I0131916

Aline Auroux, Ljiljana Damjanović-Vasilić (Eds.)
Thermal Analysis and Calorimetry

Also of interest

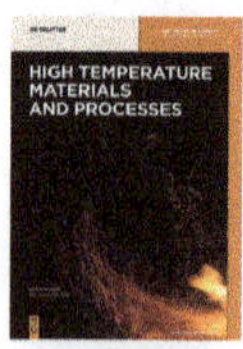

High Temperature Materials and Processes
(Open Access Journal)
Hiroyuki Fukuyama (Editor-in-Chief)
ISSN 2191-0324

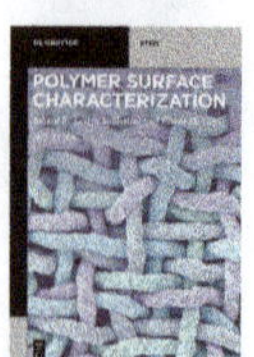

Polymer Surface Characterization
2nd Edition
Sabbatini, De Giglio (Eds.), 2022
ISBN 978-3-11-070104-3, e-ISBN (PDF) 978-3-11-070109-8,
e-ISBN (EPUB) 978-3-11-070114-2

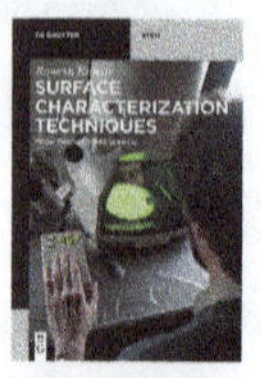

Surface Characterization Techniques.
From Theory to Research
Kumar, 2022
ISBN 978-3-11-065599-5, e-ISBN (PDF) 978-3-11-065648-0,
e-ISBN (EPUB) 978-3-11-065658-9

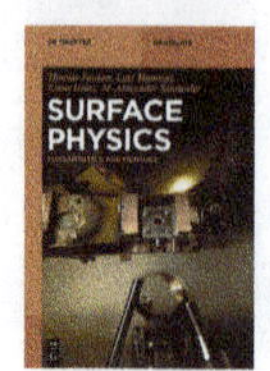

Surface Physics.
Fundamentals and Methods
Fauster, Hammer, Heinz, Schneider, 2020
ISBN 978-3-11-063668-0, e-ISBN (PDF) 978-3-11-063669-7,
e-ISBN (EPUB) 978-3-11-063699-4

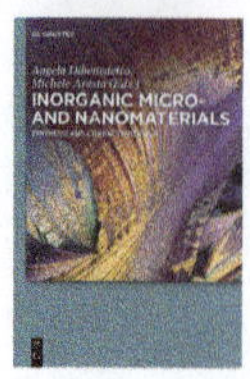

Inorganic Micro- and Nanomaterials.
Synthesis and Characterization
Dibenedetto, Aresta (Eds.), 2013
ISBN 978-3-11-030666-8, e-ISBN (PDF) 978-3-11-030687-3

Thermal Analysis and Calorimetry

Versatile Techniques

Edited by
Aline Auroux and Ljiljana Damjanović-Vasilić

DE GRUYTER

Editors
Dr. Aline Auroux
Institut de Recherches sur la Catalyse et l'Environnement de Lyon
UMR 5256 CNRS/Universite Lyon1
2 Avenue Einstein
69626 VILLEURBANNE Cedex
France
aline.auroux@outlook.com

Prof. Ljiljana Damjanović-Vasilić
Faculty of Physical Chemistry
University of Belgrade
Studentski trg 12-16
11000 Belgrade
Serbia
ljiljana@ffh.bg.ac.rs

ISBN 978-3-11-059043-2
e-ISBN (PDF) 978-3-11-059044-9
e-ISBN (EPUB) 978-3-11-059049-4

Library of Congress Control Number: 2023933617

Bibliographic information published by the Deutsche Nationalbibliothek
The Deutsche Nationalbibliothek lists this publication in the Deutsche Nationalbibliografie; detailed bibliographic data are available on the internet at http://dnb.dnb.de.

Cover image: Didier Auroux
Typesetting: Integra Software Services Pvt. Ltd.
Printing and binding: CPI books GmbH, Leck

www.degruyter.com

Abstract

This book focuses on versatile emerging applications of well-established methods of thermal analysis and calorimetry. The text covers studies with thermal and calorimetric measurements of catalysts and reactions in the field of chemical energy storage, biomass valorization, solid-state hydrogen storage, thermochemical energy storage, and DFT simulations in heterogeneous catalysis. Applications to clathrate hydrates research, characterization of polymers, the adsorption at the solid-liquid interface, and studying fluids in refrigeration systems are reviewed as the latest applications to the investigation of cultural heritage and food science. The strengths and the limitations of the different techniques are highlighted. Written by experts, it is a comprehensive reference for the scientists, engineers, and graduate students looking for a distinctive and up-to-date resource.

https://doi.org/10.1515/9783110590449-202

Preface

Thermal analysis and calorimetry are well-established methods for characterization of thermal properties of materials. These techniques are particularly important in characterization of catalysts, supports, adsorbents, and their surfaces. Alongside, the continuous interest of research community and practitioners in calorimetry and thermal analysis lead to discovery and validation of novel versatile applications.

Therefore, it is appropriate to assemble a group of texts presenting emerging and diverse applications of thermal analysis and calorimetry in different scientific areas.

The applications of thermal analysis methods and calorimetry covered in this book are:

- the study of materials for catalyst production in the field of chemical energy storage with the emphasis on precursor transformations and gas-solid interaction under reaction conditions;
- use of isothermal titration calorimetry to study the adsorption at the solid-liquid interface: the physical factors to be monitored and controlled during experiments, data processing, data analysis, and interpretation;
- the study of the hydrogen storage properties of solid-state materials;
- the characterization of the acid-base properties of the relevant catalysts for biomass conversion reactions;
- the correspondence of calorimetric studies with DFT simulations in heterogeneous catalysis;
- measuring the thermal characteristics of thermochemical energy storage materials;
- the determination of heat capacities and heat of vaporization, the key properties in refrigeration cycles;
- the characterization of polymers: accurate identification and quantification of thermodynamic transitions, thermal stability, decomposition, and chemical reactions;
- the thermo-physical properties of various gas hydrates (clathrate hydrates) which may have applications in energy recovery from marine natural hydrate resources, energy storage, hydrate structure elucidation, assessment of hydrate inhibitors/promoters, etc.;
- in the field of cultural heritage for the studies of mortars, stones, ceramics, leathers and parchments, wood, paper, and painting materials significant for the restoration and conservation of the artifacts;
- in the field of food investigation studying the effect of temperature changes on the physical and chemical properties of the raw materials and the final products important for most manufacturing protocols, transport and storage, as well as preparations for consumption.

This book will be useful for graduate students, thermal engineers, and established researchers in the fields of material science, catalysis, energy storage, and from other scientific areas.

https://doi.org/10.1515/9783110590449-203

We would like to express sincere gratitude to our colleagues who accepted invitation to write the chapters and manage to do it successfully amid Covid-19 pandemic. We are also grateful to all those who contribute to editing, redaction, and design of this book.

Villeurbanne, France
Belgrade, Serbia
September, 2022

Aline Auroux
Ljiljana Damjanović-Vasilić

Contents

Abstract — V

Preface — VII

List of contributors — XI

Andrey V. Tarasov
Chapter 1
Thermal analysis: a guide through catalyst's synthesis and reaction process — 1

Bénédicte Prélot and Jerzy Zając
Chapter 2
Contribution of isothermal titration calorimetry to elucidate the mechanism of adsorption from dilute aqueous solutions on solid surfaces: data processing, analysis, and interpretation — 47

Basile Galey, Nuno Batalha, Aline Auroux, and Georgeta Postole
Chapter 3
Thermal analysis and solid-state hydrogen storage: Mg/MgH_2 system case study — 91

Vincent Folliard and Aline Auroux
Chapter 4
Using calorimetry to study catalytic surfaces and processes for biomass valorization — 123

María-Guadalupe Cárdenas-Galindo and Brent E. Handy
Chapter 5
The correspondence of calorimetric studies with DFT simulations in heterogeneous catalysis — 159

Mohamed Zbair, Elliot Scuiller, Patrick Dutournié, and Simona Bennici
Chapter 6
Major concern regarding thermophysical parameters' measurement techniques of thermochemical storage materials — 183

Jean-Yves Coxam and Karine Ballerat-Busserolles
Chapter 7
Calorimetric methods for key properties in refrigeration cycles — 223

Rodica Chiriac, François Toche, and Olivier Boyron
Chapter 8
Calorimetry and thermal analysis for the study of polymer properties — 245

Jyoti Shanker Pandey, Asheesh Kumar, and Nicolas von Solms
Chapter 9
Role of calorimetry in clathrate hydrate research — 293

Ljiljana Damjanović-Vasilić
Chapter 10
Thermal methods as a tool for studying cultural heritage — 311

Vesna Rakić, Steva Lević, and Vladislav Rac
Chapter 11
The application of calorimetry and thermal methods of analysis in the investigation of food — 341

Index — 393

List of contributors

Andrey V. Tarasov
Fritz-Haber Institute der Max-Planck Gesellschaft
Department of Inorganic Chemistry
Faradayweg 4-6, D-14195
Berlin, Germany
tarasov@fhi-berlin.mpg.de

Bénédicte Prélot
ICGM, Univ. Montpellier,
CNRS, ENSCM, 34090 Montpellier, France
benedicte.prelot@umontpellier.fr

Jerzy Zając
ICGM, Univ. Montpellier,
CNRS, ENSCM, 34090 Montpellier, France
jerzy.zajac@umontpellier.fr

Basile Galey
Univ Lyon, University Claude Bernard Lyon 1
CNRS, IRCELYON, 2 avenue Albert Einstein
F-69626 Villeurbanne, France
basilegaley@gmail.com

Nuno Batalha
Univ Lyon, University Claude Bernard Lyon 1
CNRS, IRCELYON, 2 avenue Albert Einstein
F-69626 Villeurbanne, France
nuno.rocha-batalha@ircelyon.univ-lyon1.fr

Aline Auroux
Univ Lyon, University Claude Bernard Lyon 1
CNRS, IRCELYON, 2 avenue Albert Einstein
F-69626 Villeurbanne, France
aline.auroux@outlook.com

Georgeta Postole
Univ Lyon, University Claude Bernard Lyon 1
CNRS, IRCELYON, 2 avenue Albert Einstein
F-69626 Villeurbanne, France
georgeta.postole@ircelyon.univ-lyon1.fr

Vincent Folliard
Univ Lyon, University Claude Bernard Lyon 1
CNRS, IRCELYON, 2 avenue Albert Einstein
F-69626 Villeurbanne, France

María-Guadalupe Cárdenas-Galindo
Facultad de Ciencias Químicas
Universidad Autónoma de San Luis Potosí
Av. Manuel Nava No. 6, Zona Universitaria
C.P. 78210, San Luis Potosí
S.L.P., México
cardenas@uaslp.mx

Brent E. Handy
Facultad de Ciencias Químicas
Universidad Autónoma de San Luis Potosí
Av. Manuel Nava No. 6, Zona Universitaria
C.P. 78210, San Luis Potosí
S.L.P., México
handy@uaslp.mx

Mohamed Zbair
Université de Haute-alsace, CNRS
IS2M UMR 7361
F-68100 Mulhouse, France;
Université de Strasbourg, France
mohamed.zbair@uha.fr

Elliot Scuiller
Université de Haute-alsace, CNRS
IS2M UMR 7361
F-68100 Mulhouse, France;
Université de Strasbourg, France
elliot.scuiller@uha.fr

Patrick Dutournié
Université de Haute-alsace, CNRS
IS2M UMR 7361
F-68100 Mulhouse, France;
Université de Strasbourg, France
patrick.dutournie@uha.fr

Simona Bennici
Université de Haute-alsace, CNRS
IS2M UMR 7361
F-68100 Mulhouse, France;
Université de Strasbourg, France
simona.bennici@uha.fr

https://doi.org/10.1515/9783110590449-205

Jean-Yves Coxam
Institut de Chimie de Clermont-Ferrand
Campus Universitaire des Cézeaux
TSA 60026 - CS 60026, 24, Avenue Blaise Pascal
63178 AUBIERE, France
j-yves.coxam@uca.fr

Karine Ballerat-Busserolles
Institut de Chimie de Clermont-Ferrand
Campus Universitaire des Cézeaux
TSA 60026 - CS 60026, 24, Avenue Blaise Pascal
63178 AUBIERE, France
Karine.BALLERAT@uca.fr

Rodica CHIRIAC
Laboratoire des Multimatériaux et Interfaces (LMI)
UMR CNRS 5615, Univ Lyon
Université Claude Bernard Lyon 1
F-69622 Villeurbanne, France
rodica.chiriac@univ-lyon1.fr

François TOCHE
Laboratoire des Multimatériaux et Interfaces (LMI)
UMR CNRS 5615, Univ Lyon
Université Claude Bernard Lyon 1
F-69622 Villeurbanne, France
francois.toche@univ-lyon1.fr

Olivier BOYRON
Catalyse, Polymérisation, Procédés & Matériaux (CP2M)
UMR CNRS 5128, CPE Lyon Univ Lyon
Université Claude Bernard Lyon 1
F-69616 Villeurbanne, France
olivier.boyron@univ-lyon1.fr

Jyoti Shanker Pandey
Center for Energy Resource Engineering (CERE)
Department of Chemical Engineering
Technical University of Denmark
2800 Kgs. Lyngby, Denmark
jyshp@kt.dtu.dk

Asheesh Kumar
Upstream and Wax Rheology Division
CSIR-Indian Institute of Petroleum
Dehradun-248005, India
Center for Energy Resource Engineering (CERE)
Department of Chemical Engineering
Technical University of Denmark
2800 Kgs. Lyngby, Denmark
asheesh.kumar@iip.res.in

Nicolas von Solms
Center for Energy Resource Engineering (CERE)
Department of Chemical Engineering
Technical University of Denmark
2800 Kgs. Lyngby, Denmark
nvs@kt.dtu.dk

Ljiljana Damjanović-Vasilić
University of Belgrade-Faculty of Physical Chemistry
Studentski trg 12–16, 11158 Belgrade 118
P.O. Box 47, Serbia
ljiljana@ffh.bg.ac.rs

Vesna Rakić
Faculty of Agriculture, University of Belgrade,
Nemanjina 6, 11080 Zemun, Serbia
vesna.rakic@agrif.bg.ac.rs

Steva Lević
Faculty of Agriculture, University of Belgrade,
Nemanjina 6, 11080 Zemun, Serbia
slevic@agrif.bg.ac.rs

Vladislav Rac
Faculty of Agriculture, University of Belgrade,
Nemanjina 6, 11080 Zemun, Serbia
vladarac@agrif.bg.ac.rs

Andrey V. Tarasov

Chapter 1
Thermal analysis: a guide through catalyst's synthesis and reaction process

Abstract: Various thermal analysis methods trace all stages of catalyst formation. This chapter deals with a sequential study of the main stages of catalyst formation from widely used thermogravimetry to specially developed in situ calorimetry through precursor synthesis to the catalytic reaction. Particular focus is given to Cu, Zn-based catalytic systems for methanol synthesis and Ni, Mg-based catalysts for dry methane reforming. The calcination of hydroxocarbonates and their decomposition kinetics were investigated by the simultaneous thermal analysis–mass spectrometry (STA-MS) method. Using temperature-programmed reduction technique, the composition of oxide systems, their reduction, and activation of the metal catalyst was analyzed. Additional diffraction, spectroscopic, and microscopic methods characterized the change in metal–support interaction during successive oxidative and reducing temperature treatments. Structural–functional relationships can be identified based on thermochemical, structural, and catalytic data. High-pressure thermogravimetry was used to probe the adsorption layer on the catalyst surface under methanol synthesis and coking under dry methane reforming conditions. Finally, the application of in situ calorimetry for studying the catalyst restructuring in oxidative reactions is shown.

Keywords: calcination, Cu, Zn hydroxocarbonates, Ni catalysts, methanol synthesis, decomposition kinetics, reduction, adsorbates, deactivation, coking

1.1 Prologue

Recent discoveries in heterogeneous catalysis in the academic and industrial fields have primarily been accelerated by the advances in analytical tools and the broad application of complementary in situ methods. The instrumental progress has been achieved by developing several areas: hyphenated techniques, multiple coupling options, high-throughput experimentation, and operando methods. These techniques serve to examine the surface and bulk of the catalysts and directly measure the parameters of reaction systems under conditions of relevant catalytic performance. All of the trends above had affected the thermal methods of analysis (TA). On the other

Andrey V. Tarasov, Fritz-Haber Institute der Max-Planck Gesellschaft, Faradayweg 4-6, Berlin 14195, Germany

https://doi.org/10.1515/9783110590449-001

hand, when the methods experience technical limitations, a wide range of experimental approaches are still available, determined by the experimentalist's creativity, patience, diligence, and obstinacy. Although the growing presence of conventional machines has outnumbered the custom-designed setups, the research-specific equipment provides more profound insight than analytics with standard configuration.

Calorimetry together with thermogravimetry are techniques with a very long history. In the era of progressively developing synchrotron beam, spectroscopy, and electron microscopy in the alliance of catalysis and surface science, thermal methods have erroneously gained a reputation as supportive methods. A legitimate question arises among the catalytic community: is this still good enough? A look at the past of chemical findings reminded us that Marie Curie was literary "weighting radioactivity" in her experiments with Pierre Curie's piezoelectric balance for discovering new elements Ra and Po. Research of that level is nowadays performed on particle accelerators.

Another critical point is the definition of kilojoule. Is it a big or small value? From the experiments with energy transformation of James Joule, we know that 1 kJ will give the tennis ball a speed of 360 km/h or heat 1 L (55.5 mol) of water to 0.25 K. Given that the energy of common chemical bonds ranges from 150 to 1,000 kJ/mol, it is a very high heat evolution or consumption that is detected upon formation and decomposition of chemical substances. Even though the chemical reaction occurs on a minimal amount of active centers, the heat effects of the strong interaction, such as the formation of covalent bonds, are significant. However, weaker interactions such as vibrations of the participating molecules and long-range interactions are less available.

Even though the spectroscopy methods dominate, they experience significant limitations when quantifying specific surface sites under reaction conditions, which is imperative in catalysis research. Conventional catalysts do not consist of chemically and crystallographically homogeneous surfaces. Adsorption phenomena and thermochemical properties of the topmost layer provide necessary information for the characterization of the surface region. However, it is insufficient to build a clear picture of the chemistry of all involved processes. Even though in situ thermal analysis methods open up the possibility of studying structure–function relationships, one must keep in mind that correlation does not necessarily mean causation. Decisive information on the geometry and electronic structure of the surface layer under in situ conditions is not trivial to extract. In addition, the interplay between the flexibility of the surface and the stability of the bulk structure shapes the dynamic behavior of the catalytically active system. To bridge the gap between these elements, a high-level theory has to accompany the experimental findings. Proceeding from these basic facts, progress, and limits of instrumentation, the following chapters aim to demonstrate and remind the reader of the actual application of thermal analysis as a primary self-consistent or a secondary, complementary method for the characterization of solid catalysts.

1.2 Concerning thermal analysis in catalysis

Thermochemical information in analyzing catalysts and catalytic processes is a significant building block for understanding the active state of the catalyst. Thermogravimetry (TG), temperature-programmed reduction/oxidation (TPR/O), differential scanning calorimetry (DSC), and adsorption microcalorimetry (MC) are the main pillars of thermoanalytical methods in catalysis, and this will continue to be the case. These methods concerning the chemistry of heterogeneous reactions in the solid–gas system are applied for (i) the characterization of solid samples and surfaces and (ii) the study of heterogeneous processes. The first group includes the analysis of sample composition, determination of redox properties (TPR-TPO cycles) [1, 2], determination of temperature limits for catalyst stability (TG) [3], thermokinetics of calcination and reduction (TG, TPR/O), titration of metallic surface area, quantification of basic and acidic centers (TPD and pulse TA) [4–6], and characterization of strong and weak adsorption sites of the catalyst (MC). The second subdivision implies the application of in situ techniques to study surface adsorbates and deposits (in situ TG) [2, 7, 8], characterization of equilibrium and nonequilibrium states of solid, measurement of the heat effects of reversible phase transitions during the reaction as the catalyst structure is formed, construction of "thermochemical properties–reactivity" diagrams (in situ DSC) [9]. Instrumental basics, theoretical foundation, and numerous examples are summarized and reviewed in the literature [10].

Figure 1.1 provides a standard overview of thermochemical and adsorption processes within the catalytic cycle that an alternate application of these methods could determine. The formation of a catalyst as represented by the phase α is typically a multi-stage synthetic route involving additional calcination and reduction steps before it is exposed to reaction feed. Ex situ temperature-programmed techniques are employed on almost every stage to characterize the intermediate products in all detail and thus have a clear picture of the catalyst formation from a precursor through a precatalyst and active catalyst to a waste catalyst as represented by deactivated phase β. Once the surface is exposed to the chemical potential of the gas phase, the formation of vacancies and surface reconstructions occurs. In the current example, the active state in the reaction feed could be considered to be an endothermic formation of the defective state isostructural to the phase α. The E^* value does not include the energy of lattice reconstruction; that is, exothermic adsorption of A ($E_{ads}{}^{A}$) and desorption of product B ($E_{ads}{}^{B}$) do not modify the phase composition of the bulk and involve the formation and refill of vacancies and surface reorganization solely. The adsorbed species A overcomes the activation barrier ($E^{\#}$) and progresses through the intermediate state leading to final product formation and desorption. Meanwhile, the surface endures a row of adaptive states during the catalytic cycle. The formation of the inactive phase β is related to exothermic E_{relx} relaxation of the lattice to the more stable structure; thus, the reaction enthalpy of phase transformation could be expressed as $E_r = E_{relx} - E^*$.

Although this example is quite generalized and each specific system has to be considered separately, in the context of structure–functional relationship, these thermochemical

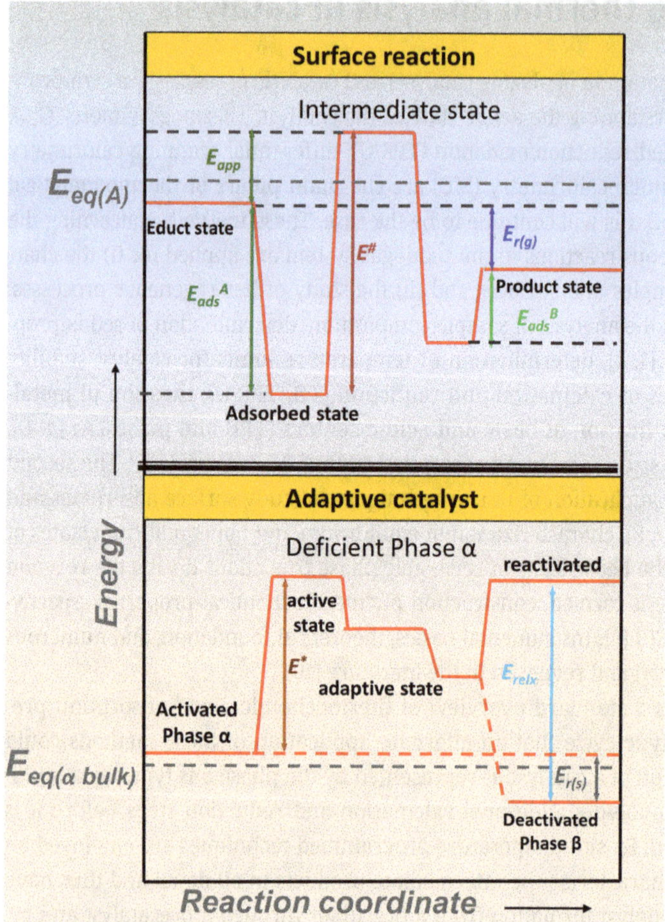

Fig. 1.1: Simplified reaction free energy diagram for a not-equilibrated exothermic reaction (adapted from [11, 12]).

parameters feature energetic nonuniformity of the properties of active surface sites. As reference values, the thermodynamic data of model systems under equilibrium conditions are used. Application of this approach for the construction of thermochemical cycle and therefore establishing probable steps in the reaction mechanism have been demonstrated in the earlier review of Pierre C. Gravelle in 1977 [13]. However, this approach is limited to relatively simple systems and reactions such as CO oxidation over stoichiometric nickel oxide. It becomes less productive for multicomponent and multiphase supported catalysts since the reference thermochemical data are unavailable for such complex systems. From the scheme (Fig. 1.1), it is seen that the active catalyst is instead in a metastable state with respect to the bulk.

Moreover, the thermochemical properties (E^* , E_{relx}) and reactivity (E_{app}, selectivity, conversion, reaction rate) as well as the adsorption properties (${E_{ads}}^A$, ${E_{ads}}^B$) undergo gradual changes in a complex manner as a response to the continuously varied chemical potential of the gas phase [11]. The catalytic property of a solid should be related to the on-the-fly state under the operation conditions, which can dramatically differ from the equilibrium state [9]. To address this dynamics of interconnected thermochemical effects and to establish a correlation of surface properties (e.g., the binding energy of lattice oxygen) with catalytic selectivity of supported catalysts, an in situ DSC in combination with pulse TA has proven to be a potent tool [9, 14–18]. Another combination of microcalorimetry with reactivity measurements is promising for determining activation barrier $E^{\#}$, which is defined as a sum of adsorption energy ${E_{ads}}^A$ and apparent activation energy E_{app} [19].

The current chapter is a review which revisits the thermoanalytical studies of the high-performance catalysts carried out over the past 10 years in the Department of Inorganic Chemistry of the Fritz-Haber Institute of the Max-Planck Society. In the following sections, we demonstrate the application of thermal analysis complementary to other methods for extensive characterization of catalysts, summarizing the findings on the nature of catalytically active material. Particular emphasis is placed on the analysis of materials evolution upon the formation of the active catalyst on each step of the synthetic path and under process conditions (phases α and β on Fig. 1.1). Much attention is paid to the catalytic Cu, Zn systems for methanol synthesis and Ni, Mg catalysts for dry methane reforming. The reactions mentioned above are widely employed in the chemical industry because of their high energetic relevance and impact on CO_2 utilization. This review-like contribution offers a perspective for elaborative thermoanalytical characterization as a primary or alternative way for catalyst examination from the synthesis to the reaction process. It aims to provide the reader with the logic behind the experiments, organizing and plotting the data to get valuable insights.

1.2.1 Acknowledgments

The author expresses sincere gratitude to the following individuals for their experimental contribution and oral advice, which laid the basis for this chapter: Prof. Dr. Malte Behrens (University Kiel/Germany), Dr. Stefani Kühl (FHI/Germany), Dr. Julia Schumann (Imperial College London/UK), Dr. Elias Frei (BASF/Germany), Dr. Marie-Mathilde Millet (Domo Chemicals/France), Dr. Katharina Mette (GNV/Berlin), Dr. Ezgi Erdem (Stanford/USA), Dr. Annette Trunschke (FHI/Germany), and Prof. Dr. Robert Schlögl (FHI/Germany). I greatly appreciate Dr. Gayana Kirakosyan (IGIC RAS/Russia) for assistance in the preparation of the manuscript.

1.3 Cu-based catalyst for industrial methanol synthesis

For several decades much attention has focused on the energy storage in chemical bonds in chemicals like methanol, dimethyl ether, and methyl-*tert*-butyl ether. These products can be applied as synthetic fuels in the transportation sector and are capable of compensating for energy losses in the energy sector during windless and cloudy weather. These products can also be used as a carbon source in the chemical industry. Methanol is among the best candidates for such an energy storage molecule. Industrial methanol synthesis relies almost exclusively on Cu-, ZnO-, and Al_2O_3-based catalysts ($Cu/ZnO/Al_2O_3$). Common $Cu/ZnO/Al_2O_3$ catalysts for methanol synthesis are often mistakenly considered as supported systems, but neither ZnO nor the low amounts of Al_2O_3 used represent classical porous oxide supports. This is apparent when considering the typical composition of modern $Cu/ZnO/(Al_2O_3)$ catalysts, characterized by a molar Cu:Zn ratio close to 70:30, while the amount of Al_2O_3 typically is significantly lower than that of ZnO. This Cu-rich composition is responsible for a specific microstructure of the industrial $Cu/ZnO/Al_2O_3$ catalyst. The TEM images reveal spherical copper nanoparticles 5–15 nm in size (Fig. 1.2D and E) and even smaller ZnO nanoparticles arranged in an alternating fashion. In the microstructure, ZnO serves as a spacer and stabilizer, preventing direct contact of the copper particles and their aggregation [20].

The preparation of the $Cu/ZnO/(Al_2O_3)$ catalyst is a multistage process and traditionally begins with coprecipitation of Cu, Zn, (Al) hydroxocarbonates, which are then aged in the mother liquor. In this first microstructure directing step (meso-structuring) the homogeneous Cu, Zn precipitate crystallizes as thin interwovèn needles of zincian malachite, which are a desirable feature determining the porosity of the final catalyst [20]. The precursor is then thermally treated in an oxidative atmosphere. At this step, the individual needles are decomposed into CuO and ZnO and pseudomorphs of the precursor needles can be still observed after mild calcination. Subsequently, the CuO component of the precatalyst is reduced in hydrogen during the activation step to afford Cu/ZnO nanoparticles with a unique microstructure exhibiting high porosity and Cu dispersion. Eventually, the catalyst is subjected to reaction conditions where the active state is finally formed. This preparation scheme of Cu/ZnO based on SEM and TEM images is shown schematically in Fig. 1.2.

The whole catalyst preparation route is very complex and responds even to slight changes in parameters. Moreover, the conditions of each thermal step are decisive for the formation of the active catalyst [22]. Numerous studies on this system have shown that the synthetic parameters during the very early stages of the catalyst synthesis have a crucial effect on the properties of the precursor and, thus, on the catalytic activity of the final catalyst. This phenomenon is referred to as “chemical memory” of the system. Therefore, calcination and activation steps are studied in detail to elucidate the chemical processes engaged during the formation of the catalytic material.

Fig. 1.2: Cartoon of the industrially applied preparation of Cu/ZnO:Al catalyst via zincian malachite precursor: (A) SEM image of the precursor before calcination. The image is colored with turquoise for visualization (B and C) TEM micrographs of the precatalyst and catalyst. The rod-like microstructure is conserved. The Cu nanoparticles of the precatalyst appear on the outer side of the nanorods (D) HRTEM image of a Cu nanoparticle with defects, decorated with ZnO overlayer, arrowheads indicate twin formation (F) HAADF-STEM image, red and yellow colors represent Cu and Zn, respectively (adapted from [21]. © 2014 WILEY-VCH Verlag GmbH & Co. KGaA, Weinheim).

The following section contains information on a specific precursor and is not unified for all types of precursors. It should not be considered a handbook on creating an active catalyst either; instead, it aims to provide readers with information and considerations relevant to their research when studying similar systems and addressing a research task to the specific method. Occasionally the author is inconsistent and switches from one type of sample structure to another: namely, from calcination of one type of Cu, Zn, (Al)-precursor to the reduction of the Cu, Zn, (Al)-precatalyst. It is done not with the aim of generalization; the samples are indeed different, but because of complete and representative existing datasets.

1.3.1 Calcination

1.3.1.1 Cu/Zn hydroxocarbonates as precursors for active catalyst

The common and industrially relevant precursors are "binary" (containing two metal ions) hydroxocarbonates – zincian malachite [$M_2(CO_3)(OH)_2$] and aurichalcite $M_5(CO_3)_2(OH)_6$ (M is $Cu_xZn_{(1-x)}$) [23, 24] – and the "ternary" hydrotalcite-like precursor

$Cu_{0.5}Zn_{0.17}Al_{0.33}(OH)_2(CO_3)_{0.17} \cdot mH_2O$. The general structure of hydrotalcite-like compounds composed of $(1-x)M^{III}M^{II}_x(OH)_2$ brucite-like layers and charge-balancing anionic species in the interlayer is shown in Fig. 1.3A. In CuZnAl hydrotalcite compounds, the metal ions are held in octahedrally coordinated sites in the hydroxide layers; such an arrangement ensures their uniform distribution over the structure. This is a prerequisite for that the reduction of these precursors will afford catalysts with a homogeneous microstructure characterized by highly dispersed metal species and strong metal–support interaction. The binary Cu/ZnO precursors allow synthesizing catalysts with a broad Cu/Zn ratio and serve as a model system for industrial catalysts usually containing 5–10 mol% Al_2O_3. The dominance of a certain hydroxocarbonate phase depends on the Cu/Zn ratio: zincian malachite dominates at Zn contents up to 31% [25], two phases – malachite and aurichalcite – are present at 30–50% Zn, and aurichalcite dominates at Zn content above 50% [26].

Thermal activation of these precursors strongly affects the porosity and structure of Cu-based catalysts. For a better understanding of solid-state transformations with heating and enabling the improvement of the calcination protocol, the knowledge of decomposition kinetics has become indispensable. This information helps design calcination protocols tailored for individual samples, which can be crucial for catalyst performance. Figure 1.3 demonstrates the precursor structures and decomposition profiles in synthetic air.

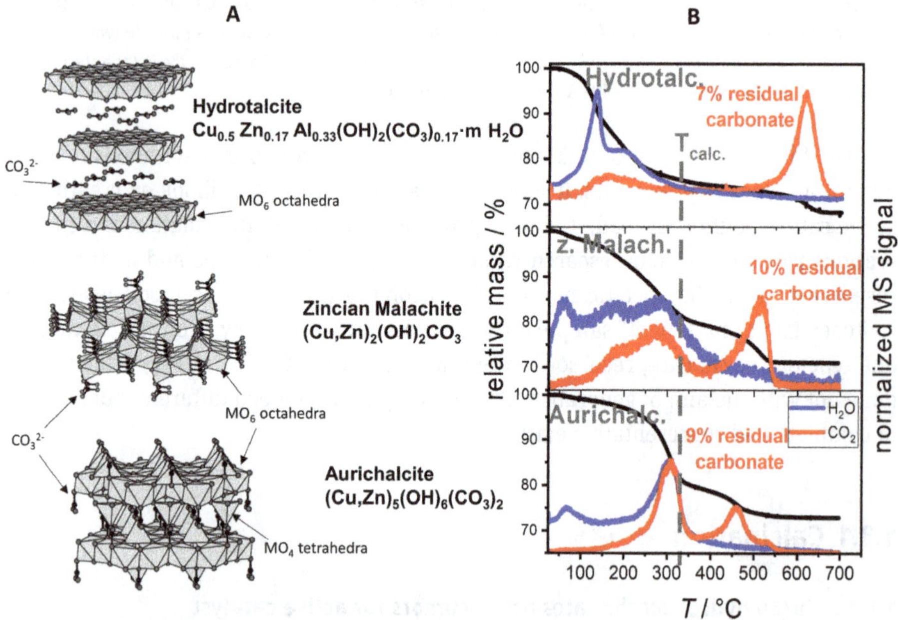

Fig. 1.3: Catalyst precursors: (A) crystal structures of the precursors (black spheres – carbons, gray spheres – oxygen, MO_6 octahedra, and MO_4 tetrahedra), hydrogen and water molecules not shown and (B) TG-MS results of calcination in synthetic air (21% O_2 in Ar), 100 mlm with 2 Kpm.

Several decomposition steps are involved for the mixed hydroxocarbonates with Cu/Zn ratios from 70/30 to 30/70 [27]:

1. Surface water molecules are removed around 70 °C
2. Interlamellar water molecules are released between 100 and 140 °C
3. Dehydroxylation and loss of interlayer carbonate ions occur at 200–300 °C

The high-temperature step (400–500 °C) is associated with the release of strongly bound CO_2. This "high-temperature carbonate" is a mixture of carbonates and oxides [26]. It is also referred to as anion-modified metal oxide $MO_{(1-y)}(CO_3)_{2y}$; it forms in the first step and then decomposes to metal oxide [28, 29]. It is worth noting that no such intermediate has been detected during the calcination of the nonmixed malachite and hydrozincite synthetic precursors [30] as well as of malachite minerals [24]. The role of this intermediate in the activity of the final catalyst has been a subject of scientific debate in the literature. It has been proposed [31] that oxygen ions in this structure are replaced by OH^- and $(CO_3)^{2-}$ groups although the existence of OH groups is doubtful. Its crystal structure is still unknown. Its abundance turns out to depend on the Cu/Zn ratio and the content of the hydroxycarbonate phase [30]. Because the catalytic precursors are usually calcined at 300–350 °C [22, 32, 33], the carbonate-modified metal oxide can remain captured in the catalyst, which can alter its catalytic performance [34].

In an in situ XRD experiment, structural changes occur under a controlled temperature program. Figure 1.4 shows the three-dimensional contour plot of combined XRD-TG-DSC data for pure Aurichalcite precursor with Cu/Zn = 40/60.

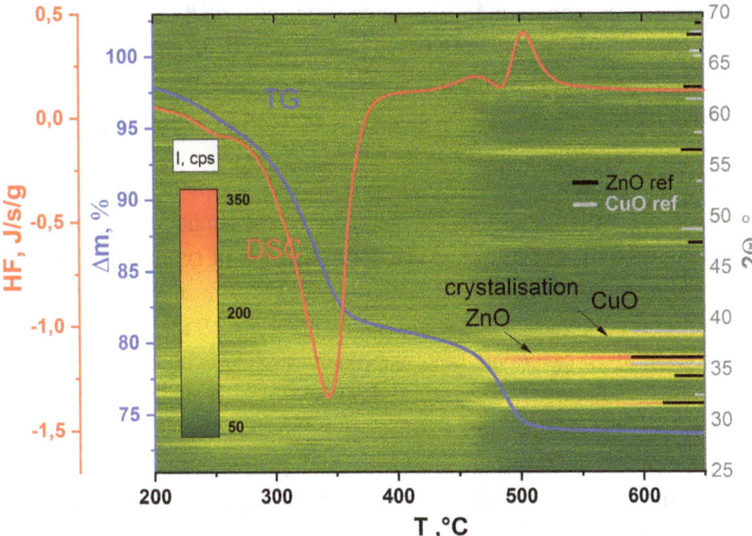

Fig. 1.4: Contour plot of in situ XRPD data combined with TG/DSC profiles for aurichalcite Cu/Zn = 40/60 in 21% O_2 in Ar with 10 Kpm. The XRD is recorded in isothermal segments in range 200–650 °C with 50 °C steps; hence, no direct kinetic comparison is possible (adapted from [35] with permission from Elsevier, copyright 2014).

The changes in reflection intensities in the XRD patterns agree with the DSC curve. Of particular interest is the temperature range 300–400 °C (Fig. 1.4), where the carbonate-modified oxide is detected. According to XRD (absence of distinct reflections/diffuse intensity signal), a phase with missing long-range order is formed. An endotherm accompanies intermediate carbonate-modified oxide formation with a maximum of about 350 °C. Its XRD pattern features a rather "XRD-amorphous" profile, which may also be associated with the inception of a nanocrystalline phase. The XRD pattern gradually starts changing from 400 °C with the decomposition of high-temperature carbonate and ends with the crystallization of oxides ZnO and CuO (yellow and red lines on XRD counter plot). The corresponding DSC curve shows an exotherm as expected for crystallization.

1.3.1.2 Nature of carbonate-modified oxide, decomposition kinetics

The crystallization of oxides was shown to be sensitive to the self-generated water vapor pressure during the reaction. The humidity of the gas atmosphere promotes decomposition and prevents the formation of the intermediate carbonate-modified oxide [35]. High partial pressures of H_2O give rise to carbonate decomposition and crystallization and, thus, to significant batch effects if not precisely controlled. The entire process is described with a formal reaction scheme (Fig. 1.5A) and could be mathematically represented by five stages, including the desorption of physisorbed water, partial dehydroxylation, and decarboxylation of interlamellar space. Aurichalcite decomposition affords an oxide (k_1) when water removal is limited, while the intermediate formation of a carbonate-modified oxide (k_2) is observed only when water is readily removed. The kinetic parameters of each step have been determined from multiple heating rate experiments using the model-free isoconversional (Flyn-Wall-Ozawa) method. They have been refined by nonlinear regression least-squares procedure for different conversion functions $f(\alpha)$ (Fig. 1.5B).

The first three stages are described by the nth-order reaction equation $f(\alpha) = (1 - \alpha)^n$. Stages 2 and 3 of aurichalcite decomposition are characterized by similar activation energies E_a of ~165 kJ/mol. The third step refers to the decomposition of crystalline aurichalcite, whereas the second step is related to the evolution of hydroxyls and weekly bound carbonate groups out of the interspace between the precursor rods (Fig. 1.2). The best-fit curve for the decomposition of carbonate-modified oxide to oxide is obtained using the three-dimensional Jander-type diffusion equation $f(\alpha) = 3(1 - \alpha)^{2/3}/2(1 -(1 - \alpha)^{1/3})$ [36]. This step has a high activation barrier $E_a = 464 \pm 71$ kJ/mol. Upon decomposition of the carbonate-modified oxide phase, ZnO starts to crystallize with the evolution of CO_2, initiating crystallization of copper oxide so that the CuO reflections appear in the XRD pattern at ~550 °C. According to Jander [36], this stage of decomposition involves diffusion-controlled reactions in the solid phase. This diffusion control can be rationalized by assuming that when carbonate-modified oxide decomposes, the zinc atoms leave their positions in the lattice and diffuse through the solid to form zinc oxide. Simultaneously,

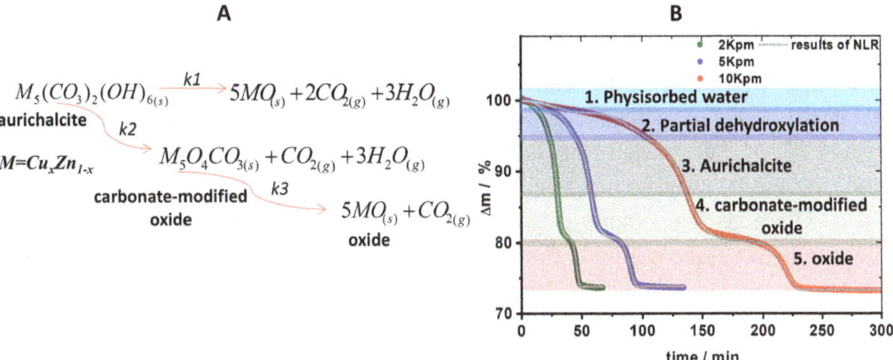

Fig. 1.5: (A) Suggested reaction scheme for decomposition of aurichalcite and (B) experimental and best-fit calculated TG curves for the aurichalcite precursor. Numbers indicate the persistent educts of decomposition (adapted from [35] with permission from Elsevier, copyright 2014).

the rest of the carbonate decomposes while copper oxide and zinc oxide crystallize. Further heat treatment of the CuO and ZnO results in crystal growth through material transport and aggregation of the oxide crystals [37]. According to the activation energy of decomposition of high-temperature carbonate for the zincian malachite (ZM) and aurichalcite (AU) precursors (297 and 464 kJ/mol, respectively), the intermediate high-temperature carbonate state is less stable in ZM. This finding is in line with the IR analysis of the calcined precursors at 330 °C (Fig. 1.6). It suggests that the carbonates in the ZM-based precursors are more like free carbonate ions, with a symmetric surrounding, whereas in the calcined AU samples, the carbonate ions are distorted because of their monodentate coordination through one oxygen atom [38].

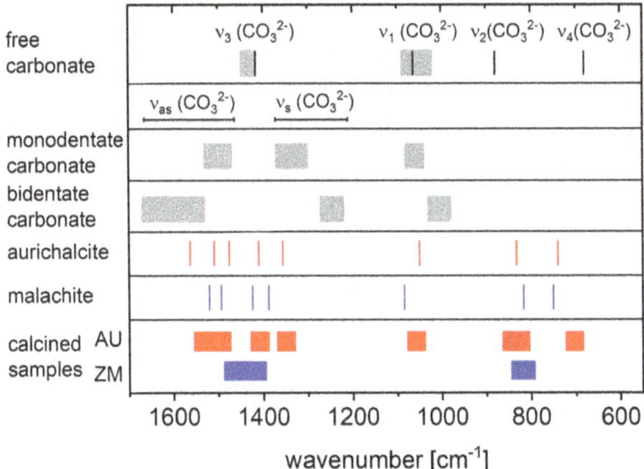

Fig. 1.6: IR vibrational frequency correlation chart of different carbonate species on metal oxides (reproduced from [38] with permission from Elsevier, copyright 2016).

In the IR spectra, strong and broad OH deformation bands in the range 1,200–500 cm^{-1} typical of the precursors vanish after calcination, indicating the complete decomposition of the hydroxyl groups. Some weak bands in this range correspond to out-of-plane and asymmetric O–C–O bending modes (ν_2 and ν_4) [38]. The calcined AU series gives rise to a band at 703 cm^{-1} and a split band around 837 cm^{-1}. The third feature of the AU series at 1,058 cm^{-1} can be assigned to a symmetric stretching vibration of carbonate (ν_1), which is expected for malachite samples at 1,085 cm^{-1} [39]. The absence of the symmetric stretching vibration bands in the spectra of the ZM precatalyst further supports the assumption that these carbonates are very weakly bound, in a symmetric environment inside the samples, like in the spectra of free carbonate-ions where this vibration is IR-inactive since it occurs without dipole changes. The bands at 820 cm^{-1} and below 650 cm^{-1} can be assigned to the bending modes ν_2 and ν_4. A broad feature below 600 cm^{-1} observed for both series (not shown in Fig. 1.6) could be ascribed to the vibrations of the M–O skeleton [40].

1.3.1.3 Effect of high-temperature carbonate on catalytic methanol synthesis

Formal kinetic analysis [35] has permitted the prediction of the content of the reaction products generated under different calcination conditions. Figure 1.7A illustrates the changes in solid degradation products of the ZM precursor with temperature. Concentration in Fig. 1.7A stands for the content of the related species in the solid product of the reaction. MO(1) and MO(2) are oxides produced via two competitive decomposition routes. They are chemically identical but differ in microstructure because of different formation paths [35]. The high-temperature carbonate accumulates at 240 °C and completely decomposes at 500 °C. Its total amount depends on the maximum calcination temperature or product transfer conditions, the latter can be changed by regulating the heating rate or gas flow. Since the carbonates are completely liberated after the subsequent activation step prior to a catalytic test, the relationship between carbonate abundancy in the precursor and the reactivity of the catalysts is somewhat indirect.

Figure 1.7B shows the simulated evolution of the HT-CO_3 concentration in solid products of ZM decomposition at different heating rates. With increasing the heating rate, more water accumulates in the atmosphere above the sample because of its faster decomposition, while H_2O cannot be removed rapidly enough. The water content in the atmosphere severely affects the hydroxycarbonate decomposition kinetics by promoting the HT-CO_3 decay and CuO and ZnO crystallization [33, 41]. This effect was also observed when water vapor was specially injected into the system during calcination [33, 35]. The increase in the content of the HT-CO_3 within one catalyst family does not improve the stability (Fig. 1.7C), while the lack of carbonate species leads to a sharper activity decay (Fig. 1.7D). At the same time, the difference in the coordination of carbonate species in the calcined samples – no coordinated carbonate in the calcined ZM sample

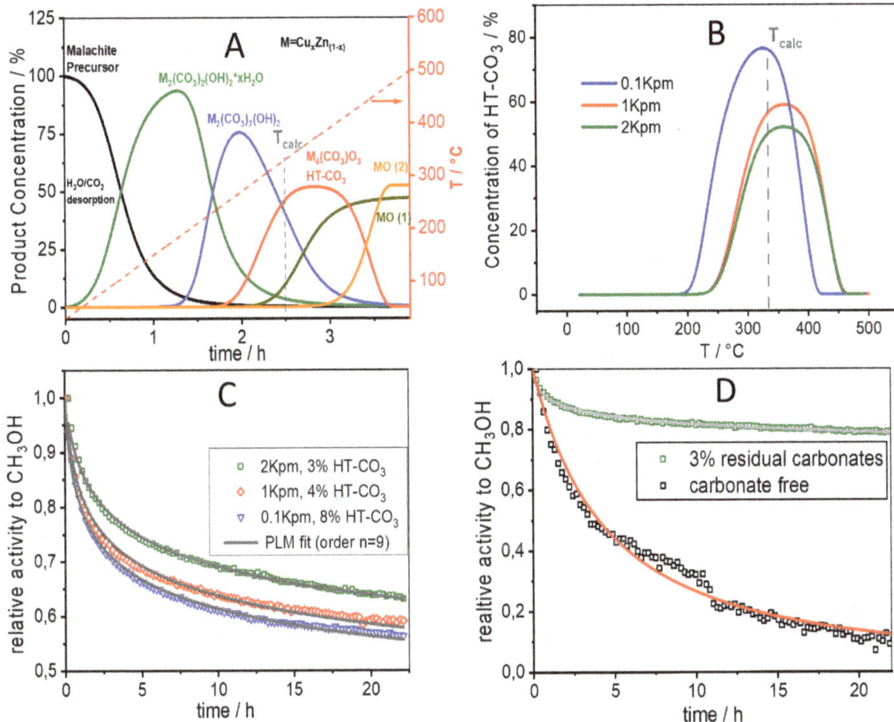

Fig. 1.7: (A) Component kinetic analysis of zincian malachite decomposition at 2 Kpm; (B) kinetic simulation of HT-CO_3 evolution at different heating rates; (C and D) relative activity $a_{rel}(t) = a(t)/a(0)$, power-law-model fit $a_{rel}(t) = ((n - 1)k^*t + 1)^{-1/(n-1)}$ for ZM-based catalyst with HT-CO_3 abundancy reaction conditions: 30 bar, 230 °C, ($8CO_2/6CO/59H_2/27He$) GHSV 3,500 h^{-1}, reduced with 20% H_2, heating with 1 Kpm to 250 °C, hold for 90 min (reproduced from [38] with permission of Elsevier, copyright 2016).

and monodentately coordinated species in the calcined AU sample – has less influence on the catalytic behavior of the final catalyst than copper content or the total amount of carbonates.

On the other hand, it should be noted that the copper content in AU is half that in ZM, and the coordinated carbonate species existing in the calcined AU sample can act as a binder between oxide components, responsible for the embedding of copper. Conversely, the free carbonate species have no binding effect and act more likely as a diluent. Thus, different factors control the stability of the microstructure: (i) the amount and (ii) the distribution of the ZnO component. The selection of the precursor phase influences both. Moreover, the choice of the calcination conditions and the resulting HT-CO_3 phase likely influence the ZnO distribution in the catalyst.

To conclude, the amount of HT-CO_3 in the calcined sample does not ensure the high intrinsic activity in methanol synthesis. However, it acts as a structural stabilizer, responsible for high surface areas and a large number of oxygen vacancies in the ZnO component of the catalysts. There is no immediate need for HT-CO_3 in the calcined or active catalyst, but, on the other hand, the HT-CO_3 in the calcined catalysts is a suitable mediator to provide mild calcination conditions ensuring no strong sintering caused by high water content [38].

1.3.2 Activation

1.3.2.1 Effect of calcination temperature on the reduction profile

A TPR using definite concentrations of hydrogen is a well-established method to study the formation of metallic nanoparticles out of oxide precursors. Notwithstanding the advantages of this analytical tool in industrial applications, the activation of the catalysts is implemented directly in the feed gas. Given this operation difference, the activation step studied under controlled TPR conditions has an academic interest in tracking the formation path of catalytically active material.

The TPR profile of a conventionally prepared catalyst (Fig. 1.8) shows a spread of the confidence interval (determined from six measurements under identical conditions) of 5–15% of the mean value in the temperature range 220–310 °C. Especially the tail of the TPR profile exhibits higher dispersion. At given conversions, the solid–gas reaction rate is usually limited by the diffusion of the reactants through the solid product layer. Therefore, a higher statistical dispersion indicates a broader particle size distribution of copper oxide in a conventional catalyst than that of the Cu, Zn, and Al-mixed oxide in a hydrotalcite-based precursor [42]. The entire reduction process for the hydrotalcite-derived catalyst occurs at higher temperatures. In CuO uniformly distributed over the Zn-Al oxide matrix, the Cu^{2+} ions are bridged through O^{2-} and/or CO_3^{2-} anions to $ZnAl_2O_4$. This firm contact should be broken before reduction for segregation of reducible Cu^{2+} species from irreducible Zn^{2+} and Al^{3+} species. Thus, the pronounced shift of the reduction temperature of the CuZnAl mixed oxide in a hydrotalcite-based precursor is associated with the well-dispersed Cu^{2+} cations, which should separate from the Zn-Al-oxide matrix and coalesce into clusters to ultimately form metallic nanoparticles. This separation into CuO and ZnO has already occurred during the calcination of commercial Cu catalysts (Fig. 1.8).

Variation of the calcination temperature for the CuZnAl-hydrotalcite precursor resulted in a saddle-like volcano activity plot for the TPR peak maximum and opposite trend for reducibility. As followed from Fig. 1.9B, the calcination temperature of maximum activity is 580 °C, which gives rise to a 242 °C reduction peak on the TPR profile Fig. 1.9A.

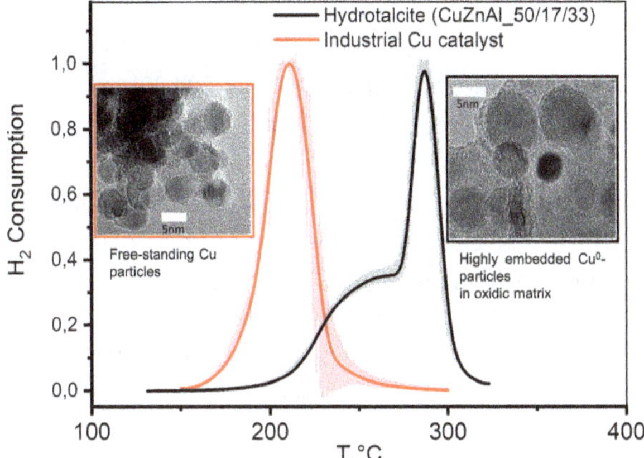

Fig. 1.8: Reduction profile of Cu, Zn, Al-mixed oxide as compared to a common calcined catalyst [42], measured at a heating rate of 6 K/min using Monti and Baiker[1] and Malet and Caballero [43] criteria. Shaded areas correspond to the confidence interval determined from six repetitions. Insets show the TEM images of the catalysts reduced in hydrogen on heating at 6 Kpm to 300 °C with 30 min holding time [44, 45] (reproduced from [42, 44], Copyright © 2014 WILEY-VCH Verlag GmbH & Co. KGaA, Weinheim, © 2010 WILEY-VCH Verlag GmbH & Co. KgaA, Weinheim).

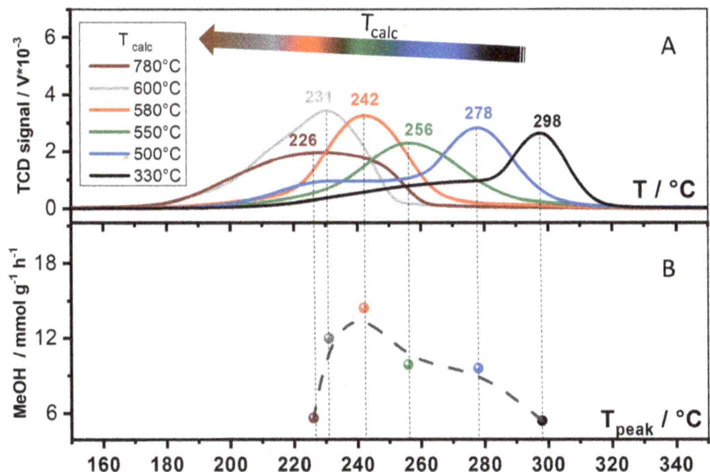

Fig. 1.9: (A) Effect of calcination temperature of Cu, Zn, Al-hydrotalcite-based precursor on the TPR profile and (B) on methanol formation rate. Reaction conditions: 30 bar, 230 °C, ($8CO_2/6CO/59H_2/27He$) GHSV 6,000 h^{-1}, reduced with 20% H_2, heating with 1 Kpm to 300 °C, hold for 90 min; (B) the arrow visualizes the increase of T_{calc}.

1.3.2.2 The course of reduction

Whereas the conventionally prepared precatalyst (calcined above 500 °C) exhibits a nearly symmetric peak shape, the TPR profile for the CuZnAl hydrotalcite-based precatalyst is bimodal (Figs. 1.8 and 1.9A). It is natural that the multistep reduction of Cu-based systems should not be characterized by a single TPR peak. The well-defined shoulder in the TPR profile can be of two origins: diffusion control of the reduction in the oxide matrix and/or the formation of a stable intermediate. Thermokinetic analysis of these systems [42, 43] has shown that the TPR curves correspond to a two-step consecutive reaction model and are best-fitted by the Prout-Tompkins *n*th-order autocatalytic equation $f(\alpha) = (1-\alpha)^n \alpha^a$ for each stage. Parameter a characterizes the autocatalytic nature of the process and is smaller than 1 for stage 1. The reaction order n is close to 1; that is, the role of autocatalysis is insignificant. Thus, the kinetic equation is close to the *n*th order function $f(\alpha) = (1-\alpha)^n$. Nevertheless, the small autocatalytic contribution in stage 1 likely results from the mechanical strain generated during the reduction with subsequent cracking of the CuO particles. Cracking increases the surface area and the number of centers where the reaction may occur. The autocatalytic character of the second stage is believed to be caused by the adsorption of molecular hydrogen onto the metal; this gives chemisorbed hydrogen atoms and further facilitates the reduction of adjacent oxide particles (spillover effect). NEXAFS spectroscopy has demonstrated that Cu_2O is formed as at least a metastable surface intermediate. The calculated kinetic models derived from TPR measurements enable the prediction of reduction-induced phase evolution, as exemplarily shown for 2 K/min in Fig. 1.10B. The NEXAFS data on the phase evolution near the surface are qualitatively consistent with the TPR-predicted bulk behavior. Both show the parallel existence of CuO, Cu_2O, and Cu during reduction and the delayed reduction of Cu_2O to metal Cu (CuO is almost consumed at the maximum of Cu_2O content). Other differences can be related to different partial pressures of hydrogen in TPR (5% H_2/Ar at 1.5 bar) and NEXAFS (0.25 bar H_2) experiments.

This resemblance is evidence that Cu(I) is actually formed as a kinetically stabilized intermediate in the course of the reduction of the hydrotalcite-derived mixed oxide. NEXAFS spectra measured during the reduction of the conventional reference catalyst point to the intermediary Cu_2O formation [42]. Nevertheless, the high "stability" of Cu_2O, responsible for the pronounced shoulder of the TPR profile of CuZnAl mixed oxide, is most likely a result of the strong interaction of Cu with the Zn-Al matrix, which hinders the reduction process. Proven for the hydrotalcite precursor, the two-stage consecutive model was vindicated for the calcined aurichalcite precursor [21]. Compared with Al-richer hydrotalcite-derived CuZnAl mixed oxide, aurichalcite-based CuO/ZnO system is characterized by a significantly lower activation barrier and a higher reduction rate, which implies freestanding Cu particles rather than the firm embedment into the oxide matrix. MS of the evolved products during TPR has revealed that the reduction involves the decomposition of residual carbonate; the latter

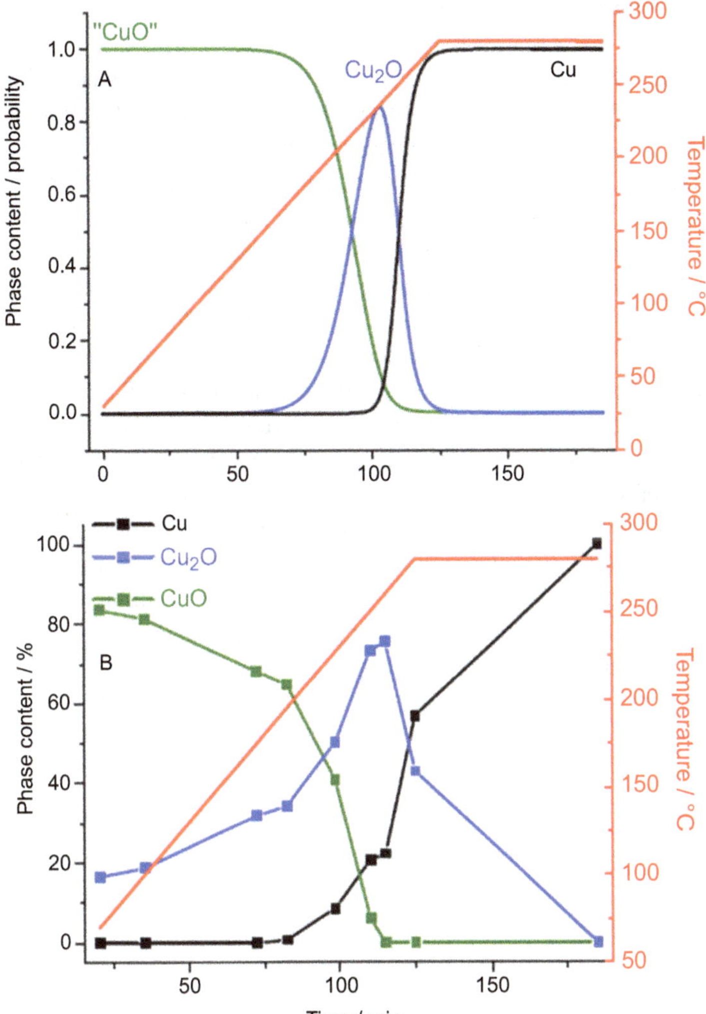

Fig. 1.10: Phase evolution of Cu, Zn, Al-calcined precursor during reduction in hydrogen with 2 Kpm (A) component kinetics-prediction form kinetic analysis of TPR data (B) quantification results of NEXAFS data – performed by the linear combination of reference spectra (reproduced from [42], Copyright © 2014 WILEY-VCH Verlag GmbH & Co. KGaA, Weinheim).

is considered to retard the reduction of CuO and to be the limiting step for the Cu(I)-Cu(0) reduction stage [43].

1.3.2.3 The impact of reduction temperature on catalytic activity

Although the Cu/ZnO catalyst withstands severe industrial conditions of methanol synthesis, the formation of an active catalyst is a sensitive process. Continuous heating in a reducing atmosphere (here called activation) promotes thermally and chemically induced structural transformation of nanoparticles (e.g., sintering, domain size growth, phase segregation, ZnO reduction, Cu/Zn diffusion, and alloy formation). Figure 1.11A displays the plot of the Cu lattice parameter versus the activation temperature in an atmosphere of 20% H_2 in He. The in situ XRD data show a gradual increase in the lattice parameter with an increase in the activation temperature beginning at 300 °C. This is consistent with the Zn diffusion into the Cu lattice in a reductive atmosphere, which starts from the surface [44]. Since a CuZn alloy (or α brass) has the same structural environment as Cu metal, the degree of CuZn alloy formation is directly reflected by the corresponding lattice parameter, which increases as Zn is incorporated. The significant amount of CuZn alloy at an activation temperature of 350 °C scales with the apparent activation energies (E_A) of the CH_3OH formation shown in Fig. 1.11B. Since the E_A increases with an increase in CuZn alloy amount, the alloy formation has a detrimental effect on the nature of the active site. The E_a at 300 °C seems unchanged, indicating that the surface alloy is unstable under industrially relevant testing conditions [44, 45]. However, when bulk alloy formed (from 350 °C), it remained persistent under the applied reaction environment. This irreversible change of the catalytic material entails the depletion of activity toward CH_3OH [46].

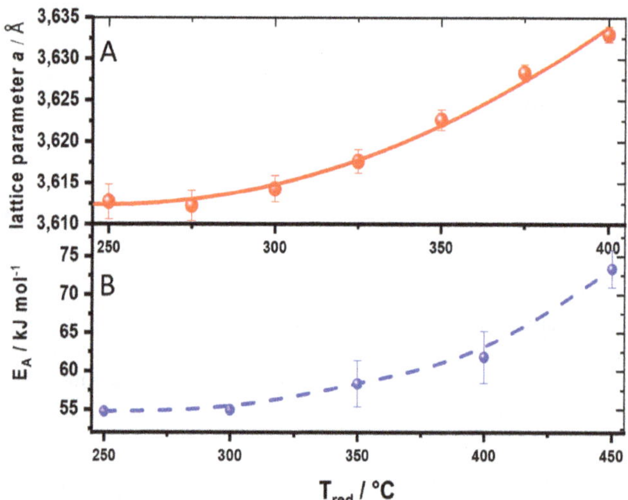

Fig. 1.11: The Cu lattice parameter versus the activation temperature under a reductive atmosphere of 20% H_2 in He, measured by in situ XRD for 3% Al promoted Cu/ZnO zincian malachite-derived catalyst: (A) apparent activation energies at different activation temperatures at 30 bar in synthesis gas ($8CO_2/6CO/59H_2/27He$) GHSV 6,000 h^{-1}, 190–250 °C; (B) the activation prior to the measurement was conducted in 20% H_2 in Ar and heating rates of 1 K/min. Splined lines are guides to the eye (adapted from [46]).

Exceeding the activation temperature over 300 °C leads to reactivity decay (alloy formation/sintering) unless the induced changes are reversible.

1.3.3 Characterization of metal–support interaction in a Cu-based catalyst

The most popular and standardized method used for studying Cu-based catalysts and quantifying the Cu surface area for more than 30 years is reactive nitrous oxide frontal (N_2O-RFC) or pulse chromatography [47–49]. A modified N_2O method has been suggested by Jensen et al. [50] for determining the Cu dispersion. The oxidations of surface and bulk Cu-species are considered as separate processes, which explain the continuous detection of N_2. This N_2O technique enables the calculation of the oxygen diffusion coefficients and, thus, the assessment of the oxygen mobility in metals (here, Cu), which is crucial for the redox dynamics of catalysts at elevated temperatures (when diffusion processes are pronounced). In other words, the catalyst resistance against oxidation can affect the catalytic performance in reactions and can be presumably used as a descriptor for assessing phase stability and deactivation.

The role of the support in the dynamics of Cu oxidation can be judged from the oxygen diffusion coefficients calculated at elevated temperatures. Figure 1.12A summarizes the diffusion coefficients of three catalyst families determined in the temperature range 30–250 °C. The diffusion of all catalysts grows continuously with temperature and significantly increases starting from 200 °C. At 220 °C, the diffusion coefficient of a malachite-derived catalyst Cu/ZnO turns out to be four times higher than that of the Al-promoted Cu/ZnO:Al. It is evident that the incorporation of Al into the ZnO lattice (as a dopant with an influence on, e.g., oxygen vacancies and SMSI) [51, 52] considerably affects the interaction with the Cu nanostructure. The higher reducibility of the ZnO:Al support at the interfacial contact with Cu leads to better incorporation [53] (ZnO overlayer) and stabilization of the Cu species (particle size and strain-induced defects), which directly influences the oxygen diffusion coefficients. The diffusion coefficient of an impregnated catalyst with 10 wt% Cu on ZnO:Al at 250 °C is 3.5-times higher than that of the coprecipitated catalyst (Fig. 1.12(A), blue-orange arrow). On the contrary, Cu/MgO shows at 250 °C a 5.5-times lower diffusion coefficient than Cu/ZnO:Al, although the Cu nanostructures are directly comparable. Since the nonreducible MgO support exhibits a strong static interaction with Cu, it stabilizes the domain size and defects without any tendencies to overgrow [54]. According to the results of single-crystal studies on MgO(100) [55–57], surface hydroxyls can be responsible for the specific stabilization of metal/oxide bonding, and this, in turn, may lead to the increase of adhesion on the Cu/MgO interface. In that respect, the overgrowth of the ZnO_x [51, 58] does not act as a protection layer preventing O diffusion; rather, it stabilizes the defective Cu nanoparticles. Moreover, the nonstatic and reducible ZnO support promotes oxygen diffusion. Comparison of the oxygen diffusion coefficient of co-precipitated Cu/MgO with that of an impregnated catalyst with 10 wt% Cu on MgO demonstrates, again, an approximately fivefold increase (Fig. 1.12(A), gray-orange arrow) due to weakening of the metal–support interaction.

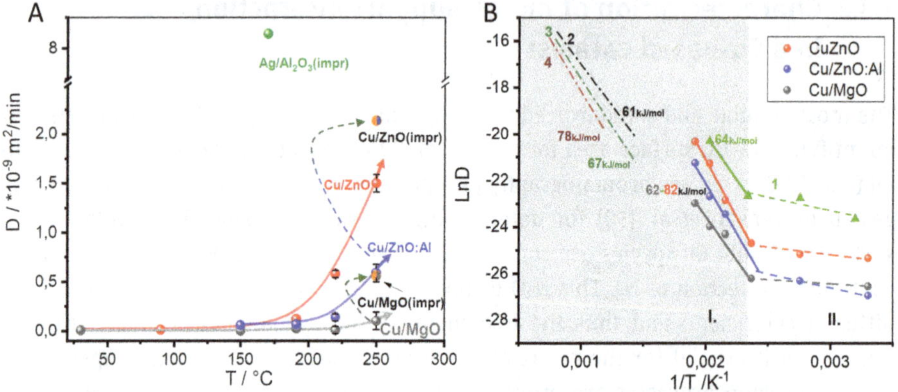

Fig. 1.12: (A) Oxygen diffusion coefficient as a function of temperature for various catalysts. Dashed arrows guide the essential change in diffusivity between co-precipitated and impregnated catalysts. Splined lines with arrows are guides to the eye; (B) Arrhenius plots of selected catalysts. Literature data indicated for supported Cu/ZnO catalyst: 1 – Jensen et al. [50]; bulk Cu: 2 – Albert et al. [59], 3 – Narula et al. [60], 4 – Ramanarayan et al. [61] (adapted from [62] with permission from Elsevier, copyright 2020).

A reference measurement with supported Ag/Al_2O_3 (Fig. 1.12A) confirms the reliability of the diffusivities. The low oxophilicity of Ag, along with its larger lattice parameter of 4.086 Å (compared to that of Cu, 3.615 Å), is responsible for a significant increase in diffusion coefficient, already at lower temperatures (Fig. 1.12(A), green point) [61–64].

Figure 1.12B illustrates the relevant Arrhenius plots for the oxygen diffusion process. The different slopes of the plot at high and low temperatures (regions I and II, respectively) indicate that different processes are involved. At temperatures below 150 °C, the formation of a thin Cu_2O film stops the oxidation [65]. The process is limited by adsorption [66] and ionic transport on the $Cu_2O_{surface}/Cu_{bulk}$ interface. The low activation energies of 10–20 kJ/mol support this since they are inherent to adsorption–desorption processes [67]. The activation energies for oxygen diffusion in solid Cu or high-temperature electrochemical studies on bulk Cu at 600–1,000 °C varies in the range of ~60–80 kJ/mol [6, 59, 60, 68] (Fig. 1.12B). These energies are consistent with the E_a calculated for Cu/ZnO (82 kJ/mol), Cu/ZnO:Al (75 kJ/mol), and Cu/MgO (62 kJ/mol), which indicate that the oxygen diffusion in a Cu–supported catalyst at moderate temperatures (150–250 °C) is likely characterized by the same kinetics as in bulk Cu at high temperatures. The increase in the E_a, inverse to the strength of the Cu–support interaction, is evidence of the probing character of the oxygen diffusion coefficient. This means that a stronger Cu-support interaction affords smaller and more defective Cu particles where oxygen diffusion occurs at a moderate rate (e.g., 59 × 10^{-11} m^2/min at

250 °C for Cu/ZnO:Al). The less defective and bigger Cu species in Cu/ZnO offer, very likely, more diffusion sites/channels (150×10^{-11} m^2/min at 250 °C). The oxidic support has a strong effect on the oxygen diffusion coefficients and the formation of a Cu_2O layer (e.g., the lack of the stabilizing character of the supports in impregnated samples leads to significantly higher diffusion coefficients). The more defective catalysts are also less prone to forming Cu_2O and have smaller diffusion coefficients (compared to Cu/ZnO). Thus, it can be suggested that the difference in metal–support interactions accounts for differences in Cu particles and their oxygen diffusion properties.

1.3.4 Reaction

Direct application of thermal analysis methods under industrially relevant conditions of methanol synthesis is challenging. Although it is technically feasible to decouple the sensing/detection unit from the reaction environment and thus exclude the effects of harsh conditions on measuring hardware, the data interpretation is rather demanding due to possible pseudothermal events (sample grains shrinkage, expansion, specimen breakdown, buoyancy force change) and difficulty to approach relevant conversion (sample holder/reactor geometry, flow dynamics). Such a measuring campaign requires essentially blank and reference measurements with substances of well-known stable properties and cross-validation with complementary techniques. High-pressure in situ thermogravimetry is able to provide quantitative insights into the adsorbed layers on an industrial-like Cu/ZnO:Al catalyst in the course of CO/CO_2 hydrogenation. Information on the composition of a catalytically formed adlayer is of great importance for revealing the relevant intermediates of the CH_3OH formation from CO- or CO_2-containing feeds.

To that end, the activated (20% H_2 in Ar) catalysts were exposed to a CO_2-containing feed (Fig. 1.13A). A direct weight gain indicates the formation of an adsorbate layer on the catalyst surface; the layer ceases to change when the mass gain reaches 1.9 wt% of the initial mass after ≈10 h in the H_2/CO_2 feed. The replacement of the feed with H_2/CO is accompanied by a reversible loss of 0.7 wt% within ≈8 h. Among carbon-based adsorbates, the water-derived species are discussed as an essential constituent of the catalytically active adlayer. The importance of H_2O/OH for the CH_3OH formation has been already demonstrated for Cu systems [69, 70]. Water molecules may have a promotional effect at finite conversion [71] or lead to product inhibition at integral conditions [72]. To elucidate the contribution of H_2O and H_2O-derived species (e.g., OH groups) to the composition of the adsorbate layer, the catalyst was saturated in a thermobalance with an H_2/Ar stream enriched with 1.5 vol% H_2O (Fig. 1.13B). A stable mass gain of 1.3 wt% was reached in ca. 10 h of TOS. Switching to a dry H_2/Ar stream resulted in a mass loss of 0.7 wt% within ≈8 h; the weight continued to decrease slowly, approaching a theoretical value of 0.4 wt% for stable hydroxyl groups on oxygen vacancies in ZnO_x [73] measured with surface titration methods [52, 74]. The mass change in an H_2O atmosphere

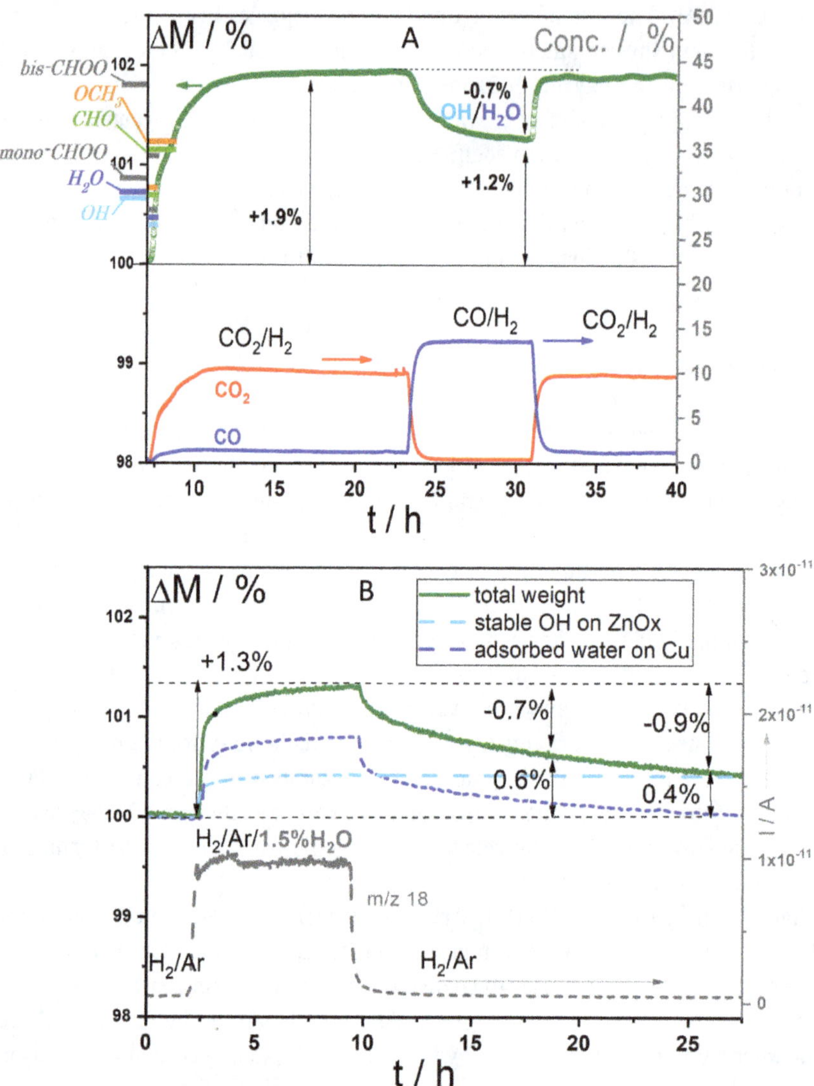

Fig. 1.13: In situ thermogravimetry of 3% Al-promoted malachite-derived catalyst Cu/ZnO:Al in different feed gases at 250 °C: (A) $CO_2/H_2/He$ (12/59/29), $CO/H_2/Ar/He$ (14/59/4/23), 120 mL/min at 30 bar, with online analysis of the gas phase; (B) introduction of 1.5 vol% H_2O vapor to 85% H_2/15% inert. Colored bars on the ΔM axis indicate the theoretical mass increase of the catalyst, provided that the surface is entirely covered with one type of species. Large bars correspond to species-oriented coverage on Cu sites, small bars – ZnO_x sites only (adapted from [73] with permission from the American Chemical Society, copyright 2020).

directly correlates with the Cu surface area and ZnO_x vacancies. These findings are visualized in Fig. 1.13B, where the mass increase is exemplarily divided into two contributions from the adsorbed H_2O and strongly bound OH groups.

From the TG curves in Fig. 1.13B, it becomes evident that the mass loss that accompanies the switching from H_2/CO_2 to H_2/CO (Fig. 1.13A) is primarily caused by the removal of H_2O-derived species. Consequently, the maximum amount of strongly bound OH-groups still present on the ZnO_x surface ranges from 0.4 to 0.8 wt% of the catalyst. In addition, it seems highly probable that carbon-based reaction intermediates also form depending on the gas feed used. According to the residual mass after switching from H_2/CO_2 to H_2/CO at 30 bar (Fig. 1.13A), a stable adlayer is 1.2 wt% of the catalyst, of which at least 0.4–0.8 wt% are non-H_2O-related adsorbates, namely, carbon-based derivatives (dependent on the stable OH-group concentration). Most of the carbon-derived adsorbates might be placed on the Cu moieties. Based on the assignments made in high-pressure in situ IR experiments [73], the mass gain in H_2/CO is rationalized mainly by the formation of methoxy groups on Cu sites, which accounts for a mass increase of 1.24%, according to the calculation of the theoretical mass change (large green bar in Fig. 1.13A). Thus, during the treatment in H_2/CO at 30 bars, the adsorbates, for example, formyl- and bis-formate-groups accumulated in H_2/CO_2 are partly replaced by methoxy groups. Consistent with previous spectroscopic analysis [75], the remaining layer may contain methoxy, formyl, and/or formate species balanced with OH groups, which in turn agrees well with the mass changes and quantitative assessments from thermogravimetric experiments. The present observations demonstrate quantification of the catalytic intermediates formed under realistic conditions (1.9 wt% or 1–2 ML). It should be noted that an exchange of accumulated intermediates lasts for hours, where H_2O-derived species are of crucial importance, being intermediates, reaction participants, and possibly rate-determining reactants.

1.3.5 Final remarks

In summary, this chapter provided fundamental insight into the thermal properties of precursors for Cu-based catalysts, solid kinetics of catalyst formation with relationship to the performance, approach for extended surface characterization and quantification of surface adsorbates under reaction conditions. Here one has to point out the thermokinetic analysis, which allows the simulation of the temperature programs and predicts the amount of high-temperature carbonate species. Changing the heating ramp from 0.1 to 2 K/min during up-scaled calcination in a rotating tube oven up to 330 °C makes it possible to adjust the amount of HT-CO_3 in calcined precursors prepared from both ZM and AU precursors. The success of this procedure is vindicated by the consecutive TG analysis of the ex situ-calcined samples for the residual carbonates, and a good agreement between the predicted and experimental mass loss has been achieved. The presence of considerable amounts of HT-CO_3 after mild calcination has pointed to the lack of significant segregation and crystallization of CuO and ZnO. However, distinct changes in crystallinity, decomposition profile, and IR absorption did not result in noticeable differences in the catalytic activity in methanol synthesis. No beneficial effect of high HT-

CO_3 content was observed. The presence of large amounts of HT-CO_3 in the calcined sample did not lead to high intrinsic activity in methanol synthesis. At the same time, HT-CO_3 acted as a structural stabilizer, ensuring high surface areas and many oxygen vacancies in the ZnO component of the catalysts. This example demonstrates the potential of thermokinetic analysis when applied systematically to elucidate thermal processes in catalyst preparation and may essentially contribute to the optimization and upscaling of the synthetic protocol. Besides, notwithstanding the conclusive character of experiments provided in this chapter, the employment of the in situ realized techniques such as XRD, TEM, IR, and NEXAFS become inevitable for the veracity of the research.

1.4 Ni-based catalyst for industrial application

Supported Ni catalysts have been widely used in various "energy-relevant" industrial processes. Renewed interest has recently arisen in catalytic processes involving CO_2 such as its hydrogenation to methane [76, 77], the production of syngas by reverse water–gas shift (WGS) [78, 79], or the dry reforming of methane (DRM) [80, 81]. Ru, Rh, or Pt metals are known to be excellent catalysts for these reactions, whereas active base metals, particularly Ni, are rapidly deactivated because of coking. However, Ni-based catalysts are economically more suitable for commercial applications than noble metal ones.

It is worth noting that a favorable combination of size, morphology, structure, and composition is crucial for designing Ni-based catalysts. For example, the catalytic activity responds to the specific surface area and the acid–base properties of the support. Because the aforementioned processes imply the adsorption and dissociation of acidic CO_2, essential supports like MgO can enhance CO_2 chemisorption, thus increasing the catalyst's coke resistance. In addition, the strong binding of nickel to the support can significantly enhance the coke resistance. Strong metal–support interaction (SMSI)) effects and surface overlayers have a significant impact on many catalyst systems like the industrial Cu/ZnO/Al_2O_3 catalyst for methanol production [82]. It should be noted that a similar technique to stabilize Ni nanoparticles by incorporating them into a stable oxide matrix has been previously used for Ni-containing perovskites [83] and spinels [84]. The same strategy has been previously applied to Cu-based catalysts for methanol synthesis, in particular, when developing Cu, Zn, Al hydrotalcite as a promising catalyst precursor [42, 85]. As described in the previous section, the resulting catalysts are characterized by a homogeneous distribution of nanosized Cu particles in an amorphous $ZnAl_2O_4$ matrix, which leads to their stabilization.

In addition, MgO shows excellent prospects since NiO and MgO can form a solid solution owing to their identical face-centered cubic (fcc) structure and very close ionic radii (0.4213 nm for MgO, 0.41769 nm for NiO) [86, 87]. The intense interaction between the Ni species and MgO affords the formation of highly dispersed Ni particles, which enhances the CO_2 conversion activity [77] and stability [88–90]. Analogously to copper-based materials, the Ni precursor chemistry is also defined by "binary" (Ni,Mg) candidates with a brucite- and hydromagnesite-type structure or "ternary" (Ni, Mg, Al) hydroxocarbonates represented by hydrotalcite (Fig. 1.14A). Similarly, the thermal preparation route of a Ni-containing catalyst is paved with several stages of temperature-programmed treatment before a catalytically active material forms. Starting from a hydroxocarbonate precursor and passing through an oxidic precatalyst to eventually afford oxide-supported metallic Ni nanoparticles as a catalyst, the complete procedure shapes the final performance and requires in-depth thermal analysis.

1.4.1 Calcination

1.4.1.1 Decomposition of NiMgAl and NiMg hydroxocarbonates

A detailed consideration of TG-EGA curves (Fig. 1.14B) has shown that the decomposition of brucite-like precursor proceeds in one step in the range 200–380 °C involving the primary water loss (signal $m/z = 18$) that accompanies the conversion of the hydroxide structure to the oxide one. Above 380 °C, the water signal shows a tail up to

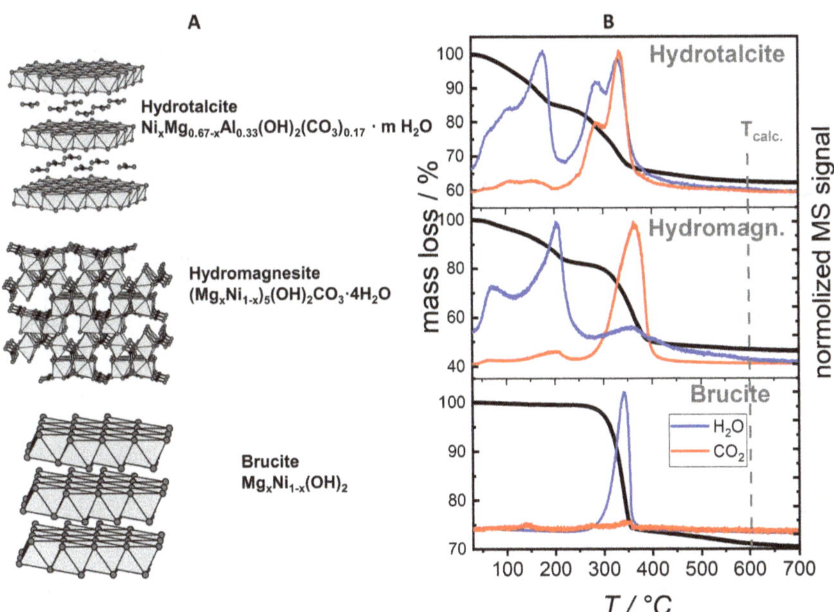

Fig. 1.14: Catalyst precursors: (A) crystalline structures of the precursors (black spheres – carbons, gray spheres – oxygen, MO_6 octahedra are filled with light gray), hydrogen and water molecules not shown) and (B) TG-EGA results of calcination in synthetic air (21% O_2 in Ar), 100 mlm with 2 Kpm.

700 °C, corresponding to the progressive water loss during the final phase transformation from the brucite to the MgO rocksalt structure.

In addition, there is a small signal ($m/z = 30$) at 500 °C due to the release of NO from the residual nitrates of the metal precursors used in synthesis (Fig. 1.14(A)). As the Ni content of the catalyst increases, the characteristic decomposition step begins at lower temperatures. The presence of a single thermal event (in contrast to two different ones) provides evidence in favor of the formation of a solid solution in a wide range of metal concentrations rather than a mere mixture of $Ni(OH)_2$ and $Mg(OH)_2$ [77]. The high phase purity of the precursors is also a prerequisite for forming the solid solution in the final catalysts.

In contrast to the single-step decomposition of the brucite-type precursor, the hydromagnesite features multiple steps on the decomposition profile. At least five steps describe the DTG curve as shown with peak deconvolution in Fig. 1.15(B). Although the DTG curve features overlapping processes, the by-phase of $Ni(OH)_2$ forms with increasing Ni content and is visible with a distinct DTG peak at 280 °C. Only partial Ni incorporation is achieved during synthesis, which is indicated again by the shift of the main thermal event to lower temperatures. Very likely, phase heterogeneity can be reduced by optimization of coprecipitation conditions.

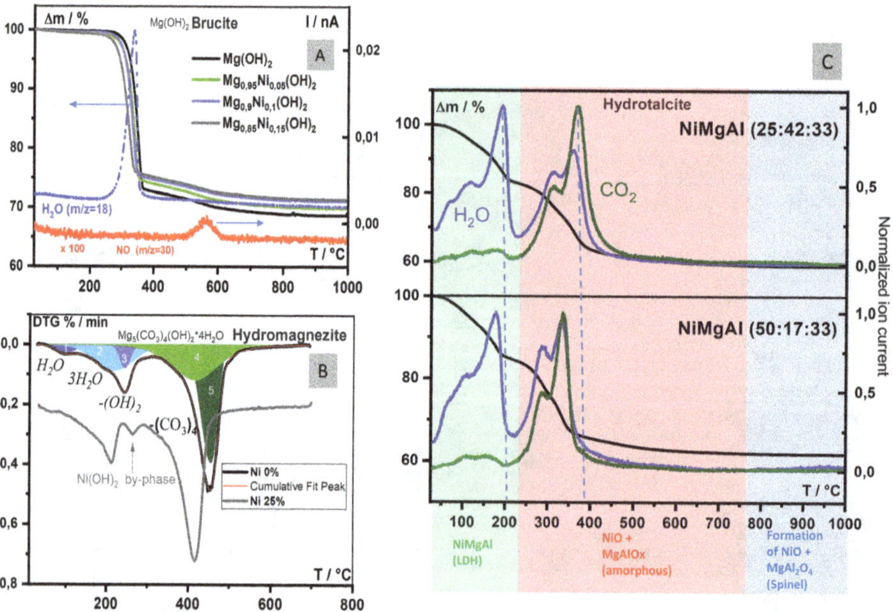

Fig. 1.15: Thermoanalytical curves of various Ni/Mg containing precursors at 2 Kpm in 21% O_2/Ar (A) TG curves for brucite series(adapted from [91, 92] with permission of American Chemical Society, copyright 2019). For more clarity, MS analysis only for a Ni-free sample is shown; (B) DTG of a hydromagnezite-based precursor. Enveloped is the peak deconvolution for decomposition curve (C) TG-MS analysis for hydrotalcite series (adapted from [81] Copyright © 2013 WILEY-VCH Verlag GmbH & Co. KGaA, Weinheim).

Similarly, the hydrotalcite-like precursor shows multiple simultaneous water and carbon dioxide evolution stages during linearly programmed heating up to 1,000 °C, as presented in Fig. 1.15(C). After drying, the decomposition can be represented by two major stages: (i) release of interlayer water in the range 125–225 °C simultaneously with removal of small quantities of CO_2 from weakly bound carbonates and (ii) bimodal dehydroxylation of the brucite layers and decarboxylation of the interlayer in the range 225–500 °C [93]. The overall mass loss measured for all samples ranges within 38–46%. It is higher for lower Ni concentrations due to a higher content of the lighter Mg. As pointed out previously, the shift of decomposition temperatures to lower values with increasing Ni content, and resemblance of the mass loss and gas-phase profiles in the Ni concentration series support the successful Mg substitution within the hydrotalcite structure. According to XRD analysis [94], the calcination product obtained in the range 250–450 °C is an amorphous, carbonate-, and hydroxyl-modified mixed Ni, Mg, and Al oxide, in which the uniform distribution of the metal species is largely inherited from the precursor upon decomposition. Further calcination at 600 °C results in a fully dehydrated and carbonate-free mixed oxide. Eventually, at 800 °C, crystallization of the spinel phase is observed. After calcination at 1,000 °C, the XRD pattern shows a mixture of rock salt- and a spinel-type phases as expected for the hydrotalcite decomposition product [93, 95]. As followed from the analysis of thermal behavior of mentioned hydroxyl-carbonate precursors, a minimum calcination temperature of 600 °C is necessary to form a pure mixed oxide precatalyst.

1.4.2 Activation

1.4.2.1 Reducibility

The reduction behavior of the materials calcined at 600 °C has been studied using TPR measurements in 5% H_2/Ar, as shown in Fig. 1.16(A). The maximum reduction temperature decreases from 914 to 663 °C with an increase in Ni content, suggesting autocatalytic reduction kinetics due to hydrogen spillover, which points to higher availability of the Ni species in the oxidic support. From general considerations, high reduction temperatures could be rationalized by the presence of $NiAl_2O_4$, which is reduced at higher temperatures than NiO because of the stronger interaction between Ni and the matrix in the former [80, 96]. Unexpectedly, excessive hydrogen consumption is observed at high temperatures for the reduction of Ni-free (Ni0) and Ni5 samples. MS has detected the evolution of CO_2, CO, and CH_4 during the reduction of the Ni-free sample. Therefore, the CO_2 release is due to the decomposition of residual interlayer carbonate ions of the hydrotalcite precursor. This is in line with the TG in air (Fig. 1.16 (C)) since for Ni-free and Ni5 samples CO_2 releases up to 900 °C. According to the observed MS traces of TPR measurement, liberated CO_2 undergoes reverse water–gas shift (rWGS) and methanation reactions at the Ni surface formed during TPR; it is precisely these processes that are responsible for a significantly higher H_2 consumption than expected and lead to the formation of CO and CH_4. For Ni5-600, the hydrogen uptake due to rWGS and methanation occurs simultaneously with the Ni^{2+} reduction. In subsequent TPOs of the Ni-free sample, oxygen consumption observed above 600 °C is accompanied by the evolution of CO_2. All this implies that CO and CH_4 formed during reduction undergo Boudouard and pyrolysis reactions, respectively, to give carbon deposited in the catalyst bed. The deposited carbon is oxidized to CO_2 at sufficiently high temperatures. Therefore, under reducing conditions, CO_2 released upon the decomposition of interlayer carbonates is partially converted to carbon even in the absence of metallic Ni and the CO_2/CH_4 gas mixture. In addition, high-resolution (HR) (S)TEM ((scanning) transmission electron microscopy) images of Ni5 and Ni50 after TPR to 1,000 °C demonstrate the existence of a crystalline overgrowth on top of the Ni particles (Fig. 1.16(B) and (C)). For the Ni50 catalyst, the overgrowth does not entirely cover the nanoparticles, so some areas of the metallic Ni surface remain exposed; this accounts for the observed lower reduction temperature because of the better availability of Ni^{2+} species. The overgrowth appears likely as nickel aluminate [97], formed as a result of moderate metal–support interaction [98].

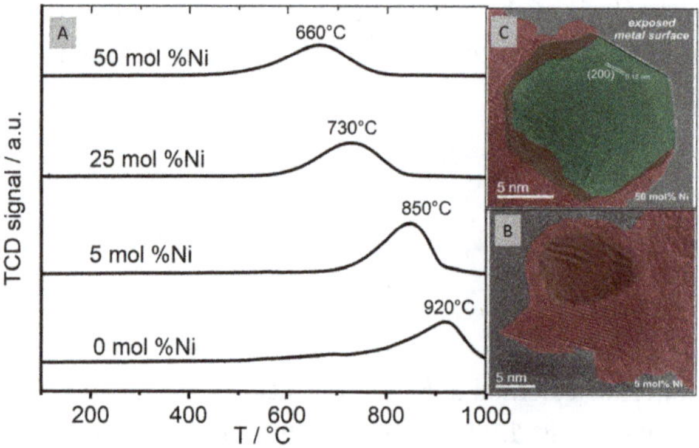

Fig. 1.16: (A) TPR profiles with the offset of the concentration series of hydrotalcite-derived oxides calcined at 600 °C using a linear heating ramp of 6 K/min in 5% H_2/Ar (60 mL/min) [94, 99]. HR-TEM images of Ni5 (B) and Ni50 (C) after TPR reduction at 1,000 °C [94]. Red and green colors represent the overgrowth and Ni particles, respectively (adapted with permission from the American Chemical Society, copyright 2016).

As mentioned in the previous section, the calcination of mixed NiMg hydroxide results in NiO–MgO solid solutions, known for their pure reducibility [88]. At higher calcination temperatures, the reducibility of the samples becomes lower (decreasing area under the TPR profiles, percentage indicated), as shown in Fig. 1.17. Following the literature, this phenomenon is attributed to the progressive diffusion of Ni atoms into the MgO matrix to form a solid solution [100–104]. The uniform spreading of the

Ni atoms over the MgO bulk hinders their reduction. The distribution of the Ni atoms in the structure is shown as the high-angle annular dark-field (HAADF) STEM image (Fig. 1.17(B)). This image demonstrates the discrete arrangement of the Ni atoms (brighter spots) in the MgO lattice. However, Ni atoms are not deposited on the surface as ad-atoms; thus, the TPR peak at 750 °C, characteristic of Ni atoms embedded in the MgO structure [100, 102, 105], underlines the homogeneous character of the $Ni_xMg_{1-x}O$ solid solution and the lack of NiO agglomerates on its surface.

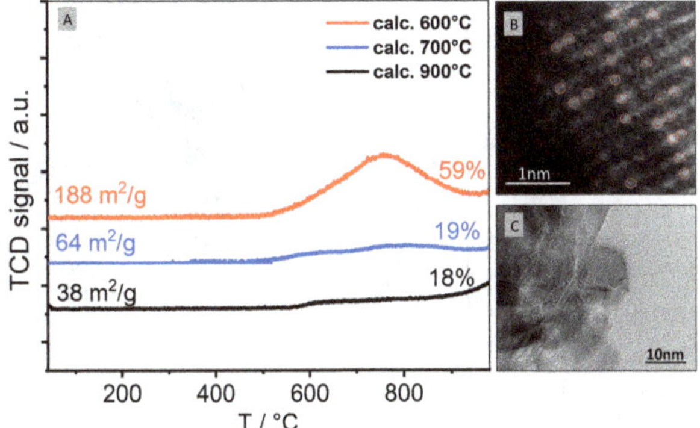

Fig. 1.17: (A) TPR profiles of the brucite-based NiO-MgO (10 at% Ni) samples calcined at different temperatures (600, 700, and 900 °C) using a linear heating ramp of 6 K/min in 5% H_2/Ar (60 mL/min). Insets indicate N_2 BET surface area of the calcined material and reducibility percentage at 1,000 °C [92]; (B) STEM-HAADF image of Ni10 calcined at 600 °C prior to TPR measurement [91]; (C) HR-TEM of a Ni particle of Ni10_600 after TPR to 1,000 °C [92] (adapted with permission from the American Chemical Society, copyright 2019).

As indicated in Fig. 1.17A, the calcination at high temperatures dramatically decreases the surface area of the samples and thus diminishes the accessibility of the Ni. A considerable fraction of Ni species is trapped inside the larger oxide particles (in contrast to the lower calcination temperature), and the contact area between the reductive gas and the material is decreased. After reduction, a well-crystallized material is formed as numerous little cubes (Fig. 1.17B). The Ni particles (≈10 nm) are uniformly dispersed [92]. Nearby well-facetted Ni particles and partly Wulff-constructed Ni particles are identified (Fig. 1.17C). The Ni particles are isolated from each other, being embedded/stabilized in the oxide support. However, in contrast to hydrotalcite-derived catalysts (Fig. 1.16), the Ni particles are not covered with the poorly reducible $Mg_{1-x}Ni_xO$ support, which is in line with the results of determining metal dispersion utilizing integral surface titration methods [106].

1.4.2.2 Redox dynamics

To study the structural and redox stability of the catalyst, single calcination, and reduction treatment of the precursor or multiple TPR/TPO (temperature-programmed oxidation) cycles are used. Subsequent TPR/TPO experiments provide not only information on the effect of redox cycling on the catalytic performance but a detailed course of structural changes (Fig. 1.18) [80].

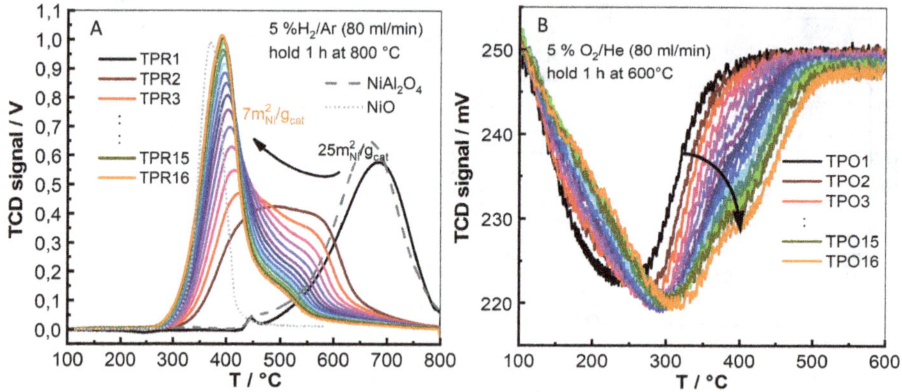

Fig. 1.18: (A) TPR cycles of hydrotalcite-based NiMgAl oxide catalyst (Ni50) calcined at 600 °C and of $NiA_{l2}O_4$ (dark gray-dashed line) and NiO (light gray-dotted line) as references; (B) TPO cycles of NiMgAl oxide catalyst. The TPR experiments of the calcined or reoxidized Ni50 catalyst are labeled TPR*n*, where *n* is the number of cycles, and the corresponding samples are named Ni50-TPR*n*. Accordingly, the reoxidation profiles are labeled TPO*n*, with *n* as the number of cycles, and the corresponding samples are denoted Ni50-TPO*n*. The TPRs were performed up to 800 °C; consequently, the TPOs up to 600 °C, analogous to the calcination process, with 16 repetitions (reproduced from [80] with permission from Elsevier, copyright 2015).

The TPR profile noticeably changed (Fig. 1.18A), whereas the total amount of consumed H_2 remained constant and corresponded to a 98% reduction of NiO. The TPR peak (temperature of maximum hydrogen consumption rate) shifted from 685 to 392 °C. Although slight changes might still be present, the system is considered stable after 16 cycles. Comparison of the profile of TPR16 with the profile of an unsupported NiO reference shows an explicit agreement, except for the slowly vanishing shoulder at higher temperatures. The (final) Ni phase after 16 reduction–reoxidation cycles consists primarily of NiO. Thus, redox cycling under the applied conditions causes the gradual transition from a strongly interacting "$NiAl_2O_4$-like" phase with considerably lower interaction to the support. Changes are also observed in the corresponding TPO profiles during reoxidation (Fig. 1.18B). Starting from a single broad peak at 220 °C, the peak maximum shifts upward to 313 °C with a shoulder arising at higher temperatures.

The slightly "bimodal" profiles after the cycling (TPR16 and TPO16) suggest more than one Ni-containing phase, with the NiO dominant. The structural transformations and temperature-induced agglomeration during cycling lead to a considerable decrease of Ni dispersion, which is indicated by a sharp decrease in metallic Ni surface area from 25 for Ni50-TPR1 to 7 m^2/gcat for Ni50-TPR16. Thus, the Ni50 catalyst undergoes structural ageing upon redox cycling: the Ni particles produced by reduction are sintered and become less dispersed. The resulting stable Ni/$MgAl_2O_4$ catalyst is characterized by a weaker interaction of the redox-active Ni phase with the crystalline support as compared to the starting material, which is demonstrated by the lower temperature of the TPR signal. This lower interaction with the crystalline support is likely responsible for the higher particle mobility leading to sintering at high temperatures. It is worth noting that the structural changes in the catalyst caused by repeated redox cycles have only a moderate influence on its activity in the DRM reaction at 900 °C.

At the same time, a significant decrease in coke deposition was observed after DRM in a fixed-bed reactor, mainly because of a lower fraction of graphitic carbon. Complementary DRM experiments carried out in a thermobalance have demonstrated that coking continues until a carbon limit of 123 wt% is reached and that redox cycling leads to a faster carbon formation [80].

1.4.3 Surface characterization

1.4.3.1 Metal dispersion and pulse TA

The catalytic properties of most supported catalysts strongly depend on their metal dispersion [107, 108]. Herefore, the design of a reliable and straightforward method for determining metal dispersion remains relevant, especially in the case of Ni. The potential of thermal analysis has been significantly extended by an improved construction of calorimetric cells and integration with new gas analysis systems. Over the past two decades, a significant breakthrough has been achieved in developing and applying pulse thermal analysis (Pulse TA®) [5, 109–113]. Pulse thermal analysis enables precise dosing of small portions of the reacting gas into an inert carrier gas stream. Simultaneous monitoring of the variations of the catalyst mass by thermogravimetric analysis (TGA), the reaction enthalpy by differential thermal analysis (DTA), and the gas composition by MS provides an insight into heterogeneous reactions. This method also has the advantage that the reaction can be interrupted between the pulses. Its completion is observed when successive pulses no longer lead to any change in the system.

In addition, under flow conditions with a short contact time, pulses better mimic the reaction environment than classical volumetric methods, turning the traditional "bulk-specific" TA into a more surface-sensitive method [49, 114]. Pulse TA® was used to oxidize the surface of metallic Ni supported on $Ni_xMg_{1-x}O$ solid solutions. The

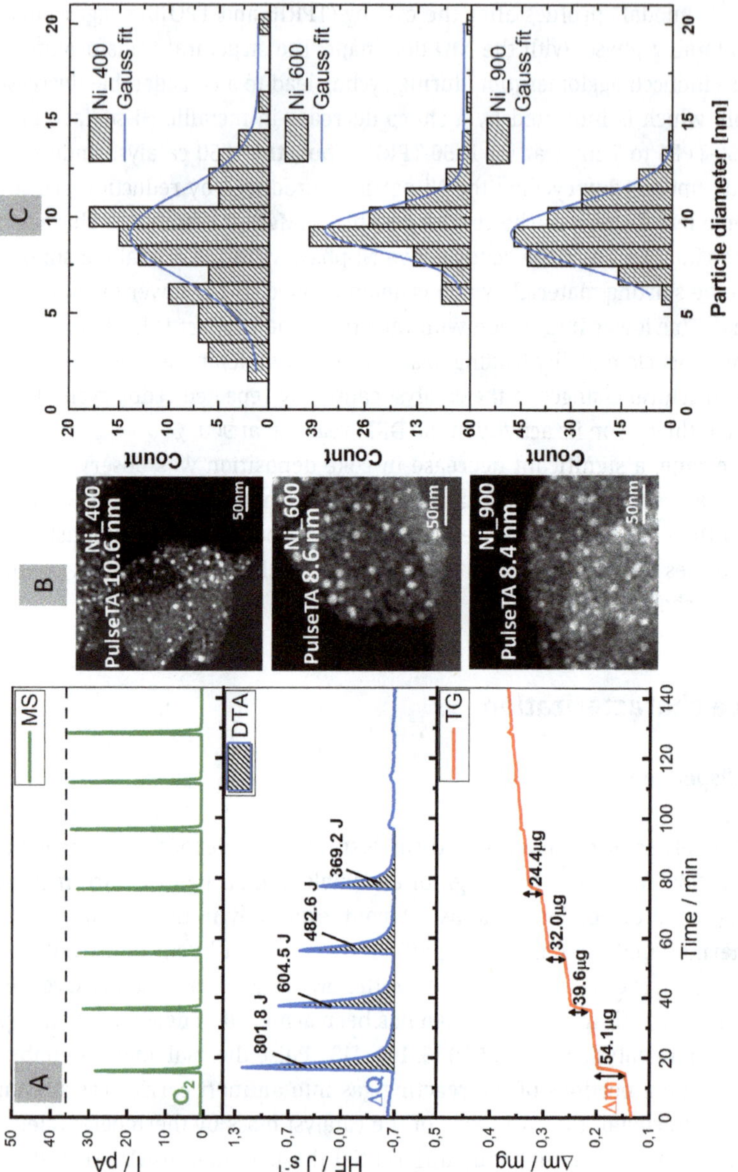

Fig. 1.19: (A) Oxygen pulse experiment at 40 °C on brucite-derived Ni5 catalyst, calcined at 600 °C and reduced in situ at 1,000 °C. (B) HAADF image of Ni10 sample calcined at 400, 600, and 900 °C after reduction at 1,000 °C. Average particle size from pulse TA indicated (C) particle size distributions on more than 110 particles of the three samples (reproduced from [106] with permission from Elsevier, copyright 2018).

amount of oxygen consumed was followed by monitoring the change in the sample mass (and using MS), while the enthalpy of the corresponding oxide formation was determined from DTA data.

After the in situ reduction of the Ni5 sample at 1,000 °C to metallic Ni, 500 μL of O_2 was pulsed into the DTA-MS chamber, and each pulse was monitored simultaneously by three complementary techniques as shown in Fig. 1.19A: the evolved oxygen ($m/z = 32$) was detected by mass spectrometry, the heat released at each oxidation was measured by DTA, and the mass gain was monitored by thermogravimetry. Seven O_2 pulses onto the reduced sample were performed, and changes were observed up to the fourth pulse for all signals, and the three last pulses caused only minor changes in the system. The O_2 uptakes decreased from pulse to pulse, as the amount of metallic Ni, still oxidizable, decreased. The chemisorption was accompanied by heat release, which decreased with a decrease in the O_2 uptake until an abrupt loss of the heat signal, implying surface saturation [109]. The peaks became stronger with each pulse as the oxygen consumption decreased. With ongoing O_2 pulses, minimal heat releases were observed, indicating consecutive subsurface oxidation. The corresponding particle size was calculated by the following equation [115]:

$$\mathrm{MCS} = \frac{F}{\left(\mathrm{Ni}^0\ \mathrm{s.a.}_{\mathrm{Ni}} * d_{\mathrm{Ni}}\right)} \tag{1.1}$$

Here, MCS stands for the metal particle size (m), F is the shape factor (6) assuming spherical particles [116, 117], d_{Ni} is the density of metallic Ni (8.908 g/cm^3) [118], and Ni^0 s.a.$_{\mathrm{Ni}}$ is the metal surface area per gram of reduced Ni (m^2/g$_{\mathrm{Ni}}$) calculated as

$$\mathrm{Ni}^0\ \mathrm{s.a.}_{\mathrm{Ni}} = \frac{\left(\left(\frac{\Delta m}{M_{\mathrm{O}}}\right) * N_{\mathrm{A}}\right)}{g_{\mathrm{metal}}} * \mathrm{RA} \tag{1.2}$$

where Δm is the mass gain (g), M_{O} is the molar mass of atomic oxygen (g/mol), N_{A} is the Avogadro constant (6.022×10^{23} mol^{-1}), RA is the Ni atomic cross-sectional area (6.49×10^{-20} m^2/atom) [119], and g_{metal} (g) corresponds to the mass of reduced nickel upon reduction treatment, assuming a dissociative chemisorption mechanism (stoichiometric factor Sf = 2; 2Ni/O_2) [120, 121].

In this case, electron microscopy was used as a complementary analysis method. Good agreement has been found between the particle size estimated from O_2 pulse TA (Fig. 1.19B) and the average and median particle size from the particle size distribution, as described by Gauss function, observed in TEM (Fig. 1.19C). The accuracy of the O_2 pulse measurements could be improved by using better gas purification systems (implying removal of moisture traces) and less porous and hydrophilic materials for the inner reactor. In addition, experimental temperatures for pulses below 10 °C are of advantage for preventing possible multilayer oxidation.

1.4.4 Reaction studies

1.4.4.1 Coking behavior in dry reforming of methane

In the DRM, a highly endothermic reaction, noble metal-based catalysts are very active and stable. Ni-based catalysts have a low cost and wide availability but are more prone to fast deactivation by coking, oxidation and poisoning. The carbon deposition occurs mainly due to the endothermic methane decomposition and the Boudouard equilibrium and leads to the deactivation of the catalyst and causes irreversible damage of reformers [122]. The combination of thermogravimetry and MS allows monitoring the carbon deposition in situ with simultaneous analysis of the gas phase. Here, great care in interpreting catalytic results should be taken because the experimental conditions in catalytic fixed-bed reactors may not be directly imposed on the sample in a measuring cell of the thermobalance. Thus, a direct comparison of the activity is, in many cases, not correct. However, the observed trends can provide better insights into carbonization kinetics during DRM. The current paragraph is devoted to comparative thermogravimetric studies of coke deposition on Ni-based and Ni-free catalysts in long-term isothermal experiments under DRM conditions.

Two conventionally prepared noble metal-based catalysts Rh/$CeZrAlO_x$ and perovskite-derived Ru/Ni/La_2O_3 were used as reference materials in the DRM reaction to compare coking behavior with hydrotalcite-derived Ni-based catalysts. Figure 1.20A demonstrates the coking dynamics of different catalysts under dry reforming conditions at 900 °C within 10 h. Because of the pronounced endothermicity of the reaction, the actual temperature of the catalyst bed during the reaction is significantly lower than that of the thermobalance reactor. According to thermodynamic calculation [99], lower temperatures enhance carbon growth. The Ni-rich catalyst strongly undergoes coking without saturation within 10 h time on stream. During the reaction, the weight of the sample drastically rises by 100% for Ni50/$MgAlO_x$ and by 70% for Ni/Ru/La_2O_3, whereas Ni5/$MgAlO_x$ and the Rh-based catalyst show stable weight gain with 1% and 3%, respectively.

An active coking process was observed during 10 h of DRM over Ni50/$MgAlO_x$ and Ni/Ru/La_2O_3. Figure 1.20B shows that the CH_4 conversion over Ni50/$MgAlO_x$ is the highest and continuously increases with the growing amount of carbon deposits. The primary form of carbonaceous deposit on this catalyst represents carbon filaments 20–30 nm thick [81]. The growing filaments move an active Ni microcrystal, and its surface remains accessible for the catalytic reaction, retaining the activity. The catalysts Ni5/$MgAlO_x$ and Rh/$CeZrAlO_x$ undergo less coking and show stable conversions within 10 h TOS. Because of unavoidable reverse water–gas shift reaction $CO_2 + H_2 \rightleftarrows CO + H_2O$, the conversion of CO_2 should be higher than that of CH_4 in the reaction. The methane conversions for both catalysts are close, whereas the samples demonstrate a pronounced difference in CO_2 conversions (Fig. 1.20C). This might be explained by a lower activity of Rh-based catalyst toward reverse water–gas shift reaction. This result agrees with post-

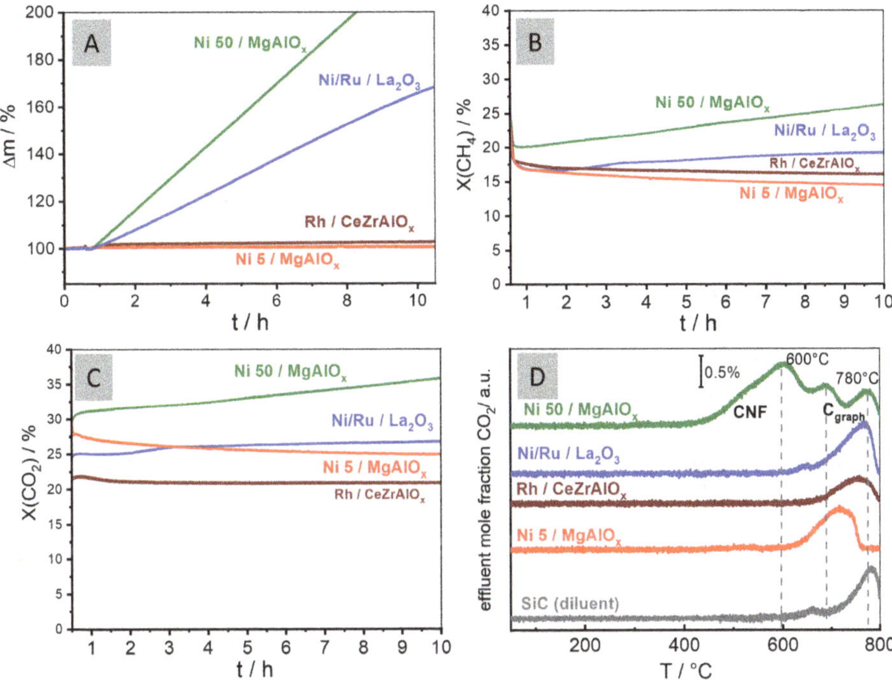

Fig. 1.20: (A) Gravimetric measurement during DRM reaction ($p = 1$ bar, 900 °C 120 Nml/min, $CO_2/CH_4 = 1.25$); (B and C) conversions versus reaction time during DRM reaction; (D) effluent mole fraction of CO_2 during the TPO experiments with 20 Nml/min, 4.5% O_2 in Ar. Assignments of TPO from [99] adapted from [123] (copyright © 2014 WILEY-VCH Verlag GmbH & Co. KGaA, Weinheim).

reaction TPO experiments in a fixed-bed reactor, which demonstrated lower carbon accumulation on Ni5/MgAlO$_x$ and Rh/CeZrAlO$_x$ catalysts (Fig. 1.20D). During the TPO experiments with catalysts, exclusively CO_2 was formed, while during the blank test with pure SiC, CO formation was observed [99]. The signals during TPO of the Ru and Rh catalysts are closely related to the profile of the SiC blank test. The noble metals Ru and Rh have a very low intrinsic coking tendency compared to the non-noble metal Ni. Therefore, only small amounts of coke should be formed on the catalyst. The primary type of carbon was graphitic. The high-temperature CO_2 peak (above 700 °C) was ascribed to the graphite-like carbon originating from methane pyrolysis. These graphitic depositions appear only during the DRM at 900 °C and present on the SiC diluent. Only for Ni50 other carbon species were obtained (Fig. 1.20D).

In summary, the present work demonstrates the ability of Ni-based and Ni-free catalysts to resist coking in the dry reforming reaction as examined by thermogravimetry at 900 °C. Ni-rich catalysts undergo strong coking; however, this severe carbon formation on Ni50/MgAlO$_x$ and Ni/Ru/La$_2$O$_3$ samples does not result in total activity loss. The Ni-free and Ni5/MgAlO$_x$ catalysts undergo less coking, featuring, on the other hand, lower conversions. Hence, the catalyst resistance against coking can be improved by using noble metal-based Ni-free precursors or decreasing the Ni content in the catalytic system on the cost of the activity.

1.4.5 Conclusive summary

This chapter presents a thermoanalytical approach to developing a thermally stable Ni/MgAl and oxide catalyst for the DRM reaction based on understanding and optimizing the catalyst synthesis. It contributes to a better fundamental comprehension of the role of both the metal Ni particles and the basic support. As for the active catalyst, particular attention has focused on observing the formation of carbon deposits through in situ TG. Detailed examination of the material at all stages of the preparation has enabled the development of a synthetic route via Ni, Mg, Al hydrotalcite-like and Ni, Mg brucite-like precursors that ensured the formation of catalytically active nanostructures. Upon the high-temperature reduction studied with TPR, Ni nanoparticles appeared strongly embedded into an oxide matrix and partially enclosed by an overlayer.

Similar to copper-based Cu, Zn, Al and Cu, Mg catalysts [56–59], this phenomenon was observed for Al-rich hydrotalcite systems. In contrast, for Al-free brucite-derived catalysts, oxidic overgrowth was not revealed. Redox TPR-TPO treatments gradually changed the interaction of the redox-active Ni phase with the oxide support to afford a crystalline Ni/$MgAl_2O_4$-type catalyst with weaker Ni–Al interactions. Furthermore, the DRM activity and the carbon deposition were enormously dependent on the Ni content: the larger the amount of incorporated Ni, the higher the activity, and the lower the coke resistance. The carbon continuously formed during the investigated time, mainly through methane pyrolysis. Based on the gained insights, it can be concluded that there should be a good balance between activity and coke resistance in a suitable catalyst, which is controlled by an interplay of Ni dispersion, embedment, and metal–support interactions. Therefore, a pulse titration method based on the oxidation of the metallic particles by the oxygen inside a TGMS setup was applied for quantifying the exposed metallic surface area. Although one could always further argue about the accuracy of the quantitative data collected, the correlations observed with the catalytic results demonstrated its relevance and the pertinence of the qualitative data obtained [92].

1.5 In situ perspective

This chapter reviewed various applications of thermal analysis methods to study materials for catalyst production in the chemical energy storage field. Particular emphasis was placed on precursor transformations and gas–solid interaction under reaction conditions. The applications considered in this chapter were chosen based on two catalyst families, which have been thoroughly examined with thermal methods at every stage of preparation. Several other examples of TGA of different catalyst precursors and thermogravimetry in reactive environments are provided by Pawelec and Fierro [2]. In this extensive review, the authors reported on elucidating the chemical processes for diverse materials during the single calcination (TG-TPO) or reduction (TG-TPR) step. Conversely, the specific focus laid here was to describe transformation kinetics and explore surface-structural phenomena along the path from a specific precursor to an active catalyst through multiple thermal steps.

In situ thermogravimetry has been proven to be a valuable tool for quantifying reaction adsorbates and deposits. However, the comparable activity to a plug flow reactor is difficult to approach due to the batch-like reactor type of the TG instruments, causing the feed gas to bypass the crucible. This drawback can be essentially resolved by in situ DSC technique, which showed great potential for thermochemical characterization of a catalyst under reaction conditions. Sinev and Bychkov [9] showed various applications of in situ DSC to investigate the catalysts' dynamic thermochemical state. In particular, the calorimetric results elucidated thermochemical identity of lattice oxygen species involved in catalytic oxidation in close relation to the factors governing the process selectivity [14–18]. The fixed-bed geometry of the Tian-Calvet DSC cell [10] can be applied in a steady (dynamic) or transient (pulsed) mode for the reaction kinetics studies, examination of performance during the operation of a catalyst, that is, caused by catalyst deactivation, and rapid screening using temperature-programmed reaction.

The unexploited possibilities of the current technique can be explored in future in studying phase transitions during catalytic operation in a steady (dynamic) mode. For example, aerosil-supported potassium pentavanadate was investigated under oxidative propane dehydrogenation (ODP) conditions in the temperature range of 300–500 °C (Fig. 1.21). The melting of the supported phase in inert feed at 409 °C (Fig. 1.21A) correlates directly with the loss of activity and selectivity increase under reaction conditions [124] (Fig. 1.21B, C). First, the heat flow increases with the growing temperature due to the exothermic reaction. The product-based heat flow equals the experimental heat flow up to the liquefaction. Once the phase transformation occurs, the discrepancy between the heat flows (highlighted area on plot B) becomes pronounced, not expected solely from product spectra. Heat flux decreases during the phase transition from solid to liquid (plot B). The conversion profile (plot C) perfectly follows the heat profile, which first increases with growing temperature and then levels off when endothermicity starts. A remarkable increase in selectivity to the propylene coincides with the activity loss. This observation suggests that the mobility of the potassium vanadate phase due to melting may inhibit the reactivity of surface species.

Interesting information on isotope effects in silver-catalyzed ethylene oxidation to ethylene oxide (EO) can be obtained using in situ DSC in a transient mode combined with pulse thermal analysis. First, the Ag/α-Al_2O_3 [125, 126] catalyst was conditioned inside the in situ DSC Tian-Calvet cell [9] under ethylene epoxidation conditions until a steady state was reached (200 °C, 1 bar, $C_2H_4/O_2/Ar$ (25/5/70) 30 mlm (GHSV ~ 14,300 h^{-1}).

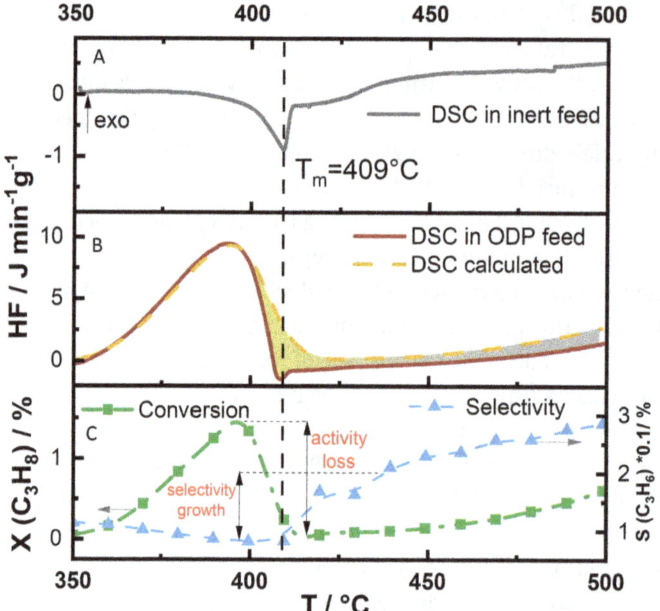

Fig. 1.21: In situ DSC of the aerosil-supported $K_3V_5O_{14}$ (10 wt%) in propane oxidation with 0.42 Kpm and 9 mlm total flow, (A) melting on $K_3V_5O_{15}$ in inert feed (B) DSC at ODP conditions $C_3H_8/O_2/N_2$ = 7.5/7.5/85 vol%, GHSV ≈ 4,300 h^{-1}. The dashed line shows the heat flow calculated from standard formation enthalpies of products and their concentrations. The area under the curve highlights the region where the catalyst transforms; (C) conversion and selectivity plots during the temperature-programmed DSC experiment.

Then a continuous 500 ppm C_2H_4 flow, maintained over a catalyst, is interrupted with short pulses of $^{16}O_2$ or $^{18}O_2$ (Fig. 1.22). Each pulse results in 420 kJ/mol(O_2) reaction heat, with X(C_2H_4) 35% and S(EO) 10%. The resemblance of the heat flow profile for $^{16}O_2$ pulses with one for $^{18}O_2$ pulses and the similarity of the total heat per pulse indicate on minor impact of kinetic and thermodynamic isotope effects on the course of reaction at 200 °C. The oxygen from the gas phase ($^{18}O_2$) gradually exchanges subsurface oxygen species (^{16}O). The prestored oxygen ^{16}O diffuses from the subsurface region to the surface and is mainly incorporated into CO_2 ($C^{16}O^{18}O$ trace).

On the contrary, EO ($C_2H_4{}^{16}O$) levels off quickly and contains only ^{18}O isotope ($C_2H_4{}^{18}O$ trace). These results demonstrate that silver catalyst serves as an oxygen reservoir in their active state. The surface and subsurface oxygen atoms exchange is an ongoing reversible process. Adsorbed oxygen atoms (^{18}O) react with ethylene to give EO. However, subsurface atoms (^{16}O) migrate to the surface to exchange with adsorbed oxygen and likely attack EO molecules to form scrambled CO_2. This example shows the crucial role of oxygen diffusion and counter diffusion on catalytic ethylene epoxidation.

Analysis of the heat balance may provide an idea about the active state of the catalyst. The difference between directly measured heat ~420 kJ/mol (Fig. 1.22) and the product-based reaction heat, deduced from gas-phase composition and thermodynamic data, shows 10–30 kJ/mol. This energy difference can be ascribed to the mean Ag–O bond strength, namely the energy of oxygen participating in catalytic conversion, which may differ from the adsorption energy acquired by direct admission of oxygen dosages over the oxygen-deficient catalyst. This range of energies may suggest molecular species, dissolved subsurface oxygen, and oxygen adsorbed on grain boundaries [127]. However, the reactive environment of the current transient experiment does not correspond to the chemical potential of typically applied EO synthesis conditions. Thus, the thermochemical state of the catalyst will be different when the reaction conditions are applied, and the catalyst may accommodate other O species converting ethylene. Comparing the direct calorimetric heat and product-based heat may be helpful to spot side reactions in the gas phase and discover the state of solid bulk.

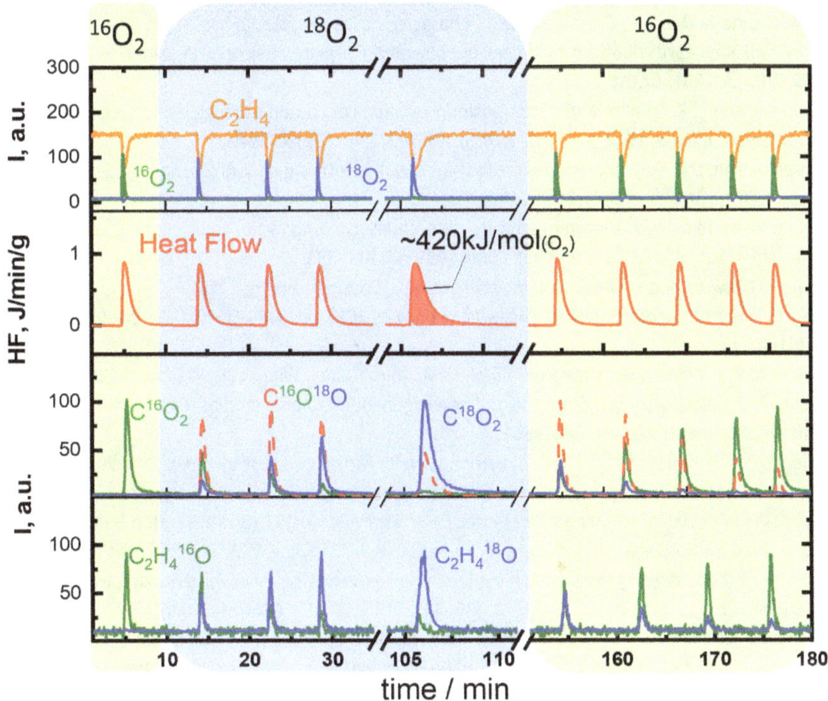

Fig. 1.22: In situ DSC-pulse isotope exchange experiment on Ag/α-Al_2O_3 catalyst at 200 °C.

These two cases briefly highlight the possibility of in situ catalyst characterization and expand the known applications [9] of in situ studies of the overall process using the DSC method. Among traditional bulk-specific TA methods, in situ DSC becomes an indispensable tool for investigating the evolution of the active surface.

References

[1] Monti, D.; Baiker, A. Temperature-programmed reduction. Parametric sensitivity and estimation of kinetic parameters. *J. Catal.* **1983**, 323–335, DOI:10.1016/0021-9517(83)90058-1.

[2] Pawelec, B.; Fierro, J. L. G. Applications of thermal analysis in the preparation of catalyst and in catalysis. In *Handbook of Thermal Analysis and Calorimetry*; Brown, M. E.; Gallagher, P. K., Eds.; *Elsevier*; Amsterdam; **2003**; *Vol. 2*: *Applications to Inorganic and Miscellaneous Materials*, pp. 119–189. DOI:10.1016/S1573-4374(03)80008-0

[3] Maciejewski, M.; Baiker, A. Potential of thermal analysis in preparation and characterization of solid catalyst. *J. Therm. Anal.* **1997**, *48*, 611–622, DOI:10.1007/BF01979507.

[4] Maciejewski, M.; Müller, C. A.; Tschan, R.; Emmerich, W. D.; Baiker, A. Novel pulse thermal analysis method and its potential for investigating gas-solid reactions. *Thermochim. Acta* **1997**, *295*, 167–182, DOI:10.1016/S0040-6031(97)00105-6.

[5] Feist, M.; König, R.; Bäßler, S.; Kemnitz, E. Adsorption properties of various forms of aluminium trifluoride investigated by Pulseta®. *Thermochim. Acta* **2010**, 498, 100–105, DOI:10.1016/j.tca.2009.10.008.

[6] Feist, M.; Teinz, K.; Manuel, S. R.; Kemnitz, E. Characterization of surfacial basic sites of sol gel-prepared alkaline earth fluorides by means of Pulseta®. *Thermochim. Acta* **2011**, *524*, 170–178, DOI:10.1016/j.tca.2011.07.010.

[7] Rostrup-Nielsen, J. R. Equilibria of decomposition reactions of carbon monoxide and methane over nickel catalysts. *J. Catal.* **1972**, 343–356, DOI:10.1016/0021-9517(72)90170-4.

[8] Rostrup-Nielsen, J. R. Coking on nickel catalyst for steam reforming of hydrocarbons. *J. Catal.* **1974**, 184–201, DOI:10.1016/0021-9517(74)90263-2.

[9] Sinev, M. Y.; Bychkov, V. Y. High-temperature differential scanning in-situ calorimetric study of the mechanism of catalytic processes. *Kinet. Catal.* **1999**, *40*, 819–835.

[10] Auroux, A. *Calorimetry and Thermal Methods in Catalysis*; Springer; Berlin; **2013**.

[11] Schlogl, R. Heterogeneous Catalysis. *Angewandte Chemie.* **2015**, *54*, 3465–3520, DOI:10.1002/anie.201410738.

[12] van Santen, R. A. *Modern Heterogeneous Catalysis: An Introduction*; Wiley-VCH; Weinheim; **2017**.

[13] Gravelle, P. C. Calorimetry in adsorption and heterogeneous catatysis studies. *Catal. Rev.* **1977**, *16*, 37–110, DOI:10.1080/03602457708079634.

[14] Gordienko, Y.; Usmanov, T.; Bychkov, V.; Lomonosov, V.; Fattakhova, Z.; Tulenin, Y.; Shashkin, D.; Sinev, M. Oxygen availability and catalytic performance of Nawmn/Sio2 mixed oxide and its components in oxidative coupling of methane. *Catal. Today* **2016**, DOI:10.1016/j.cattod.2016.04.021.

[15] Lomonosov, V. I.; Gordienko, Y. A.; Sinev, M. Y.; Rogov, V. A.; Sadykov, V. A. Thermochemical properties of the lattice oxygen in W,Mn-containing mixed oxide catalysts for the oxidative coupling of methane. *Russ. J. Phys. Chem. A* **2018**, *92*, 430–437, DOI:10.1134/s0036024418030147.

[16] Sinev, M. Y.; Fattakhova, Z. T.; Bychkov, V. Y.; Lomonosov, V. I.; Gordienko, Y. A. Dynamics and thermochemistry of oxygen uptake by a mixed Ce–Pr oxide. *Russ. J. Phys. Chem. A* **2018**, *92*, 424–429, DOI:10.1134/s0036024418030263.

[17] Sadykov, V., et al. Mechanism of Ch4 dry reforming by pulse microcalorimetry: Metal nanoparticles on perovskite/fluorite supports with high oxygen mobility. *Thermochim. Acta* **2013**, *567*, 27–34, DOI:10.1016/j.tca.2013.01.034.

[18] Simonov, M. N.; Rogov, V. A.; Smirnova, M. Y.; Sadykov, V. A. Pulse microcalorimetry study of methane dry reforming reaction on Ni/Ceria-Zirconia catalyst. *Catalysts* **2017**, *7*, 268, DOI:10.3390/catal7090268.

[19] Kube, P.; Frank, B.; Wrabetz, S.; Kröhnert, J.; Hävecker, M.; Velasco-Vélez, J.; Noack, J.; Schlögl, R.; Trunschke, A. Functional analysis of catalysts for lower alkane oxidation. *ChemCatChem* **2017**, *9*, 573–585, DOI:10.1002/cctc.201601194.

[20] Behrens, M.; Schlögl, R. How to prepare a good Cu/Zno catalyst or the role of solid state chemistry for the synthesis of nanostructured catalysts. *Z. Anorg. Allg. Chem.* **2013**, *639*, 2683–2695, DOI:10.1002/zaac.201300356.
[21] Schumann, J.; Lunkenbein, T.; Tarasov, A.; Thomas, N.; Schlögl, R.; Behrens, M. Synthesis and characterisation of a highly active Cu/Zno:Al catalyst. *ChemCatChem.* **2014**, *6*, 2889–2897, DOI:10.1002/cctc.201402278.
[22] Baltes, C.; Vukojevic, S.; Schuth, F. Correlations between synthesis, precursor, and catalyst structure and activity of a large set of Cuo/Zno/Al2o3 catalysts for methanol synthesis. *J. Catal.* **2008**, *258*, 334–344, DOI:10.1016/j.jcat.2008.07.004.
[23] Reddy, B. J.; Frost, R. L.; Locke, A. Synthesis and spectroscopic characterisation of aurichalcite (Zn, Cu2+)5(Co3)2(Oh)6; Implications for Cu–Zno catalyst precursors. *Transition Met. Chem.* **2007**, *33*, 331–339, DOI:10.1007/s11243-007-9044-9.
[24] Behrens, M.; Girgsdies, F.; Trunschke, A.; Schlögl, R. Minerals as model compounds for Cu/Zno catalyst precursors: Structural and thermal properties and ir spectra of mineral and synthetic (zincian) malachite, rosasite and aurichalcite and a catalyst precursor mixture. *Eur. J. Inorg. Chem.* **2009**, *2009*, 1347–1357, DOI:10.1002/ejic.200801216.
[25] Zwiener, L.; Girgsdies, F.; Brennecke, D.; Teschner, D.; Machoke, A. G. F.; Schlögl, R.; Frei, E. Evolution of zincian malachite synthesis by low temperature co-precipitation and its catalytic impact on the methanol synthesis. *Appl. Catal. B Environ.* **2019**, *249*, 218–226, DOI:10.1016/j.apcatb.2019.02.023.
[26] Frost, R. L.; Locke, A. J.; Hales, M. C.; Martens, W. N. Thermal stability of synthetic aurichalcite implications for making mixed metal oxides for use as catalysts. *J. Therm. Anal. Calorim.* **2008**, *94*, 203–208, DOI:10.1007/s10973-007-8634-2.
[27] Günter, M. M.; Ressler, T.; Jentoft, R. E.; Bems, B. Redox behavior of copper oxide/zinc oxide catalysts in the steam reforming of methanol studied by in situ X-ray diffraction and absorption spectroscopy. *J. Catal.* **2001**, *203*, 133–149, DOI:10.1006/jcat.2001.3322.
[28] Shin-ichiro, F.; Shuhei, M.; Yoshinori, K.; Nobutsune, T. Effects of the calcination and reduction conditions on a Cu/Zno methanol synthesis catalyst. *React. Kinet. Catal. Lett.* **2000**, *70*, 11–16.
[29] Millar, G. J.; Holm, I. H.; Uwins, P. J. R.; Drennan, J. Characterization of precursors to methanol synthesis catalysts Cu/Zno system. *J. Chem. Soc., Faraday Trans.* **1998**, *94*, 593–600, DOI:10.1039/a703954i.
[30] Bems, B.; Schur, M.; Dassenoy, A.; Junkes, H.; Herein, D.; Schlogl, R. Relations between synthesis and microstructural properties of copper/zinc hydroxycarbonates. *Chemistry* **2003**, *9*, 2039–2052, DOI:10.1002/chem.200204122.
[31] Yurieva, T. M. Catalyst for methanol synthesis: Preparation and activation. *React. Kinet. Catal. Lett.* **1995**, *55*, 513–521, DOI:10.1007/BF02073088.
[32] Behrens, M.; Kasatkin, I.; Kühl, S.; Weinberg, G. Phase-pure Cu,Zn,Al hydrotalcite-like materials as precursors for copper rich Cu/Zno/Al2o3 catalysts. *Chem. Mater.* **2010**, *22*, 386–397, DOI:10.1021/cm9029165.
[33] Fujita, S.-I.; Moribe, S.; Kanamori, Y.; Kakudate, M.; Takazawa, N. Preparation of a coprecipitated Cu/Zno catalyst for the methanol synthesis from Co2 – Effects of the calcination and reduction conditions on the catalytic performance. *Appl. Catal. A: Gen.* **2001**, *207*, 121–128, DOI:10.1016/S0926-860X(00)00616-5.
[34] Minyukova, T. P.; Khassin, A. A.; Khasin, A. V.; Yurieva, T. M. Formation of effective copper-based catalysts of methanol synthesis. *Kinet. Catal.* **2020**, *61*, 886–893, DOI:10.1134/S0023158420060087.
[35] Tarasov, A.; Schumann, J.; Girgsdies, F.; Thomas, N.; Behrens, M. Thermokinetic investigation of binary Cu/Zn hydroxycarbonates as precursors for Cu/Zno catalysts. *Thermochim. Acta* **2014**, *591*, 1–9, DOI:10.1016/j.tca.2014.04.025.

[36] Jander, W. *Reaktionen Im Festen Zustände Bei Höheren Temperaturen*; Chemische Institut der Universität: Würzburg; **1927**.

[37] Markov, L.; Ioncheva, R. Synthesis and thermal decomposition of Cu(Ii)-Zn(Ii) hydroxide nitrate mixed crystals. *Mater. Chem. Phys.* **1990**, *26*, 493–504, DOI:10.1016/0254-0584(90)90059-J.

[38] Schumann, J.; Tarasov, A.; Thomas, N.; Schlögl, R.; Behrens, M. C. Zn-based catalysts for methanol synthesis: On the effect of calcination conditions and the part of residual carbonates. *Appl. Catal. A: Gen.* **2016**, *516*, 117–126, DOI:10.1016/j.apcata.2016.01.037.

[39] Schmidt, M.; Lutz, H. D. Hydrogen bonding in basic copper salts: A spectroscopic study of malachite, Cu2(Oh)2co3, and brochantite, Cu4(Oh)6so4. *Phys. Chem. Miner.* **1993**, *20*, 27–32, DOI:10.1007/BF00202247.

[40] Goldsmith, J. A.; Ross, S. D. The infra-red spectra of azurite and malachite. *Spectrochim. Acta Part A: Mol. Spectrosc.* **1968**, *24*, 2131–2137, DOI:10.1016/0584-8539(68)80273-9.

[41] Koga, N.; Criado, J. M.; Tanaka, H. Apparent kinetic behaviour of the thermal decomposition of synthetic malachite. *Thermochim. Acta* **1999**, *341*, 387–394, DOI:10.1016/S0040-6031(99)00289-0.

[42] Kühl, S.; Tarasov, A.; Zander, S.; Kasatkin, I.; Behrens, M. Cu-based catalyst resulting from a Cu,Zn,Al hydrotalcite-like compound: A microstructural, thermoanalytical, and in situ XAS study. *Chem. – Eur. J.* **2014**, *20*, 3782–3792, DOI:10.1002/chem.201302599.

[43] Tarasov, A.; Kühl, S.; Schumann, J.; Behrens, M. Thermokinetic study of the reduction process of a Cuo/Znal2o4 catalyst. *High Temp. High Press.* **2013**, *42*, 377–386.

[44] Chinchen, G. C.; Waugh, K. C.; Whan, D. A. The activity and state of the copper surface in methanol synthesis catalysts. *Appl. Catal.* **1986**, *25*, 101–107, DOI:10.1016/S0166-9834(00)81226-9.

[45] Lunkenbein, T.; Girgsdies, F.; Kandemir, T.; Thomas, N.; Behrens, M.; Schlögl, R.; Frei, E. Bridging the time gap: A copper/zinc oxide/aluminum oxide catalyst for methanol synthesis studied under industrially relevant conditions and time scales. *Angewandte Chemie* **2016**, *128*, 12900–12904, DOI:10.1002/ange.201603368.

[46] Frei, E.; Gaur, A.; Lichtenberg, H.; Zwiener, L.; Scherzer, M.; Girgsdies, F.; Lunkenbein, T.; Schlögl, R. Cu–Zn alloy formation as unfavored state for efficient methanol catalysts. *ChemCatChem* **2020**, *12*, 4029–4033, DOI:10.1002/cctc.202000777.

[47] 66136-1, D., Determination of the Degree of Dispersion of Metals Using Gas Chemisorption. In *Part:1 Principles*, DIN: **2004**; Vol. 66136-1.

[48] 66136-3, D., Detrmination of the Degree of Dispersion of Metals Using Chemisorption. In *Part 3: Flow method*, DIN: **2007**; Vol. 66136-3.

[49] Hinrichsen, O.; Genger, T.; Muhler, M. Chemisorption of N2o and H2 for the surface determination of copper catalysts. *Chem. Eng. Technol.* **2000**, *23*, 956–959, DOI:10.1002/1521-4125(200011)23:11<956::AID-CEAT956>3.0.CO;2-L.

[50] Jensen, J. R.; Johannessen, T.; Livbjerg, H. An improved N2o-method for measuring Cu-dispersion. *Appl. Catal. A: Gen.* **2004**, *266*, 117–122, DOI:10.1016/j.apcata.2004.02.009.

[51] Lunkenbein, T.; Schumann, J.; Behrens, M.; Schlogl, R.; Willinger, M. G. Formation of a Zno overlayer in industrial Cu/Zno/Al2 O3 catalysts induced by strong metal-support interactions. *Angewandte Chemie* **2015**, *54*, 4544–4548, DOI:10.1002/anie.201411581.

[52] Schumann, J.; Eichelbaum, M.; Lunkenbein, T.; Thomas, N.; Álvarez Galván, M. C.; Schlögl, R.; Behrens, M. Promoting strong metal support interaction: Doping Zno for enhanced activity of Cu/Zno:M (M = Al, Ga, Mg) catalysts. *ACS Catal.* **2015**, *5*, 3260–3270, DOI:10.1021/acscatal.5b00188.

[53] Frei, E.; Gaur, A.; Lichtenberg, H.; Heine, C.; Friedrich, M.; Greiner, M.; Lunkenbein, T.; Grunwaldt, J.-D.; Schlögl, R. Activating a Cu/Zno : Al catalyst – much more than reduction: Decomposition, self-doping and polymorphism. *ChemCatChem* **2019**, *11*, 1587–1592, DOI:10.1002/cctc.201900069.

[54] Zander, S.; Kunkes, E. L.; Schuster, M. E.; Schumann, J.; Weinberg, G.; Teschner, D.; Jacobsen, N.; Schlogl, R.; Behrens, M. The role of the oxide component in the development of copper composite

catalysts for methanol synthesis. *Angewandte Chemie* **2013**, *52*, 6536–6540, DOI:10.1002/anie.201301419.

[55] Brown, M. A.; Carrasco, E.; Sterrer, M.; Freund, H.-J. Enhanced stability of gold clusters supported on hydroxylated Mgo(001) surfaces. *J. Am. Chem. Soc.* **2010**, *132*, 4064–4065, DOI:10.1021/ja100343m.

[56] Brown, M. A.; Fujimori, Y.; Ringleb, F.; Shao, X.; Stavale, F.; Nilius, N.; Sterrer, M.; Freund, H.-J. Oxidation of Au by surface Oh: Nucleation and electronic structure of gold on hydroxylated Mgo (001). *J. Am. Chem. Soc.* **2011**, *133*, 10668–10676, DOI:10.1021/ja204798z.

[57] Starr, D. E.; Diaz, S. F.; Musgrove, J. E.; Ranney, J. T.; Bald, D. J.; Nelen, L.; Ihm, H.; Campbell, C. T. Heat of adsorption of Cu and Pb on hydroxyl-covered Mgo(100). *Surf. Sci.* **2002**, *515*, 13–20, DOI:10.1016/S0039-6028(02)01915-5.

[58] Lunkenbein, T.; Girgsdies, F.; Kandemir, T.; Thomas, N.; Behrens, M.; Schlogl, R.; Frei, E. Bridging the time gap: A copper/zinc oxide/aluminum oxide catalyst for methanol synthesis studied under industrially relevant conditions and time scales. *Angewandte Chemie* **2016**, DOI:10.1002/anie.201603368.

[59] Albert, E.; Kircheim, R. Diffusivity of oxygen in copper. *Scripta Metallurgica* **1981**, *15*, 673–677.

[60] Narula, M. L.; Tare, V. B.; LWorrell, W. L. Diffusivity and solubility of oxygen in solid copper using potentiostatic and potentiometric techniques. *Metall. Trans. B* **1983**, *14B*, 673–677.

[61] Ramanarayanan, T. A.; Rapp, R. A. The diffusivity and solubility of oxygen in liquid tin and solid silver and the diffusivity. *Metall. Trans.* **1972**, *3*, 3239–3246, DOI:10.1007/bf02661339.

[62] Tarasov, A. V.; Klyushin, A. Y.; Friedrich, M.; Girgsdies, F.; Schlögl, R.; Frei, E. Oxygen diffusion in Cu-based catalysts: A probe for metal support interactions. *Appl. Catal. A: Gen.* **2020**, *594*, 117460, DOI:10.1016/j.apcata.2020.117460.

[63] Ramanarayanan, T. A.; Worrell, W. L. Overvoltage phenomena in electrochemical cells with oxygen-saturated copper electrodes. *Metall. Trans.* **1974**, *5*, 1773–1777, DOI:10.1007/bf02644140.

[64] Gryaznov, V. M.; Gul'yanova, S. G.; Kanizius, S. Diffusion of oxygen through a silver mambrane. *Russ. J. Phys. Chem.* **1973**, *47*, 1517–1518.

[65] Greiner, M. T.; Jones, T. E.; Johnson, B. E.; Rocha, T. C.; Wang, Z. J.; Armbruster, M.; Willinger, M.; Knop-Gericke, A.; Schlogl, R. The oxidation of copper catalysts during ethylene epoxidation. *Phys. Chem. Chem. Phys.* **2015**, *17*, 25073–25089, DOI:10.1039/c5cp03722k.

[66] Dell, R. M.; Stone, F. S.; Tiley, P. F. The adsorption of oxygen and other gases on copper. *Trans. Faraday Soc.* **1953**, *49*, 195–201, DOI:10.1039/TF9534900195.

[67] Fromhold, A. T. Introduction to gas-solid reactions resulting in the formation of thin-film barrier layers. In *Theory of Metal Oxidation*; Amelinckx, S.; Gevers, R.; Nihoul, J., Eds.; North-Holland Publishing Company: Netherlands; **1976**; Vol. 1, pp. 3–37.

[68] Pastorek, R. L.; Rapp, R. A. *Trans AIME* **1969**, *245*, 1711.

[69] Zhao, Y.-F.; Yang, Y.; Mims, C.; Peden, C. H. F.; Li, J.; Mei, D. Insight into methanol synthesis from Co2 hydrogenation on Cu(111): Complex reaction network and the effects of H2o. *J. Catal.* **2011**, *281*, 199–211, DOI:10.1016/j.jcat.2011.04.012.

[70] Yang, Y.; Mims, C. A.; Mei, D. H.; Peden, C. H. F.; Campbell, C. T. Mechanistic studies of methanol synthesis over Cu from Co/Co2/H2/H2o mixtures: The source of C in methanol and the role of water. *J. Catal.* **2013**, *298*, 10–17, DOI:10.1016/j.jcat.2012.10.028.

[71] Martin, O.; Pérez-Ramírez, J. New and revisited insights into the promotion of methanol synthesis catalysts by Co2. *Catal. Sci. Technol.* **2013**, *3*, 3343, DOI:10.1039/c3cy00573a.

[72] Sahibzada, M.; Metcalfe, I. S.; Chadwick, D. Methanol synthesis from Co/Co2/H2 over Cu/Zno/Al2o3 at differential and finite conversions. *J. Catal.* **1998**, *174*, 111–118, DOI:10.1006/jcat.1998.1964.

[73] Tarasov, A. V.; Seitz, F.; Schlögl, R.; Frei, E. In situ quantification of reaction adsorbates in low-temperature methanol synthesis on a high-performance Cu/Zno:Al catalyst. *ACS Catal.* **2019**, *9*, 5537–5544, DOI:10.1021/acscatal.9b01241.

[74] Kuld, S.; Thohauge, M.; Falsig, H.; Elkjaer, C. F.; Helveg, S.; Chorkendorff, I.; Sehested, J. Quantifying the promotion of Cu catalysts by Zno for methanol synthesis. *Science* **2016**, *352*, 969–974.
[75] Kandemir, T.; Friedrich, M.; Parker, S. F.; Studt, F.; Lennon, D.; Schlogl, R.; Behrens, M. Different routes to methanol: Inelastic neutron scattering spectroscopy of adsorbates on supported copper catalysts. *Phys. Chem. Chem. Phys.* **2016**, *18*, 17253–17258, DOI:10.1039/C6CP00967K.
[76] Van Herwijnen, T.; Van Doesburg, H.; De Jong, W. A. Kinetics of the methanation of Co and Co2 on a nickel catalyst. *J. Catal.* **1973**, *28*, 391–402, DOI:10.1016/0021-9517(73)90132-2.
[77] Li, Y.; Lu, G.; Ma, J. Highly active and stable nano Nio–Mgo catalyst encapsulated by silica with a core–shell structure for Co2 methanation. *RSC Adv.* **2014**, *4*, 17420–17428, DOI:10.1039/C3RA46569A.
[78] Sun, F.-M.; Yan, C.-F.; Wang, Z.-D.; Guo, C.-Q.; Huang, S.-L. Ni/Ce–Zr–O catalyst for high Co2 conversion during reverse water gas shift reaction (RWGS). *Int. J. Hydrogen Energy* **2015**, *40*, 15985–15993, DOI:10.1016/j.ijhydene.2015.10.004.
[79] Wang, L.; Zhang, S.; Liu, Y. Reverse water gas shift reaction over co-precipitated Ni-Ceo2 catalysts. *J. Rare Earths* **2008**, *26*, 66–70, DOI:10.1016/S1002-0721(08)60039-3.
[80] Mette, K.; Kühl, S.; Tarasov, A.; Düdder, H.; Kähler, K.; Muhler, M.; Schlögl, R.; Behrens, M. Redox dynamics of Ni catalysts in Co2 reforming of methane. *Catal. Today* **2015**, *242*, 101–110, DOI:10.1016/j.cattod.2014.06.011.
[81] Mette, K.; Kühl, S.; Düdder, H.; Kähler, K.; Tarasov, A.; Muhler, M.; Behrens, M. Stable performance of Ni catalysts in the dry reforming of methane at high temperatures for the efficient conversion of Co_2 into syngas. *ChemCatChem* **2014**, 1000–1004, DOI:10.1002/cctc.201300699.
[82] Behrens, M., et al. The active site of methanol synthesis over Cu/Zno/Al2o3 industrial catalysts. *Science* **2012**, *336*, 893–897, DOI:10.1126/science.1219831.
[83] Choudhary, V. R.; Uphade, B. S.; Belhekar, A. A. Oxidative conversion of methane to syngas over LaNiO3perovskite with or without simultaneous steam and Co2 reforming reactions: Influence of partial substitution of La and Ni. *J. Catal.* **1996**, *163*, 312–318, DOI:10.1006/jcat.1996.0332.
[84] Guo, J.; Lou, H.; Zhao, H.; Chai, D.; Zheng, X. Dry reforming of methane over nickel catalysts supported on magnesium aluminate spinels. *Appl. Catal. A: Gen.* **2004**, *273*, 75–82, DOI:10.1016/j.apcata.2004.06.014.
[85] Kühl, S.; *Synthesis and Characterization of Cu-Based Catalysts Resulting from Cu,Zn,Xhydrotalcite-Like Compounds.* Doctoral Thesis, Technische Universität Berlin, Berlin, **2012**, DOI:10.14279/depositonce-3140.
[86] Kuzmin, A.; Mironova, N. Composition dependence of the lattice parameter in Nicmg1-Co solid solutions. *J. Phys.* **1998**, *10*, 7937–7944.
[87] Kuzmin, A.; Mironova, N.; Purans, J.; Rodionov, A. X-ray absorption spectroscopy study of Nicmg1-Co solid solutions on the Ni K edge. *J. Phys.: Condens. Matter* **1995**, *7*, DOI:10.1088/0953-8984/7/48/023.
[88] Yan, Y.; Dai, Y.; He, H.; Yu, Y.; Yang, Y. A novel W-doped Ni-Mg mixed oxide catalyst for Co2 methanation. *Appl. Catal. B Environ.* **2016**, *196*, 108–116, DOI:10.1016/j.apcatb.2016.05.016.
[89] Nakayama, T.; Ichikuni, N.; Sato, S.; Nozaki, F. Ni/Mgo catalyst prepared using citric acid for hydrogenation of carbon dioxide. *Appl. Catal. A: Gen.* **1997**, *158*, 185–199, DOI:10.1016/S0926-860X(96)00399-7.
[90] Tomishige, K.; Himeno, Y.; Matsuo, Y.; Yoshinaga, Y.; Fujimoto, K. Catalytic performance and carbon deposition behavior of a Nio–Mgo solid solution in methane reforming with carbon dioxide under pressurized conditions. *Ind. Eng. Chem. Res.* **2000**, *39*, 1891–1897, DOI:10.1021/ie990884z.
[91] Millet, M. M., et al. Ni single atom catalysts for Co2 activation. *J. Am. Chem. Soc.* **2019**, *141*, 2451–2461, DOI:10.1021/jacs.8b11729.
[92] Millet, -M.-M.; Tarasov, A. V.; Girgsdies, F.; Algara-Siller, G.; Schlögl, R.; Frei, E. Highly dispersed Ni0/Nixmg1–Xo catalysts derived from solid solutions: How metal and support control the Co2 hydrogenation. *ACS Catal.* **2019**, 8534–8546, DOI:10.1021/acscatal.9b02332.

[93] Cavani, F.; Trifirò, F.; Vaccari, A. Hydrotalcite-type anionic clays: Preparation, properties and applications. *Catal. Today* **1991**, *11*, 173–301, DOI:10.1016/0920-5861(91)80068-K.

[94] Mette, K., et al. High-temperature stable Ni nanoparticles for the dry reforming of methane. *ACS Catal.* **2016**, *6*, 7238–7248, DOI:10.1021/acscatal.6b01683.

[95] Zhang, F.; Xiang, X.; Li, F.; Duan, X. Layered double hydroxides as catalytic materials: Recent development. *Catal. Surv. Asia* **2008**, *12*, 253, DOI:10.1007/s10563-008-9061-5.

[96] Becerra, A. M.; Castro-Luna, A. E. An investigation on the presence of Nial2o4 in a stable Ni on alumina catalyst for dry reforming. *J. Chil. Chem. Soc.* **2005**, *50*, 465–469, DOI:10.4067/s0717-97072005000200005.

[97] Ross, J. R. H.; Steel, M. C. F.; Zeini-Isfahani, A. Evidence for the participation of surface nickel aluminate sites in the steam reforming of methane over nickel/alumina catalysts. *J. Catal.* **1978**, *52*, 280–290, DOI:10.1016/0021-9517(78)90142-2.

[98] Ruppert, A. M.; Weckhuysen, B. M. Metal–support interactions. In *Handbook of Heterogeneous Catalysis*; Ertl, G.; Knözinger, H.; Schüth, F; Weitkamp, J., Eds.; Wilei-VCH; Weinheim; 2008; Vol. 2; pp. 1178–1188.

[99] Düdder, H.; Kähler, K.; Krause, B.; Mette, K.; Kühl, S.; Behrens, M.; Scherer, V.; Muhler, M. The role of carbonaceous deposits in the activity and stability of Ni-based catalysts applied in the dry reforming of methane. *Catal. Sci. Technol.* **2014**, *4*, 3317–3328, DOI:10.1039/c4cy00409d.

[100] Parmaliana, A.; Arena, F.; Frusteri, F.; Giordano, N. Temperature-programmed reduction study of Nio–Mgo interactions in magnesia-supported Ni catalysts and Nio–Mgo physical mixture. *J. Chem. Soc. Faraday Trans.* **1990**, *86*, 2663–2669, DOI:10.1039/FT9908602663.

[101] Arena, F.; Frusteri, F.; Parmaliana, A.; Plyasova, L.; Shmakov, A. N. Effect of calcination on the structure of Ni/Mgo catalyst: An X-ray diffraction study. *J. Chem. Soc. Faraday Trans.* **1996**, *92*, 469–471, DOI:10.1039/FT9969200469.

[102] Arena, F.; Frusteri, F.; Parmaliana, A.; Giordano, N. On the reduction of Nio forms in magnesia supported catalysts. *React. Kinet. Catal. Lett.* **1990**, *42*, 121–126, DOI:10.1007/BF02137627.

[103] Arena, F.; Licciardello, A.; Parmaliana, A. The role of Ni2+ diffusion on the reducibility of Nio/Mgo system: A combined Trp-Xps study. *Catal. Lett.* **1990**, *6*, 139–149, DOI:10.1007/BF00764063.

[104] Bond, G. C.; Sarsam, S. P. Reduction of nickel/magnesia catalysts. *Appl. Catal.* **1988**, *38*, 365–377, DOI:10.1016/S0166-9834(00)82839-0.

[105] Jafarbegloo, M.; Tarlani, A.; Mesbah, A. W.; Muzart, J.; Sahebdelfar, S. Nio–Mgo solid solution prepared by sol–gel method as precursor for Ni/Mgo methane dry reforming catalyst: Effect of calcination temperature on catalytic performance. *Catal. Lett.* **2016**, *146*, 238–248, DOI:10.1007/s10562-015-1638-9.

[106] Millet, -M.-M.; Frei, E.; Algara-Siller, G.; Schlögl, R.; Tarasov, A. Surface titration of supported Ni catalysts by O2-pulse thermal analysis. *Appl. Catal. A: Gen.* **2018**, *566*, 155–163, DOI:10.1016/j.apcata.2018.08.023.

[107] Chareonpanich, M.; Teabpinyok, N.; Kaewtaweesub, S. *Effect of Nickel Particle Size on Dry Reforming Temperature.* WCECS 2008 **2008**.

[108] Garbarino, G.; Riani, P.; Magistri, L.; Busca, G. A study of the methanation of carbon dioxide on Ni/Al2o3 catalysts at atmospheric pressure. *Int. J. Hydrogen Energy* **2014**, *39*, 11557–11565, DOI:10.1016/j.ijhydene.2014.05.111.

[109] Feist, M.; Teinz, K.; Manuel, S. R.; Kemnitz, E. Characterization of surfacial basic sites of sol gel-prepared alkaline earth fluorides by means of Pulseta®. *Thermochim. Acta* **2011**, *524*, 170–178, DOI:10.1016/j.tca.2011.07.010.

[110] Maciejewski, M.; Baiker, A. Chapter 4 – Pulse thermal analysis. In *Handbook of Thermal Analysis and Calorimetry*; Gallagher, M. E. B.; Patrick, K., Eds.; Elsevier Science B.V.; Amsterdam; **2008**; Vol. 5, pp. 93–132.

[111] Maciejewski, M.; Müller, C. A.; Tschan, R.; Emmerich, W. D.; Baiker, A. Novel pulse thermal analysis method and its potential for investigating gas-solid reactions. *Thermochim. Acta* **1997**, *295*, 167–182, DOI:10.1016/S0040-6031(97)00105-6.

[112] Eigenmann, F.; Maciejewski, M.; Baiker, A. Gas adsorption studied by pulse thermal analysis. *Thermochim. Acta* **2000**, *359*, 131–141, DOI:10.1016/S0040-6031(00)00516-5.

[113] Maciejewski, M.; Baiker, A. Quantitative calibration of mass spectrometric signals measured in coupled Ta-Ms system. *Thermochim. Acta* **1997**, *295*, 95–105, DOI:10.1016/S0040-6031(97)00100-7.

[114] Xia, X.; d'Alnoncourt, R. N.; Muhler, M. Entropy of adsorption of carbon monoxide on energetically heterogeneous surfaces. *J. Therm. Anal. Calorim.* **2008**, *91*, 167–172, DOI:10.1007/s10973-007-8440-x.

[115] Geyer, R.; Hunold, J.; Keck, M.; Kraak, P.; Pachulski, A.; Schödel, R. Methods for determining the metal crystallite size of Ni supported catalysts. *Chemie Ingenieur Technik* **2011**, *84*, 160–164, DOI:10.1002/cite.201100101.

[116] Hoang-Van, C.; Kachaya, Y.; Teichner, S. J.; Arnaud, Y.; Dalmon, J. A. Characterization of nickel catalysts by chemisorption techniques, X-ray diffraction and magnetic measurements: Effects of support, precursor and hydrogen pretreatment. *Appl. Catal.* **1989**, *46*, 281–296, DOI:10.1016/S0166-9834(00)81123-9.

[117] Webb, P. A.; Introduction to Chemical Adsorption Analytical Techniques | and Their Applications to Catalysis. Technical publications, Micromeritics instrument corp.; **2003**.

[118] Nakai, K.; Nakamura, K. Pulse Chemisorption Measurement <Metal>. *BEL-CAT Application note* **2003**, *CAT-APP-002*.

[119] *Catalyst Deactivation;* Bartholomew, C. H.; Butt, J. B., Eds.; Elsevier: Amsterdam; **1991**; Vol. 68, p. 532.

[120] Yang, H.; Whitten, J. L. Dissociative adsorption of H2 on Ni(111). *J. Chem. Phys.* **1993**, *98*, 5039–5049, DOI:10.1063/1.464958.

[121] Hamza, A. V.; Madix, R. J. Dynamics of the dissociative adsorption of hydrogen on nickel(100). *J. Phys. Chem.* **1985**, *89*, 5381–5386, DOI:10.1021/j100271a014.

[122] Moulijn, J. A.; Annelies, E. D.; Kapteijn, F. Activity loss. In *Handbook of Hetarogeneous Catalysis*; Ertl, G.; Knötzinger, H.; Schüth, F.; Weitkamp, J., Eds.; WILEY-VCH Verlag GmbH & Co. KGaA: Weinheim; **2008**; Vol. 4, pp. 1829–1846.

[123] Tarasov, A.; Düdder, H.; Mette, K.; Kühl, S.; Kähler, K.; Schlögl, R.; Muhler, M.; Behrens, M. Investigation of coking during dry reforming of methane by means of thermogravimetry. *Chemie Ingenieur Technik* **2014**, *86*, 1916–1924, DOI:10.1002/cite.201400092.

[124] Ahi, H. *Supported Liquid Phase Catalysts.* Doctoral Thesis, Technische Universität Berlin, Berlin, **2018**, DOI:10.14279/depositonce-6711.

[125] Lamoth, M., et al. Nanocatalysts unravel the selective state of Ag. *ChemCatChem* **2020**, *12*, 2977–2988, DOI:10.1002/cctc.202000035.

[126] Lamoth, M.; Plodinec, M.; Scharfenberg, L.; Wrabetz, S.; Girgsdies, F.; Jones, T.; Rosowski, F.; Horn, R.; Schlögl, R.; Frei, E. Supported Ag nanoparticles and clusters for Co oxidation: Size effects and influence of the silver–oxygen interactions. *ACS Appl. Nano Mater.* **2019**, *2*, 2909–2920, DOI:10.1021/acsanm.9b00344.

[127] Jones, T. E.; Rocha, T. C. R.; Knop-Gericke, A.; Stampfl, C.; Schlögl, R.; Piccinin, S. Insights into the electronic structure of the oxygen species active in alkene epoxidation on silver. *ACS Catal.* **2015**, *5*, 5846–5850, DOI:10.1021/acscatal.5b01543.

Bénédicte Prélot and Jerzy Zając

Chapter 2 Contribution of isothermal titration calorimetry to elucidate the mechanism of adsorption from dilute aqueous solutions on solid surfaces: data processing, analysis, and interpretation

Abstract: This chapter describes the use of isothermal titration calorimetry (ITC) to monitor the enthalpy changes accompanying the adsorption of ionic species from dilute aqueous solutions on charged solid surfaces. From the application point of view, it thus covers a broad range of interfacial phenomena of great interest in environmental remediation and catalysis but also in medicinal and pharmaceutical research. The incremental titration procedure is thoroughly detailed together with the subsequent data processing on the basis of appropriate thermodynamic analysis so as to evaluate the cumulative enthalpy of displacement and its changes along the adsorption isotherm in single-solute and two-solute systems. Some illustrative examples of data analysis and processing in selected adsorption systems are presented to highlight the main challenges in correlating the enthalpy balance recorded upon dilution and adsorption calorimetry runs with the measurements of individual adsorption isotherms. It is explained how to exploit the comparison between the measurements carried out in single-solute and two-solute systems in order to get insight into competitive or cooperative effects in ion adsorption at a charged solid-liquid interface. The most important parameters to be controlled or, at least, monitored carefully during the measurements are identified, and their impact on the adsorption mechanisms is discussed.

Keywords: isothermal titration calorimetry, experimental procedures, thermal data processing, ion adsorption, charged solid surfaces, dilute aqueous solutions

2.1 Introduction

Despite great complexity of the mechanism of adsorption from solution at the solid-liquid interface, the tendency in thermodynamic description of the phenomenon has often been to refer to the formalism of gas adsorption on solid surfaces. The most spectacular example of this approach is the use of Langmuir-type equation to fit the

Bénédicte Prélot and Jerzy Zając, ICGM, Univ. Montpellier, CNRS, ENSCM, Montpellier, France

https://doi.org/10.1515/9783110590449-002

individual adsorption isotherm for a solute which is preferentially adsorbed at the interface from a dilute aqueous solution. In a further step, researchers take advantage of this simplified modeling of the experimental adsorption curves to calculate the standard Gibbs free energy of adsorption, and ultimately also the related standard enthalpy and entropy (e.g., see references [1–9]).

The Langmuir isotherm equation, originally derived to describe the monolayer gas adsorption on a homogeneous solid surface, is usually transposed to the case of adsorption from solution by replacing the equilibrium partial pressure by the solute concentration in the equilibrium bulk solution [10]. Even though the individual adsorption isotherms usually adopt a Langmuir-type shape and thus the generalized Langmuir equation fits the experimental data well, the thermodynamic description of the phenomenon is at least incomplete. The main reason for this is related to the competitive character of adsorption from solution, the driving force of which is the macroscopic outcome of surface and molecular interactions of the solution components with the solid surface and between themselves both in the interfacial region and in the bulk solution [11–13].

Contrary to gas adsorption, the solid-liquid interface is never empty and always packed with solvent molecules and solute molecules or ions. In consequence, the preferential adsorption of a solute is inevitably accompanied by desorption of solvent molecules, other molecules or ions preadsorbed at the interface. In the case of dilute aqueous solutions, changes in the structure of the vicinal water or in the hydration layers of the adsorbing and desorbing species may also occur. When the adsorption takes place at a charged interface from an ionic solution, transfer of individual ionic species between the interface and the supernatant solution results in an interface which has the structure of ionic double layer being, at equilibrium, electrically neutral as a whole [14–17]. Depending on the chemical specificity of the solid-solute interactions involved, ionic species may form various surface complexes located at different distances of approach to the charged surface [18–21]. In numerous adsorption systems, the composition and structure of the interfacial region vary from one point on the adsorption isotherm to another as a result of charge compensating mechanism. Furthermore, changes in the electrical charge and potential of the solid surface can be also observed [16].

In light of the above consideration, the total enthalpy balance accompanying adsorption from solution is composed of several hardly separable contributions which may be exothermic or endothermic. For an adsorption system under particular experimental conditions, this enthalpy balance, measured directly by isothermal titration calorimetry (ITC), takes either positive or negative values depending on the interactions operating among various components in a given adsorption range. It may even be that the change between exothermic and endothermic adsorption occurs for various degrees of surface coverage by the adsorbed species in the same system, for example, when passing from one adsorption range to another [22, 23]. It is thus not surprising that the ITC measurements yield enthalpy values which are so often different from the isosteric heats of adsorption calculated from the temperature dependence of the individual adsorption isotherms by following the so-called Van't Hoff procedure [24, 25]. By the way, the latter not only requires the knowledge of experimental adsorption curves for, at least, three different temperatures, which may sometimes pose technical problems but also some crude approximations are usually attempted to the activity coefficients within the equilibrium constant so as to facilitate calculations. The possibility of quantifying, in a model-independent manner, the enthalpy changes for selected quantities of adsorption along the adsorption isotherm at a fixed temperature constitutes the major asset of the ITC technique.

This chapter deals with the processing of thermal data obtained in an ITC adsorption experiment, their thermodynamic analysis, and possible interpretation to shed more light on the adsorption mechanism, thereby going beyond the conclusions drawn on the basis of individual adsorption isotherms. For practical reasons, the analysis is restricted to single-solute and two-solute dilute aqueous solutions. This usually suffices to clearly distinguish between competitive and cooperative effects in adsorption, which is one of the priority areas of the present report. In a typical ITC experiment, a relatively concentrated stock solution is injected into a calorimetric cell containing a suspension of solid particles in a liquid. Dilution of the stock solution precedes the adsorption of solute species at the boundary between the solid particles and the liquid phase. This particular contribution should be assessed accordingly and subtracted from the total enthalpy change so as to extract the "pure" effect of adsorption. Another difficulty with the ITC technique is that the determination of adsorption isotherms cannot be done experimentally throughout the duration of the calorimetry run. Instead, the adsorption curves are measured in a separate adsorption experiment. It thus seems obvious that the ITC adsorption and dilution experiments, as well as the measurements of adsorption isotherms, should be strictly correlated with one another. For this reason, the thermodynamic analysis of the ITC dilution experiment and that of the isotherm measurements have been also included in this chapter.

It should be emphasized here that the analysis of the thermal signal is based on the simplified operating principles of a typical ITC system, without referring to a particular model of calorimeter. A detailed description of the measuring principles and procedures, technical data and specifications, as well as dedicated software for control and data collection, or report creation assigned to a given ITC setup can be usually found in the manual provided with the purchased equipment. Nevertheless, studies with the use of ITC equipment have been mostly devoted either to phenomena at liquid-liquid interfaces or to complexation reactions, protein-ligand interactions, and self-assembly processes occurring in a homogeneous liquid phase. The potential user of an ITC calorimeter who wants to study the adsorption at the solid-liquid interface may be disappointed when consulting such manuals since experimental information specific to adsorption systems, the physical factors to be monitored and controlled during each calorimetry run, or postmeasurement processing of the measured data are either highly incomplete or even unavailable. Therefore, the main motivation for writing the present chapter has been to fill this gap and help the reader carry out and interpret the calorimetry measurements in an adequate way.

2.2 Typical ITC operating principles and implementation of the ITC adsorption experiment

2.2.1 General operating mode

Schematic representation of an isothermal titration calorimeter is given in Fig. 2.1. A typical twin ITC instrument contains two cells: a sample cell, where a given reaction or phenomenon occurs, and a reference one, the content of which does not change upon measurement. Depending on the type of experiment, the reference cell is usually filled with the pure solvent or buffer solution in order to balance the heat capacity of the sample cell. A constant electric power is supplied to both cells continuously. The two cells are connected to each other through a thermoelectric device or Peltier device, which monitors the difference of temperature between the cells. Such a device is, in turn,

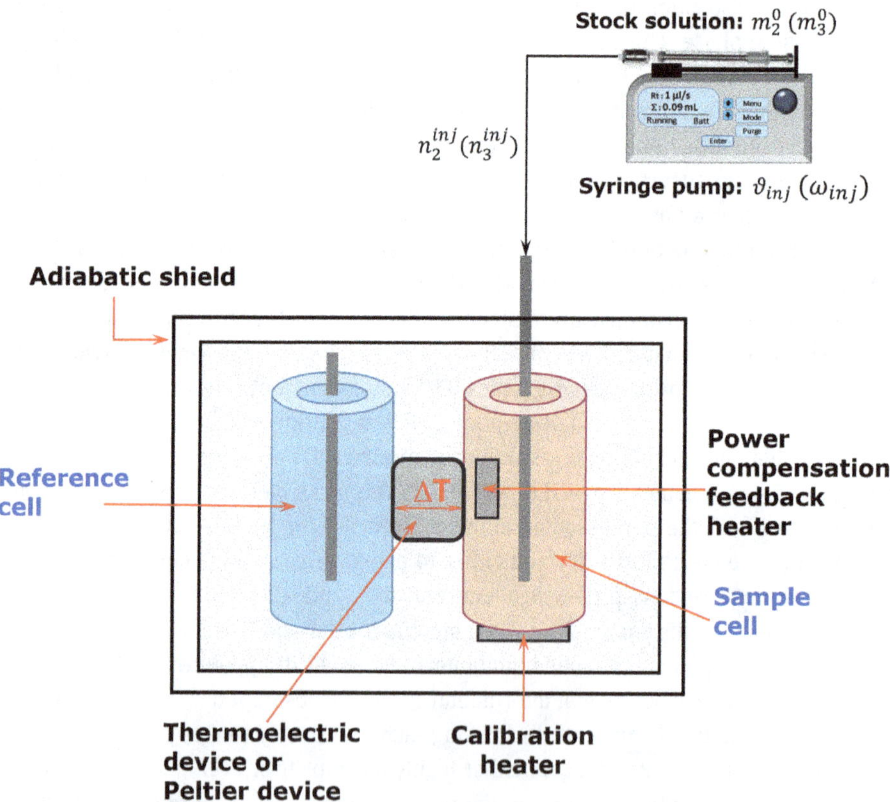

Fig. 2.1: Simplified drawing of the main parts of a typical twin ITC calorimetry system with some control elements.

connected to a feedback power supply and the temperature difference is thus calibrated to a power level. In consequence, the power supplied to the sample cell heater is reduced or increased, according to the exothermic or endothermic character of the experiment, to maintain zero temperature difference between both cells. The cell feedback power (i.e., differential power) represents the heat released or absorbed as a result of the exothermic or endothermic reaction or phenomenon that occurs in the sample cell.

Since the reacting system is at constant pressure (i.e., it is open to the atmosphere), this heat, Q_P, is equal to the change in enthalpy, ΔH, accompanying the reactions or phenomena studied:

$$Q_P = \Delta H \tag{2.1}$$

Note that the sign convention used in eq. (2.1) means that the heat absorbed by the system takes positive values ($\Delta H > 0$), whereas the heat evolved by the system is negative ($\Delta H < 0$).

A typical ITC device is equipped with an automated syringe pump allowing aliquots of a given stock solution of titrant to be injected in a controlled manner into the sample cell. The volume of the aliquot injected in a single injection step, ϑ_{inj}, also called the delivery volume, is related to the delivery time, t_{inj}. They both are the primary parameters of the pump to be set prior to each calorimetry run. They are obviously linked with each other through the flow rate of the pump, d_{pump}, which is not considered as a directly modifiable parameter in recent calorimetry systems.

The quantities m_j^0 appearing in Fig. 2.1 represent the molality (expressed in mol kg^{-1}) of the j th solute in the stock solution; namely, the number of moles of this solute in a given solution corresponding to 10^3 g of solvent. In a more general case considered in this chapter, the stock solution in the syringe may contain simultaneously solute 2 of molar mass M_2 and solute 3 of molar mass M_3. The molality of solute 2 in this stock solution is m_2^0 and that of solute 3 is m_3^0. With dilute aqueous solutions, the delivery volume may be interchanged with the delivery mass, ω_{inj}, expressed in grams.

It is very important to understand why the use of molality units is often preferred to molar concentration in adsorption studies by calorimetry. First, the determination of solution composition is achieved with greater precision, since any solution may be prepared by weighing the solute and solvent, or the stock solution and solvent (dilution of the stock solution). Second, the molality is independent of temperature contrary to the molarity. Furthermore, this approach may be easily extended to the case of dilute aqueous solutions of strong electrolytes which dissociate completely into free ions of opposite electrical charges and are subject to the condition of electrical neutrality [26].

In the case of single-solute solutions, the quantity of solute introduced into the cell during one injection step can be calculated from the following expression:

$$n_2^{\text{inj}} = \frac{m_2^0 \cdot \omega_{\text{inj}}}{10^3 + m_2^0 \cdot M_2} \tag{2.2}$$

To investigate the competitive or cooperative effects in adsorption from two-solute solutions, the simplest way is to mix up the two solutes in stoichiometric proportions within the same stock solution. Therefore, the mass of solvent, M_1^{inj}, number of moles of solute 2, n_2^{inj}, and solute 3, n_3^{inj}, injected into the sample cell of the ITC calorimeter during a single injection are expressed as follows:

$$M_1^{\text{inj}} = \frac{\omega_{\text{inj}}}{1 + \frac{m_2^0}{10^3} \cdot M_2 + \frac{m_3^0}{10^3} \cdot M_3}, \quad n_2^{\text{inj}} = \frac{m_2^0}{10^3} \cdot M_1^{\text{inj}}, \quad n_3^{\text{inj}} = \frac{m_3^0}{10^3} \cdot M_1^{\text{inj}} \tag{2.3}$$

Very often equimolar solutions of both solutes are used and, consequently, $n_2^{\text{inj}} = n_3^{\text{inj}}$.

In practice, it is a requirement for accurate measurements to control the stability and regularity of the delivery process. If it is possible to easily remove the calorimetric cell from the device, the best way to determine the real value of ω_{inj} (or ϑ_{inj}) is by weighing the sample cell before and after the calorimetry run. An alternative way is to check the mass of the syringe in the beginning and at the end of the ITC experiment. However, this obviously depends on the type of syringe used.

2.2.2 Individual adsorption isotherms and adsorption kinetics

When adsorbing on the solid surface, the adsorbate units displace a certain number of solvent molecules or other solute species because of the limited extent of the adsorption space. Therefore, the phenomenon of adsorption induces an uneven partition of solvent and solute molecules between the solid-liquid interface and the bulk solution. The determination of the amount of solute adsorbed is usually done by following the so-called solution depletion procedure which consists in comparing the composition of the bulk solution before and after the attainment of adsorption equilibrium [13]. In the case of dilute aqueous solutions covered by the present chapter, the amount of solute j adsorbed onto solid adsorbent is determined from one of the following expressions, which can be used interchangeably:

$$\Delta_{\text{ads}} n_j = \frac{V^{\odot}}{m_S \cdot A_S} \cdot (C_j^{\odot} - C_j) \quad \text{or} \quad \Delta_{\text{ads}} n_j = \frac{M_1^{\odot}}{m_S \cdot A_S} \cdot (m_j^{\odot} - m_j) \tag{2.4}$$

where m_S and A_S are, respectively, the mass and the specific surface area of the solid sample, $C_j^{\odot}$ ($m_j^{\odot}$) is the molarity (molality) of the initial solution (i.e., before adsorption), C_j (m_j) is the molarity (molality) of the supernatant solution (i.e., after the attainment of adsorption equilibrium), $V^{\odot}$ denotes the initial volume of the solution and $M_1^{\odot}$ is the initial mass of the solvent. Note that the initial values of volume, mass, and concentration do not have the same meaning as those used in further description of the calorimetry measurements and therefore, they have been denoted with a different superscript symbol.

In a strictly formal approach to adsorption, the determination of the amount of solute adsorbed leads to an excess of number of moles for this component present in the interfacial region [12, 26]. Nevertheless, when the solute is preferentially retained by the solid from a dilute aqueous solution, the number of moles of solute adsorbed, $\Delta_{ads}n_j$, is simply calculated from the difference between the total amount of this component introduced into the adsorption system and its amount remaining in the bulk solution after the attainment of the adsorption equilibrium. The plot of $\Delta_{ads}n_j$ as a function of the solute content in the supernatant solution at a constant temperature represents the individual adsorption isotherm for this component.

Note that in each of eq. (2.4), the product $m_S \cdot A_S$ in the denominator constitutes a general normalization factor introduced to overcome the effects associated with the total surface area of the solid sample. Indeed, when the overall surface of the adsorbent is accessible to adsorbed species, the maximum quantity of adsorption will be enhanced for solids possessing higher specific surface areas. This version of eq. (2.4) is thus used whenever one wants to compare the adsorption performance of various solid materials on a per unit surface area basis. The value of A_S to put into eqs. (2.4) is routinely determined from the measurements of gaseous nitrogen adsorption at 77 K by applying the surface area analysis based on the well-known Brunauer, Emmett, and Teller isotherm model [27]. The immersion calorimetry measurements at room temperature combined with a modified Harkins and Jura procedure provide an alternative method for determining specific surface areas of powders, even though the experimental procedures are long and fastidious [28].

It is rather common practice in adsorption studies that the right-hand terms in eq. (2.4) are divided only by the mass of the solid sample, m_S, and the quantity of adsorption is expressed as moles of the adsorbate per gram of the adsorbent [29]. This is recommended especially when the access of adsorbed species to the solid surface is restrained in some porous materials or just when the specific surface area of the solid sample is unknown. Of course, it is still possible to use the accessible surface area (e.g., pore surface or external surface) as determined by following one of the available methods for analysis of gas adsorption isotherms: for example, *t*-plot or its extension to micropore analysis referred to as the MP method, model-independent α_S-plot or a more sophisticated approach based on the density functional theory and its successive developments [27, 30–34].

The knowledge of individual adsorption isotherms for all adsorbing species, that is, $\Delta_{ads}n_j = \Delta_{ads}n_j(m_j), j = 2, 3$, is necessary to obtain a more complete description of the competitive or cooperative effects in adsorption from two-solute solutions. Within the framework of the present chapter, these curves are compulsory for designing the ITC adsorption experiment prior to each calorimetry run, as well as for the final data processing. It is worthwhile mentioning here that the batch approach to the adsorption isotherm based on the solution depletion method is by far the most commonly used in calorimetry studies for a wide range of sorption systems (suffice it to analyze the selected examples of studies reported during the last decade [4, 9, 35–45]).

The time necessary for a given adsorption system to reach the thermodynamic (adsorption) equilibrium is another important parameter required at the preliminary design stage. Adsorption kinetics includes information about the transfer mode of adsorbing solute species from the liquid phase to the solid surface and the nature of interactions responsible for adsorption [29]. The kinetic curves represent the variation of the adsorbate retention at the interface along time at a constant temperature corresponding to a given initial value of $C_j^{\odot}$ or $m_j^{\odot}$, namely:

$$\Delta_{\text{ads}} n_j = \Delta_{\text{ads}} n_j(t) \qquad \text{for given } C_j^{\odot} \ (m_j^{\odot}) \tag{2.5}$$

The objective of kinetic studies is usually to identify the slowest adsorption step and thus to decide the equilibration time to be applied in the ITC experiment.

2.2.3 Incremental titration procedure, the resulting thermal effects, and mass balance

The incremental titration is by far the most common method applied in the ITC measurements. It consists of injecting small aliquots of titrant into the sample cell containing either less concentrated solution (dilution experiment) or solid suspension (adsorption experiment). Complexation or affinity experiments can also be carried out by following similar titration procedures to evaluate binding interactions between two species in ternary systems (e.g., metal-ligand bonding).

The necessity of ensuring an even distribution of the injected titrant portion and a homogeneous mixing of the reactants within the sample cell is one of the most important challenges of the ITC systems. It is usually realized through a special stirring mechanism: for example, flatted needle as the outlet nozzle of the stirring syringe or a paddle stirrer with a slow speed drill. The latter solution is usually recommended at least in the case of adsorption onto powdered solids from liquid solutions, although too high rotation speeds and feed rates may lead to nonnegligible frictional heat effects (being exothermic). The system should be left to reach thermal equilibrium for a sufficiently long period before next injection to ensure that a baseline is established for accurate heat determination. With a satisfactory return of the thermal signal to the baseline, each injection of the stock solution results in a heat pulse that is integrated with respect to time.

It is important to realize that the injected species are collected inside the device. One consequence of this accumulation of matter is that the overall heat capacity of the sample cell varies upon successive injections, thus perturbing the conversion of measured voltage to heat flow. The accurate determination of the heat quantities will thus require a calibration of the device, which is usually performed using in-built calibration heaters, namely precision resistors with known calibration power. For more details about the calibration procedures (e.g., static or dynamic calibration, validation procedures making use of external calibration heaters, or chemical test reactions), the interested reader should consult the manual of the particular calorimetry equipment.

Ultimately, the successive injections of a stock solution into the sample cell result in a raw heat flow record representing several thermal peaks equally spaced in time (whenever an automatic injection mode is employed). The integration of the areas under these thermal peaks should lead to the determination of the individual heat effects, $\Delta_{inj}H_i$, $i=1,2,\ldots$, accompanying a finite number of injections carried out during a given calorimetry run. When the volume (mass) of the stock solution injected into the sample cell and the related heat pulses are sufficiently small, the enthalpy change $\Delta_{inj}H_i$ recorded upon each i th step is considered as a good approximation of the differential enthalpy effect (i.e., the so-called pseudo-differential enthalpy values). The thermal effects of successive injections can be also summed up to obtain the cumulative enthalpy change; after k injections, one obtains

$$\Delta_{inj}H^k_{cum} = \sum_{i=1}^{k} \Delta_{inj}H_i \tag{2.6}$$

The simplified model of ITC adsorption experiment presented in Fig. 2.2 allows appropriate mass balance calculations to be made. Prior to injection sequence, there are M_1^0 grams of pure solvent and a sample m_S of powdered solid in the sample cell. Small aliquots of the stock solution containing both solutes dissolved in deionized water are injected by a high-precision syringe pump. In practice, the delivery time and volume (mass), t_{inj} and ϑ_{inj} (ω_{inj}), are most frequently kept constant throughout the duration of the experiment. Therefore, n_2^{inj} moles of solute 2 and n_3^{inj} moles of solute 3 are simultaneously introduced into the sample cell during successive injection steps. They are easily calculated from eq. (2.3); again, m_2^0 and m_3^0 represent the molalities of solute 2 and solute 3 in the stock solution placed in the injection syringe.

This model represents the simplest case where the solid sample behaves as a neutral component which does not dissolve in the aqueous phase and its role is limited only to providing the interfacial area and the force field for adsorbing species. Therefore, the mass of the solid sample used, m_S, and its specific bulk enthalpy, h_S, do not change during the adsorption process.

Consider first the case of adsorption from single-solute solutions. With the above simplification in mind, the quantity of solute adsorbed from the beginning of the ITC experiment, where the adsorbent is immersed only in M_1^0 grams of pure solvent, to a final state, corresponding to the formation of a solid-liquid interface in equilibrium with a bulk solution of molality m_2^k remaining in the cell, can be evaluated directly on the basis of the mass balance for the calorimetric device:

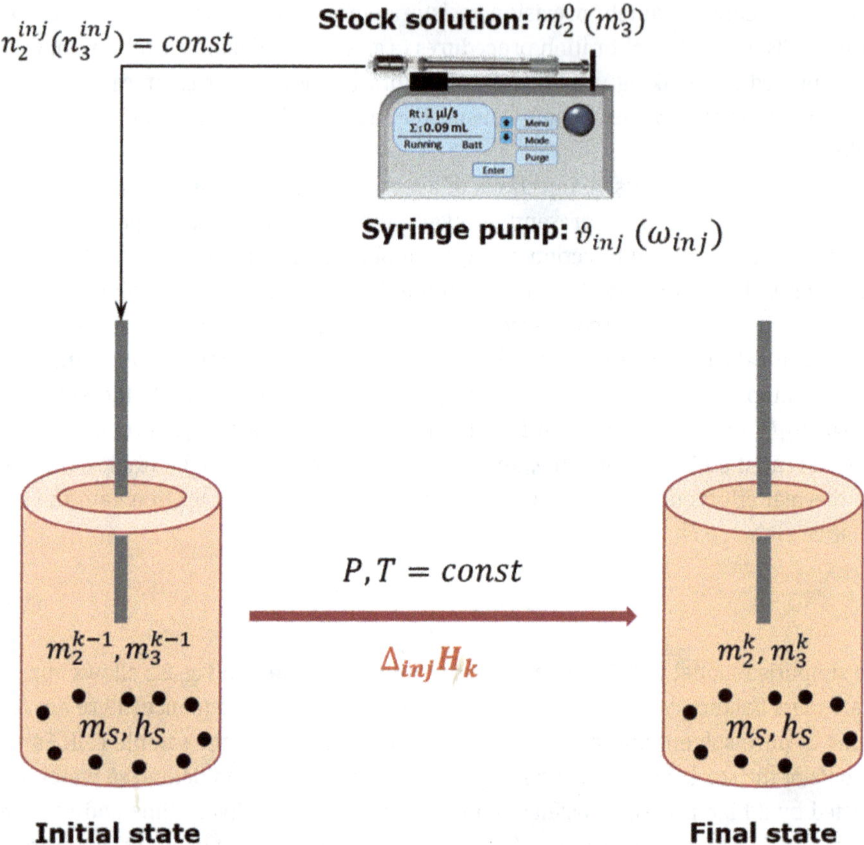

Fig. 2.2: Schematic representation of the *k*th injection step in the ITC adsorption experiment.

$$\Delta_{\text{ads}} n_2^k \left(0 \to m_2^k\right) = k \cdot n_2^{\text{inj}} - m_2^k \cdot \frac{M_1^k}{10^3} = k \cdot n_2^{\text{inj}} - m_2^k \cdot \frac{1}{10^3} \cdot \left(M_1^0 + \frac{10^3 \cdot k \cdot n_2^{\text{inj}}}{m_2^0} \right) \tag{2.7}$$

According to eq. (2.7) for a given series of injections, 1, ..., k, the plot of $\Delta_{\text{ads}} n_2$ against m_2 represents a straight line passing through two characteristic points P1 and P2. They correspond to hypothetical situations in which the whole amount of solute injected into the sample cell either had been retained at the interface (P1) or had remained in the supernatant solution (P2). At the end of each calorimetry run, when the thermogram has been recorded and the real values of parameters M_1^0, m_2^0, ω_{inj} have been determined with precision, the coordinates of points P1 and P2 are calculated on the basis of eqs. (2.2) and (2.7) for all injection steps.

It is the particularity of the ITC adsorption measurements that simultaneous determination of quantity $\Delta_{\text{ads}} n_2^k$ defined by eq. (2.7) cannot be done during the calorimetry run. Instead, a specific procedure for data processing can be applied in association with the use of individual adsorption isotherm for the solute determined in a separate adsorption experiment (i.e., solution depletion method). Finally, the real equilibrium values $\Delta_{ads}n_2^k$ and m_2^k are determined from the intersection between the calorimetric line, that is, plot of $\Delta_{ads}n_2$ versus m_2, and the experimental adsorption isotherm. The coordinates of intersection points can be found numerically first by fitting, to the experimental adsorption data, either any model isotherm equation or a sum of polynomial functions and then by setting the resulting equation and that of calorimetric line (i.e., eq. (2.7)) equal to each other. Alternatively, a continuous adsorption curve can be constructed on the basis of spline interpolation method in a way to capture the trend in the experimental data along the entire adsorption range and the intersection determined graphically, as exemplified in Fig. 2.3.

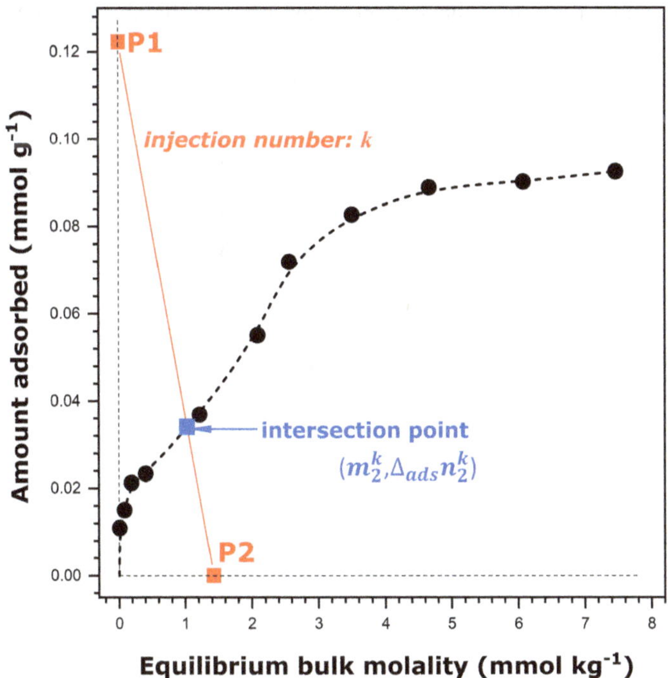

Fig. 2.3: Adsorption of ammonium citrate tribasic onto γ-alumina from single-solute aqueous solutions at 298 K (adapted from [46]). Schematic representation of the data processing procedure leading to the determination of the equilibrium values for m_2^k and $\Delta_{ads}n_2^k$; experimental adsorption isotherm (black circles) and calorimetric straight line corresponding to a series of k successive injections (red line). The dashed black curve has been constructed by spline interpolation method to capture the trend in the experimental data along the entire adsorption range.

A treatment analogous to that presented above may be applied to the case of competitive or cooperative adsorption from two-solute solutions. Therefore, the amount of each solute adsorbed at the solid-liquid interface is calculated as follows:

$$\Delta_{\text{ads}} n_2^k(0 \to m_2^k) = k \cdot n_2^{\text{inj}} - m_2^k \cdot \frac{1}{10^3} \cdot (M_1^0 + k \cdot M_1^{\text{inj}}) \tag{2.8a}$$

$$\Delta_{\text{ads}} n_3^k(0 \to m_3^k) = k \cdot n_3^{\text{inj}} - m_3^k \cdot \frac{1}{10^3} \cdot (M_1^0 + k \cdot M_1^{\text{inj}}) \tag{2.8b}$$

The numerical or graphical procedures for determining the amount adsorbed and the related equilibrium molality of the solute, described previously, also apply in the present case separately for each solute. This means that it is necessary to measure the individual adsorption isotherms for solute 2 and solute 3 from the same two-solute solution. The next step is to find the points of intersection between these experimental adsorption curves and the calorimetry lines defined by eqs. (2.8a) and (2.8b). In the general case, the characteristic points P1 and P2 for both solutes are different from one another. However, when the solutes have been mixed up in equimolar amounts

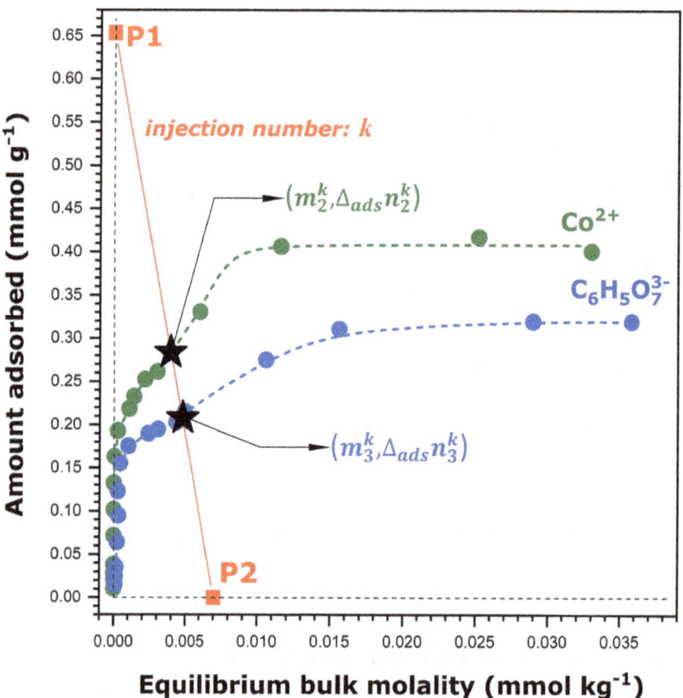

Fig. 2.4: Simultaneous adsorption of cobalt(II) nitrate hexahydrate (green circles) and ammonium citrate tribasic (blue circles) onto γ-alumina from two-solute aqueous solutions at 298 K (adapted from [46]). Schematic representation of the data processing procedure leading to the determination of the equilibrium values for m_2^k, $\Delta_{\text{ads}} n_2^k$ and m_3^k, $\Delta_{\text{ads}} n_3^k$; experimental adsorption isotherms (circles) and calorimetric straight line corresponding to a series of k injections (red line). The dashed green and blue curves have been constructed by spline interpolation method to capture the trend in the experimental data along the entire adsorption ranges. The calorimetry line is the same for both solutes since the stock solution in the syringe contains equimolar quantities of Co(II) and citrate species.

in the stock solution, that is, $m_2^0 = m_3^0$, and the injection mass, ω_{inj}, is kept constant, the two couples of points P1 and P2 are identical and they define one unique calorimetry line. An example of data processing for such a case is illustrated in Fig. 2.4.

Finally, it is of high importance to carry out the measurements of adsorption isotherms under experimental conditions identical to those applied in the ITC adsorption experiment (e.g., temperature, ratio of mass of solid to volume of solution, pH, and ionic strength). Only if this condition is fulfilled, the partition of the solute between the adsorbed and bulk phases, being strictly determined by the adsorption equilibrium at a given temperature, will not depend on the path by which the adsorption system passes from its initial state to the equilibrium. Further comments on this aspect of ITC measurements will be made in the next section.

2.2.4 Designing ITC adsorption experiment: operating conditions and baseline stability

There is absolutely no doubt that careful design and optimization of the operational procedures applied during the ITC adsorption experiment is crucial for the quality of the resulting data and their correct interpretation. Such parameters as the volume (mass) and time, ϑ_{inj} (ω_{inj}) and t_{inj}, of titrant delivery from the syringe to the sample cell, equilibration time, Δt_{inj}, between two successive injections sufficient for the thermal signal to return to the baseline should be optimized in a way to ensure the stability of the thermal baseline over time.

Usually, preliminary studies of adsorption kinetics provide important indications on the optimal equilibration time, Δt_{inj}. For a given type of adsorption system, Δt_{inj} also may be experimentally adjusted in some blank runs. On this occasion, it should not be forgotten that it is closely connected with the delivery parameters and that it decides, together with the number of injections, the total duration of the experiment. Indeed, the ITC adsorption experiments can last for several hours or more and, consequently, some deviations of the baseline from a perfectly horizontal line are often observed. The probability of such deviations increases when certain heat pulses are greater than the others recorded in the same calorimetry run. Fully automated algorithms of baseline correction and data smoothing may be included in the software package purchased with the commercial calorimetry device. Nevertheless, to prevent great routine errors, it is strongly recommended to utilize software dedicated to subtract the baseline and integrate the peaks which allow better control of the smoothing procedures (e.g., the peak analyzer software offered by OriginPro). An example of raw and processed titration thermograms is presented in Fig. 2.5.

It is important to mention here that the raw thermograms may be of significantly lower quality than those shown in Fig. 2.5. In such a case, much attention should be paid not to markedly alter the thermogram and preserve all its features when applying the baseline correction and peak integration procedures. Sometimes, better optimization of

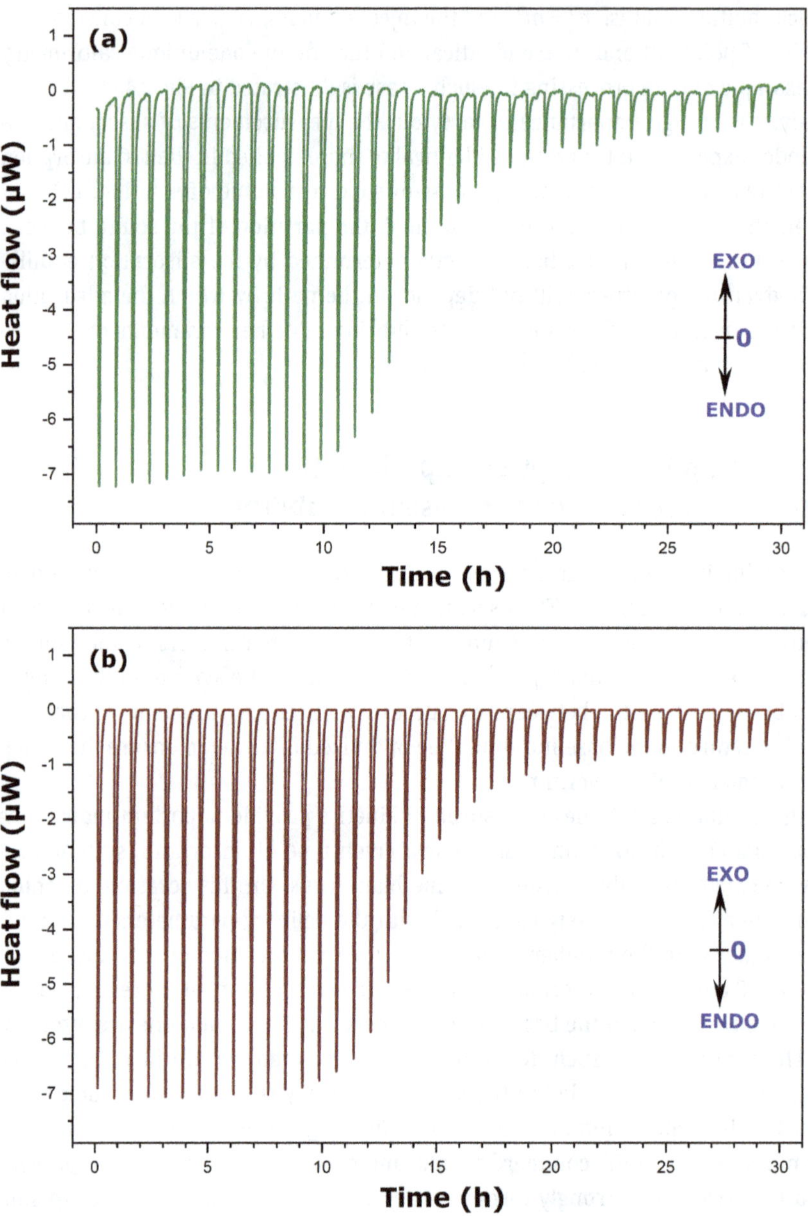

Fig. 2.5: Raw (a) and processed (b) thermal profiles for dilution of a 60 mmol kg^{-1} aqueous solution of dodecylguanidinium chloride in deionized water at 298 K (adapted from [47]). Records of 40 successive injections of 5 μL aliquots into a 1 mL glass ampoule containing initially 800 μL of deionized water (the equilibration time between two injections was set at 45 min). Peak analyzer procedures available in OriginPro software were applied to subtract the baseline (panel b).

the injection sequence may improve the stability of the baseline. Another possibility is to divide the long-lasting ITC experiment into shorter runs, with the final state of one run being taken as the initial state for a subsequent run.

When deciding the operating conditions to be applied, it is important to remember that the number of injections, the delivery volume and time, the equilibration time, and the concentration of the stock solution are linked among themselves, as well as with the technical specifications of the calorimetric device (e.g., heat flow detection limit and precision, precision and regularity of the automated syringe pump, and volume capacity of the syringe and the calorimetric cell). Already, caution should be paid to changing the delivery parameters. When increasing the amount of stock solution injected into the calorimetric cell, and especially if the related thermal effects are significant, it will probably be necessary to prolong the lapse of time between two successive injections for the thermal signal to return completely to the baseline. Another problem, whose negative impact is not always well understood, relates to the temperature difference between the exterior and the interior of the calorimeter, especially when the enthalpy effects at higher temperatures are to be measured. Since the stock solution contained in the syringe is usually at room temperature, its injection to the calorimetric cell thermostated at a higher temperature perturbs the thermal equilibrium of the calorimetric device and may give a hardly predictable contribution to the overall thermal effect. Improved thermostating of the syringe and the injection circuit sticking out of the calorimeter is a good way of avoiding this undesired effect of injection. Furthermore, added to the initial quantity of solvent, M_1^0, and mass of the solid sample, m_S, the total amount of the titrant introduced during the ITC adsorption experiment must not exceed the total volume capacity of the calorimetric cell.

In consideration of the above discussion, it may be concluded that the total number of injections, the equilibration time Δt_{inj}, and, to a lesser extent, the delivery volume and time, will be rather regarded as fixed parameters for a given type of adsorption system in connection with its adsorption and kinetic behavior.

The approach adopted in the previous section to calculate the amount adsorbed and the related equilibrium molality of the solute achieved after successive titrant injections (cf. discussion referring to Figs. 2.3 and 2.4) can be helpful for predicting the injection sequence necessary to attain a given quantity of adsorption, $\Delta_{ads}n_2$. For this purpose, in the very first stage of designing the calorimetry experiment, the calorimetry straight lines are simulated on the basis of eq. (2.7) or (2.8) with a given set of changeable parameters, namely, M_1^0, m_S, m_2^0, m_3^0. These simulated lines are subsequently confronted with the individual adsorption isotherms determined experimentally prior to the ITC experiment. Figure 2.6 illustrates how the range of adsorption quantities to be covered in a calorimetry run would change when the molality of a single-solute stock solution of the titrant were diminished.

Full equivalence of the experimental conditions applied in the measurements of adsorption isotherms and the ITC adsorption experiment may pose a particular problem in the case of ion adsorption on charged surfaces. In the case of incremental titration

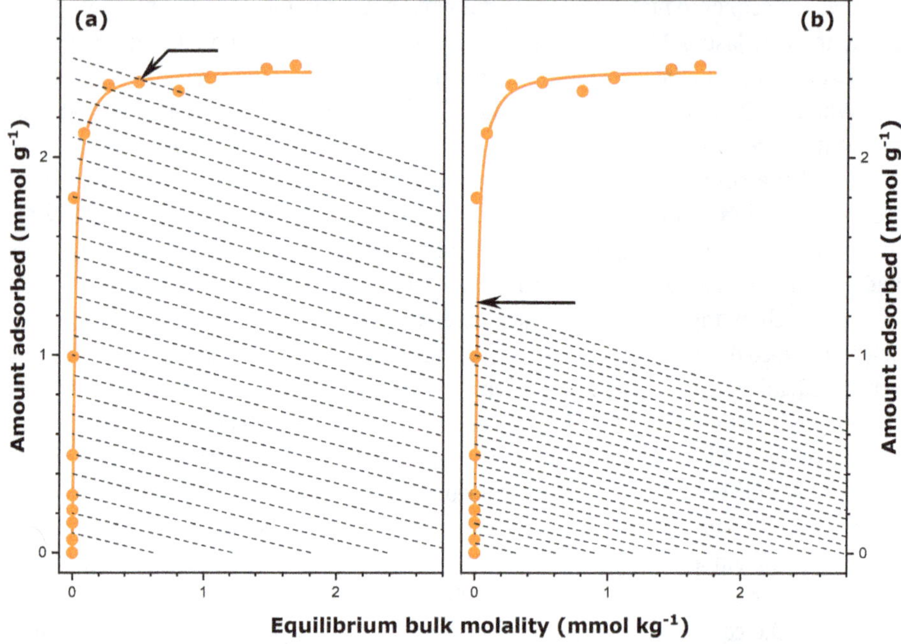

Fig. 2.6: Designing of an ITC adsorption experiment: graphic representation of the data processing showing how to predict the range of adsorption quantities (marked with an arrow) to be covered in a calorimetry run. Adsorption of chromate anion from aqueous solution of potassium chromate (K_2CrO_4) onto porous ionosilica at 298 K (orange circles) (adapted from [48]). The orange curve has been constructed by spline interpolation method to capture the trend in the experimental data along the entire adsorption range. The dashed straight lines represent the theoretical $\Delta_{ads}n_2$ versus m_2 plots obtained for 25 successive injections from eqs. (2.2) and (2.7) with the following set of parameters programmed for the future calorimetry run: M_1^0 = 800 mg, ω_{inj} = 10 mg or ϑ_{inj} = 10 µL, t_{inj} = 10 s, m_S = 5 mg; m_2^0 = 50 mmol kg^{-1} (panel a) and m_2^0 = 25 mmol kg^{-1} (panel b).

method, the injected species are collected inside the device. Therefore, the composition of the reacting system in the sample cell changes concomitantly upon successive injections, in line with the uneven distribution of the solute species between the interfacial region and the bulk solution according to the adsorption mechanism. In the case of two-solute systems, the competition between both solutes to adsorb at the interface often depletes the equilibrium bulk solution of that solute which is preferentially adsorbed. In consequence, pH and ionic strength of ionic solutions are not maintained constant. This feature is of high importance in charged interfacial systems where the mechanisms studied depend strongly on these two parameters (e.g., mineral oxide surfaces with a pH-dependent electrical charge).

One solution quite commonly proposed in experimental and modeling studies consists in considering the adsorption phenomenon under constant pH and ionic strength conditions [16]. In this respect, the solution depletion method offers the possibility of maintaining a constant pH value within the solid suspension during the course of adsorption process. For example, when the solid sample and the stock solution of a given composition are equilibrated in stoppered flasks or tubes to determine separate experimental points on the adsorption isotherm, this may be done in the following way. First, the initial pH in each suspension has to be measured by means of a pH-meter and adjusted to a desired value by adding small doses of an acidic or basic standard solution. Then, during the equilibration period to attain the adsorption equilibrium, the tube shaking (rotation) will be stopped periodically and the pH of the suspension be measured and, if necessary, re-adjusted to the initial value. This procedure of pH adjustment is thus quite complicated and obviously it needs to be correlated with the kinetic pattern recorded for the system studied.

While taking into account the actual progress in the development of calorimetric devices, it is practically impossible to readjust the pH of the solid suspension during or after each injection step in a way that it is done in solution depletion method.

The use of a buffer solution or a background electrolyte to weaken the effects of pH and ionic strength is another possibility [43, 45]. Nevertheless, it should be stressed that the addition of new species to the aqueous phase may greatly perturb the mechanism of adsorption if these species enter into competition with the main solute components. Therefore, this possibility should be always considered with caution.

2.3 Analysis of enthalpy balance in ITC experiments and correlation between dilution and adsorption calorimetry runs

The differential or cumulative enthalpy values measured during the successive injections in the ITC device (see Section 2.2.3) inevitably include dilution effects. In order to extract the net effect of adsorption, an appropriate dilution experiment has to be performed with the use of an incremental titration procedure which follows exactly the one applied in the main adsorption run. From this point of view, the description of dilution experiment in this section has a subsidiary character.

The model of ITC adsorption experiment introduced in Section 2.2.3 (cf. Fig. 2.2) may be still utilized in calculations of the enthalpy balance but it should be completed by adding adequate enthalpy quantities chosen to represent the thermodynamic state of the solution and the interfacial region. The same formalism of incremental titration will be applied in the analysis of enthalpy changes upon dilution of a stock solution, the only exception being the absence of the solid sample in the calorimetric cell.

2.3.1 Enthalpy quantities adequate for the implementation of incremental titration model

In the thermodynamic treatment of the solid-liquid interface, it is necessary to introduce the so-called excess thermodynamic functions [12, 13, 26]. As mentioned earlier, only the cases of single-solute and two-solute solutions are considered here. Accordingly, the interfacial enthalpy for a solid-liquid interface, H^{σ}_{SL}, which remains in thermodynamic equilibrium with a ternary solution containing solutes 2 and 3 dissolved in pure water as a solvent, is defined as follows:

$$H^{\sigma}_{SL}(m_2, m_3) = H - m_S \cdot h_S - H_L(m_2, m_3) \tag{2.9}$$

where H is the total enthalpy of the heterogeneous system, m_S is the mass of the solid, h_S is the specific (per unit mass) enthalpy of the bulk of the solid phase, H_L is the enthalpy of the equilibrium bulk solution of which the composition is given by molalities m_2 and m_3.

From a rigorous point of view, the total enthalpy, H_L, of a single-solute or two-solute solution, as a function of the composition, is expressed by means of partial molal enthalpies of the components, $\overline{h_j}, j = 1, 2, 3$ [26]. The quantity $\overline{h_j}$ for a given solute, $j = 2$ or $j = 3$, plays an important role in the thermodynamics of solutions since it is an intensive state function representing the rate at which the enthalpy of solution changes with the amount of this solute added as the temperature, pressure, and amounts of the other components are kept constant. Therefore [26],

$$H_L = \sum_{j=1}^{2\,\text{or}\,3} n_j \cdot \overline{h_j}, \qquad T, P = \text{const} \tag{2.10}$$

where n_j is the number of moles of solvent (component 1) or solute (components 2 and 3) in this solution.

In the case of dilute binary aqueous solutions, this enthalpy H_L may be conveniently expressed in terms of the apparent molal enthalpy of the unique solute, Φ_H, as

$$H_L = \frac{M_1}{10^3} \cdot m_2 \cdot \Phi_H(m_2) + M_1 \cdot h_1^* \tag{2.11}$$

where M_1 is the total mass of the solvent in the solution, h_1^* is the specific enthalpy (per unit mass) of the pure solvent, m_2 is the molality of the solute, and $\Phi_H(m_2)$ represents the value of Φ_H corresponding to a particular value of m_2. In the case of dilute aqueous solutions of strong electrolytes, eqs. (2.10) and (2.11) are still used with n_2, $\overline{h_2}$, m_2, and Φ_H referring to the whole electrolyte component.

The partial molar, $\overline{h_2}$, and apparent molal, Φ_H, enthalpies of solute in a binary solution are related by the following expressions:

$$\overline{h_2} = \Phi_H + m_2 \cdot \frac{d\Phi_H}{dm_2}, \qquad \text{at infinite dilution: } \overline{h_2}(\infty) = \Phi_H(\infty) \tag{2.12}$$

The apparent molal enthalpy of solute in a binary solution represents the change in the solution enthalpy when all of the solute is added to the solution, per mole of solute added, and it may be potentially determined from the calorimetric measurements of the enthalpy of dilution to produce infinitely dilute solution according to the following reaction pattern:

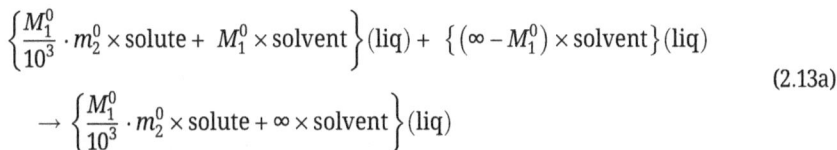

$$\begin{aligned} &\left\{\frac{M_1^0}{10^3} \cdot m_2^0 \times \text{solute} + M_1^0 \times \text{solvent}\right\}(\text{liq}) + \left\{\left(\infty - M_1^0\right) \times \text{solvent}\right\}(\text{liq}) \\ &\quad \rightarrow \left\{\frac{M_1^0}{10^3} \cdot m_2^0 \times \text{solute} + \infty \times \text{solvent}\right\}(\text{liq}) \end{aligned} \tag{2.13a}$$

The corresponding change in the enthalpy is equal to

$$\Delta_{\text{dil}} H\left(m_2^0 \rightarrow 0\right) = -\frac{M_1^0}{10^3} \cdot m_2^0 \cdot \left[\Phi_H\left(m_2^0\right) - \Phi_H(\infty)\right] \tag{2.13b}$$

The difference in square brackets on the right-hand side of eq. (2.13b), called the relative apparent molal enthalpy of the solute, is thus an extremely useful quantity because it relates the partial molar function to the experimentally observed enthalpy of dilution.

In the case of two-solute solutions, the thermodynamic treatment of the dilution experiment is less evident. Thermodynamically rigorous eq. (2.10) is still valid but it has limited usefulness in describing the dilution effects in a simple way. In order to adapt eq. (2.11) to the present case, one of the possibilities is to treat the system as a pseudo-binary one by defining the weighted mean apparent molal enthalpy, $\overline{\Phi_H^{23}}$, of the two solutes [49]; therefore,

$$H_L = \frac{M_1}{10^3} \cdot (m_2 + m_3) \cdot \overline{\Phi_H^{23}}(m_2, m_3) + M_1 \cdot h_1^* \tag{2.14}$$

2.3.2 Enthalpy changes accompanying adsorption from single-solute solutions

Considering the above definitions and in accordance with eqs. (2.7), (2.9), and (2.11), the enthalpy balance in the initial (init) and final (fin) states of the model system presented in Fig. 2.2 can be written in the following way:

$$H_{\text{init}} = \frac{10^3 \cdot n_2^{\text{inj}}}{m_2^0} \cdot h_1^* + n_2^{\text{inj}} \cdot \Phi_H(m_2^0) + H_{\text{SL}}^\sigma(m_2^{k-1}) + m_S \cdot h_S + M_1^{k-1} \cdot h_1^* + \\ + \left[(k-1) \cdot n_2^{\text{inj}} - \Delta_{\text{ads}} n_2^{k-1}\right] \cdot \Phi_H(m_2^{k-1}) \tag{2.15a}$$

$$H_{\text{fin}} = H_{\text{SL}}^\sigma(m_2^k) + m_S \cdot h_S + M_1^k \cdot h_1^* + \left(k \cdot n_2^{\text{inj}} - \Delta_{\text{ads}} n_2^k\right) \cdot \Phi_H(m_2^k) \tag{2.15b}$$

According to eq. (2.6), the cumulative enthalpy change upon k successive injections takes the following form:

$$\Delta_{\text{inj}} H_{\text{cum}}^k = \left[H_{\text{SL}}^\sigma(m_2^k) - H_{\text{S1}}^\sigma\right] - \Delta_{\text{ads}} n_2^k \cdot \Phi_H(m_2^k) - n_2^{\text{inj}} \cdot k \cdot \left[\Phi_H(m_2^0) - \Phi_H(m_2^k)\right] \tag{2.16}$$

where H_{S1}^σ denotes the interfacial enthalpy for the boundary between the solid and the pure solvent (before the first injection step, the solid sample is immersed in M_1^0 grams of solvent). The first two terms on the right-hand side of eq. (2.16) describe the adsorption of $\Delta_{\text{ads}} n_2^k$ moles of solute from the solution of molality m_2^k displacing the equivalent amount of solvent in the reverse direction so as to form a new solid-liquid interface described by an interfacial enthalpy of $H_{\text{SL}}^\sigma(m_2^k)$. It is worth noting that, according to this expression, the solid-liquid interfacial enthalpy is considered with respect to the equilibrium bulk solution of molality m_2^k taken as the reference state [50, 51]. Therefore, the expressions such as $\left[H_{\text{SL}}^\sigma(m_2^k) - H_{\text{S1}}^\sigma\right] - \Delta_{\text{ads}} n_2^k \cdot \Phi_H(m_2^k)$ can be finally interpreted as describing the enthalpy of progressive displacement of the solvent by the solute adsorbed from a dilute binary solution, that is, $\Delta_{\text{dpl}} H = \Delta_{\text{dpl}} H(\Delta_{\text{ads}} n_2^k, m_2^k)$.

In line with eq. (2.13), the last term on the right-hand side of eq. (2.16) represents the dilution effects in the bulk solution when $n_2^{\text{inj}} \cdot k$ moles of the solute are diluted from molality m_2^0 to molality m_2^k. Finally, the cumulative enthalpy change measured in the ITC adsorption experiment represents the sum of the enthalpy of displacement and the enthalpy of dilution:

$$\Delta_{\text{inj}} H_{\text{cum}}^k = \Delta_{\text{dpl}} H_{\text{cum}}(\Delta_{\text{ads}} n_2^k, m_2^k) + \Delta_{\text{dil}} H_{\text{cum}}(m_2^k) \tag{2.17}$$

The data analysis appears much more complicated in the case of adsorption from aqueous solution containing strong electrolytes as solutes potentially adsorbing at a charged solid-liquid interface. Initially, the solid particles, even those immersed in pure water, may develop around them an ionic double layer which contains counterions issued from dissociation of surface functional groups or partial dissolution of the solid phase. In consequence, the interfacial enthalpy in the initial state, H_{S1}^σ, is strictly related to this particular interfacial structure. The interfacial enthalpy after the kth injection, $H_{\text{SL}}^\sigma(m_2^k)$, is the resultant of ion exchange pathway and it remains in equilibrium with the bulk solution containing not only the main ionic component, m_2^k, but also the one which has been displaced from the interfacial region during the experiment. Therefore, the expressions $\Delta_{\text{dpl}} H_{\text{cum}}(\Delta_{\text{ads}} n_2^k, m_2^k)$ must include, at least, contributions due to: (i) adsorption of $\Delta_{\text{ads}} n_2^k$ moles of an ion from the bulk solution of molality

m_2^k, (ii) desorption of an equivalent amount of another ion with the same charge (but not always the same valence), (iii) changes in the environment of the solvent molecules remaining adsorbed [3, 4, 37, 41, 52]. With ions possessing large hydrophobic moieties, the partial dewetting of the solid surface can also provide an endothermic contribution to the total enthalpy of displacement [22, 44, 45].

An important conclusion from the above discussion is that the enthalpy quantity measured in the ITC adsorption experiment always represents the enthalpy of displacement of the solvent by the adsorbed solute species, and additionally in ionic systems, the displacement of some ionic species initially present at the interface by an ionic solute which is preferentially adsorbed. It is thus recommended to preserve the term "differential or cumulative enthalpy of displacement" in studies of adsorption from solutions.

It is appropriate to mention at the end of this section that the choice of the reference state is an important issue for the thermodynamic treatment of solid-liquid systems, especially when the standard thermodynamic parameters of adsorption are to be derived from the simultaneous measurements of adsorption isotherms and enthalpies of adsorption [1, 3, 9, 41, 42]. The values of the Gibbs free energy change upon adsorption with respect to reference states corresponding to arbitrary chosen points on the adsorption isotherm are determined from the equilibrium adsorption curves [41, 53]. The differential molar enthalpy changes can be also measured by ITC technique as a function of the quantity of solute retained by a given adsorbent, thus enabling determination of the corresponding entropy changes with respect to the chosen reference points. Sometimes, the values of the thermodynamic functions relevant to the reference state of equilibrium solution at infinite dilution would be more useful (e.g., when one seeks to reduce the impact of solute-solute interactions [41]). The related enthalpy change can be calculated from the plots of the experimental enthalpy of displacement as a function of the equilibrium solute concentration, by extrapolating them to zero concentration.

2.3.3 Correlation between dilution and adsorption experiments in single-solute systems

The dilution terms $\Delta_{\text{dpl}}H_{\text{cum}}\left(m_2^k\right)$ in eq. (2.17) are usually determined in an appropriate ITC dilution experiment by following the same incremental titration sequence. The model of dilution experiment in the ITC measurements thus corresponds to that presented in Fig. 2.2 from which only the solid sample has been removed. All other parameters remain unchanged. The mass balance leads to the following expressions for the total amount of solute, n_2^k, and the mass of solvent, M_1^k inside the sample cell after the kth injection step:

$$n_2^k = k \cdot n_2^{\text{inj}} \text{ and } M_1^k = M_1^0 + \frac{10^3 \cdot k \cdot n_2^{\text{inj}}}{m_2^0} \tag{2.18a}$$

Therefore, the equilibrium molality of the solution, m_2^k, in the cell after this injection is calculated as follows:

$$m_2^k = \frac{10^3 \cdot k \cdot n_2^{\mathrm{inj}}}{M_1^0 + \frac{10^3 \cdot k \cdot n_2^{\mathrm{inj}}}{m_2^0}} \tag{2.18b}$$

According to eq. (2.11), the enthalpy balance in the initial (init) and final (fin) states of the model system can be written as follows:

$$H_{\mathrm{init}} = \frac{10^3 \cdot n_2^{\mathrm{inj}}}{m_2^0} \cdot h_1^* + n_2^{\mathrm{inj}} \cdot \Phi_H(m_2^0) + M_1^{k-1} \cdot h_1^* + (k-1) \cdot n_2^{\mathrm{inj}} \cdot \Phi_H(m_2^{k-1}) \tag{2.19a}$$

$$H_{\mathrm{fin}} = M_1^k \cdot h_1^* + k \cdot n_2^{\mathrm{inj}} \cdot \Phi_H(m_2^k) \tag{2.19b}$$

This leads to the cumulative enthalpy change of k successive injections (cf. eq. (2.6)):

$$\Delta_{\mathrm{inj}} H_{\mathrm{cum}}^k = -k \cdot n_2^{\mathrm{inj}} \cdot \left[\Phi_H(m_2^0) - \Phi_H(m_2^k)\right] \tag{2.20}$$

which again represents the enthalpy of dilution of a binary solution containing $k \cdot n_2^{\mathrm{inj}}$ moles of solute from molality m_2^0 to molality m_2^k.

Despite the fact that the expression on the right-hand side of eq. (2.20) has exactly the same form as the enthalpy of dilution term in eq. (2.16), the correlation between the two types of ITC experiment is not evident. Although n_2^{inj}, m_2^0, and M_1^0 are the same, the solute molality, m_2^k, after kth injection in the dilution experiment (calculated directly from eq. (2.18b)) does not correspond to the solute molality in the supernatant solution remaining in the calorimetric cell after attaining the adsorption equilibrium. Indeed, the latter m_2^k value is determined from the intersection between the experimental adsorption isotherm and the calorimetric straight line derived from the mass balance inside the calorimetric cell, as shown in Fig. 2.3 (see the discussion in Section 2.2.3).

Therefore, the calculation procedure to be applied is as follows. The curves representing the cumulative enthalpy of dilution, $\Delta_{\mathrm{dil}} H_{\mathrm{cum}}$, as a function of the equilibrium molality of the solute in the calorimetric cell, m_2, are obtained by processing the results of the ITC dilution experiment with the injection sequence identical to that applied in the ITC adsorption experiment (i.e., the same n_2^{inj}, m_2^0, and M_1^0 values). The continuous $\Delta_{\mathrm{dil}} H_{\mathrm{cum}}$ versus m_2 plots can be constructed on the basis of spline or polynomial interpolation method applied to experimental points given by eqs. (2.18b) and (2.20). An example of the data processing procedure is shown in Fig. 2.7.

In the next stage, the results of ITC adsorption experiment are processed to determine the amount adsorbed, $\Delta_{\mathrm{ads}} n_2^k$, after successive k injections and the related molality of the equilibrium bulk solution, m_2^k, from the intersection between the experimental adsorption isotherm and the calorimetric straight lines. Ultimately, each dilution contribution, $\Delta_{\mathrm{dil}} H_{\mathrm{cum}}^k$, to the enthalpy change measured in the ITC adsorption experiment will be read directly from the $\Delta_{\mathrm{dil}} H_{\mathrm{cum}}$ versus m_2 plot for a given m_2^k value.

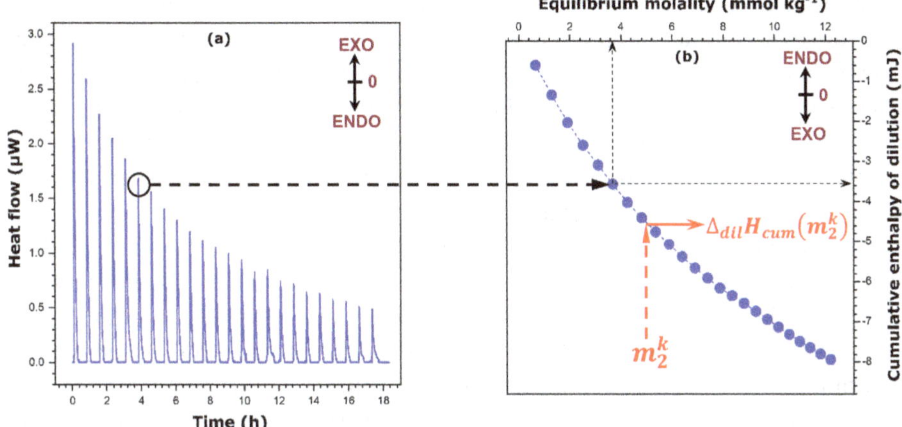

Fig. 2.7: Schematic representation of the data processing leading to evaluate the dilution contribution, $\Delta_{\text{dil}}H_{\text{cum}}(m_2^k)$, to the cumulative enthalpy change, $\Delta_{\text{inj}}H_{\text{cum}}^k$, accompanying adsorption of ammonium citrate tribasic onto γ-alumina from single-solute aqueous solutions at 298 K (adapted from [46]): (a) records of 24 successive injections of 10 µL aliquots of a 50 mmol L^{-1} stock solution into a 1 mL glass ampoule containing initially 800 µL of deionized water (the equilibration time between 2 injections was set at 45 min); (b) the enthalpy of dilution curve obtained from these heat flow records. The equilibrium molality of the bulk solution, m_2^k, after k injections of the same stock solution into a 1 mL glass ampoule containing initially 800 µL of deionized water and 10 mg of γ-Al_2O_3, was obtained from the intersection between the experimental isotherm and the corresponding calorimetric straight line (see Section 2.2.3).

When adsorption of an ion from ionic aqueous solutions at a charged interface is accompanied by desorption of another ion initially present at the interface, the progressive release of the outgoing ion makes the equilibrium solution in the calorimetric cell vary in composition (in relation with both ions) from injection to injection. In consequence, the background dilution experiment cannot be carried out any longer by making use of a single-solute solution but it should be programmed in a way to ensure that the real composition of the equilibrium solution in the cell follows the same trend as that observed in the ITC adsorption experiment. In numerous adsorption systems where the adsorption is due to strong solute-surface interactions, the dilution term is either small or sufficiently weak to be negligible [23]. Otherwise, the problem may be tackled by performing several tests of dilution with aqueous solutions containing, in addition to the main solute, the other ion at different concentrations. This would at least give an idea of the importance of the effect. Finally, another approach is to use a background electrolyte at a higher concentration to impose the ionic strength, always at the risk of blurring the main mechanism of competitive adsorption.

Other difficulties are encountered in adsorption studies when the solute species undergo self-assembly processes in aqueous solutions. This is the case of adsorption of surface-active compounds (surfactants) which can form micelles if the solution composition exceeds that of a particular molarity value known as the critical micelle concentration (CMC) [47, 54–60]. Depending on the type of surfactant and its aqueous environment, micelle formation may be an exothermic, endothermic, or athermal process and the enthalpy of micellization is usually of significant magnitude. In calorimetry studies of surfactant adsorption, concentrated stock solutions are often used (e.g., 10× CMC [61, 62]) to cover a large adsorption range. When such a stock solution is injected into the calorimetric cell, the micellar aggregates contained in it are first destroyed upon dilution and the resulting monomer species are subsequently adsorbed at the solid-liquid interface. Therefore, the dilution contribution to the cumulative enthalpy change measured in the ITC adsorption experiment includes also the enthalpy of micellization taken with the opposite sign. For the purpose of designing the ITC adsorption experiment, both the enthalpy of micellization and the CMC should be thus determined. This can easily be done in the ITC dilution experiment since the self-aggregation phenomenon is accompanied by a change in the slope of the partial molal enthalpy of the solute with respect to concentration. Indeed, calorimetry measurements of the $\Delta_{\text{inj}}H^k_{\text{cum}}$ terms in eq. (2.20) corresponding to various values of $\Phi_H\left(m_2^k\right)$, when normalized using the moles of solute injected, offer an experimental tool for determining the apparent molal enthalpy of solute, Φ_H, and, in turn, the related partial molal enthalpy, $\overline{h_2}$ (see eq. (2.12)). As can be seen in Fig. 2.8, the plots of $\frac{\Delta_{\text{inj}}H^k_{\text{cum}}}{n_2^{\text{inj}}}$ against m_2^k and k are often linear.

In Fig. 2.8, clear breaks in slope mark the change in the state of the diluted species leading to the formation of micellar aggregates at the CMC. This CMC parameter is thus obtained from the intersection between the two linear plots of $\frac{\Delta_{\text{inj}}H^k_{\text{cum}}}{n_2^{\text{inj}}}$ versus m_2^k. According to eqs. (2.12) and (2.20), the values of the partial molal enthalpy for surfactant solute before and after the slope break are constant and their difference provides the standard molar enthalpy change accompanying the process, $\Delta_{\text{mic}}h^0$. It is thus evaluated by plotting the terms $\frac{\Delta_{\text{inj}}H^k_{\text{cum}}}{n_2^{\text{inj}}}$ as a function of injection number k.

2.3.4 Enthalpy changes accompanying adsorption from two-solute solutions

In the case of two-solute solutions, the procedures for determining the enthalpy balance after k successive injections, analogous to those used in Sections 2.3.2 and 2.3.3, lead to adequate expressions for the cumulative enthalpy change upon dilution and adsorption.

In the case of the ITC dilution experiment

$$\Delta_{\text{inj}}H^k_{\text{cum}} = -k\cdot\left(n_2^{\text{inj}} + n_3^{\text{inj}}\right)\cdot\left[\overline{\Phi_H^{23}}\left(m_2^0, m_3^0\right) - \overline{\Phi_H^{23}}\left(m_2^k, m_3^k\right)\right] \tag{2.21}$$

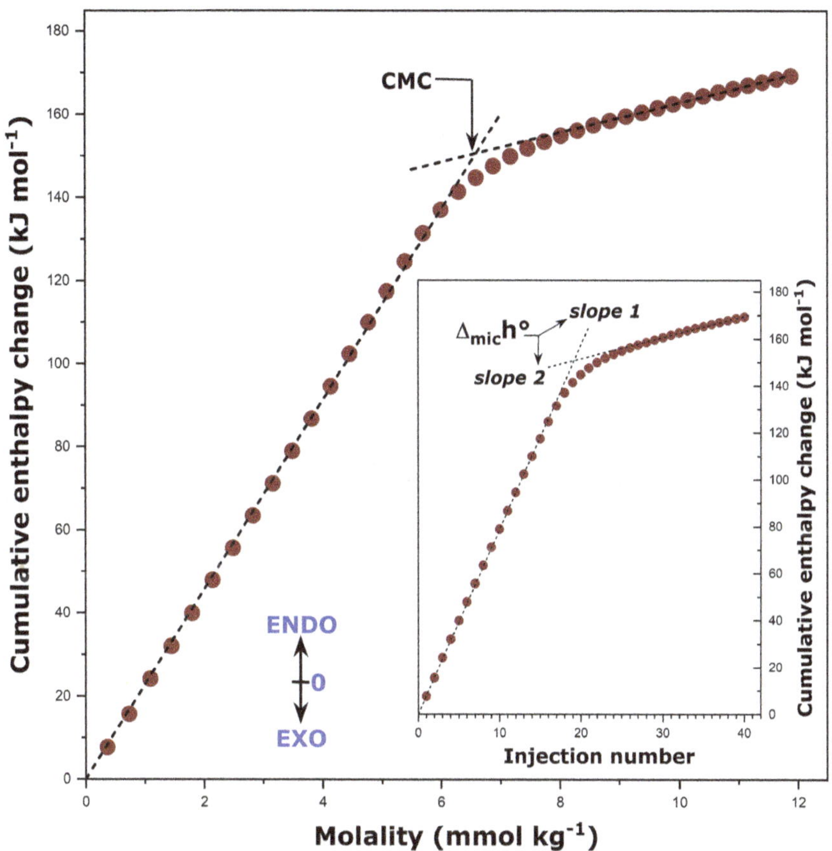

Fig. 2.8: Cumulative enthalpy of dilution of a 60 mmol kg^{-1} aqueous solution of dodecylguanidinium chloride in deionized water at 298 K (adapted from [47]). The plots of $\frac{\Delta_{inj}H^k_{cum}}{n_2^{inj}}$ against the equilibrium molality m_2^k (main panel) and injection number k (inset).

where $\overline{\Phi_H^{23}}$ stands for the weighted mean apparent molal enthalpy of the two solutes in aqueous solutions, according to eq. (2.14). With this formalism in mind, the experimental enthalpy change $\Delta_{inj}H^k_{cum}$ represents the cumulative enthalpy of dilution, $\Delta_{dil}H_{cum}(m_2^k, m_3^k)$, of a ternary solution containing $k \cdot (n_2^{inj} + n_3^{inj})$ moles of both solutes from composition (m_2^0, m_3^0) to composition (m_2^k, m_3^k). Therefore, the calorimetric measurements of terms $\frac{\Delta_{inj}H^k_{cum}}{n_2^{inj}+n_3^{inj}}$ for various values of m_2^k, m_3^k, and k allows the relative apparent enthalpy, $\overline{\Phi_H^{23}}(m_2^k, m_3^k) - \overline{\Phi_H^{23}}(m_2^0, m_3^0)$, to be determined as a function of $m_2^k + m_3^k$.

With the interfacial enthalpy H^{σ}_{S1} before the first injection step where the solid sample is immersed in M_1^0 grams of solvent, as well as the enthalpy $H^{\sigma}_{SL}(m_2^k, m_3^k)$ of the boundary formed after the k th injection and remaining in equilibrium with the bulk solution of composition m_2^k and m_3^k, the cumulative enthalpy change measured in the ITC adsorption experiment can be expressed in the following form:

$$\Delta_{\text{inj}}H_{\text{cum}}^k = \left[H_{\text{SL}}^{\sigma}\left(m_2^k\right) - H_{\text{S1}}^{\sigma}\right] - \left(\Delta_{\text{ads}}n_2^k + \Delta_{\text{ads}}n_3^k\right) \cdot \overline{\Phi_H^{23}}\left(m_2^k, m_3^k\right) + \\ - \left(n_2^{\text{inj}} + n_3^{\text{inj}}\right) \cdot k \cdot \left[\overline{\Phi_H^{23}}\left(m_2^0, m_3^0\right) - \overline{\Phi_H^{23}}\left(m_2^k, m_3^k\right)\right] \quad (2.22)$$

Following the same reasoning as in the case of single-solute systems (cf. Section 2.3.2), one obtains

$$\Delta_{\text{inj}}H_{\text{cum}}^k = \Delta_{\text{dpl}}H_{\text{cum}}\left(\Delta_{\text{ads}}n_2^k, \Delta_{\text{ads}}n_3^k, m_2^k, m_3^k\right) + \Delta_{\text{dil}}H_{\text{cum}}\left(m_2^k, m_3^k\right) \quad (2.23)$$

The cumulative enthalpy of displacement in the two-solute system, that is, $\Delta_{\text{dpl}}H_{\text{cum}}$ $\left(\Delta_{\text{ads}}n_2^k, \Delta_{\text{ads}}n_3^k, m_2^k, m_3^k\right)$, may be thus inferred from the ITC adsorption experiment if the corresponding cumulative enthalpy of dilution, $\Delta_{\text{dil}}H_{\text{cum}}\left(m_2^k, m_3^k\right)$, has been determined in a separate ITC dilution run. In the case of adsorption at a charged interface from an ionic aqueous solution, the enthalpy of displacement obviously includes the terms due to desorption of the equivalent number of ions initially present within the interfacial region to neutralize the charge of the solid surface. Following the discussion in Section 2.2.3 (cf. Fig. 2.4), the values of $\Delta_{\text{ads}}n_2^k$, $\Delta_{\text{ads}}n_3^k$, m_2^k, and m_3^k required for evaluating the dilution terms are inferred from the points of intersection between the experimental adsorption curves for both solutes and the calorimetry lines defined by eqs. (2.8a) and (2.8b).

2.3.5 Correlation between dilution and adsorption experiments in two-solute systems

Determining the correction term due to dilution of a two-solute aqueous solution is often a real issue for extraction of the enthalpy of displacement term from eq. (2.23), especially when the two solutes compete against each other in the interfacial region. In consequence, the proportion between them in the supernatant within the calorimetric cell changes constantly and it is different from that in the stock solution having a constant composition $\left(m_2^0, m_3^0\right)$. This problem may be solved in different manners depending on the specificity of the adsorption system studied.

When the cumulative enthalpy change after k successive injections, $\Delta_{\text{inj}}H_{\text{cum}}^k$, is markedly dominated by the displacement term due to strong solute-surface interactions, the dilution contribution can be simply neglected. Otherwise, it is highly recommended to study and thoroughly compare the dilution phenomena in single-solute and two-solute systems by following the ITC procedures described below.

Even though the quantity $\overline{\Phi_H^{23}}(m_2, m_3)$ has no clear thermodynamic significance, it may constitute the basis on which to decide whether the two types of solute species interact with one another and/or compete against one another for interactions with the solvent in the ternary solution. For this purpose, the results of the ITC dilution experiment performed by making use of a two-solute solution will be compared with those obtained in two separate ITC dilution experiments in which a binary solution of either solute 2 or solute 3 is employed. Based on these three types of calorimetry experiment, it is possible to construct the following excess function of variable $m_2 + m_3$:

$$\Delta_{23}H(m_2 + m_3) = (n_2 + n_3) \cdot \left[\overline{\Phi_H^{23}}(m_2, m_3) - \overline{\Phi_H^{23}}\left(m_2^0, m_3^0\right)\right] + \\ - n_2\left[\Phi_H(m_2) - \Phi_H\left(m_2^0\right)\right] - n_3\left[\Phi_H(m_3) - \Phi_H\left(m_3^0\right)\right] \quad (2.24)$$

where the first term on the right-hand side represents the cumulative enthalpy change in the ternary system; the second and third terms correspond to the cumulative enthalpy of dilutions measured separately in both binary systems for the same values of m_2, m_3, m_2^0, and m_3^0. Note that the values of n_2 and n_3 appearing in eq. (2.24) are taken as corresponding to the local composition, $m_2 + m_3$, of the ternary system. When the function $\Delta_{23}H$ takes zero values in a given $m_2 + m_3$ range, the mixing of both solutes in water does not involve any cooperative or competitive interactions. Two opposite situations are depicted in Fig. 2.9.

According to Fig. 2.9a, the dilution of single-solute solutions of cadmium and strontium is exothermic, whereas that of barium is endothermic. In the case of two-solute solution of strontium and cadmium (Fig. 2.9b), the experimental $\Delta_{\text{dil}}H_{\text{cum}}$ versus $(m_2 + m_3)$ curve and the theoretical $\Delta_{23}H$ versus $(m_2 + m_3)$ one practically overlap with each other over a wide molality range, which means that their dilution pattern resembles that of ideal mixture of the two solutes. On the contrary, the mixing of strontium and barium is far from being ideal as the dilution is much more endothermic than what can be deduced from the theoretical values given by eq. (2.24).

In particular situations where the two solutes or some of their moieties bind together in a reversible way to form complexes upon mixing, carrying out the ITC complexation experiments is necessary in providing in-depth knowledge about the binding stoichiometry, n, the equilibrium dissociation constant, K_d, related to the Gibbs energy of binding, ΔG_{bind}, by the formula $K_d = \exp(\Delta G_{\text{bind}}/R \cdot T)$, as well as the related enthalpy change upon binding, ΔH_{bind} [63–66].

Given the subject matter of the present chapter, the chelation of metal ions with a variety of ligands is a notable example representing also the case of cooperative effects in adsorption.

Here, the main difficulty lies in determining the concentration of metal-ligand complex in addition to free metal or ligand species in the equilibrium bulk solution remaining in the calorimetric cell after each injection of a new portion of the stock solution. Therefore, the knowledge of dissociation constant in the systems studied is mandatory to extract the concentration of free and complex species from the overall concentrations of metal and ligand in various solutions. The dissociation of a complex between a metal, M, and a ligand, L, can be represented by the following chemical reaction, in

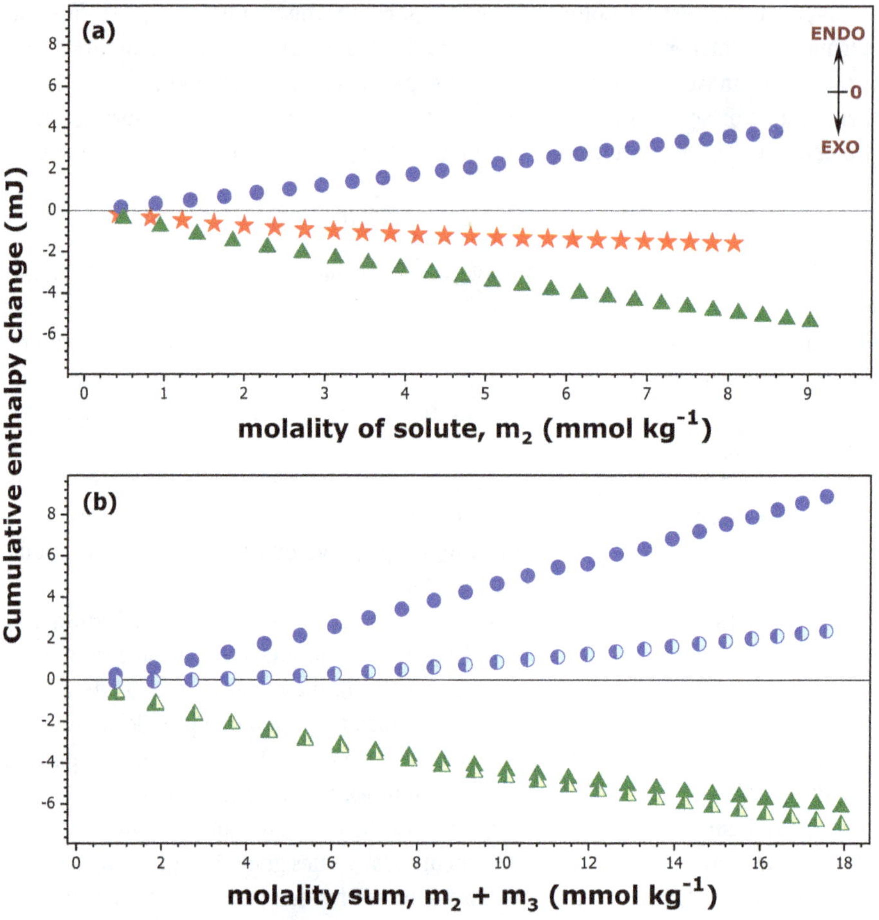

Fig. 2.9: Variations of the cumulative enthalpy change accompanying the dilution of single-solute and two-solute aqueous solutions at 298 K (adapted from [23]): (a) the enthalpy curves for dilution of 40 mmol kg^{-1} solutions of strontium nitrate (red stars), barium nitrate (blue circles), and cadmium nitrate (green triangles); (b) comparison between the cumulative enthalpy change accompanying the dilution of two-solute aqueous solutions (filled symbols) and the sum of the second and third terms on the right-hand side of eq. (2.24) (half-filled symbols) for ternary solutions containing 40 mmol kg^{-1} of strontium nitrate and 40 mmol kg^{-1} of barium nitrate (blue circles) or cadmium nitrate (green triangles).

which the electrical charges of the ionic species have been omitted for the sake of simplicity:

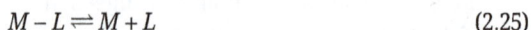

$$M - L \rightleftharpoons M + L \tag{2.25}$$

The case of dilute aqueous solutions usually justifies the use of molarities in the mass-action expressions (to be thermodynamically correct, the activities of the substances should have been compared). With the equilibrium molar concentrations of free metal,

$[M]_k$, free ligand, $[L]_k$, and complex, $[M-L]_k$, species after the kth injection into the calorimetric cell, the dissociation constant may be expressed as follows according to the independent one-site binding model [64]:

$$K_d = \frac{[M]_k \cdot [L]_k}{[M-L]_k} = \frac{\{C_2^k - [M-L]_k\} \cdot \{C_3^k - [M-L]_k\}}{[M-L]_k} \tag{2.26}$$

where C_2^k and C_3^k represent the total molarities, respectively, of metal and ligand in the equilibrium solution. Since the molar concentrations of aqueous solutions employed in the ITC experiment are usually low, it is possible to apply the thermodynamic formalism based on the molality units by replacing C_2^k and C_3^k with m_2^k and m_3^k, respectively.

In a typical ITC experiment, the thermal peaks corresponding to successive injections of one binding partner (e.g., metal ion) in an aqueous stock solution from the syringe into the sample cell containing an aqueous solution of the other binding partner (e.g., ligand species) are integrated and plotted against the mole ratio between the binding partner injected and the one in the sample cell to generate a binding isotherm (Fig. 2.10). The binding isotherm is then fitted to the theoretical isotherm expression derived on the basis of one-site or multi-site binding models to obtain the related dissociation constant [67–74].

Nevertheless, it should be realized that the K_d values inferred from the ITC binding curves strongly depend on the experimental conditions applied and may not always be comparable to analogous values determined from complexometric titrations performed by means of such other methods as potentiometry or spectrophotometry. A comprehensive account of multiple complexation equilibria and parasitic effects accompanying the metal and ligand binding (e.g., dilution of the titrant, precipitation, hydrolysis, and redox reactions) may be found in the literature (e.g., [65, 75]).

2.4 Possible interpretation of the results of ITC adsorption measurements with a view to improving the understanding of adsorption mechanism

The description of adsorption process is usually inferred from the analysis of variations of the differential molar enthalpy of displacement as a function of the quantity of solute adsorption since it is the differential enthalpy which is sensitive to any change in the partial mechanism. On the contrary, all the interactions and effects involved are merged together into the cumulative enthalpy change where they may be masked to a great extent, with a risk of being hardly identifiable. The direct determination of the differential enthalpy effects in the ITC adsorption experiment is still

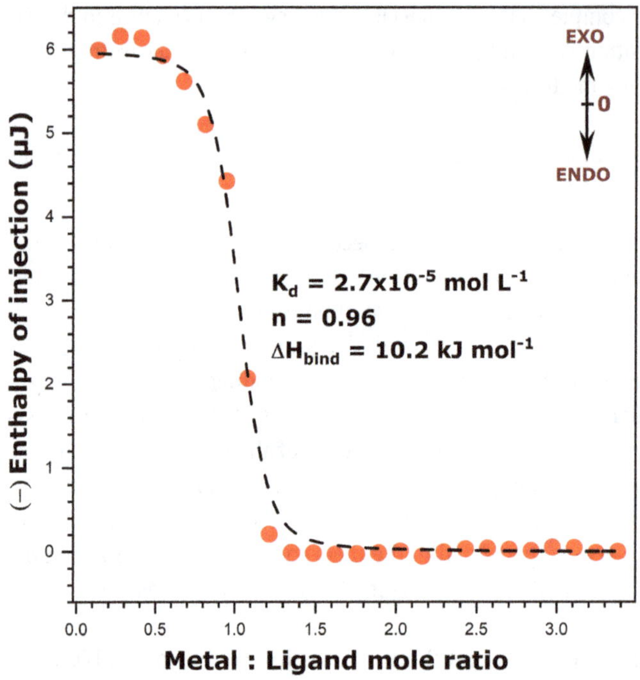

Fig. 2.10: ITC heat data (enthalpy of injection taken with the opposite sign) corresponding to successive injections of a co-containing stock solution into a citrate-containing aqueous solution placed in the calorimetric cell at 298 K (adapted from [46]): the solute concentrations, respectively, in the syringe and in the sample cell are as follows: 2 mol L^{-1} cobalt(II) nitrate hexahydrate and 5 mmol L^{-1} ammonium citrate tribasic. The dissociation constant, K_d, binding stoichiometry, n, and binding enthalpy, ΔH_{bind}, for $Co(C_6H_5O_7)^-$ complex have been obtained by fitting the heat experimental values (red circles) to the independent site binding model (dashed line).

possible (i.e., by calculating the $\Delta_{inj}H_k$ values for individual k th injections) but more complex than that of the cumulative ones [51]. This is all the more serious when the pseudo-differential enthalpy curve obtained directly from the ITC measurements does not represent a regular analytical function [37, 41].

In the case of single-solute system, a better solution to this problem proposed within the framework of this chapter should rather be to exploit the full potential of the cumulative enthalpy curve, that is, $\Delta_{dpl}H_{cum} = \Delta_{dpl}H_{cum}(\Delta_{ads}n_2)$, and differentiate this function with respect to the amount of solute, $\Delta_{ads}n_2$, adsorbed at the solid-liquid interface so as to compute the differential molar enthalpy of displacement. This approach is particularly simple when the cumulative enthalpy curve contains linear portions.

As far as the competitive or cooperative adsorption in two-solute systems is concerned, the values of $\Delta_{dil}H_{cum}(\Delta_{ads}n_2, \Delta_{ads}n_3,\ m_2,\ m_3)$ should be plotted against the total amount adsorbed which is a sum of the adsorption quantities for the two solutes, that is, $\Delta_{ads}n_2 + \Delta_{ads}n_3$. This representation is highly recommended, particularly in the case of strong competitive adsorption between the two solutes when the individual adsorption isotherms may be decreasing functions on some concentration intervals.

It is possible to gain more insight about the competitive or cooperative character of the phenomenon by comparing the experimental enthalpy values, $\Delta_{dpl}H_{cum}$, with the theoretical ones, $\Delta_{dpl}H^{23}_{cum}$, calculated based on the assumption of full independence of adsorption mechanisms for both adsorbate species. Indeed, it may be initially assumed that solutes 2 and 3 adsorb in an independent manner on surface sites specific for each of them when mixed up together in the two-solute system. Consequently, each solute could follow the same adsorption mechanism it had in the single-solute system. The additivity of enthalpy contributions given individually by each solute component to the total displacement effect leads to the following expression:

$$\Delta_{dpl}H^{23}_{cum} = \Delta_{ads}n^{23}_2 \cdot \Delta_{dpl}h^*_2 + \Delta_{ads}n^{23}_3 \cdot \Delta_{dpl}h^*_3 \tag{2.27}$$

where $\Delta_{ads}n^{23}_2$ and $\Delta_{ads}n^{23}_3$ correspond to the quantities of individual adsorption for solutes 2 and solute 3, respectively, as measured in the two-solute system (the superscript 23), $\Delta_{dpl}h^*_2$ and $\Delta_{dpl}h^*_3$ denote the molar enthalpies of displacement accompanying the adsorption of each solute from an appropriate single-solute solution. Such an enthalpy should be determined based on appropriate adsorption and calorimetry experiments by following the procedures described in Section 2.3 separately for each solute. The measurements of adsorption from single-solute solutions can thus be regarded as mandatory pre-requisites for a deeper comprehension of the adsorption mechanisms in two-solute systems.

Two selected examples illustrating the proposed approach with regard to interpretation of calorimetry results are discussed briefly below.

2.4.1 Example of competitive ion adsorption

Figure 2.11 presents an example of the cumulative enthalpy of displacement measured upon simultaneous Cd(II) and Sr(II) adsorption from two-metal nitrate solutions onto zeolite 4A, together with the corresponding adsorption isotherms determined individually for each of the metal components [23]. Following the recommendations given throughout this chapter, the two metal solutes are mixed up in equimolar proportions and the enthalpy of displacement, $\Delta_{dpl}H_{cum}$, is plotted as a function of the sum of adsorption quantities, $\Delta_{ads}n_2 + \Delta_{ads}n_3$. The experimentally measured $\Delta_{dpl}H_{cum}$ values are also compared with the theoretical ones, $\Delta_{dpl}H^{23}_{cum}$, generated from eq. (2.27) on the basis of adsorption and calorimetry measurements performed in both single metal systems.

First and foremost, there is always an obligation to identify, during the programming phase prior to the experimental studies, the molecular nature of the solute species to be adsorbed and evaluate the sorption performance of the adsorbent, with the

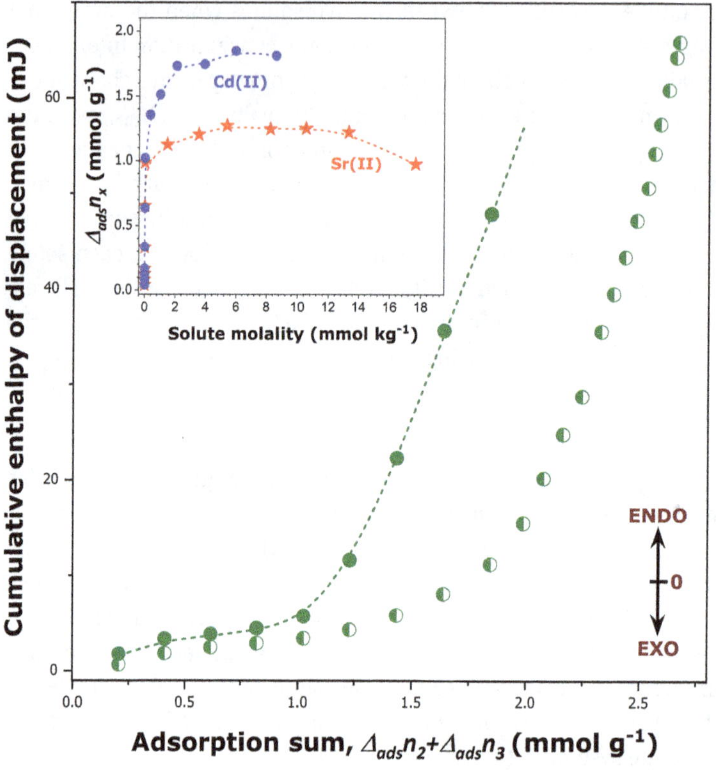

Fig. 2.11: Variations of the cumulative enthalpy of displacement (green circles) accompanying simultaneous adsorption of cadmium and strontium nitrates onto 4A zeolite from two-solute solutions in water at 298 K (adapted from [23]). The dashed line represents the uncorrected enthalpy curve obtained without subtracting the dilution effects. The half-filled symbols refer to the theoretical enthalpy values calculated on the basis of eq. (2.27). The inset shows the individual adsorption isotherms for both solutes from two-solute solutions.

view of shedding light on potential adsorbate-adsorbent interactions responsible for adsorption.

As far as the present solute species are concerned, it is mandatory to analyze the speciation diagram of cadmium and strontium nitrates under the conditions of pH and ionic strength applied in adsorption and calorimetry measurements to be sure that free metal cations represent the predominant species. Such a verification procedure is particularly important when the pH and ionic strength of the aqueous phase may change upon adsorption at risk of formation of hydrolyzed or complex metal species. Appropriate speciation diagrams may be sometimes found in the literature. Nevertheless, it is always more appropriate to establish such diagrams under real conditions (e.g., concentrations and counter-ions) with the aid of dedicated chemical speciation software (e.g., Medusa-Hydra, Mineql+, Minsorb, Fiteql, Minteq, Minfit, Phreeqc, Ecosat, Orchestra, Geosurf, CHEAQS Next, and Geochemist's Workbench [76–85]).

Concerning the adsorbent used, zeolites A (Linde Type A) with a permanent charge are known to possess extra-framework compensating cations which are located in various crystallographic sites, depending on the cation nature and its crystal content, the hydration state of the zeolite structure, the Si-to-Al ratio in the framework, or the degree of cation exchange [86–90]. Therefore, it seems likely that the adsorption of free Cd(II) or/and Sr(II) cations follows the cation exchange pathway in stoichiometric proportion so as to neutralize this permanent charge along the adsorption isotherm. By the way, the release of Na^+ counter-ions from the zeolite structure and dehydration of Cd^{2+} ions upon adsorption are supposed to represent potential endothermic mechanisms which overcome the principal exothermic contribution due to the metal uptake, thus rendering the overall displacement process in the cadmium-zeolite system endothermic and entropy-driven in the whole adsorption range studied [23, 52, 91].

Without going into too much detail about the mechanism of adsorption from single-metal solutions, the comparison between $\Delta_{dpl}H_{cum}$ and $\Delta_{dpl}H^{23}_{cum}$ in Fig. 2.11 provides strong indications for the existence of two adsorption regimes. There is little difference between the experimental and theoretical enthalpy values in an adsorption interval (i.e., $\Delta_{ads}n_2 + \Delta_{ads}n_3$) below 1 mmol g^{-1}, corresponding to half of the quasi-vertical segments on the individual adsorption isotherms. It is thus reasonable to conclude that both types of metal cation adsorb in a quite independent manner in this domain. Beyond this range, there is evidence to suggest that the direct competition between Cd^{2+} and Sr^{2+} cations modifies the sequence in which the zeolite crystallographic sites are occupied. This conclusion is additionally supported by the decrease in the strontium uptake at high equilibrium concentrations (cf. inset to Fig. 2.11).

It is important to realize that the enthalpy measurements correlated with the individual adsorption isotherms provide a macroscopic account of the exchange pathway but they fail to identify the crystallographic sites in zeolite 4A on which the cation exchange between heavy metal cations and Na^+ ions occurs at every stage of adsorption. To rationalize the cation exchange pathway and provide a microscopic description of the adsorption phenomenon, it could be relevant to account for this variety of crystallographic sites, their various strength of interaction, possible migration of the compensating cations of different type. One way to accomplish this goal would be through the Rietveld analysis of X-ray powder diffraction patterns widely used for obtaining the structural information about minerals [92–96]. Nevertheless, the presence of liquid water is not sufficiently taken into account in the structure refinement procedures, the consequence being that the full understanding of the driving force of ion exchange cannot be really attained.

2.4.2 Example of cooperative ion adsorption from dilute aqueous solutions

ITC technique offers also the possibility of monitoring cooperative effects in adsorption at the macroscopic level through measurements of the enthalpy of displacement and its changes along the adsorption isotherm. To illustrate this point, consider Fig. 2.12, which shows the experimental enthalpy curves obtained in the case of individual and simultaneous adsorption of Co(II) cations, $[Co(H_2O)_6]^{2-}$, and ligand acetate anions, CH_3COO^-, from single-solute and two-solute aqueous solutions onto powdered γ-Al_2O_3 possessing a pH-dependent surface charge [46]. To quantify the attachment of metal and ligand species to the solid support under the unadjusted pH condition, the isotherms of individual adsorption of the components have to be measured from a single-solute and two-solute aqueous phase and interpreted according to the observed changes in the pH of the equilibrium solid-liquid suspension and the results of potentiometric titrations [97].

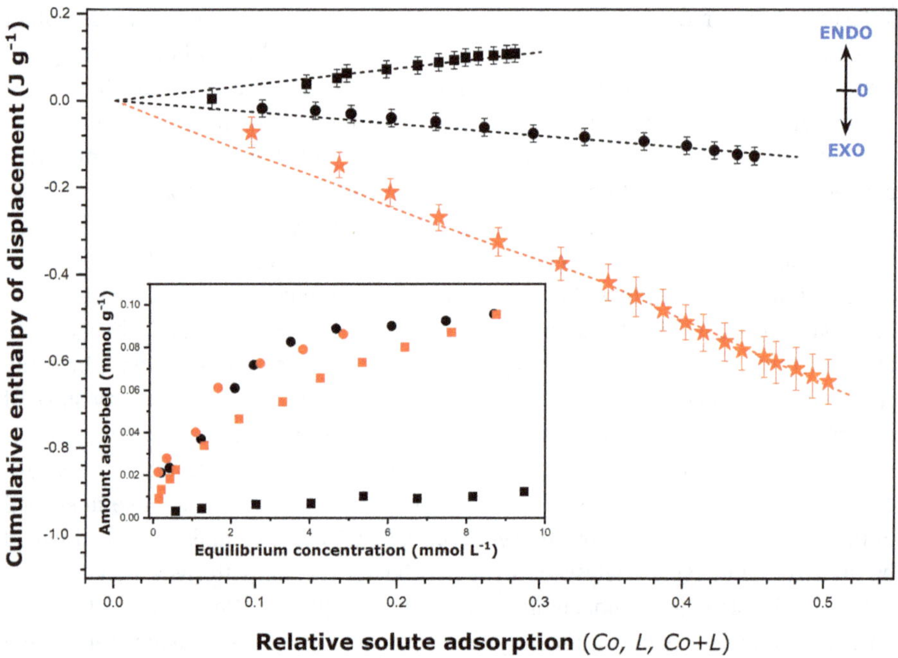

Fig. 2.12: Variations of the cumulative enthalpy of displacement accompanying the solute adsorption onto γ-alumina from single-solute (black symbols) and two-solute aqueous solutions (red symbols) at 298 K under the unadjusted pH condition: cobalt alone (circles), acetate alone (squares), and cobalt-acetate (stars). The dashed lines represent the theoretical enthalpy curves obtained by assuming the formation of inner-sphere (cobalt) or outer-sphere (acetate) complexes in the single-solute systems, as well as type A (solid-metal-ligand) complexes through cooperative adsorption of metal and ligand species in the two-solute system. The inset shows the individual isotherms of cobalt (circles) and acetate (squares) adsorption from single-solute (black symbols) and two-solute (red symbols) solutions (adapted from [46, 97]).

The linearity of $\Delta_{dpl}H_{cum}$ versus $\Delta_{ads}n_2$ plots in the single-solute systems is a strong argument for supporting the conclusions about the nature of binary surface complexes, previously drawn on the basis of other experimental studies [97]. This linearity indicates that the differential enthalpy of displacement should be a constant function of $\Delta_{ads}n_2$ and the adsorption mechanism remains unchanged in the adsorption range studied (restricted in Fig. 2.12 to the initial range of low $\Delta_{ads}n_2$ values). The exothermic character of the displacement process for cobalt cations in the absence of ligand species argues in favor of strong solute-solid interactions dominating all other contributions to the overall enthalpy effect. Hence the attachment of metal cations to the alumina surface can be modeled in the form of mononuclear and mono-dentate inner-sphere complexes. On the contrary, the formation of outer-sphere complexes is rather postulated in the case of acetate anions adsorbed from single-solute solutions so as to account for the endothermic character of the displacement process. When the two solutes are mixed up in equimolar proportions, simple modeling of the enthalpy of displacement data on the basis of eq. (2.27) suggests that both metal and ligand likely preserve their modes of attachment to γ-Al_2O_3 in the ternary system. Indeed, the formation of inner-sphere cobalt complexes yields in turn positively charged sites for further ligand binding, thus giving rise to a great increase in the acetate adsorption as observed in the inset to Fig. 2.12. This allows for validation of the hypothesis of cooperative attachment to the alumina surface, likely in accordance with the formation of ternary solid-metal-ligand complexes of type A.

2.4.3 Main limitation of ITC technique in adsorption studies

The above analysis of experimental results obtained by direct calorimetry measurements, as well as the opinions formulated by other ITC users [36, 98–101], clearly confirm that the ITC technique provides some specific information and rather limited perception of the adsorbed species. Therefore, some additional experiments are always necessary to complete the information. Certain surface-sensitive microscopic or spectroscopic techniques, for example, scanning electron microscopy, nonlinear optics, X-ray photoelectron spectroscopy, extended X-ray absorption fine structure spectroscopy, X-ray absorption near edge spectroscopy, energy-dispersive X-ray spectroscopy, infrared spectroscopy by attenuated total reflection, Raman spectroscopy, UV–vis diffuse reflectance spectroscopy, or nuclear magnetic resonance spectroscopy (NMR) may give a microscopic insight into the nature of surface complexes formed on solid surfaces [35–39, 102–110]. However, they give reliable results in the range of sufficiently high concentrations of adsorbed species on the solid surface. Since there is always a risk that the necessity of drying or freezing the sample for the purpose of spectroscopic studies perturbs the mode of adsorbate binding to the surface, the challenge nowadays is to work under in situ conditions [36, 106, 108, 111, 112].

2.5 Concluding remarks

Far from representing all cases of the ITC use in the literature of the subject, the examples of solid-liquid adsorption studies analyzed within this chapter testify that isothermal microcalorimetry is now established as a useful tool for continuous monitoring and assessment of the main interactions involved in adsorption phenomena and, in consequence, for better understanding of the adsorption mechanisms at a molecular level. Thermodynamic studies of the interactions between deoxyribonucleic acid or peptide hormones and hydrophilic or hydrophobic solid substrates are evidence of the growing importance of the use of ITC technique in medicinal and pharmaceutical applications.

From the practical point of view, ITC can provide essential information on the adsorption energetics always required in industrial applications to establish the energy balance of technological processes.

A summary of experimental procedures and the subsequent data processing is presented with a view to draw the attention of the reader to the complex thermodynamic interpretation of the relevant observables. It is recommended that further methodical efforts should be made in this respect. The most important parameters to be monitored are highlighted since they can become important sources of systematic errors. It is also clear that further technological progress in designing new accessories to better control the experimental conditions (e.g., pH and ionic strength) is needed beyond the constant search for devices allowing ultra-sensitive heat flow measurements.

It is important to realize that calorimetry alone is not capable of solving satisfactorily many detailed problems concerning the resulting interfacial mechanisms. On a macroscopic level, it certainly cannot provide much specific information on entropy changes accompanying the adsorption process. The formalism based on the treatment of experimental adsorption isotherms is model-dependent and, therefore, it cannot be applied to analyze, in routinely manner, a great number of adsorption systems. By the way, it does not necessarily reflect correctly the competitive character of the phenomenon which manifests itself even in the single-solute systems. On a microscopic level, the knowledge of the nature of the adsorbed species, their specific orientation, and their correlation at the solid-liquid interface, or their distribution over the adsorbent surface constitute the greatest challenge that cannot be solved without developing adequate in situ microscopy or spectroscopy techniques capable of detecting local changes in the surface properties under real operating conditions.

References

[1] Teodoro, F. S.; Nhandeyara Do Carmo Ramos, S.; Elias, M. M. C.; Mageste, A. B.; Ferreira, G. M. D.; da Silva, L. H. M.; Gil, L. F.; Gurgel, L. V. A. Synthesis and application of a new carboxylated cellulose derivative. Part I: Removal of Co^{2+}, Cu^{2+} and Ni^{2+} from monocomponent spiked aqueous solution. *J. Colloid Interface Sci.* **2016**, *483*, 185–200. DOI:10.1016/j.jcis.2016.08.004.

[2] Nhandeyara Do Carmo Ramos, S.; Xavier, A. L. P.; Teodoro, F. S.; Elias, M. M. C.; Gonçalves, F. J.; Gil, L. F.; de Freitas, R. P.; Gurgel, L. V. A. Modeling mono- and multi-component adsorption of cobalt (II), copper(II), and nickel(II) metal ions from aqueous solution onto a new carboxylated sugarcane bagasse. Part I: Batch adsorption study. *Ind Crops Prod* **2015**, *74*, 357–371. DOI:10.1016/j.indcrop.2015.05.022.

[3] Xavier, A. L. P.; Adarme, O. F. H.; Furtado, L. M.; Ferreira, G. M. D.; da Silva, L. H. M.; Gil, L. F.; Gurgel, L. V. A. Modeling adsorption of copper(II), cobalt(II) and nickel(II) metal ions from aqueous solution onto a new carboxylated sugarcane bagasse. Part II: Optimization of monocomponent fixed-bed column adsorption. *J. Colloid Interface Sci.* **2018**, *516*, 431–445. DOI:10.1016/j.jcis.2018.01.068.

[4] Lanas, S. G.; Valiente, M.; Tolazzi, M.; Melchior, A. Thermodynamics of Hg^{2+} and Ag^{+} adsorption by 3-mercaptopropionic acid-functionalized superparamagnetic iron oxide nanoparticles. *J. Therm. Anal. Calorim.* **2019**, *136(3)*, 1153–1162. DOI:10.1007/s10973-018-7763-0.

[5] Fonseca, M. G.; Airoldi, C. Thermodynamics data of interaction of copper nitrate with native and modified chrysotile fibers in aqueous solution. *J. Colloid Interface Sci.* **2001**, *240(1)*, 229–236. DOI:10.1006/jcis.2001.7581.

[6] Lima, I. S.; Airoldi, C. A thermodynamic investigation on chitosan–divalent cation interactions. *Thermochim. Acta* **2004**, *421 (1-2)*, 133–139. DOI:10.1016/j.tca.2004.03.012.

[7] Machado, M. O.; Lopes, E. C. N.; Sousa, K. S.; Airoldi, C. The effectiveness of the protected amino group on crosslinked chitosans for copper removal and the thermodynamics of interaction at the solid/liquid interface. *Carbohydr. Polym.* **2009**, *77(4)*, 760–766. DOI:10.1016/j.carbpol.2009.02.031.

[8] Badshah, S.; Airoldi, C. Layered inorganic–organic hybrid with talc-like structure for cation removal at the solid/liquid interface. *Thermochim. Acta* **2013**, *552*, 28–36. DOI:10.1016/j.tca.2012.11.005.

[9] Oliveira, M.; Simoni, J. A.; Airoldi, C. Chitosan metal-crosslinked beads applied for n-alkylmonoamines removal from aqueous solutions – A thermodynamic study. *J. Chem. Thermodyn.* **2014**, *73*, 197–205. DOI:10.1016/j.jct.2013.12.030.

[10] Azizian, S.; Eris, S.; Wilson, L. D. Re-evaluation of the century-old Langmuir isotherm for modeling adsorption phenomena in solution. *Chem. Phys.* **2018**, *513*, 99–104. DOI:10.1016/j.chemphys.2018.06.022.

[11] Kipling, J. J. Adsorption of non-electrolytes from solution. *Quart. Rev. Chem. Soc., Lond.* **1951**, *5(1)*, 60–74. DOI:10.1039/QR9510500060.

[12] Defay, R.; Prigogine, I.; Bellemans, A.; Everett, D. H. *Surface Tension and Adsorption;* London: Longmans, 1966.

[13] Everett, D. H. Reporting data on adsorption from solution at the solid/solution interface (Recommendations 1986). *Pure Appl. Chem.* **1986**, *58(7)*, 967–984. DOI:10.1351/pac198658070967.

[14] Sparnaay, M. J. *The Electrical Double Layer*; Glasgow: Pergamon Press, 1972.

[15] Kohler, H.-H. Surface Charge and Surface Potential. In *Coagulation and Flocculation: Theory and Applications*, Dobias, B.; Surfactant Science Series, Vol. 47; Marcel Dekker, 1993; pp 37–56.

[16] Koopal, L. K. Adsorption of Ions and Surfactants. In *Coagulation and Flocculation : Theory and Applications*, Dobias, B. Ed.; Surfactant Science Series, Vol. 47; Marcel Dekker, 1993; pp 101–207.

[17] Hiemstra, T.; Van Riemsdijk, W. H. A surface structural approach to ion adsorption: The charge distribution (CD) model. *J. Colloid Interface Sci.* **1996**, *179(2)*, 488–508. DOI:10.1006/jcis.1996.0242.

[18] Sposito, G. *The Surface Chemistry of Natural Particles*; Oxford: Oxford University Press, 2004.

[19] Bourikas, K.; Kordulis, C.; Vakros, J.; Lycourghiotis, A. Adsorption of cobalt species on the interface, which is developed between aqueous solution and metal oxides used for the preparation of supported catalysts: A critical review. *Adv. Colloid Interface Sci.* **2004**, *110(3)*, 97–120. DOI:10.1016/j.cis.2004.04.001.

[20] Schindler, P. W. Co-adsorption of Metal Ions and Organic Ligands; Formation of Ternary Surface Complexes. In *Mineral-water Interface Geochemistry*, M. F. Hochella, J., A. F. White Ed.; Reviews in Mineralogy and Geochemistry, Mineralogical Society of America, 1990; pp 281–307.

[21] Fein, J. B. The Effects of Ternary Surface Complexes on the Adsorption of Metal Cations and Organic Acids on Mineral Surfaces. In *Water-Rock Interactions, Ore Deposits, and Environmental Geochemistry - A Tribute to David A. Crerar*, Hellmann, R., S. A. Wood Ed.; The Geochemical Society, 2002; pp 365–378.

[22] Trompette, J. L.; Zajac, J.; Keh, E.; Partyka, S. Scanning of the cationic surfactant adsorption on a hydrophilic silica surface at low surface coverages. *Langmuir* **1994**, *10(3)*, 812–818. DOI:10.1021/la00015a036.

[23] Prelot, B.; Araïssi, M.; Gras, P.; Marchandeau, F.; Zajac, J. Contribution of calorimetry to the understanding of competitive adsorption of calcium, strontium, barium, and cadmium onto 4A type zeolite from two-metal aqueous solutions. *Thermochim. Acta* **2018**, *664*, 39–47. DOI:10.1016/j.tca.2018.04.006.

[24] Builes, S.; Sandler, S. I.; Xiong, R. Isosteric heats of gas and liquid adsorption. *Langmuir* **2013**, *29(33)*, 10416–10422. DOI:10.1021/la401035p.

[25] Anastopoulos, I.; Kyzas, G. Z. Are the thermodynamic parameters correctly estimated in liquid-phase adsorption phenomena? *J. Mol. Liq.* **2016**, *218*, 174–185. DOI:10.1016/j.molliq.2016.02.059.

[26] Guggenheim, E. A. *Thermodynamics*; Amsterdam: North Holland Publishing Co., 1967.

[27] Gregg, S. J.; Sing, K. S. W. *Adsorption, Surface Area, and Porosity*; London: Academic Press, 1982.

[28] Partyka, S.; Rouquerol, F.; Rouquerol, J. Calorimetric determination of surface areas: Possibilities of a modified Harkins and Jura procedure. *J. Colloid Interface Sci.* **1979**, *68(1)*, 21–31. DOI:10.1016/0021-9797(79)90255-8.

[29] Sahoo, T. R.; Prelot, B. Adsorption Processes for the Removal of Contaminants from Wastewater: The Perspective Role of Nanomaterials and Nanotechnology. In *Nanomaterials for the Detection and Removal of Wastewater Pollutants*, Bonelli, B., Freyria, F. S., Rossetti, I., Sethi, R. Eds.; Elsevier, 2020; pp 161–222.

[30] de Boer, J. H.; Linsen, B. G.; Osinga, T. J. Studies on pore systems in catalysts: VI. The universal t curve. *J. Catal.* **1965**, *4(6)*, 643–648. DOI:10.1016/0021-9517(65)90263-0.

[31] Lippens, B. C.; Linsen, B. G.; Boer, J. H. d. Studies on pore systems in catalysts I. The adsorption of nitrogen; apparatus and calculation. *J. Catal* **1964**, *3(1)*, 32–37. DOI:10.1016/0021-9517(64)90089-2.

[32] Brunauer, S.; Mikhail, R. S.; Bodor, E. E. Pore structure analysis without a pore shape model. *J. Colloid Interface Sci.* **1967**, *24(4)*, 451–463. DOI:10.1016/0021-9797(67)90243-3.

[33] Occelli, M. L.; Olivier, J. P.; Petre, A.; Auroux, A. Determination of pore size distribution, surface area, and acidity in fluid cracking catalysts (FCCs) from nonlocal density functional theoretical models of adsorption and from microcalorimetry methods. *J. Phys. Chem. B* **2003**, *107(17)*, 4128–4136. DOI:10.1021/jp022242m.

[34] Schlumberger, C.; Thommes, M. Characterization of hierarchically ordered porous materials by physisorption and mercury porosimetry – A tutorial review. *Adv. Mater. Interfaces* **2021**, *8 (4)*, 2002181. DOI:10.1002/admi.202002181.

[35] Lyngsie, G.; Penn, C. J.; Hansen, H. C.; Borggaard, O. K. Phosphate sorption by three potential filter materials as assessed by isothermal titration calorimetry. *J. Environ. Manage.* **2014**, *143*, 26–33. DOI:10.1016/j.jenvman.2014.04.010.

[36] Hong, Z.-N.; Yan, J.; Jiang, J.; Li, J.-y.; Xu, R.-k. Isothermal titration calorimetry as a useful tool to examine adsorption mechanisms of phosphate on gibbsite at various solution conditions. *Soil Sci. Soc. Am. J.* **2020**, *84(4)*, 1110–1124. DOI:10.1002/saj2.20101.

[37] Alam, M. S.; Gorman-Lewis, D.; Chen, N.; Flynn, S. L.; Ok, Y. S.; Konhauser, K. O.; Alessi, D. S. Thermodynamic analysis of nickel(II) and zinc(II) adsorption to biochar. *Environ. Sci. Technol.* **2018**, *52(11)*, 6246–6255. DOI:10.1021/acs.est.7b06261.

[38] Alam, M. S.; Gorman-Lewis, D.; Chen, N.; Safari, S.; Baek, K.; Konhauser, K. O.; Alessi, D. S. Mechanisms of the removal of U(VI) from aqueous solution using biochar: A combined spectroscopic and modeling approach. *Environ. Sci. Technol.* **2018**, *52(22)*, 13057–13067. DOI:10.1021/acs.est.8b01715.

[39] Darmograi, G.; Prelot, B.; Geneste, A.; De Menorval, L.-C.; Zajac, J. Removal of three anionic orange-type dyes and Cr(VI) oxyanion from aqueous solutions onto strongly basic anion-exchange resin. The effect of single-component and competitive adsorption. *Colloids Surf. A Physicochem. Eng. Asp.* **2016**, *508*, 240–250. DOI:10.1016/j.colsurfa.2016.08.063.

[40] Darmograi, G.; Prelot, B.; Geneste, A.; Martin-Gassin, G.; Salles, F.; Zajac, J. How does competition between anionic pollutants affect adsorption onto Mg–Al layered double hydroxide? Three competition schemes. *J. Phys. Chem. C* **2016**, *120(19)*, 10410–10418. DOI:10.1021/acs.jpcc.6b01888.

[41] Ferreira, G. M. D.; Ferreira, G. M. D.; Hespanhol, M. C.; de Paula Rezende, J.; Dos Santos Pires, A. C.; Gurgel, L. V. A.; da Silva, L. H. M. Adsorption of red azo dyes on multi-walled carbon nanotubes and activated carbon: A thermodynamic study. *Colloids Surf. A Physicochem. Eng. Asp.* **2017**, *529*, 531–540. DOI:10.1016/j.colsurfa.2017.06.021.

[42] Fideles, R. A.; Ferreira, G. M. D.; Teodoro, F. S.; Adarme, O. F. H.; da Silva, L. H. M.; Gil, L. F.; Gurgel, L. V. A. Trimellitated sugarcane bagasse: A versatile adsorbent for removal of cationic dyes from aqueous solution. Part I: Batch adsorption in a monocomponent system. *J. Colloid Interface Sci.* **2018**, *515*, 172–188. DOI:10.1016/j.jcis.2018.01.025.

[43] Chen, W.-Y.; Matulis, D.; Hu, W.-P.; Lai, Y.-F.; Wang, W.-H. Studies of the interactions mechanism between DNA and silica surfaces by isothermal titration calorimetry. *J. Taiwan Inst. Chem. Eng.* **2020**, *116*, 62–66. DOI:10.1016/j.jtice.2020.11.019.

[44] Pinholt, C.; Hostrup, S.; Bukrinsky, J. T.; Frokjaer, S.; Jorgensen, L. Influence of acylation on the adsorption of insulin to hydrophobic surfaces. *Pharm. Res.* **2011**, *28(5)*, 1031–1040. DOI:10.1007/s11095-010-0349-6.

[45] Pinholt, C.; Kapp, S. J.; Bukrinsky, J. T.; Hostrup, S.; Frokjaer, S.; Norde, W.; Jorgensen, L. Influence of acylation on the adsorption of GLP-2 to hydrophobic surfaces. *Int. J. Pharm.* **2013**, *440(1)*, 63–71. DOI:10.1016/j.ijpharm.2012.01.040.

[46] Ali Ahmad, M.; Prelot, B.; Zajac, J. Calorimetric screening of co-operative effects in adsorption of Co (II) on γ-alumina surface in the presence of Co-complexing anions in aqueous solution. *Thermochim. Acta* **2020**, *694*, 178800. DOI:10.1016/j.tca.2020.178800.

[47] Bouchal, R.; Hamel, A.; Hesemann, P.; In, M.; Prelot, B.; Zajac, J. Micellization behavior of long-chain substituted alkylguanidinium surfactants. *Int. J. Mol. Sci.* **2016**, *17 (2)*, 223. DOI:10.3390/ijms17020223.

[48] Thach, U. D.; Prelot, B.; Pellet-Rostaing, S.; Zajac, J.; Hesemann, P. Surface properties and chemical constitution as crucial parameters for the sorption properties of ionosilicas: The Case of chromate adsorption. *ACS Appl. Nano Mater.* **2018**, *1(5)*, 2076–2087. DOI:10.1021/acsanm.8b00020.

[49] Corea, M.; Grolier, J.-P. E.; Del Río, J. M. Determination of Thermodynamic Partial Properties in Multicomponent Systems by Titration Techniques. In *Advances in Titration Techniques*, Hoang, V. D. Ed.; IntechOpen, 2017; p 69706.

[50] Denoyel, R.; Rouquerol, F.; Rouquerol, J. Thermodynamics of adsorption from solution: Experimental and formal assessment of the enthalpies of displacement. *J. Colloid Interface Sci.* **1990**, *136(2)*, 375–384. DOI:10.1016/0021-9797(90)90384-Z.

[51] Zajac, J. Calorimetry at the Solid–Liquid Interface. In *Calorimetry and Thermal Methods in Catalysis*, Auroux, A. Ed.; Springer Berlin Heidelberg, 2013; pp 197–270.
[52] Prelot, B.; Lantenois, S.; Chorro, C.; Charbonnel, M.-C.; Zajac, J.; Douillard, J. M. Effect of nanoscale pore space confinement on cadmium adsorption from aqueous solution onto ordered mesoporous silica: A combined adsorption and flow calorimetry study. *J. Phys. Chem. C* **2011**, *115(40)*, 19686–19695. DOI:10.1021/jp2015885.
[53] Blaschke, T.; Varon, J.; Werner, A.; Hasse, H. Microcalorimetric study of the adsorption of PEGylated lysozyme on a strong cation exchange resin. *J. Chromatogr. A* **2011**, *1218(29)*, 4720–4726. DOI:10.1016/j.chroma.2011.05.063.
[54] Chaghi, R.; de Ménorval, L.-C.; Charnay, C.; Derrien, G.; Zajac, J. Interactions of phenol with cationic micelles of hexadecyltrimethylammonium bromide studied by titration calorimetry, conductimetry, and 1H NMR in the range of low additive and surfactant concentrations. *J. Colloid Interface Sci.* **2008**, *326(1)*, 227–234. DOI:10.1016/j.jcis.2008.07.035.
[55] Hou, Y.; Han, Y.; Deng, M.; Xiang, J.; Wang, Y. Aggregation behavior of a tetrameric cationic surfactant in aqueous solution. *Langmuir* **2010**, *26(1)*, 28–33. DOI:10.1021/la903672r.
[56] Hait, S.; Majhi, P.; Blume, A.; Moulik, S. A critical assessment of micellization of sodium dodecyl benzene sulfonate (SDBS) and its interaction with poly(vinyl pyrrolidone) and hydrophobically modified polymers, JR 400 and LM 200. *J. Phys. Chem. B* **2003**, *107*, 3650–3658. DOI:10.1021/jp027379r.
[57] Tsamaloukas, A. D.; Beck, A.; Heerklotz, H. Modeling the micellization behavior of mixed and pure n-alkyl-maltosides. *Langmuir* **2009**, *25(8)*, 4393–4401. DOI:10.1021/la8033935.
[58] Medeiros, M.; Marcos, X.; Velasco-Medina, A. A.; Perez-Casas, S.; Gracia-Fadrique, J. Micellization and adsorption modeling of single and mixed nonionic surfactants. *Colloids Surf. A Physicochem. Eng. Asp.* **2018**, *556*, 81–92. DOI:10.1016/j.colsurfa.2018.08.005.
[59] Stodghill, S. P.; Smith, A. E.; O'Haver, J. H. Thermodynamics of micellization and adsorption of three alkyltrimethylammonium bromides using isothermal titration calorimetry. *Langmuir* **2004**, *20(26)*, 11387–11392. DOI:10.1021/la047954d.
[60] Mezzetta, A.; Łuczak, J.; Woch, J.; Chiappe, C.; Nowicki, J.; Guazzelli, L. Surface active fatty acid ILs: Influence of the hydrophobic tail and/or the imidazolium hydroxyl functionalization on aggregates formation. *J. Mol. Liq.* **2019**, *289*. DOI:10.1016/j.molliq.2019.111155.
[61] Zajac, J.; Chorro, C.; Lindheimer, M.; Partyka, S. Thermodynamics of micellization and adsorption of zwitterionic surfactants in aqueous media. *Langmuir* **1997**, *13(6)*, 1486–1495. DOI:10.1021/la960926d.
[62] Zajac, J. Adsorption microcalorimetry used to study interfacial aggregation of quaternary ammonium surfactants (zwitterionic and cationic) on powdered silica supports in dilute aqueous solutions. *Colloids Surf. A Physicochem. Eng. Asp.* **2000**, *167 (1-2)*, 3–19. DOI:10.1016/S0927-7757(99)00479-3.
[63] Velazquez-Campoy, A.; Freire, E. Isothermal titration calorimetry to determine association constants for high-affinity ligands. *Nat. Protoc.* **2006**, *1(1)*, 186–191. DOI:10.1038/nprot.2006.28.
[64] Grossoehme, N. E.; Spuches, A. M.; Wilcox, D. E. Application of isothermal titration calorimetry in bioinorganic chemistry. *J. Biol. Inorg. Chem.* **2010**, *15(8)*, 1183–1191. DOI:10.1007/s00775-010-0693-3.
[65] Wilcox, D. E. Isothermal titration calorimetry of metal ions binding to proteins: An overview of recent studies. *Inorganica Chim. Acta* **2008**, *361(4)*, 857–867. DOI:10.1016/j.ica.2007.10.032.
[66] Wyrzykowski, D.; Chmurzyński, L. Thermodynamics of citrate complexation with Mn2+, Co2+, Ni2+ and Zn2+ ions. *J. Therm. Anal. Calorim.* **2010**, *102(1)*, 61–64. DOI:10.1007/s10973-009-0523-4.
[67] Cheema, M. A.; Taboada, P.; Barbosa, S.; Juárez, J.; Gutiérrez-Pichel, M.; Siddiq, M.; Mosquera, V. Human serum albumin unfolding pathway upon drug binding: A thermodynamic and spectroscopic description. *J. Chem. Thermodyn.* **2009**, *41(4)*, 439–447. DOI:10.1016/j.jct.2008.11.011.

[68] Ouimet, C. M.; Shao, H.; Rauch, J. N.; Dawod, M.; Nordhues, B.; Dickey, C. A.; Gestwicki, J. E.; Kennedy, R. T. Protein cross-linking capillary electrophoresis for protein–protein interaction analysis. *Anal. Chem.* **2016**, *88(16)*, 8272–8278. DOI:10.1021/acs.analchem.6b02126.

[69] Johnson, R. A.; Manley, O. M.; Spuches, A. M.; Grossoehme, N. E. Dissecting ITC data of metal ions binding to ligands and proteins. *Biochim. Biophys. Acta Gen. Subj.* **2016**, *1860*(5), 892–901. DOI:10.1016/j.bbagen.2015.08.018.

[70] Bou-Abdallah, F.; Arosio, P.; Santambrogio, P.; Yang, X.; Janus-Chandler, C.; Chasteen, N. D. Ferrous ion binding to recombinant human H-chain ferritin. An isothermal titration calorimetry study. *Biochemistry* **2002**, *41(37)*, 11184–11191. DOI:10.1021/bi020215g.

[71] Terpstra, T.; McNally, J.; Han, T.-H.-L.; Ha-Duong, N.-T.; El-Hage-Chahine, J.-M.; Bou-Abdallah, F. Direct thermodynamic and kinetic measurements of Fe^{2+} and Zn^{2+} binding to human serum transferrin. *J. Inorg. Biochem.* **2014**, *136*, 24–32. DOI:10.1016/j.jinorgbio.2014.03.007.

[72] Sacco, C.; Skowronsky, R. A.; Gade, S.; Kenney, J. M.; Spuches, A. M. Calorimetric investigation of copper(II) binding to Aβ peptides: Thermodynamics of coordination plasticity. *J. Biol. Inorg. Chem.* **2012**, *17(4)*, 531–541. DOI:10.1007/s00775-012-0874-3.

[73] Black, C. B.; Cowan, J. A. A critical evaluation of metal-promoted Klenow 3′-5′ exonuclease activity: Calorimetric and kinetic analyses support a one-metal-ion mechanism. *J. Biol. Inorg. Chem.* **1998**, *3(3)*, 292–299. DOI:10.1007/s007750050234.

[74] Arranz, P.; Bianchi, A.; Cuesta, R.; Giorgi, C.; Godino, M. L.; Gutiérrez, M. D.; López, R.; Santiago, A. Binding and removal of sulfate, phosphate, arsenate, tetrachloromercurate, and chromate in aqueous solution by means of an activated carbon functionalized with a pyrimidine-based anion receptor (HL). Crystal structures of $[H_3L(HgCl_4)]\cdot H_2O$ and $[H_3L(HgBr_4)]\cdot H_2O$ showing anion–π interactions. *Inorg. Chem.* **2010**, *49(20)*, 9321–9332. DOI:10.1021/ic100929f.

[75] Connors, K. A. *Binding Constants: The Measurement of Molecular Complex Stability*; New York: Wiley, 1987.

[76] Keizer, M.; Van Riemsdijk, *W. ECOSAT: A Computer Program for the Calculation of Speciation and Transport in Soil-water Systems;* Wageningen Agricultural University, 1994.

[77] Bradbury, M. H.; Baeyens, B. A mechanistic description of Ni and Zn sorption on Na-montmorillonite Part II: Modelling. *J Contam Hydrol* **1997**, *27* (*3-4*), 223–248. DOI:10.1016/S0169-7722(97)00007-7.

[78] Sahai, N.; Sverjensky, D. A. GEOSURF: A computer program for modeling adsorption on mineral surfaces from aqueous solution. *Comput. Geosci.* **1998**, *24*(*9*), 853–873. DOI:10.1016/S0098-3004(97)00142-8.

[79] Herbelin, A. L.; Westall, J. C. *FITEQL: A Computer Program for Determination of Chemical Equilibrium Constants from Experimental Data*; Oregon State University, 1999.

[80] Schecher, W. D.; McAvoy, D. C. *MINEQL+: A Chemical Equilibrium Modeling System; Version 4.5 For Windows Workbook*; Environmental Research Software, 2001.

[81] Meeussen, J. C. ORCHESTRA: An object-oriented framework for implementing chemical equilibrium models. *Environ. Sci. Technol.* **2003**, *37*(*6*), 1175–1182. DOI:10.1021/es025597s.

[82] Bethke, C. M.; Yeakel, S. *The Geochemist's Workbench, Release 8.0 GWB Essentials Guide*; University of Illinois, 2010.

[83] Gustafsson, J. P. *Visual MINTEQ 3.0 User Guide*; Royal Institute of Technology, 2011.

[84] Parkhurst, D. L.; Appelo, C. Description of Input and Examples for PHREEQC Version 3: A Computer Program for Speciation, Batch-reaction, One-dimensional Transport, and Inverse Geochemical Calculations. In *U.S. Geological Survey Techniques and Methods, Book 6, Modeling Techniques*, US Geological Survey, 2013.

[85] Xie, X.; Giammar, D. E.; Wang, Z. Minfit: A spreadsheet-based tool for parameter estimation in an equilibrium speciation software program. *Environ. Sci. Technol.* **2016**, *50*(*20*), 11112–11120. DOI:10.1021/acs.est.6b03399.

[86] Reed, T. B.; Breck, D. W. Crystalline zeolites. II. Crystal structure of synthetic zeolite, type A. *J. Am. Chem. Soc.* **1956**, *78*(*23*), 5972–5977. DOI:10.1021/ja01604a002.

[87] Van Tassel, P. R.; Phillips, J. C.; Davis, H. T.; McCormick, A. V. Zeolite adsorption site location and shape shown by simulated isodensity surfaces. *J. Mol. Graph.* **1993**, *11*(*3*), 180–184. DOI:10.1016/0263-7855(93)80070-8.

[88] Lin, R.; Ladshaw, A.; Nan, Y.; Liu, J.; Yiacoumi, S.; Tsouris, C.; DePaoli, D. W.; Tavlarides, L. L. Isotherms for Water adsorption on molecular sieve 3A: Influence of cation composition. *Ind. Eng. Chem. Res.* **2015**, *54*(*42*), 10442–10448. DOI:10.1021/acs.iecr.5b01411.

[89] Julbe, A.; Drobek, M. Zeolite A type. In *Encyclopedia of Membranes*, Drioli, E., Giorno, L. Eds.; Springer Berlin Heidelberg, 2016; pp 2055–2056.

[90] Kocevski, V.; Zeidman, B. D.; Henager, C. H.; Besmann, T. M. Communication: First-principles evaluation of alkali ion adsorption and ion exchange in pure silica LTA zeolite. *J. Chem. Phys.* **2018**, *149 13*, 131102.

[91] Prelot, B.; Lantenois, S.; Charbonnel, M.-C.; Marchandeau, F.; Douillard, J. M.; Zajac, J. What are the main contributions to the total enthalpy of displacement accompanying the adsorption of some multivalent metals at the silica–electrolyte interface? *J. Colloid Interface Sci.* **2013**, *396*, 205–209. DOI:10.1016/j.jcis.2012.12.049.

[92] Wang, X.; Liao, L. Rietveld structure refinement of Cu-Trien exchanged nontronites. *Front. Chem.* **2018**, *6*, Original Research. DOI:10.3389/fchem.2018.00558.

[93] Ufer, K.; Kleeberg, R.; Bergmann, J.; Dohrmann, R. Rietveld refinement of disordered illite-smectite mixed-layer structures by a recursive algorithm. II: Powder-pattern refinement and quantitative phase analysis. *Clays Clay Miner.* **2012**, *60*(*5*), 535–552. DOI:10.1346/CCMN.2012.0600508.

[94] Gualtieri, A. F.; Ferrari, S.; Galli, E.; Di Renzo, F.; van Beek, W. Rietveld structure refinement of zeolite ECR-1. *Chem. Mater.* **2006**, *18*(*1*), 76–84. DOI:10.1021/cm051985s.

[95] Ikeda, T.; Izumi, F.; Kodaira, T.; Kamiyama, T. Structural study of sodium-type zeolite LTA by combination of Rietveld and maximum-entropy methods. *Chem. Mater.* **1998**, *10*(*12*), 3996–4004. DOI:10.1021/cm980442y.

[96] Howell, P. A refinement of the cation positions in synthetic zeolite type A. *Acta Crystallographica*. **1960**, *13*(*9*), 737–741. DOI:10.1107/S0365110X60001758.

[97] Ali Ahmad, M.; Zajac, J.; Prelot, B. The effect of chelating anions on the retention of Co(II) by γ-alumina from aqueous solutions under the unadjusted pH condition of supported catalyst preparation. *J. Colloid Interface Sci.* **2019**, *535*, 182–194. DOI:10.1016/j.jcis.2018.09.091.

[98] Penn, C. J.; Warren, J. G. Investigating phosphorus sorption onto kaolinite using isothermal titration calorimetry. *Soil Sci. Soc. Am. J.* **2009**, *73*(*2*), 560–568. DOI:10.2136/sssaj2008.0198.

[99] Penn, C. J.; Zhang, H. Isothermal titration calorimetry as an indicator of phosphorus sorption behavior. *Soil Sci. Soc. Am. J.* **2010**, *74*(*2*), 502–511. DOI:10.2136/sssaj2009.0199.

[100] Penn, C.; Heeren, D.; Fox, G.; Kumar, A. Application of isothermal calorimetry to phosphorus sorption onto soils in a flow-through system. *Soil Sci. Soc. Am. J.* **2014**, *78*(*1*), 147–156. DOI:10.2136/sssaj2013.06.0239.

[101] Penn, C. J.; Gonzalez, J. M.; Chagas, I. Investigation of atrazine sorption to biochar with titration calorimetry and flow-through analysis: Implications for design of pollution-control structures. *Front. Chem.* **2018**, *6*, Original Research. DOI:10.3389/fchem.2018.00307.

[102] Manceau, A.; Schlegel, M.; Nagy, K. L.; Charlet, L. Evidence for the formation of trioctahedral clay upon sorption of Co(2+) on quartz. *J. Colloid Interface Sci.* **1999**, *220*(*2*), 181–197. DOI:10.1006/jcis.1999.6547.

[103] Sahai, N.; Carroll, S. A.; Roberts, S.; O'Day, P. A. X-Ray absorption spectroscopy of Strontium(II) coordination. *J. Colloid Interface Sci.* **2000**, *222*(*2*), 198–212.

[104] Elzinga, E. J.; Sparks, D. L. X-ray absorption spectroscopy study of the effects of pH and ionic strength on Pb(II) sorption to amorphous silica. *Environ. Sci. Technol.* **2002**, *36(20)*, 4352–4357. DOI:10.1021/es0158509.

[105] Chen, CC; Coleman, M. L.; Katz, L. E. Bridging the gap between macroscopic and spectroscopic studies of metal ion sorption at the oxide/water interface: Sr(II), Co(II), and Pb(II) sorption to quartz. *Environ. Sci. Technol.* **2006**, *40(1)*, 142–148. DOI:10.1021/es050356g.

[106] Müller, K.; Gröschel, A.; Rossberg, A.; Bok, F.; Franzen, C.; Brendler, V.; Foerstendorf, H. In situ spectroscopic identification of neptunium(V) inner-sphere complexes on the hematite–water interface. *Environ. Sci. Technol.* **2015**, *49(4)*, 2560–2567. DOI:10.1021/es5051925.

[107] Fuller, A. J.; Shaw, S.; Peacock, C. L.; Trivedi, D.; Burke, I. T. EXAFS study of Sr sorption to illite, goethite, chlorite, and mixed sediment under hyperalkaline conditions. *Langmuir* **2016**, *32(12)*, 2937–2946. DOI:10.1021/acs.langmuir.5b04633.

[108] Davantès, A.; Schlaup, C.; Carrier, X.; Rivallan, M.; Lefèvre, G. In situ cobalt speciation on γ-Al2O3 in the presence of carboxylate ligands in supported catalyst preparation. *J. Phys. Chem. C* **2017**, *121(39)*, 21461–21471. DOI:10.1021/acs.jpcc.7b06559.

[109] Wang, X.; Wang, Z.; Peak, D.; Tang, Y.; Feng, X.; Zhu, M. Quantification of coexisting inner- and outer-sphere complexation of sulfate on hematite surfaces. *ACS Earth Space Chem.* **2018**, *2*. DOI:10.1021/acsearthspacechem.7b00154.

[110] Grégoire, B.; Bantignies, J.-L.; Le-Parc, R.; Prélot, B.; Zajac, J.; Layrac, G.; Tichit, D.; Martin-Gassin, G. Multiscale mechanistic study of the adsorption of methyl orange on the external surface of layered double hydroxide. *J. Phys. Chem. C* **2019**, *123(36)*, 22212–22220. DOI:10.1021/acs.jpcc.9b04705.

[111] Jürgensen, A.; Raschke, H.; Esser, N.; Hergenröder, R. An in situ XPS study of L-cysteine co-adsorbed with water on polycrystalline copper and gold. *Appl. Surf. Sci.* **2018**, *435*, 870–879. DOI:10.1016/j.apsusc.2017.11.150.

[112] Assaf, M.; Martin-Gassin, G.; Prelot, B.; Gassin, P.-M. Driving forces of cationic dye adsorption, confinement, and long-range correlation in zeolitic materials. *Langmuir* **2022**, *38(3)*, 1296–1303. DOI:10.1021/acs.langmuir.1c03280.

Basile Galey, Nuno Batalha, Aline Auroux, and Georgeta Postole

Chapter 3
Thermal analysis and solid-state hydrogen storage: Mg/MgH_2 system case study

Abstract: Through the Mg/MgH_2 system case study, this chapter presents the primary interest of thermal analysis and calorimetry to study the hydrogen storage properties of solid-state materials in detail. The dehydrogenation temperature and the hydrogen storage capacity can be obtained by performing temperature programmed desorption and thermogravimetric analysis experiments. These two techniques allow fast experiments with only a few milligrams of sample and are generally used in the literature to study the potential of a new additive, the influence of the preparation method, and to test the performances of new storage systems.

For a deeper characterization of the dehydrogenation properties, specifically the kinetics (apparent activation energy) and the thermodynamics (enthalpy), DSC (differential scanning calorimetry) technique can be used. The Sieverts (volumetric) technique is a powerful tool to investigate the hydrogenation/dehydrogenation properties (kinetics and thermodynamics) and the reversibility of the storage system under isothermal conditions. Finally, the high-pressure DSC technique is of particular interest as it allows to couple calorimetric experiments with a volumetric analysis and study all the important hydrogen storage properties, including storage capacity, hydrogenation and dehydrogenation temperatures, apparent activation energies, and enthalpies, by using only one technique. In this chapter, the strengths and weaknesses of the different instruments are highlighted as well as the working principles and the measured storage properties.

Keywords: thermal analysis, H_2 storage, H_2 production, Mg/MgH_2 system

3.1 Introduction

The world's global development is directly related to the availability and consumption of energy [1]. The constant population growth and the living standards improvement induce an important increase in the global energy use rates [2]. Because of climate change [3] and fossil resources starvation [4], one of the major challenges of the coming decades will be decreasing greenhouse gases emissions and following the growing energy demand simultaneously [5]. The existing energy system, mostly using oil for

Basile Galey, Nuno Batalha, Aline Auroux, Georgeta Postole, Univ Lyon, Université Claude Bernard Lyon 1, CNRS, IRCELYON, F-69626, Villeurbanne, France

https://doi.org/10.1515/9783110590449-003

mobility and coal for electricity production, is not sustainable, and a major transition through renewable resources is necessary [6].

Sun and wind are universally recognized as promising renewable substitutes for electricity production [7]. However, their large-scale development is limited, among others, by their intermittent nature [8, 9]. Therefore, the coupling with an electricity storage system is mandatory to globalize their utilization for stationary and mobile applications [10]. Batteries, compressed air, flywheels, and capacitors can be used for short-term purposes. However, for long-term electricity storage, a new system is necessary [11]. In many global sustainable energy scenarios, hydrogen has been identified as tomorrow's energy vector [12 –15]. Indeed, worldwide production, development, and utilization of hydrogen could not only overcome the renewable electricity storage issue but also limit the mobility sector's reliance on fossil fuels [16].

To use hydrogen as tomorrow's energy vector, developing a reliable, competitive, and secure way for its storage is essential. For mobile applications, the ideal storage system should be a light material able to maintain hydrogen compactly and safely [17]. Hydrogen can be stored as pressurized gas, in liquid-state and in solid-state. Liquid-state storage is hardly practicable as cryogenic temperatures are required to maintain hydrogen under this physical form [18]. For gas-state storage, very high pressures are necessary to reduce the volume, e.g., 700 bars, increasing the cost and leading to safety issues [19, 20]. In opposition, solid-state systems can reversibly store a significant amount of hydrogen under moderate pressure and temperature conditions, thus being highly desirable [21]. The key parameters of such systems are (i) the reversible hydrogen storage capacity, (ii) the hydrogenation/dehydrogenation temperatures (controlled by thermodynamics and kinetics), and (iii) the operational cycle life span [22]. Table 3.1 shows the ultimate technical targets given by the U.S. DOE (United States Department of Energy) for onboard hydrogen storage for light-duty vehicles [23].

Tab. 3.1: Technical targets given by the DOE for onboard hydrogen storage for light-duty vehicles.

Hydrogen storage parameters	2020	2025	Ultimate
Reversible system gravimetric capacity (wt%)	4.5	5.5	6.5
Min/max delivery temperature (°C)	−40/80	−40/80	−40/80
Operational cycle life span (cycles)	1,500	1,500	1,500
System fill time (min)	3–5	3–5	3–5
Hydrogen storage system cost ($/kg H_2)	333	300	266

In most solid-state storage systems, hydrogen is either adsorbed on porous materials (MOF, carbon nanotubes, etc.) or absorbed as a chemical compound (metal hydrides, chemical hydrides, etc.) [24–27]. Table 3.2 summarizes the hydrogen storage properties of the most reported solid-state systems, divided into five families: (1) microporous

Tab. 3.2: Literature review of the hydrogen storage properties of the main solid-state systems.

Solid-state hydrogen storage systems			H_2 capacity (wt%)	De-hydrogenation temperature (°C)	Hydrogenation conditions	$\Delta H_{formation}$ (kJ/mol H_2)	Reversibility (wt% of H_2)	Main asset	Main limits	Ref
Microporous adsorbents	Carbon nanotubes		5–10[c]	/	−140 °C – 0.4 bar[d]	1–10	5–10[c]	Thermodynamics, storage capacity	Temperature, pressure	[26, 28]
	MOF-5		7.1[d]	/	−196 °C – 70 bar[d]	2–5[b]	7.1[d]			[26, 29]
Intermetallic compounds	AB5 – $LaNi_5$		1.37[d]	73[b]	28–10 bar[d]	-30[d]	1.37[b, d]	Temperature, reversibility	Storage capacity	[30, 31]
	AB – TiFe		1.89[d]	27	20–10 bar[d]	-30[d]	1.89[d]			[32]
Metal hydrides	MgH_2		7.67[a, c, d]	> 400[a, b, c]	>300 °C – 15 bar[b, d]	−74.7[b, d]	7.67[b, d]	Reversible storage capacity	Thermodynamics, kinetics	[33, 34]
Complex	Alanates	$NaAlH_4$	7.5[c, d]	185–230[c]	125 °C – 86 bar[d]	−113[d]	5.5[d]	Temperature, storage capacity	Re-hydrogenation pressure, safety	[35, 36]
		$LiAlH_4$	10.6[d]	150 – 180 – 400[b]	180 °C – 80 bar[d]	−119	1[d]			[37, 38]
hydrides	Boro-hydrides	$LiBH_4$	18.4[a, c]	>500[a, c]	600 °C – 155 bar[d]	−190[d]	8.3[d]	Storage capacity	Re-hydrogenation conditions	[39, 40]
		$NaBH_4$	10.6[a, c, d]	>565[a, b, c]	420 °C – 32 bar[d]	−188[d]	2.3[d, c]			[41, 42]

(continued)

Tab. 3.2 (continued)

Solid-state hydrogen storage systems		H_2 capacity (wt%)	De-hydrogenation temperature (°C)	Hydrogenation conditions	ΔH formation (kJ/mol H_2)	Reversibility (wt% of H_2)	Main asset	Main limits	Ref
Chemical hydrides	NH_3BH_3	19.6[a, d]	100 – 150 – 500[a, b]	Off-board regeneration	−21[b]	Off-board regeneration	Storage capacity, temperature	Regeneration, H_2 purity	[43]
	$LiNH_2BH_3$	10.6[c, d]	90[b, c]		−3 to −5[d]				[44]

Thermal analysis instruments used in the references to determine the hydrogen storage properties of the main solid-state systems:
[a] TGA (thermogravimetric analysis).
[b] DSC (differential scanning calorimetry).
[c] TPD (temperature programmed desorption).
[d] Sieverts (volumetric) apparatus.

adsorbents, (2) intermetallic compounds, (3) simple metal hydrides, (4) complex hydrides, and (5) chemical hydrides.

A simple comparison between the technical targets given by the U.S. DOE in Tab. 3.1, and the hydrogen storage properties of the different materials presented in Tab. 3.2, shows that no system has reached the targets for onboard hydrogen storage for light-duty vehicles [25–27]. Either the storage capacity is too low (intermetallic compounds, microporous adsorbents at room temperature), or the hydrogenation/dehydrogenation conditions are too hard (borohydrides, microporous adsorbents, metal hydrides), or the systems suffer from reversibility issues (alanates, chemical hydrides) [28–44]. Intensive research work is thus still needed to achieve the DOE goals and develop a reliable, competitive, and secure solid-state system to store hydrogen.

Thermal analysis and calorimetry techniques are widely used to study the hydrogen storage properties of the different solid-state systems [45]. Indeed, numerous information can be obtained with a simple TPD (temperature programmed desorption) experiment, like the hydrogen storage capacity, the hydrogen release temperatures, and the apparent activation energy for dehydrogenation [46]. Volumetric techniques like Sievert instruments are even more versatile and can give information about hydrogen absorption/release temperatures, thermodynamics (ΔH and ΔS of hydrogenation and dehydrogenation), and kinetics (global) together with the cycle life span behavior [47]. The present chapter focuses on using these techniques for studying the hydrogen storage properties of magnesium/magnesium hydride-based systems, including their strengths and weaknesses. Finally, the differential scanning calorimetry (DSC) technique coupled to volumetry (high-pressure DSC) will be used as a powerful tool to supply all information related to the storage system performances.

3.2 Hydrogen storage in magnesium: mechanisms, interests, and limits

Magnesium can reversibly absorb 7.67 wt% of hydrogen to form a stable compound under moderate conditions of pressure and temperature, i.e., magnesium hydride (MgH_2). The hydrogenation of Mg is exothermic and thermodynamically favorable at ambient temperature but kinetically unfavorable due to the slow diffusion process of hydrogen in magnesium. However, increasing temperature permits overcoming the energy barrier, increasing the kinetics of the diffusion process. The steps of the hydrogenation mechanism, as presented by Bérudé et al. in [48] are:

1) H_2 adsorption – at the surface of Mg.
2) H_2 dissociation – at the surface of Mg.
3) H absorption – at the surface of Mg.
4) H diffusion in Mg – the H atoms move to subsurface sites and diffuse through Mg. This solution of H in Mg is called the α-phase.

5) β-Phase nucleation and growth – with the increase of H concentration in α phase, H–H interactions become important, and a more stable phase nucleates, β phase (hydride phase).

The dehydrogenation of MgH_2 is endothermic and not thermodynamically possible at ambient temperature. Consequently, energy input is mandatory, as temperatures higher than 350 °C are required for bulk MgH_2 conversion. The mechanisms of hydrogen desorption in MgH_2 are the followings, the reverse of Mg hydrogenation:

1) Thermodynamic barrier – input of energy necessary for Mg-H bond dissociation
2) α phase nucleation and growth – corresponding to the formation of the solution of H in Mg
3) H diffusion in Mg and MgH_2 – the H atoms move to the surface of the material
4) H_2 recombination – at the surface of Mg/MgH_2
5) H_2 physidesorption

A comparison of the hydrogen storage properties of bulk Mg/MgH_2 with the technical targets given by the U.S. DOE (Tabs. 3.1 and 3.2) show that, despite an interestingly high reversible H_2 storage capacity, the system is limited by its thermodynamic properties ($\Delta H = \pm 74.7$ kJ/mol H_2) and slow hydrogenation/dehydrogenation kinetics. Different approaches have been explored in the last decades, e.g., nanostructuration, catalytic activators addition, and alloys formation, to overcome the drawbacks of Mg/MgH_2. The way these strategies impact the hydrogen storage properties of MgH_2 is summarized in Tab. 3.3 [49–55]. The best systems, so far, can reversibly store 5–6 wt% of hydrogen at temperatures as low as 200 °C [56–59]. However, the progress made up to date is not enough to meet the U.S. DOE requirements concerning the dehydrogenation temperature and the cycle life span behavior of the hydrogen storage systems. One of the other main issues of the Mg/MgH_2 system is its lack of stability in the air [60]. Indeed, magnesium can be easily oxidized by O_2 and H_2O, and important precautions to avoid air contact must be taken.

Tab. 3.3: Impact of the approaches to overcome the drawbacks of Mg/MgH_2: comparison with bulk MgH_2.

Approach explored	Storage capacity	H_2 release T	Kinetics	ΔH	Life span
Nanostructuration	No change	Decreased	Increased	No change	Highly decreased
Catalytic activators addition	Decreased[a]	Highly decreased	Highly increased	No change	Decreased
Alloys formation	Highly decreased[b]	Decreased[b]	Increased	Decreased[b]	Increased

[a] Depending on the amount of additive.
[b] Depending on the formed alloy.

3.3 Thermal analysis to perform screening processes: TPD and TGA

Dehydrogenation of MgH_2 is the limiting step preventing the design of a system based on Mg/MgH_2 technology within targets (Tab. 3.1). Therefore, it is of primary interest to quickly identify the impact of a material, an additive, or a preparation method on the decomposition temperature of magnesium hydride to assess its potential. To do so, thermal analysis techniques like TPD, TGA (thermogravimetric analysis), and DSC are interesting. These techniques can be used to make a screening process and select only a few systems for further characterization and require a few milligrams of a sample. Furthermore, the experiments can be performed under an inert atmosphere (nitrogen, helium, or argon) to avoid oxidation. The principle of such techniques is to heat the sample at a controlled heating rate and follow either the mass change, the hydrogen release signal, or the heat flow as a function of the temperature for TGA, TPD, and DSC, respectively. Of the three thermal analysis technics, TPD and TGA are more suitable for screening experiments as they can determine both the dehydrogenation temperature and the hydrogen storage capacity (only the temperature for DSC).

The quantification of the hydrogen storage capacity with TPD requires a calibration factor. Such calibration can be obtained by analyzing the hydrogen consumption during the reduction, under hydrogen atmosphere, of different masses of copper(II) oxide, Fig. 3.1a. The areas under the curves are correlated with the amount of hydrogen necessary to reduce copper(II) oxide into copper, and a calibration curve is obtained, Fig. 3.1b. In the example shown in Fig. 3.1, the calibration factor value is 5.29×10^{-8} mmol H_2/mV s. This factor allows the correlation between the area under a dehydrogenation peak with the amount of hydrogen released, i.e., the H_2 storage capacity of the sample. Other materials referred to in the literature as standards for the calibration factor determination include hydrogenated PdGd alloys [61], TiH_2 [61, 62], and palladium hydride [63].

The first important thing in the screening process design with MgH_2-based samples is to be sure that the experiments are done under the same conditions for the results to be comparable. Below is a list of important factors to be considered when doing a screening process through TPD or TGA:

- The analyses must be performed in the same thermal analysis instrument/technique. It is hard to compare results obtained by TPD with TGA as it is not the same phenomena that are measured, i.e., change in the gas flow versus change in the mass.
- The same mass of sample should be used, as a difference in mass will change the dehydrogenation temperature (peak maximum – Fig. 3.1a).
- The same atmosphere (Fig. 3.2a) and gas flow [64] must be used as these impact the measured dehydrogenation temperature.
- The same thermal procedure should be used (Fig. 3.2b), i.e., starting temperature, ramp, end temperature, presence of isotherms, etc. [65].
- Identical crucibles should be used [66].

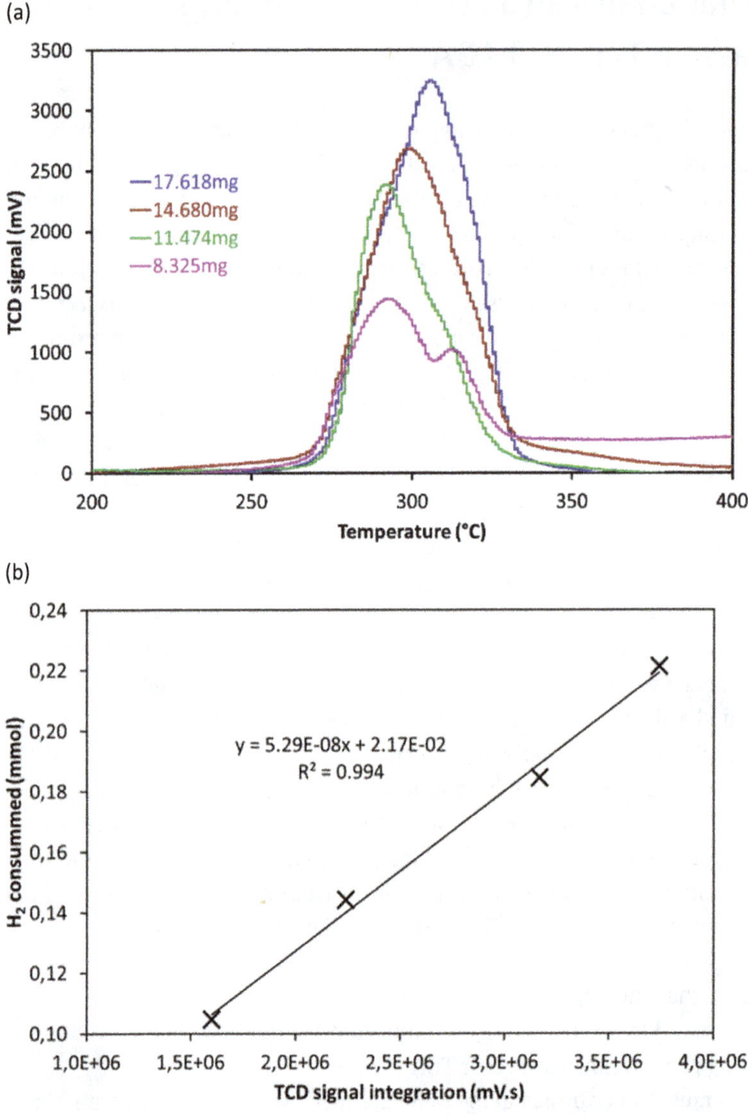

Fig. 3.1: (a) TPR experiments performed with different masses of copper(II) oxide, and (b) example of calibration curve allowing to correlate the area under a dehydrogenation peak with the amount of H_2 released.

Figure 3.2 presents experiments performed by Galey et al. (unpublished data) for the same sample (commercial MgH_2 from McPhy company) but under a different atmosphere (Fig. 3.2a) and heating rate (Fig. 3.2b). The difference between the results is significant, thus highlighting the importance of the experimental conditions in the screening process design. In Fig. 3.2a, oxidation of magnesium at high temperature is noticeable because of the presence of a few ppm of O_2 and/or H_2O, which come in the gas used for the experiment (even in helium). Such weight gain attributed to the exposure of highly reactive Mg metal to trace amounts of O_2 present in the flow inert gas with Mg oxides formation was also observed by Beattie et al. [67]. By doing experiments under Ar flow and using hermetic pans with a pinhole to evacuate the desorbed H_2, Galey et al. [68–70] succeeded to avoid the contact between the sample and the surrounding gas, thus preventing oxidation while Beattie et al. [67], although also loading samples into hermetically sealed pans in an inert atmosphere glove box, could not preclude trace O_2 from the N_2 stream and/or diffusion of O_2 from the ambient atmosphere. When competition between MgH_2 dehydrogenation and Mg oxidation processes is observed, care must be taken in reporting the mass loss. In Fig. 3.2b, the temperature of MgH_2 dehydrogenation is dislocated to higher temperatures by increasing the TPD heating rate. Similar results were obtained by Perejón et al. [71] when measuring the kinetics of MgH_2 dehydrogenation through TGA.

Figure 3.3 includes an example of a screening process performed through TPD to study the impact of adding 5 wt% of different transition metal nanoparticles, e.g., nickel, titanium, iron, and vanadium oxide, to MgH_2. The samples were prepared by planetary ball milling under argon for 5 h, using balls of zirconium dioxide of 1 mm diameter and a milling frequency of 300 revolutions per minute. The ball to powder mass ratio was 100:1. A sample of magnesium hydride milled under the same conditions without nanoparticles was also prepared for comparison. For the screening process, 2–3 mg of powder was heated from 40 to 450 °C, at a heating rate of 2 °C/min and under 40 mL/min of argon. The results show that nickel allows the lowest dehydrogenation temperature (260 °C), followed by iron, V_2O_5, and titanium (Fig. 3.3a). With a high hydrogen storage capacity of 6.6 wt% (Fig. 3.3b), doping MgH_2 with nickel nanoparticles seems an interesting option. Now further characterization can be performed on this sample, for instance, to understand how the nickel nanoparticles impact the dehydrogenation of magnesium hydride.

Another example of a screening process through TPD includes selecting the best amount of additives to design an effective hydrogen storage system. Indeed, when a compound shows potential as a doping agent to MgH_2, the optimal quantity to add has to be found to have a low dehydrogenation temperature and a high hydrogen storage capacity. Figure 3.4 presents the results obtained by TPD for doping with a nickel-based complex, $NiHCl(PCy_3)_2$, where PCy_3 is $P(C_6H_{11})_3$ [68]. Different amounts (5, 10, 20, 25, and 50 wt%) were added to MgH_2 by planetary ball milling under the same conditions described earlier. The experiments show that when the amount of complex increases in the mixtures, the dehydrogenation temperature decreases until the optimal amount of 20 wt% of the $NiHCl(PCy_3)_2$ is reached. At higher amounts, i.e., 25 and 50 wt%, the decomposition temperature shifted toward higher temperatures.

Finally, screening processes through TPD and TGA are also widely used to determine the best preparation conditions, especially when ball-milling is used to obtain "MgH_2 + additive" mixtures. For example, Xie et al. [72] studied the role of the milling time on the dehydrogenation behavior of MgH_2/Ni composite by using DSC and TGA for the screening process. Alternatively, Hanada et al. [73] used TGA to study the correlation between structural characteristics and the amount of hydrogen desorbed in the mechanical milling process. Indeed, the simple, fast, and inexpensive nature of TPD and TGA makes these

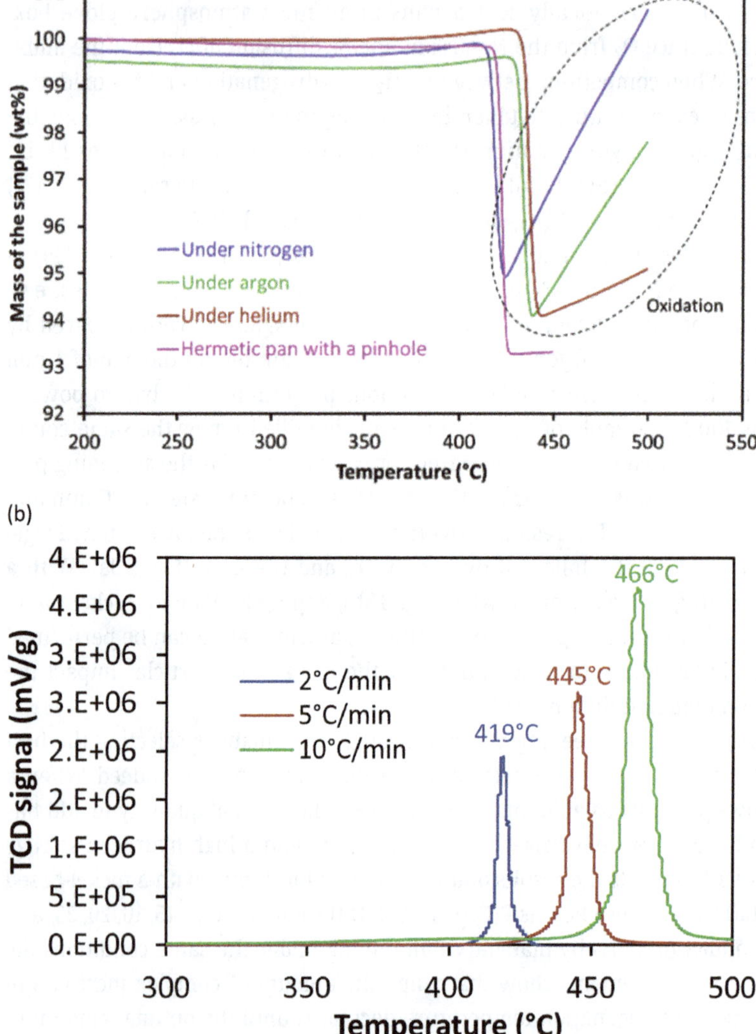

Fig. 3.2: (a) TGA experiments performed with ≈ 2 mg of commercial MgH_2, from 20 to 500 °C, at a heating rate of 2 °C/min and under different atmospheres (gas flow = 20 mL/min) and (b) TPD experiments performed with ≈ 2 mg of commercial MgH_2 (from McPhy company) heated from 20 to 500 °C, under 40 mL/min of argon and at different heating rates.

(a)

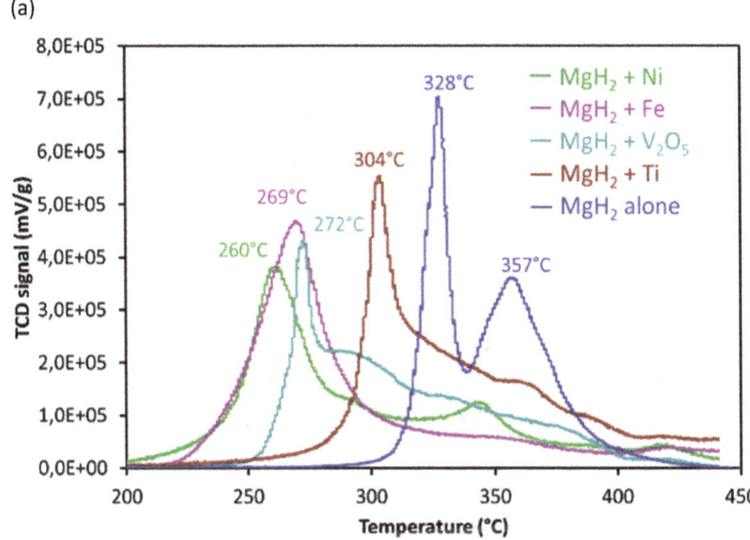

(b)

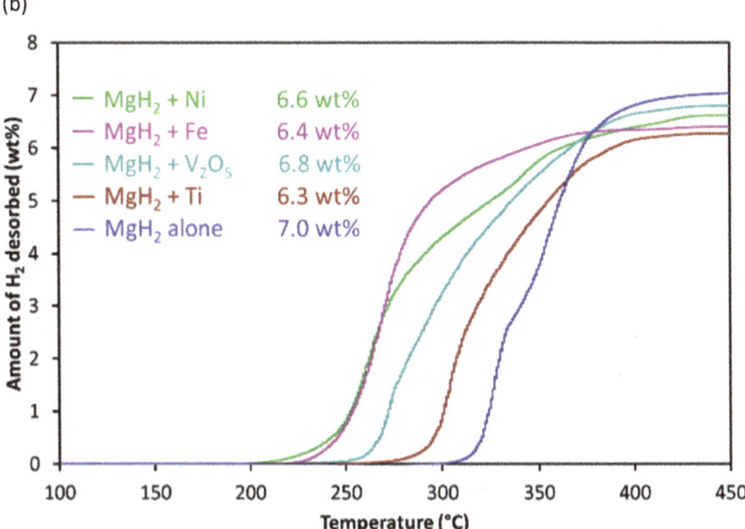

Fig. 3.3: Screening process performed by TPD to study the impact of transition metal nanoparticles on the dehydrogenation properties of MgH_2. (a) Dehydrogenation profiles and (b) H_2 desorption capacity curves.

the techniques of choice for selecting promising hydrogen storage samples to be further studied, hence providing a quick strategy for material screening.

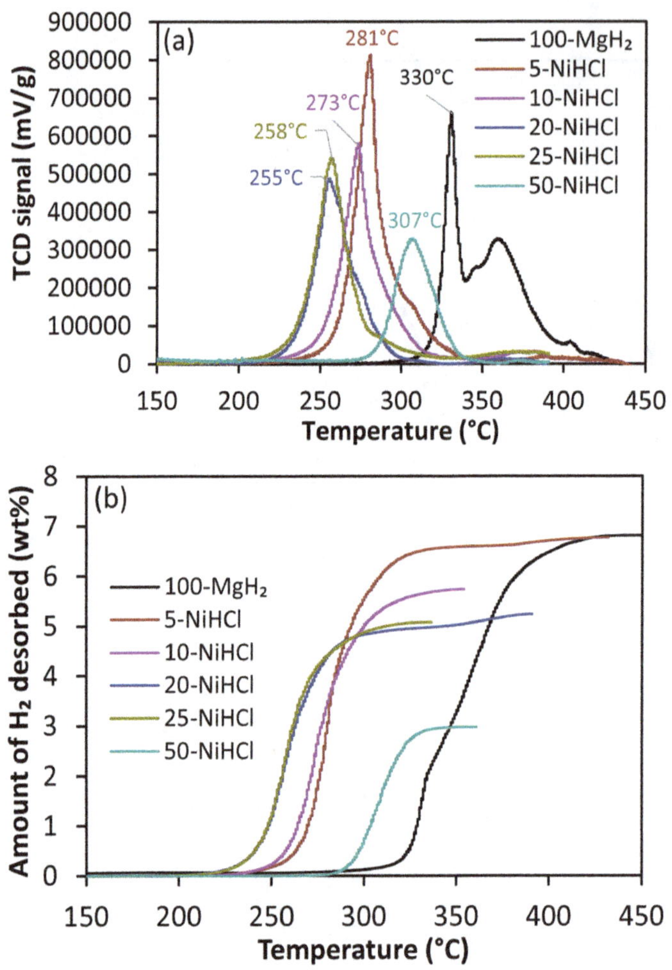

Fig. 3.4: Screening process performed by TPD to study the impact of the amount of $NiHCl(PCy_3)_2$ on the dehydrogenation properties of MgH_2. (a) Desorption profiles and (b) H_2 desorption capacity curves (reproduced from [68] with permission from Elsevier, copyright 2019).

3.4 Thermal analysis to study in detail the dehydrogenation properties of MgH_2: DSC

The calorimetry techniques (DSC) are widely used to characterize the dehydrogenation properties of MgH_2-based systems deeply. While TPD and TGA techniques allow investigating the impact of additives on the MgH_2 decomposition temperature, by using DSC the enthalpy of the dehydrogenation reaction and the apparent activation energy can also be obtained. By DSC, the impact of additives on the dehydrogenation properties of MgH_2 can be studied from both a thermodynamic and a kinetic point of view. It is of particular interest to know if an alloy is formed during the composite preparation, leading to a change in the enthalpy of dehydrogenation. Much like for TPD and TGA, the experimental conditions (mass, heating rate, gas flow, crucible use, etc.) have to be strictly controlled for repeatable and quantitative measurements.

The amount of heat involved in the dehydrogenation can be obtained with a single DSC experiment by integrating the decomposition peak (Fig. 3.5). Decomposition of MgH_2 is an endothermic phenomenon. Consequently, heat consumption is observed during the process, corresponding to the enthalpy of the reaction. As reported by Bogdanović et al. [34], the theoretical value for the enthalpy of MgH_2 dehydrogenation is 74.7 kJ/mol H_2. Hence, the value of 74 kJ/mol H_2 STP found in Fig. 3.5 agrees with the theoretical one.

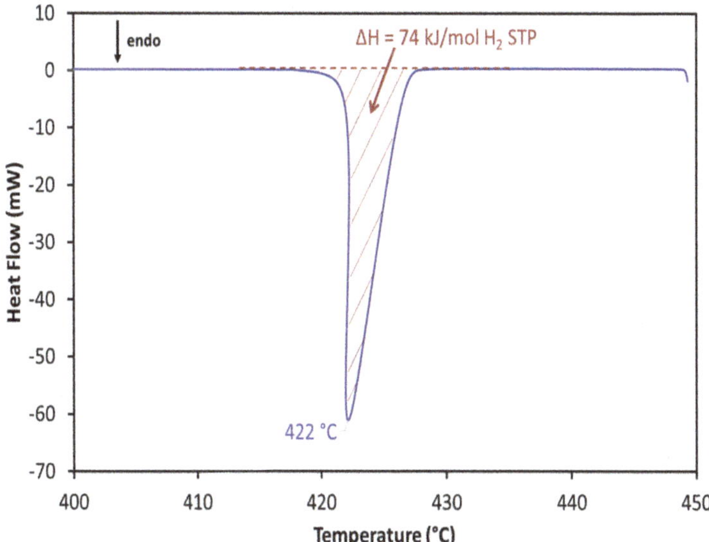

Fig. 3.5: DSC analysis performed on commercial MgH_2. The powder (1.94 mg) was heated from 25 to 450 °C (heating rate of 2 °C/min) under argon flow (40 mL/min). A hermetic aluminum pan with a pinhole sealed inside a glove box was used to avoid the sample oxidation.

The calculation of the dehydrogenation reaction's apparent activation energy (E_a) can be performed through the Kissinger method [74, 75]. For this, three DSC experiments are enough, although a higher accuracy is obtained with four or more analyses performed at different heating rates. It is worth noting that the apparent activation energy can also be obtained by applying the Kissinger method to TPD and TGA experiments. The Kissinger equation (eqn. 3.1) allows determining the apparent activation energy for a decomposition reaction regardless of the reaction order [76]. The linearization of the Kissinger equation [74, 75] in logarithmic form is reported as follows:

$$\ln\left(\frac{\beta}{T^2}\right) = \frac{-E_a}{RT} + \ln\left(\frac{RC_0}{T^2 E_a}\right) \tag{3.1}$$

where R is the gas constant, β is the heating rate, C_0 is the reaction order, and T is the temperature of the DSC peak. The apparent activation energy values are obtained from the slope of the straight lines interpolating the data obtained by plotting $\ln(\beta/T^2)$ versus $1,000/T$. It is important to mention that the calculated activation energy depends on the experimental technique used.

In Fig. 3.6, the data obtained by TPD and DSC techniques are used to calculate the apparent activation energy (E_a) of the dehydrogenation for the commercial MgH_2 and MgH_2 doped with 5 wt% of $RuH_2(H_2)_2(PCy_3)_2$ complex (5-RuPCy3) [70]. The values obtained are significantly different, with 231 versus 134 kJ/mol H_2 and 139 versus 77 kJ/mol H_2 for MgH_2 and 5-RuPCy$_3$, respectively, with higher E_a being obtained by DSC. The discrepancies are inherent to the different methods used for signal acquisition in each technique, which is impacted by different phenomena. For instance, the heat exchange rate variations can affect the DSC signal, while the gas flow rate can significantly change for TPD (Fig. 3.2b). However, it is essential to note that a similar decrease in the activation energy was obtained in both techniques when MgH_2 was doped with Ru-based complex, i.e., 60% and 57% E_a reduction measured by DSC and TPD, respectively, when comparing 5-RuPCy$_3$ with MgH_2.

In Fig. 3.3, the screening process showed that adding 5 wt% of nickel nanoparticles to MgH_2 by milling allows for the lowest dehydrogenation temperature among the studied samples. However, though TPD alone, the impact of Ni nanoparticles on the dehydrogenation of magnesium hydride cannot be explained, i.e., if the reason is linked to kinetics and/or thermodynamic phenomena. DSC experiments performed at different heating rates (2, 5, 7.5, and 10 °C/min) for the pure milled MgH_2 and MgH_2 milled with 5 wt% of nickel nanoparticles provided new insights into the thermodynamic changes in the composite containing Ni. The DSC obtained for these samples are presented in Fig. 3.7a and 3.7b, along with the Kissinger plots in Fig. 3.7c. When comparing Fig. 3.7a with Fig. 3.7b, a slight change in the dehydrogenation enthalpy can be observed with MgH_2 doped with Ni displaying 67 ± 4 kJ/mol H_2 against 72 ± 4 kJ/mol H_2 for pure milled MgH_2. This change in enthalpy can be attributed to the formation of a Mg_2NiH_4 phase during the milling preparation. In addition to the changes in the heat of reaction, a significant change was observed in the apparent activation energy determined by the Kissinger method (Fig. 3.7c). As it can be seen in Fig. 3.7c, the powder with nickel has an activation energy of 97 kJ/mol H_2 against 176 kJ/mol H_2 for pure-milled MgH_2. The difference is significant and indicates that nickel greatly enhances the dehydrogenation kinetics of magnesium hydride. Indeed, similar conclusions were obtained by Q. Huo et al. [77] when mixing NiO/C with MgH_2.

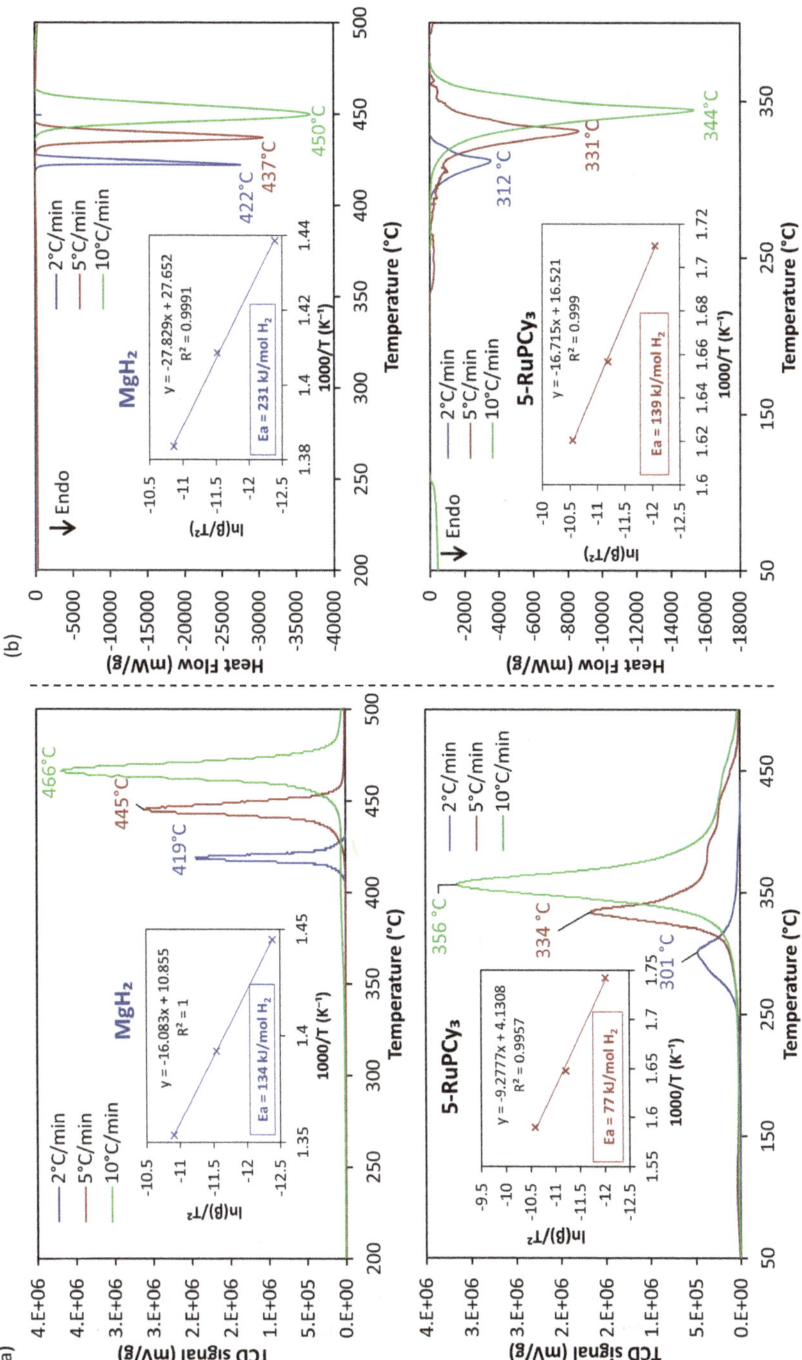

Fig. 3.6: Determination of the dehydrogenation apparent activation energy for commercial MgH_2 and MgH_2 milled with 5 wt% of $RuH_2(H_2)_2(PCy_3)_2$ complex (5-$RuPCy_3$). (a) TPD and (b) DSC results.

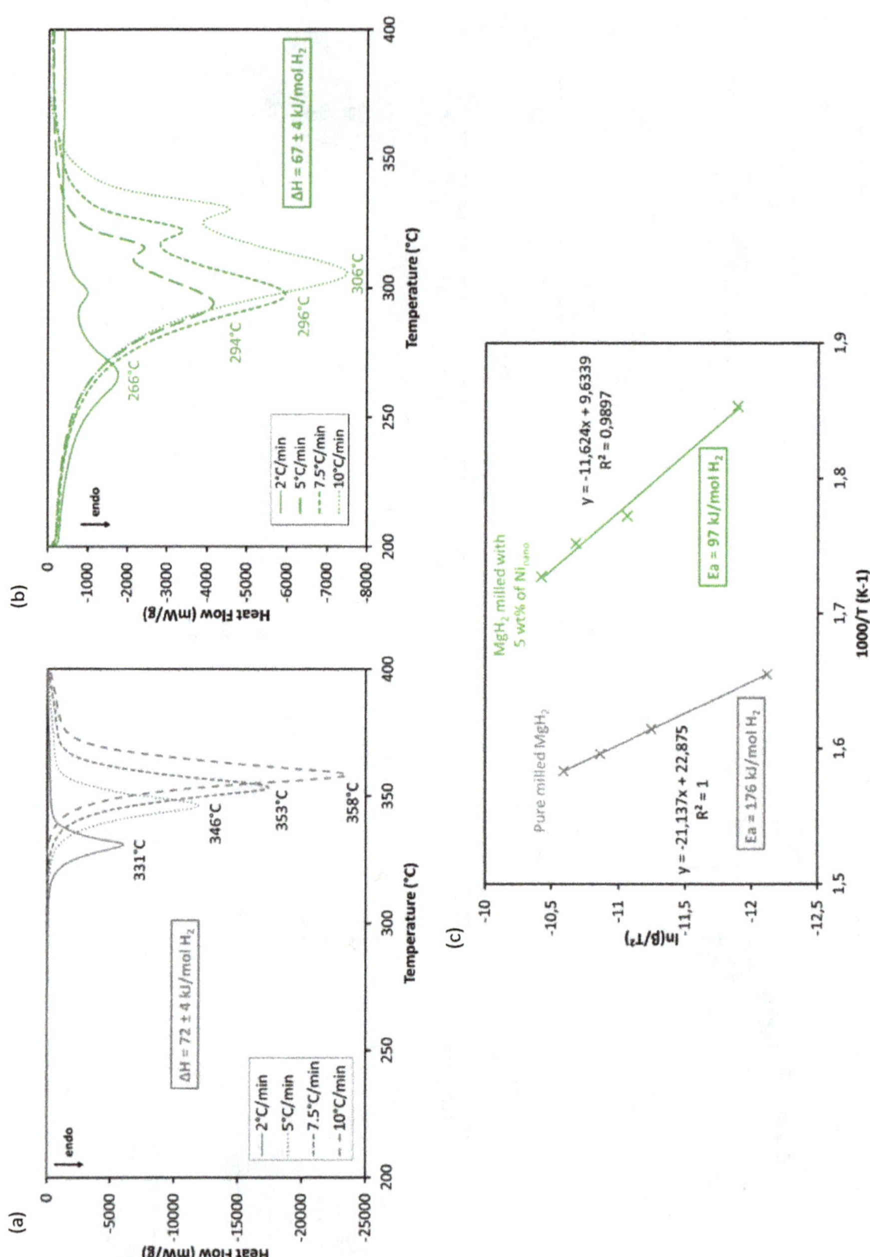

Fig. 3.7: DSC analysis performed on (a) pure-milled MgH_2 and (b) MgH_2 milled with 5 wt% of nickel nanoparticles. Experimental conditions: ≈ 2 mg of sample, under argon flow (40 mL/min), from 25 to 450 °C, at different heating rates (2, 5, 7.5, and 10 °C/min), hermetic aluminum pans with a pinhole. (c) Kissinger plot, $\ln(\beta/T^2)$ versus $1,000/T$.

Other recent examples of the use of DSC for understanding kinetic and thermodynamic phenomena in the storage of H_2 by MgH_2 materials include the work done by the C. An and Q. Deng [78] on core-shell Ni@C additives to MgH_2 and the work by C. Zhou et al. [79] on the effect of CuTi additives on MgH_2. In both cases, the authors reported increased kinetics of MgH_2 dehydrogenation in the presence of an additive. In addition to assessing the impact of MgH_2 additives, DSC has also been extensively used in understanding the effect of specific material synthesis in the thermodynamics and kinetics of hydrogen storage [80–82]. Therefore, DSC is a powerful tool that can significantly help understanding how modifications to the material influence the hydrogen storage process.

3.5 Volumetric techniques to study the hydrogenation of Mg: Sieverts apparatus

The Sieverts (volumetric) technics are widely used to study the hydrogenation of magnesium [45, 69, 70, 83]. In a Sieverts volumetric instrument, Fig. 3.8, a calibrated reference volume is filled with H_2 up to a given pressure and then opened in the sample cell (whose volume is also calibrated). Depending on the system used, i.e., pressure and temperature conditions, H_2 is either absorbed or desorbed by the sample, and the amount involved is calculated from the change of gas pressure in the cell.

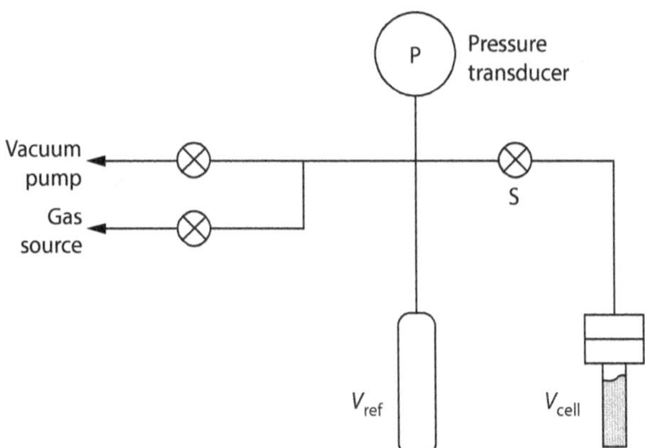

Fig. 3.8: Schematic representation of a simplified Sieverts apparatus (reproduced from [83] with permission from Elsevier, copyright 2007).

The Sievert instruments work most of the time under isothermal conditions and require ~100 mg of sample to perform the analysis. Despite the multiple advantages, the primary limitation of these instruments is the duration of the experiments, which can take between a few hours and several weeks. However, the Sieverts technique is widely used due to its versatility. Indeed, this volumetric method permits obtaining multiple types of information by doing different sets of experiments, such as:

- The determination of hydrogenation and dehydrogenation thermodynamic parameters and equilibrium pressure by making a pressure-composition isotherm (PCI). Duration – from a few days to a week.
- The global and direct kinetics of hydrogenation and dehydrogenation thanks to absorption/desorption isotherms. Duration – from a few minutes to a few days.
- The cycle life span behavior by performing hydrogenation/dehydrogenation cycles. Duration – several weeks.

For PCI (or PCT for pressure-composition-temperature) experiments, magnesium is placed in the sample chamber of the apparatus, and the cell is heated at a given temperature. Afterward, the hydrogen pressure in the cell is increased step by step until equilibrium is reached (Fig. 3.9a), and the PCI curve (Fig. 3.9b) can be built. The choice of the pressure steps depends on the properties of the sample used. Finally, the same kind of curve can be built for desorption by decreasing the hydrogen pressure step by step instead of increasing it.

The PCI curves give a good insight onto the mechanisms of hydrogenation and dehydrogenation, Fig. 3.9b. As proposed by Wei et al. in [45], in the A-B section, the hydrogen molecules firstly dissociate into hydrogen atoms at the Mg surface and then diffuse into the sub-surface and occupy the interstitial sites to form a solid solution, the α phase. With the increasing hydrogen pressure, the hydrogen concentration in the interstitial site reaches a critical point (point B), where a hydride β phase with a larger interstitial concentration and a lower density form. A plateau can be found in section B-C, which shows the coexistence of α and β phases, and the length of this plateau represents the reversible hydrogen storage capacity of the material. The pressure starts to rise again at point C, which signals the end of the α to β phase transformation. The metal hydride becomes a single hydride phase in section C-D, and the maximum storage capacity of the host material is reached. The dehydrogenation of MgH_2 follows a reverse pathway with, most of the time, a small hysteresis (difference of equilibrium pressure between hydrogenation and dehydrogenation, Fig. 3.9b).

Two important storage properties can be obtained with an absorption and a desorption PCI. Firstly, the reversible hydrogen storage capacity, corresponding to the length of the equilibrium plateau, i.e., 6.8 wt% of H_2 for the example given in Fig. 3.9. Secondly, the equilibrium pressure for both absorption and desorption. At pressures higher than the equilibrium plateau for absorption, the hydrogenation process will be favorable and the other way around for desorption. For example, in the case of pure milled MgH_2 heated at 350 °C (Fig. 3.9b), pressures higher than 6.5 bars are needed for hydrogenation, while pressures lower than 5.5 bars will favor dehydrogenation. These equilibrium pressures depend on the chosen temperature and the storage material.

With minimum 3 PCI curves obtained at different temperatures, the thermodynamic parameters of the hydrogenation/dehydrogenation can be calculated thanks to the Van't Hoff method. It allows to determine the enthalpy and entropy for both H_2 absorption and desorption based on the following equation:

$$\ln\left(\frac{P_{eq}}{P_0}\right) = \frac{\Delta H_0}{RT} - \frac{\Delta S_0}{R} \tag{3.2}$$

(a)

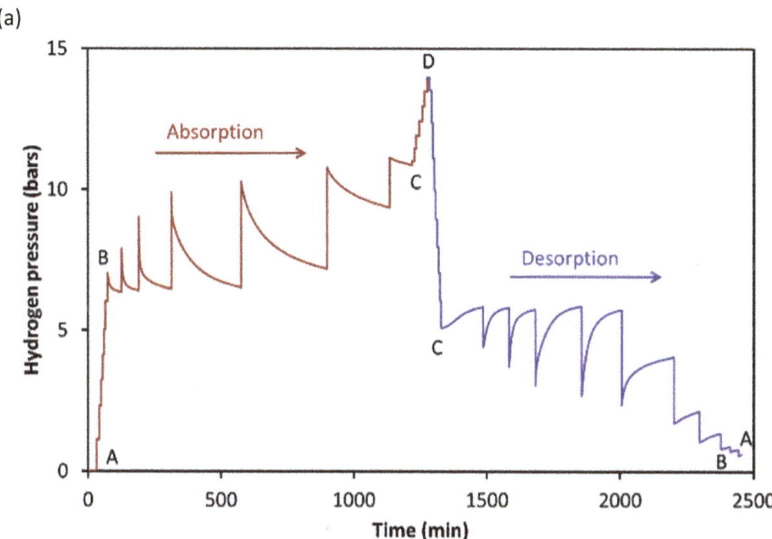

(b)

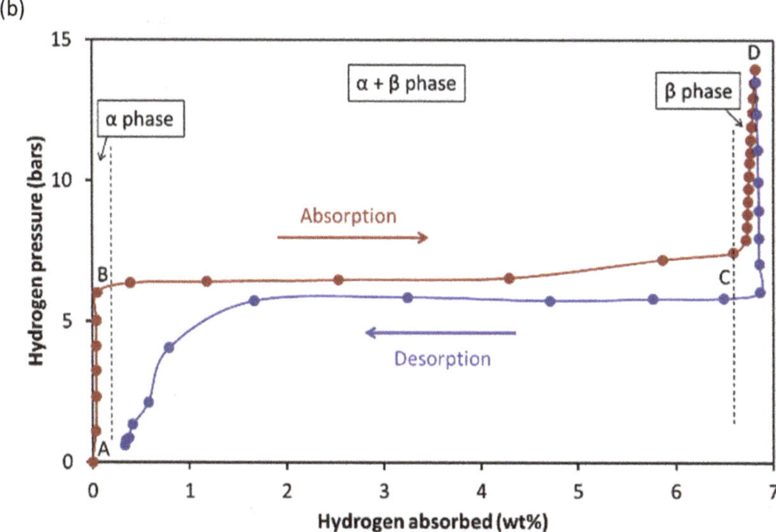

Fig. 3.9: Building of a PCI curve obtained with pure-milled MgH_2 heated at 350 °C: (a) equilibrium curve and (b) PCI curve.

where: R is the ideal gas constant, P_{eq} is the pressure at the equilibrium plateau, P_0 is the atmospheric pressure, and T is the temperature of the isotherm. The enthalpies and entropies values are obtained from the slope of the straight lines interpolating the data obtained by plotting $\ln(P_{eq}/P_0)$ versus $1/T$ for each sample.

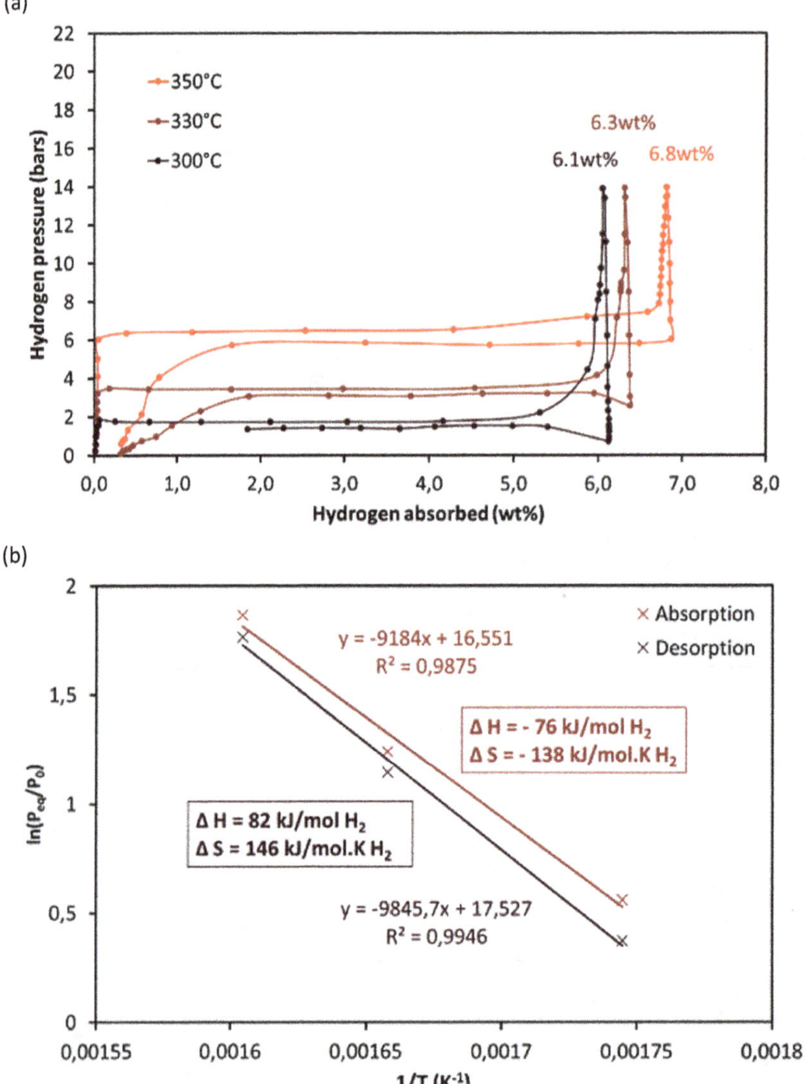

Fig. 3.10: (a) PCI curves at 300, 330, and 350 °C obtained for pure-milled MgH_2 and (b) corresponding Van't Hoff plots (adapted from [70] with permission from RSC, copyright 2018).

Fig. 3.10 presents the results obtained for pure-milled MgH_2 [70]. Three PCI at 300, 330, and 350 °C were performed (Fig. 3.10a) and the corresponding Van't Hoff plots are given in Fig. 3.10b. According to Aguey-Zinsou et al. [33] the formation enthalpy and entropy of the Mg/MgH_2 system are ~75 kJ/mol H_2 and ~130 kJ/mol H_2.K, respectively. The values obtained in Fig. 3.10 are close to the theoretical one for hydrogenation but slightly higher for dehydrogenation. This difference, often reported in the literature, can be caused by a certain degree of hysteresis, probably because of powder sintering during hydrogen absorption at high temperatures [70, 84, 85].

As referred previously, the main limitation of the PCI curves and the Van't Hoff method is the duration of the experiments. For example, the PCI curve obtained in Fig. 3.9 was performed in 2 days. As 3 PCI are at least required, nearly a week is needed to determine the thermodynamic properties of the Mg/MgH_2 system through the Van't Hoff method. However, the duration can be shortened if the hydrogenation and dehydrogenation kinetics are very high.

The Sieverts method can also be used to study the hydrogenation and dehydrogenation kinetics of Mg/MgH_2. While several increasing or decreasing pressure steps are required for building a PCI, a single one is necessary in this case. The temperature must be firstly set at the desired value before sending a given hydrogen pressure in the sample cell and follow the evolution of the pressure versus time until equilibrium. The evolution of the H_2 pressure can be related to the amount of hydrogen absorbed or desorbed by the sample, depending on if the pressure increases (dehydrogenation) or decreases (hydrogenation). Figure 3.11 presents hydrogenation isotherms obtained for the commercial MgH_2 sample, as received (commercial MgH_2, Fig. 3.11a) and after milling (pure-milled MgH_2, Fig. 3.11b). The two powders were firstly dehydrogenated under vacuum at 350 °C. The isotherms were performed under 30 bars of H_2 at 300 °C for as received commercial MgH_2 and at 200 and 300 °C for the milled sample. For as received commercial MgH_2, more than 40 h were needed to absorb 5 wt% of H_2 at 300 °C, while it takes only 40 min for the milled MgH_2 to absorb 6.6 wt% of hydrogen at the same temperature. The impact of the milling preparation on the hydrogenation kinetics is therefore considerable. When the temperature is decreased, Fig. 3.11b, the kinetics are also decreased, and the milled MgH_2 sample requires ~10 h at 200 °C, to absorb 6 wt% of H_2. The same kind of curve can be obtained for dehydrogenation, as presented by Aguey-Zinsou et al. [86].

With hydrogenation or dehydrogenation isotherms obtained at different temperatures (minimum 3), kinetic models can be used to calculate the apparent activation energy of the reaction. Numerous different models are available, helping to determine the limiting kinetic step of the hydrogenation or dehydrogenation process [87]. For example, Chen et al. [88] demonstrated that the rate-limiting step for hydrogen absorption in MgH_2-ZrO_2 composite is hydrogen diffusion in magnesium. A low value of 13 kJ/mol H_2 was obtained for the apparent activation energy of hydrogenation. Barkhordarian et al. [89] showed that the rate-limiting step for hydrogen desorption in MgH_2-Nb_2O_5 composites changes with the temperature and the Nb_2O_5 content, leading

(a)

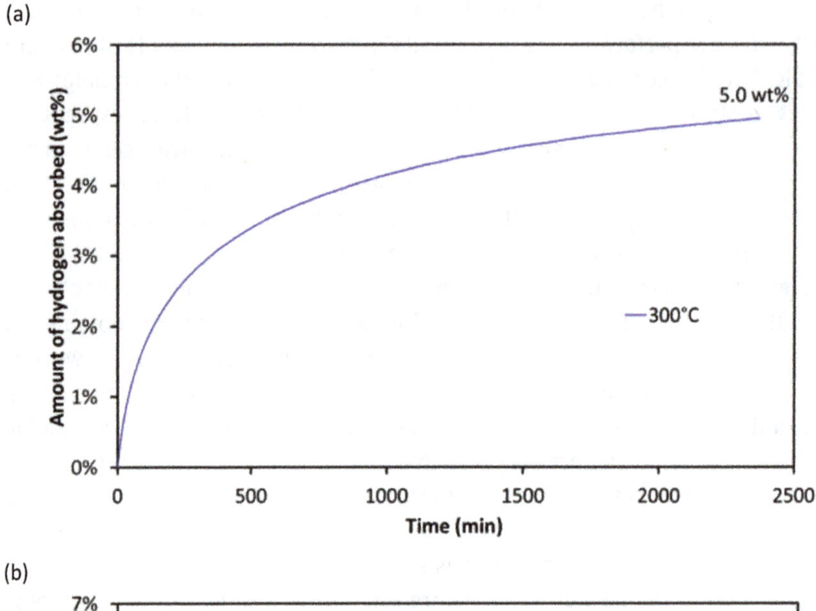

(b)

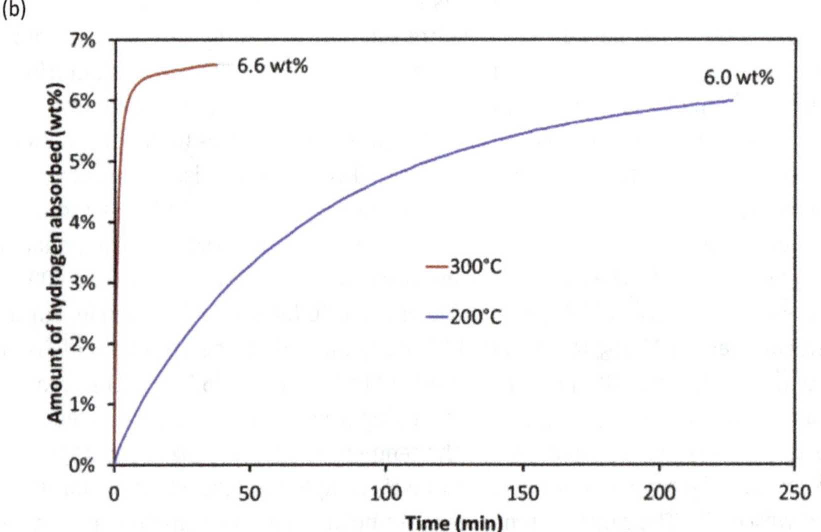

Fig. 3.11: (a) Hydrogenation isotherms under 30 bars of H_2 at different temperatures, obtained for (a) commercial MgH_2 and (b) pure-milled MgH_2.

to an activation energy of 61 kJ/mol H_2 when 0.2 mol% catalysts were present. Finally, Milosevic et al. [90] used the Johnson-Mehl-Avrami (JMA) nucleation and growth model and found that the dehydrogenation of MgH_2-VO_2 composites is limited by the nucleation rate or the dimensionality of nuclei, depending on the catalyst content. However, the use of kinetic models to study the hydrogen storage properties of

Mg/MgH$_2$ is difficult as the results can be considerably different depending on the samples, the temperature/pressure conditions, and the chosen models [33].

Finally, the Sieverts technique is interesting for studying the cycle life span behavior of the Mg/MgH$_2$ system. For this, several following hydrogenation dehydrogenation isotherms at constant temperature must be performed. As presented in Fig. 3.12, Xia et al. [91] performed 100 hydrogenation-dehydrogenation cycles on Mg nanoparticles supported on graphene and doped with nickel nanoparticles. The results show a high cycle life span for the two samples studied. Crivello et al. [92] also investigated in detail the cycle life span of magnesium hydride by Sieverts volumetric method. Crivello et al. [92] reported that MgH$_2$ under pellet form shows better cycling stability than MgH$_2$ as a powder.

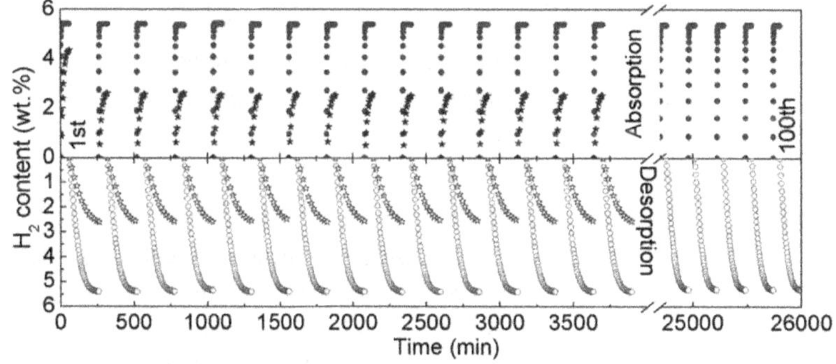

Fig. 3.12: Example of cycle life span study performed on Mg nanoparticles supported on graphene and doped with nickel nanoparticles with a Sievert apparatus [91].

Such cycle life span studies are time-consuming, and several weeks are needed when the storage system behavior after extensive cycling is required [93]. A compromise between the temperature and the duration of the experiment is most of the time necessary (high temperature for higher kinetics). However, the differences between the cycles at high temperatures are less visible than at low temperatures. As shown in Fig. 3.13, at 300 °C, the difference between the hydrogenation properties of pure-milled MgH$_2$ (m-MgH$_2$) and MgH$_2$ doped with a ruthenium complex (5-RuPCy3 [70]) is negligible, while at 200 °C this is significant. The choice of the temperature conditions for an accurate cycle life span study can therefore be difficult. If the temperature is too high, the difference in behavior during cycling will not be visible. If the temperature is too low, the kinetics will be too slow, and the experiments will take too much time.

To make a life span study under nonisothermal conditions and thus avoid the issue of the compromise between the temperature and the duration of the experiments, HP-DSC is a tool of particular interest. This technique allows to couple a volumetric line with a calorimeter. The Mg sample is heated under a controlled hydrogen

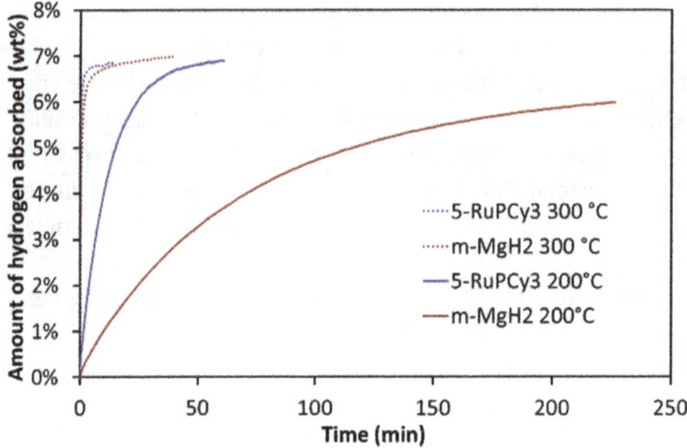

Fig. 3.13: Hydrogenation isotherms under 30 bars of H_2 at two temperatures 200 and 300 °C and for two samples pure-milled MgH_2 (m-MgH_2) and MgH_2 milled with 5 wt% of $RuPCy_3$ complex (5-RuPCy3) (adapted from [70] with permission from RSC, copyright 2018).

pressure leading to its hydrogenation. Both the heat released and the hydrogen consumed are measured. The sample is then cooled under vacuum (or a given H_2 pressure) until its complete dehydrogenation. Again, the heat consumed and the hydrogen released are measured, and several hydrogenation-dehydrogenation cycles can be performed. Figure 3.14 shows results obtained by Li et al. [94] and Zhang et al. [95]. They used an HP-DSC technique to study the life span behavior upon the cycling of their samples. No major differences could be observed between the first and last cycle, meaning that the systems formed are stable.

Although its use is scarce, the HP-DSC technique seems particularly versatile for studying all the hydrogen storage properties of the Mg/MgH_2 system:

- The hydrogenation and dehydrogenation temperatures by DSC
- The hydrogenation and dehydrogenation kinetics, thanks to the coupling with a volumetric line
- The hydrogenation and dehydrogenation thermodynamics – volumetric line (Van't Hoff) and DSC (Kissinger, direct measurement of heat)
- The hydrogen storage capacity – volumetric line
- The behavior after extensive cycling – coupling

Regarding the difficulties exposed earlier when it comes to comparing the results obtained with different instruments (see Tab. 3.4), the advantages of an all-in-one apparatus like HP-DSC are significant.

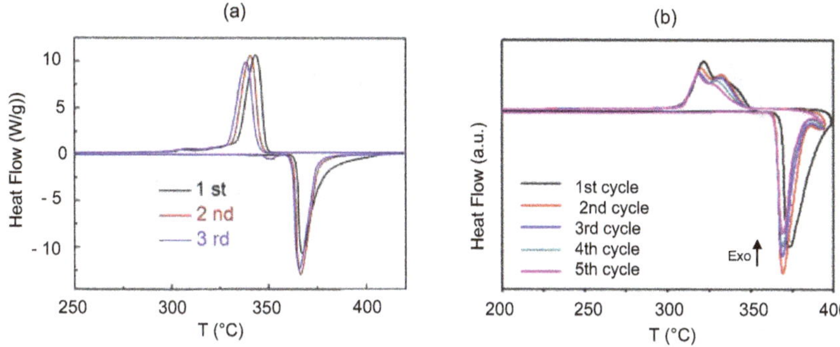

Fig. 3.14: Example of cycle life span study performed by HP-DSC by (a) Li et al. (reproduced from [94] with permission from Elsevier, copyright 2016) and (b) Zhang et al. (adapted from [95] with permission from Elsevier, copyright 2017).

3.6 Conclusion

Despite the significant number of solid-state hydrogen storage materials studied and reported in the literature, none of the investigated systems can answer the requirements for onboard applications yet. They still suffer from low reversible hydrogen storage capacity, low hydrogen uptake/release kinetics, high thermodynamic stability, and/or safety issues. For the extensive use of these materials, it is of primary interest to understand the thermal phenomena associated with the hydrogenation and dehydrogenation processes.

Through the case study of the Mg/MgH$_2$ system, the considerable interest in thermal analysis techniques and calorimetry was detailed. Table 3.4 presents an overview of the methods used in the literature to study the hydrogen storage properties of solid-state systems. The strengths and weaknesses of these techniques are highlighted as well as the working principles and the measured storage properties. The duration of the experiments performed with the different techniques and the properties measured are also indicated. HP-DSC is the most versatile of all the presented techniques and can be used to study the hydrogenation and dehydrogenation kinetics and thermodynamics in detail. Furthermore, this technique permits quantifying the hydrogen storage capacity and the cycle life span behavior of the different solid-state systems.

Tab. 3.4: overview of the thermal analysis apparatus used to study the hydrogen storage properties of the Mg/MgH_2 system.

Apparatus		Principle	H_2 capacity	Hydrogenation: $Mg + H_2 \rightarrow MgH_2$			Dehydrogenation: $MgH_2 \rightarrow Mg + H_2$			Cycle life	Strengths	Weaknesses
				Conditions	Thermo	Kinetics	Conditions	Thermo	Kinetics			
Thermal analysis	DSC	Measure of heat	/	/	/	/	T	E_a, Q	Rate, model	/	Short experiments, 1–10 mg scale	Dehydrogenation only, no H_2 capacity
	ATG	Measure of mass loss	Direct value	/	/	/	T	E_a	Rate, model	/	Short experiments, 1–10 mg scale	Dehydrogenation only
	TPD	Measure of H_2 released	Calibration	/	/	/	T	E_a	Rate, model	/	Short experiments, 1–10 mg scale	Dehydrogenation only
Sieverts technic	PCT	Measure of H_2 pressure	Calculation	T, P	ΔH, ΔS	Rate, model	T, P	ΔH, ΔS	Rate, model	Yes	Versatility	Isothermal conditions, long experiments, 100–500 mg scale
Coupled apparatus	HP-DSC	Measure of heat and H_2 pressure	Calculation	T, P	ΔH, ΔS, E_a, Q,	Rate, model	T	E_a, Q, ΔH, ΔS	Rate, model	Yes	All-in-one apparatus	Cost of the apparatus

Green: Value obtained with a single experiment.
Yellow: Value obtained in more than one experiment, but in less than a day.
Orange: Value obtained in more than one day, but in less than a week.
Red: Value obtained in more than 1 week.
/: Value not available with the apparatus.

References

[1] Chow, J.; Kopp, R. J.; Portney P. R. Energy resources and global development. *Science* **2003**, *302*, 1528–1531, DOI:10.1126/science.1091939.

[2] Arto, I.; Capellán-Pérez, I.; Lago, R.; Bueno, G.; Bermejo, R. The energy requirements of a developed world. *Energy Sustain. Dev.* **2016**, *33*, 1–13, DOI:10.1016/j.esd.2016.04.001.

[3] Graves, P. E. Implications of global warming: Two eras. *World Dev. Perspect.* **2017**, *7–8*, 9–14, DOI:10.1016/j.wdp.2017.10.002.

[4] Bockris, J. O. Will lack of energy lead to the demise of high-technology countries in this century? *Int. J. Hydrog. Energy* **2007**, *32*, 153–158, DOI:10.1016/j.ijhydene.2006.08.025.

[5] Kramer, G. J.; Haigh, M. No quick switch to low-carbon energy. *Nature* **2009**, *462*, 568–569, DOI:10.1038/462568a.

[6] Bauer, C.; Treyer, K.; Heck, T.; Hirschberg, S. Greenhouse Gas Emissions from Energy Systems, Comparison, and Overview. In *Encyclopedia of the Anthropocene.* Elsevier: Oxford. 2018, pp. 473–484, DOI:10.1016/B978-0-12-809665-9.09276-4.

[7] Benson, S. M.; Orr, F. M. Sustainability and energy conversions. *MRS Bull.* **2008**, *33*, 297–302, DOI:10.1557/mrs2008.257.

[8] Baranes, E.; Jacqmin, J.; Poudou, J.-C. Non-renewable and intermittent renewable energy sources: Friends and foes? *Energy Policy* **2017**, *111*, 58–67, DOI:10.1016/j.enpol.2017.09.018.

[9] Can Şener, Ş. E.; Sharp, J. L.; Anctil, A. Factors impacting diverging paths of renewable energy: A review. *Renew. Sust. Energ. Rev.* **2018**, *81*, 2335–2342, DOI:10.1016/j.rser.2017.06.042.

[10] Olabi, A. G. Renewable energy and energy storage systems. *Energy* **2017**, *136*, 1–6, DOI:10.1016/j.energy.2017.07.054.

[11] Dutta, S. A review on production, storage of hydrogen and its utilization as an energy resource. *J. Ind. Eng. Chem.* **2014**, *20*, 1148–1156, DOI:10.1016/j.jiec.2013.07.037.

[12] Andrews, J.; Shabani, B. The role of hydrogen in a global sustainable energy strategy. *WIREs Energy Environ.* **2014**, *3*, 474–489, DOI:10.1002/wene.103.

[13] Barreto, L.; Makihira, A.; Riahi, K. The hydrogen economy in the twenty-first century: A sustainable development scenario. *Int. J. Hydrog. Energy* **2003**, *28*, 267–284, DOI:10.1016/S0360-3199(02)00074-5.

[14] Bakenne, A.; Nuttall, W.; Kazantzis, N. Sankey-Diagram-based insights into the hydrogen economy of today. *Int. J. Hydrog. Energy* **2016**, *41*, 7744–7753, DOI:10.1016/j.ijhydene.2015.12.216.

[15] Winter, C.-J. Hydrogen energy – Abundant, efficient, clean: A debate over the energy-system-of-change. *Int. J. Hydrog. Energy* **2009**, *34*, S1–S52, DOI:10.1016/j.ijhydene.2009.05.063.

[16] Momirlan, M.; Veziroglu, T. N. The properties of hydrogen as fuel tomorrow in sustainable energy system for a cleaner planet. *Int. J. Hydrog. Energy* **2005**, *30*, 795–802, DOI:10.1016/j.ijhydene.2004.10.011.

[17] Durbin, D. J.; Malardier-Jugroot, C. Review of hydrogen storage techniques for on board vehicle application. *Int. J. Hydrog. Energy* **2013**, *38*, 14595–14,617, DOI:10.1016/j.ijhydene.2013.07.058.

[18] Zhang, F.; Zhao, P.; Niu, M.; Maddy, J. The survey of key technologies in hydrogen energy storage. *Int. J. Hydrog. Energy* **2016**, *41*, 14535–14552, DOI:10.1016/j.ijhydene.2016.05.293.

[19] Zhou, L. Progress and problems in hydrogen storage methods. *Renew. Sust. Energ. Rev.* **2005**, *9*, 395–408, DOI:10.1016/j.rser.2004.05.005.

[20] Shen, C.; Ma, L.; Huang, G.; Wu, Y.; Zheng, J.; Liu, Y.; Hu, J. Consequence assessment of high-pressure hydrogen storage tank rupture during fire test. *J. Loss. Prev. Process Ind.* **2018**, *55*, 223–231, DOI:10.1016/j.jlp.2018.06.016.

[21] Ren, J.; Musyoka, N. M.; Langmi, H. W.; Mathe, M.; Liao, S. Current research trends and perspectives on materials-based hydrogen storage solutions: A critical review. *Int. J. Hydrog. Energy* **2017**, *42*, 289–311, DOI:10.1016/j.ijhydene.2016.11.195.

[22] Klebanoff, L. E.; Keller, J. O. 5 Years of hydrogen storage research in the U.S. DOE Metal Hydride Center of Excellence (MHCoE). *Int. J. Hydrog. Energy* **2013**, *38*, 4533–4576, DOI:10.1016/j.ijhydene.2013.01.051.

[23] DOE. Technical Targets for Onboard Hydrogen Storage for Light-Duty Vehicles | Department of Energy, (n.d.). https://www.energy.gov/eere/fuelcells/doe-technical-targets-onboard-hydrogen-storage-light-duty-vehicles (accessed April 5, 2022).

[24] Lee, S.-Y.; Lee, J.-H.; Kim, Y.-H.; Kim, J.-W.; Lee, K.-J.; Park, S.-J. Recent progress using solid-state materials for hydrogen storage: A short review. *Process.* **2022**, *10*, *304*, DOI:10.3390/pr10020304.

[25] Chen, Z.; Kirlikovali, K. O.; Idrees, K. B.; Wasson, M. C.; Farha, O. K. Porous materials for hydrogen storage. *Chemistry.* **2022**, *8*, 693–716, DOI:10.1016/j.chempr.2022.01.012.

[26] Lototskyy, M.; Yartys, V. A. Comparative analysis of the efficiencies of hydrogen storage systems utilising solid state H storage materials. *J. Alloys Compd.* **2015**, *645*, S365–S373, DOI:10.1016/j.jallcom.2014.12.107.

[27] Lim, K. L.; Kazemian, H.; Yaakob, Z.; Daud, W. R. W. Solid-state materials and methods for hydrogen storage: A critical review. *Chem. Eng. Technol.* **2010**, *33*, 213–226, DOI:10.1002/ceat.200900376.

[28] Dillon, A. C.; Jones, K. M.; Bekkedahl, T. A.; Kiang, C. H.; Bethune, D. S.; Heben, M. J. Storage of hydrogen in single-walled carbon nanotubes. *Nature* **1997**, *386*, 377–379, DOI:10.1038/386377a0.

[29] Liu, D.; Purewal, J. J.; Yang, J.; Sudik, A.; Maurer, S.; Mueller, U.; Ni, J.; Siegel, D. J. MOF-5 composites exhibiting improved thermal conductivity. *Int. J. Hydrog. Energy* **2012**, *37*, 6109–6117, DOI:10.1016/j.ijhydene.2011.12.129.

[30] An, X. H.; Gu, Q. F.; Zhang, J. Y.; Chen, S. L.; Yu, X. B.; Li, Q. Experimental investigation and thermodynamic reassessment of La–Ni and LaNi5–H systems. *Calphad.* **2013**, *40*, 48–55, DOI:10.1016/j.calphad.2012.12.002.

[31] Liang, G.; Huot, J.; Schulz, R. Hydrogen storage properties of the mechanically alloyed LaNi5-based materials. *J. Alloys Compd.* **2001**, *320*, 133–139, DOI:10.1016/S0925-8388(01)00929-X.

[32] Davids, M. W.; Lototskyy, M. Influence of oxygen introduced in TiFe-based hydride forming alloy on its morphology, structural and hydrogen sorption properties. *Int. J. Hydrog. Energy* **2012**, *37*, 18155–18162, DOI:10.1016/j.ijhydene.2012.09.106.

[33] Aguey-Zinsou, K.-F.; Ares-Fernández, J.-R. Hydrogen in magnesium: New perspectives toward functional stores. *Energy Environ. Sci.* **2010**, *3*, 526–543, DOI:10.1039/B921645F.

[34] Bogdanović, B.; Bohmhammel, K.; Christ, B.; Reiser, A.; Schlichte, K.; Vehlen, R.; Wolf, U. Thermodynamic investigation of the magnesium–hydrogen system. *J. Alloys Compd.* **1999**, *282*, 84–92, DOI:10.1016/S0925-8388(98)00829-9.

[35] Wang, J.; Ebner, A. D.; Ritter, J. A. Preparation of a new Ti catalyst for improved performance of NaAlH4. *Int. J. Hydrog. Energy* **2012**, *37*, 11650–11655, DOI:10.1016/j.ijhydene.2012.05.101.

[36] Orimo, S.; Nakamori, Y.; Eliseo, J. R.; Züttel, A.; Jensen, C. M. Complex hydrides for hydrogen storage. *Chem. Rev.* **2007**, *107*, 4111–4132, DOI:10.1021/cr0501846.

[37] Xueping, Z.; Ping, L.; Xuanhui, Q. Effect of additives on the reversibility of lithium alanate (LiAlH4). *Rare Metal Mat. Eng.* **2009**, *38*, 766–769, DOI:10.1016/S1875-5372(10)60034-3.

[38] Varin, R. A.; Zbroniec, L. Decomposition behavior of unmilled and ball milled lithium alanate (LiAlH4) including long-term storage and moisture effects. *J. Alloys Compd.* **2010**, *504*, 89–101, DOI:10.1016/j.jallcom.2010.05.059.

[39] Züttel, A.; Wenger, P.; Rentsch, S.; Sudan, P.; Mauron, Ph.; Emmenegger, Ch. LiBH4 a new hydrogen storage material. *J. Power Sources* **2003**, *118*, 1–7, DOI:10.1016/S0378-7753(03)00054-5.

[40] Mauron, P.; Buchter, F.; Friedrichs, O.; Remhof, A.; Bielmann, M.; Zwicky, C. N.; Züttel, A. Stability and reversibility of LiBH4. *J. Phys. Chem. B.* **2008**, *112*, 906–910, DOI:10.1021/jp077572r.

[41] Urgnani, J.; Torres, F. J.; Palumbo, M.; Baricco, M. Hydrogen release from solid state NaBH4. *Int. J. Hydrog. Energy* **2008**, *33*, 3111–3115, DOI:10.1016/j.ijhydene.2008.03.031.

[42] Chong, L.; Zou, J.; Zeng, X.; Ding, W. Study on reversible hydrogen sorption behaviors of a 3NaBH4/HoF3 composite. *Int. J. Hydrog. Energy* **2014**, *39*, 14275–14281, DOI:10.1016/j.ijhydene.2014.03.051.
[43] Wolf, G.; Baumann, J.; Baitalow, F.; Hoffmann, F. P. Calorimetric process monitoring of thermal decomposition of B–N–H compounds. *Thermochim. Acta.* **2000**, *343*, 19–25, DOI:10.1016/S0040-6031(99)00365-2.
[44] Xiong, Z.; Yong, C. K.; Wu, G.; Chen, P.; Shaw, W.; Karkamkar, A.; Autrey, T.; Jones, M. O.; Johnson, S. R. Edwards, P. P.; David, W. I. F. High-capacity hydrogen storage in lithium and sodium amidoboranes. *Nat. Mater.* **2008**, *7*, 138–141, DOI:10.1038/nmat2081.
[45] Wei, T. Y.; Lim, K. L.; Tseng, Y. S.; Chan, S. L. I. A review on the characterization of hydrogen in hydrogen storage materials. *Renew. Sust. Energ. Rev.* **2017**, *79*, 1122–1133, DOI:10.1016/j.rser.2017.05.132.
[46] Yu, H.; Bennici, S.; Auroux, A. Hydrogen storage and release: Kinetic and thermodynamic studies of MgH_2 activated by transition metal nanoparticles. *Int. J. Hydrog. Energy* **2014**, *39*, 11633–11641, DOI:10.1016/j.ijhydene.2014.05.069.
[47] Liang, G.; Huot, J.; Boily, S.; Van Neste, A.; Schulz, R. Catalytic effect of transition metals on hydrogen sorption in nanocrystalline ball milled MgH_2–Tm (Tm=Ti, V, Mn, Fe and Ni) systems. *J. Alloys Compd.* **1999**, *292*, 247–252, DOI:10.1016/S0925-8388(99)00442-9.
[48] Bérubé, V.; Radtke, G.; Dresselhaus, M.; Chen, G. Size effects on the hydrogen storage properties of nanostructured metal hydrides: A review. *Int. J. Energy Res.* **2007**, *31*, 637–663, DOI:10.1002/er.1284.
[49] Callini, E.; Aguey-Zinsou, K.-F.; Ahuja, R.; Ares, J. R.; Bals, S.; Biliškov, N.; Chakraborty, S.; Charalambopoulou, G.; Chaudhary, A.-L.; Cuevas, F.; Dam, B.; de Jongh, P.; Dornheim, M.; Filinchuk, Y.; Grbović Novaković, J.; Hirscher, M.; Jensen, T. R.; Jensen, P. B.; Novaković, N.; Lai, Q.; Leardini, F.; Gattia, D. M.; Pasquini, L.; Steriotis, T.; Turner, S.; Vegge, T.; Züttel, A.; Montone, A. Nanostructured materials for solid-state hydrogen storage: A review of the achievement of COST Action MP1103. *Int. J. Hydrog. Energy* **2016**, *41*, 14404–14428, DOI:10.1016/j.ijhydene.2016.04.025.
[50] Sadhasivam, T.; Kim, H.-T.; Jung, S.; Roh, S.-H.; Park, J.-H.; Jung, H.-Y. Dimensional effects of nanostructured Mg/MgH_2 for hydrogen storage applications: A review. *Renew. Sust. Energ. Rev.* **2017**, *72*, 523–534, DOI:10.1016/j.rser.2017.01.107.
[51] Schneemann, A.; White, J. L.; Kang, S.; Jeong, S.; Wan, L. F.; Cho, E. S.; Heo, T. W.; Prendergast, D.; Urban, J. J.; Wood, B. C.; Allendorf, M. D.; Stavila, V. Nanostructured metal hydrides for hydrogen storage. *Chem. Rev.* **2018**, *118*, 10775–10839, DOI:10.1021/acs.chemrev.8b00313.
[52] Li, J.; Li, B.; Shao, H.; Li, W.; Lin, H. Catalysis and downsizing in Mg-based hydrogen storage materials. *Catalysts.* **2018**, *8*, 89, DOI:10.3390/catal8020089.
[53] Webb, C. J. A review of catalyst-enhanced magnesium hydride as a hydrogen storage material. *J. Phys. Chem. Solids* **2015**, *84*, 96–106, DOI:10.1016/j.jpcs.2014.06.014.
[54] Wang, Y.; Wang, Y. Recent advances in additive-enhanced magnesium hydride for hydrogen storage. *Prog. Nat. Sci: Mater. Int.* **2017**, *27*, 41–49, DOI:10.1016/j.pnsc.2016.12.016.
[55] Sun, Y.; Shen, C.; Lai, Q.; Liu, W.; Wang, D.-W.; Aguey-Zinsou, K.-F. Tailoring magnesium based materials for hydrogen storage through synthesis: Current state of the art. *Energy Storage Mater.* **2018**, *10*, 168–198, DOI:10.1016/j.ensm.2017.01.010.
[56] Kumar, S.; Singh, A.; Tiwari, G. P.; Kojima, Y.; Kain, V. Thermodynamics and kinetics of nano-engineered Mg-MgH_2 system for reversible hydrogen storage application. *Thermochim. Acta* **2017**, *652*, 103–108, DOI:10.1016/j.tca.2017.03.021.
[57] Zhang, T.; Isobe, S.; Jain, A.; Wang, Y.; Yamaguchi, S.; Miyaoka, H.; Ichikawa, T.; Kojima, Y.; Hashimoto, N. Enhancement of hydrogen desorption kinetics in magnesium hydride by doping with lithium metatitanate. *J. Alloys Compd.* **2017**, *711*, 400–405, DOI:10.1016/j.jallcom.2017.03.361.
[58] Zhang, J.; Zhu, Y.; Yao, L.; Xu, C.; Liu, Y.; Li, L. State of the art multi-strategy improvement of Mg-based hydrides for hydrogen storage. *J. Alloys Compd.* **2019**, *782*, 796–823, DOI:10.1016/j.jallcom.2018.12.217.

[59] Shinde, S. S.; Kim, D.-H.; Yu, J.-Y.; Lee, J.-H. Self-assembled air-stable magnesium hydride embedded in 3-D activated carbon for reversible hydrogen storage. *Nanoscale* **2017**, *9*, 7094–7103, DOI:10.1039/C7NR01699A.

[60] Vincent, S. D.; Huot, J. Effect of air contamination on ball milling and cold rolling of magnesium hydride. *J. Alloys Compd.* **2011**, *509*, L175–L179, DOI:10.1016/j.jallcom.2011.02.147.

[61] von Zeppelin, F.; Haluška, M.; Hirscher, M. Thermal desorption spectroscopy as a quantitative tool to determine the hydrogen content in solids. *Thermochim. Acta.* **2003**, *404*, 251–258, DOI:10.1016/S0040-6031(03)00183-7.

[62] Fernández, J. F.; Cuevas, F.; Sánchez, C. Simultaneous differential scanning calorimetry and thermal desorption spectroscopy measurements for the study of the decomposition of metal hydrides. *J. Alloys Compd.* **2000**, *298*, 244–253, DOI:10.1016/S0925-8388(99)00620-9.

[63] Panella, B.; Hirscher, M.; Ludescher, B. Low-temperature thermal-desorption mass spectroscopy applied to investigate the hydrogen adsorption on porous materials. *Microporous Mesoporous Mater.* **2007**, *103*, 230–234, DOI:10.1016/j.micromeso.2007.02.001.

[64] Swallowe, G. M. The effect of purge gas flow rate on thermogravimetric experiments. *Thermochim. Acta* **1985**, *65*, 151–154, DOI:10.1016/0040-6031(83)80017-3.

[65] Falconer, J. L.; Schwarz, J. A. Temperature-programmed desorption and reaction: Applications to supported catalysts. *Catal. Rev.* **1983**, *25*, 141–227, DOI:10.1080/01614948308079666.

[66] Loganathan, S.; Valapa, R. B.; Mishra, R. K.; Pugazhenthi, G.; Thomas, S. Chapter 4. Thermogravimetric Analysis for Characterization of Nanomaterials. In Thomas, S.; Thomas, R.; Zachariah, A. K.; Mishra R. K. (Eds.). *Thermal and Rheological Measurement Techniques for Nanomaterials Characterization*. Elsevier. 2017, pp. 67–108, DOI:10.1016/B978-0-323-46139-9.00004-9.

[67] Beattie, S. D.; Setthanan, U.; McGrady, G. S. Thermal desorption of hydrogen from magnesium hydride (MgH_2): An in situ microscopy study by environmental SEM and TEM. *Int. J. Hydrog. Energy* **2011**, *36*, 6014–6021, DOI:10.1016/j.ijhydene.2011.02.026.

[68] Galey, B.; Auroux, A.; Sabo-Etienne, S.; Dhaher, S.; Grellier, M.; Postole, G. Improved hydrogen storage properties of Mg/MgH_2 thanks to the addition of nickel hydride complex precursors. *Int. J. Hydrog. Energy* **2019**, *44*, 28848–28862, DOI:10.1016/j.ijhydene.2019.09.127.

[69] Galey, B.; Auroux, A.; Sabo-Etienne, S.; Grellier, M.; Postole, G. Enhancing hydrogen storage properties of the Mg/MgH_2 system by the addition of bis(tricyclohexylphosphine)nickel(II) dichloride. *Int. J. Hydrog. Energy* **2019**, *44*, 11939–11,952, DOI:10.1016/j.ijhydene.2019.03.114.

[70] Galey, B.; Auroux, A.; Sabo-Etienne, S.; Grellier, M.; Dhaher, S.; Postole, G. Impact of the addition of poly-dihydrogen ruthenium precursor complexes on the hydrogen storage properties of the Mg/MgH_2 system. *Sustain. Energy Fuels* **2018**, *2*, 2335–2344, DOI:10.1039/C8SE00170G.

[71] Perejón, A.; Sánchez-Jiménez, P. E.; Criado, J. M.; Pérez-Maqueda, L. A. Magnesium hydride for energy storage applications: The kinetics of dehydrogenation under different working conditions. *J. Alloys Compd.* **2016**, *681*, 571–579, DOI:10.1016/j.jallcom.2016.04.191.

[72] Xie, L.; LI, J.; Zhang, T.; Kou, H. Role of milling time and Ni content on dehydrogenation behavior of MgH_2/Ni composite. *Trans. Nonferrous Met. Soc. China.* **2017**, *27*, 569–577, DOI:10.1016/S1003-6326(17)60063-3.

[73] Hanada, N.; Ichikawa, T.; Orimo, S.-I.; Fujii, H. Correlation between hydrogen storage properties and structural characteristics in mechanically milled magnesium hydride MgH_2. *J. Alloys Compd.* **2004**, *366*, 269–273, DOI:10.1016/S0925-8388(03)00734-5.

[74] Kissinger, H. E. Variation of peak temperature with heating rate in differential thermal analysis. *J. Res. Natl. Bur. Stand.* **1956**, *57*, 217–221.

[75] Kissinger, H. E. Reaction Kinetics in Differential Thermal Analysis. *Anal. Chem.* **1957**, *29*, 1702–1706, DOI:10.1021/ac60131a045.

[76] Vyazovkin, S.; Kissinger method in kinetics of materials: Things to beware and be aware of. *Molecules* **2020**, *25*, 2813, DOI:10.3390/molecules25122813.

[77] Hou, Q.; Yang, X.; Zhang, J.; Yang, W.; Lv, E. Catalytic effect of NiO/C derived from Ni-UMOFNs on the hydrogen storage performance of magnesium hydride. *J. Alloys Compd.* **2022**, *899*, 163314, DOI:10.1016/j.jallcom.2021.163314.
[78] An, C.; Deng, Q. Improvement of hydrogen desorption characteristics of MgH_2 with core-shell Ni@C composites. *Molecules* **2018**, *23*, 3113, DOI:10.3390/molecules23123113.
[79] Zhou, C.; Bowman, R. C.; Fang, Z. Z.; Lu, J.; Xu, L.; Sun, P.; Liu, H.; Wu, H.; Liu, Y. Amorphous TiCu-based additives for improving hydrogen storage properties of magnesium hydride. *ACS Appl. Mater. Interfaces* **2019**, *11*, 38868–38879, DOI:10.1021/acsami.9b16076.
[80] Li, Q.; Lu, Y.; Luo, Q.; Yang, X.; Yang, Y.; Tan, J.; Dong, Z.; Dang, J.; Li, J.; Chen, Y.; Jiang, B.; Sun, S.; Pan, F. Thermodynamics and kinetics of hydriding and dehydriding reactions in Mg-based hydrogen storage materials. *J. Magnes. Alloys* **2021**, *9*, 1922–1941, DOI:10.1016/j.jma.2021.10.002.
[81] Mao, J.; Huang, T.; Panda, S.; Zou, J.; Ding, W. Direct observations of diffusion controlled microstructure transition in Mg-In/Mg-Ag ultrafine particles with enhanced hydrogen storage and hydrolysis properties. *Chem. Eng. J.* **2021**, *418*, 129301, DOI:10.1016/j.cej.2021.129301.
[82] Zhang, J.; Li, Z.; Wu, Y.; Guo, X.; Ye, J.; Yuan, B.; Wang, S.; Jiang, L. Recent advances on the thermal destabilization of Mg-based hydrogen storage materials. *RSC Adv.* **2019**, *9*, 408–428, DOI:10.1039/C8RA05596C.
[83] Blach, T. P.; Gray, E. M. A. Sieverts apparatus and methodology for accurate determination of hydrogen uptake by light-atom hosts. *J. Alloys Compd.* **2007**, *446–447*, 692–697, DOI:10.1016/j.jallcom.2006.12.061.
[84] Moretto, P.; Zlotea, C.; Dolci, F.; Amieiro, A.; Bobet, J.-L.; Borgschulte, A.; Chandra, D.; Enoki, H.; De Rango, P.; Fruchart, D.; Jepsen, J.; Latroche, M.; Jansa, I. L.; Moser, D.; Sartori, S.; Wang, S. M.; Zan, J. A. A round Robin test exercise on hydrogen absorption/desorption properties of a magnesium hydride based material. *Int. J. Hydrog. Energy.* **2013**, *38*, 6704–6717, DOI:10.1016/j.ijhydene.2013.03.118.
[85] Flanagan, T. B.; Park, C.-N.; Oates, W. A. Hysteresis in solid state reactions. *Prog. Solid State Chem.* **1995**, *23*, 291–363, DOI:10.1016/0079-6786(95)00006-G.
[86] Aguey-Zinsou, K.-F.; Ares Fernandez, J. R.; Klassen, T.; Bormann, R. Effect of Nb2O5 on MgH_2 properties during mechanical milling. *Int. J. Hydrog. Energy* **2007**, *32*, 2400–2407, DOI:10.1016/j.ijhydene.2006.10.068.
[87] Pang, Y.; Li, Q. A review on kinetic models and corresponding analysis methods for hydrogen storage materials. *Int. J. Hydrog. Energy* **2016**, *41*, 18072–18087, DOI:10.1016/j.ijhydene.2016.08.018.
[88] Chen, B.-H.; Chuang, Y.-S.; Chen, C.-K. Improving the hydrogenation properties of MgH_2 at room temperature by doping with nano-size ZrO2 catalyst. *J. Alloys Compd.* **2016**, *655*, 21–27, DOI:10.1016/j.jallcom.2015.09.163.
[89] Barkhordarian, G.; Klassen, T.; Bormann, R. Effect of Nb2O5 content on hydrogen reaction kinetics of Mg. *J. Alloys Compd.* **2004**, *364*, 242–246, DOI:10.1016/S0925-8388(03)00530-9.
[90] Milošević, S.; Kurko, S.; Pasquini, L.; Matović, L.; Vujasin, R.; Novaković, N.; Novaković, J. G. Fast hydrogen sorption from MgH_2–VO2(B) composite materials. *J. Power Sources* **2016**, *307*, 481–488, DOI:10.1016/j.jpowsour.2015.12.108.
[91] Xia, G.; Tan, Y.; Chen, X.; Sun, D.; Guo, Z.; Liu, H.; Ouyang, L.; Zhu, M.; Yu, X. Monodisperse magnesium hydride nanoparticles uniformly self-sssembled on graphene. *Adv. Mater.* **2015**, *27*, 5981–5988, DOI:10.1002/adma.201502005.
[92] Crivello, J.-C.; Dam, B.; Denys, R. V.; Dornheim, M.; Grant, D. M.; Huot, J.; Jensen, T. R.; de Jongh, P.; Latroche, M.; Milanese, C.; Milčius, D.; Walker, G. S.; Webb, C. J.; Zlotea, C.; Yartys, V. A. Review of magnesium hydride-based materials: Development and optimisation. *Appl. Phys. A* **2016**, *122*, *97*, DOI:10.1007/s00339-016-9602-0.

[93] El-Eskandarany, M. S.; Shaban, E.; Al-Halaili, B. Nanocrystalline β-γ-β cyclic phase transformation in reacted ball milled MgH_2 powders. *Int. J. Hydrog. Energy* **2014**, *39*, 12727–12740, DOI:10.1016/j.ijhydene.2014.06.097.
[94] Li, L.; Zhang, Z.; Jiao, L.; Yuan, H.; Wang, Y. In situ preparation of nanocrystalline Ni@C and its effect on hydrogen storage properties of MgH_2. *Int. J. Hydrog. Energy* **2016**, *41*, 18121–18129, DOI:10.1016/j.ijhydene.2016.07.170.
[95] Zhang, Q.; Wang, Y.; Zang, L.; Chang, X.; Jiao, L.; Yuan, H.; Wang, Y. Core-shell Ni3N@Nitrogen-doped carbon: Synthesis and application in MgH_2. *J. Alloys Compd.* **2017**, *703*, 381–388, DOI:10.1016/j.jallcom.2017.01.224.

Vincent Folliard and Aline Auroux*

Chapter 4
Using calorimetry to study catalytic surfaces and processes for biomass valorization

Abstract: Biomass conversion reactions offer a promising way to reduce the environmental impact of chemical processes. Many of these reactions are acid-catalyzed. This chapter presents adsorption microcalorimetry as an interesting tool to characterize the acid-base properties of the relevant catalysts. After a brief introduction to the technique, most of the discussion is devoted to case studies concerning various applications to gas- and liquid-phase reactions such as 5-HMF synthesis and production of acrolein. These show how adsorption microcalorimetry can be used to study and explain the catalytic activity and consequently help to improve the reaction yield and selectivity.

Keywords: Acid base properties of catalysts, Adsorption microcalorimetry, Fructose dehydration, Cellobiose hydrolysis, Glycerol dehydration, oxidative coupling of alcohols, Acrolein production

4.1 Introduction

Currently, a majority of chemicals and fuels are produced from non-renewable fuels such as oil, coal, and natural gas. While this has allowed for rapid economic development, a decline in reserves and growing concerns about environmental impact have led to the need for alternatives to replace fossil fuels. The use of biomass has been raised as a promising approach toward a bio-based economy. As an abundant carbon-neutral resource, biomass can be used as feedstock for the synthesis of fuels and various value-added chemicals. Moreover, advances in processes and engineering have led to the development of new manufacturing methodologies for the use of renewable biomass. Catalysis is a method of choice to transform biomass into value-added chemicals or fuels, and acid-catalyzed processes constitute a very important area for the application of heterogeneous catalysis in this context. In this field, the need to replace liquid acids (which cause corrosion, toxic waste, and separation problems) with solid acids is essential. Besides, water-tolerant properties are required for catalytic processes involving biomass transformation.

***Corresponding author: Aline Auroux,** Univ Lyon, Université Claude Bernard Lyon 1, CNRS, IRCELYON, 2 avenue Albert Einstein, F-69626, Villeurbanne, France, e-mail: aline.auroux@outlook.com
Vincent Folliard, Univ Lyon, Université Claude Bernard Lyon 1, CNRS, IRCELYON, 2 avenue Albert Einstein, F-69626, Villeurbanne, France

https://doi.org/10.1515/9783110590449-004

Among biomass components, sugars are often considered ideal feedstocks for the production of both fuel and value-added chemicals [1]. A 2004 report of the United States Department of Energy (DoE) identified multiple top value-added chemicals that can be produced from biomass carbohydrates [2].

Among building blocks derived from renewable resources, 5-hydroxymethylfurfural (5-HMF) is a very promising one. Indeed, 5-hydroxymethylfurfural can be hydrogenated to produce 2,5-dimethylfuran (DMF), an attractive biofuel [3], or oxidized to synthesize 2,5-furandicarboxylic acid (FDCA), an interesting alternative to terephthalic acid in the production of polyethylene terephthalate and polybutylene terephthalate [4]. Moreover, 5-HMF can also be reduced to obtain 2,5-bis(hydroxymethyl)furan, and 2,5-bis(hydroxymethyl)tetrahydrofuran, which lead to alcohol components for the production of polyesters, thus providing completely biomass-derived polymers when combined with FDCA [5].

Fructose and glucose are particularly interesting as precursor carbohydrates. It is known that 5-HMF can be obtained by dehydration of fructose. This process has been extensively studied using acid catalysts such as mineral acids [6, 7], organic acids, phosphates [8, 9], zeolites [10], ion-exchanged resins [11], or ionic liquids [12]. Another way to synthesize 5-HMF is the hydrolysis of cellobiose which is a disaccharide consisting of two molecules of d-glucose linked by a β-1,4′-glycosidic bond. This reaction has already been studied over different catalysts such as niobium phosphates [13], niobic acid [14], WO_3/ZrO_2 [15], metals supported on alumina [16], TiO_2-based catalysts [17], acidic resins [14], or ionic liquids [12].

Another widely studied value-added chemical is acrolein. Known as the simplest unsaturated aldehyde, acrolein is widely used as an intermediate for the synthesis of numerous industrial compounds such as acrylic acid, resins, superabsorbent polymers, detergents, biocides, or in feed applications such as methionine [18, 19]. Commercially synthesized for the first time in 1942 by Degussa using aldol condensation of formaldehyde and acetaldehyde, acrolein is nowadays mainly produced by propylene oxidation over multicomponent bismuth-iron-molybdate catalysts [19]. In this case again, due to the need to reduce greenhouse gas emissions, environmentally friendly processes have been developed. Among biomass-based alternative processes, gas-phase dehydration of glycerol over various acidic catalysts such as tungstated titania, tungstated zirconia, zeolites, or supported heteropolyacids is probably the best known [20–24]. Widely available due to its generation as a coproduct of biodiesel production [20], glycerol constitutes a rather inexpensive source of carbon. Besides, acrolein synthesis from glycerol can be done on-site, reducing the environmental footprint and minimizing risks related to transportation and storage [25].

Another alternative to the fossil fuel-based synthesis of acrolein, oxidative coupling of alcohols (OCA), constitutes a new promising way to transform biomass into sustainably produced acrolein. Based on the transformation of a mixture of methanol and ethanol, which can be biobased, this two-step process is performed over an iron molybdate catalyst (first step) and an aldolization catalyst (second step). Catalysts that have been studied for the second (aldolization) step include mixed oxides supported on silica [25, 26], hydrotalcites [26], alumina [26], magnesia [26] or spinel [27–29].

All of the above biomass transformation processes are driven by the acid-base properties of the catalyst. Using appropriate catalytic material is key to achieving high conversion and high selectivity. Thus, it is necessary to characterize the acid-base properties of catalysts and eventually modulate them to optimize processes.

4.2 The calorimetry technique

Among the many techniques that can be used to probe the acidity and basicity of solids, adsorption microcalorimetry is a versatile technique to characterize the number, the strength, and the strength distribution of surface sites. We will discuss two different implementations of the technique: the first one allows the determination of the acidity and basicity of a given solid, the so-called "intrinsic" acidity, in the gas phase, while the second one in the liquid phase gives access to the "effective" acidity [30].

4.2.1 Gas-phase calorimetry and "intrinsic" acidity and basicity

Adsorption microcalorimetry of probe molecules in the gas phase is particularly suitable to determine the acid-base properties of a given solid (intrinsic properties). To this end, gas-phase microcalorimetry requires two types of equipment for the simultaneous determination of the strength of adsorption and the adsorbed amount. Adsorption strength is usually determined using an isothermal and differential microcalorimeter, for example, a Tian-Calvet heat-flow microcalorimeter, which allows the entire reaction to being performed at a constant fixed temperature independently of the amount of gas introduced. The amount is commonly determined using a manometric line, coupled to the calorimeter. A detailed description of the technique and apparatus can be found elsewhere [31, 32].

The experimental procedure consists in the introduction of successive small doses of probe molecules (commonly ammonia as basic probe to titrate the acidic sites and SO_2 or CO_2 as acidic probes to titrate basic sites) and the measurement of evolved heats until thermal equilibrium is attained. The manometric line is equipped with a capacitance manometer and can be used to determine the amount of adsorbed gas probe molecule, which can be plotted against the equilibrium gas pressure to obtain isotherms of adsorption (Fig. 4.1).

The first run of adsorption is performed over the activated sample (pretreated under high vacuum) to arrive at the total adsorbed amount (V_{tot}) of the probe molecule. For each dose, a thermal signal is recorded and used to determine the evolved heat, while the adsorbed amount is measured by manometry; this gives access to differential heats of adsorption (q_{diff}) (Fig. 4.1).

After the first run is complete, desorption of the sample is performed by evacuation under vacuum to desorb any reversibly adsorbed amount of the probe. A second adsorption run is then performed to determine the physisorption amount (V_{reads}), which can be then subtracted from the total amount measured during the first adsorption to evaluate the irreversibly adsorbed amount (V_{irr}) of the probe molecule.

By studying how the differential heats of adsorption vary with the adsorbed amount, gas phase adsorption microcalorimetry experiments give access to not only the amounts and overall strength but also the strength distribution of the active surface sites.

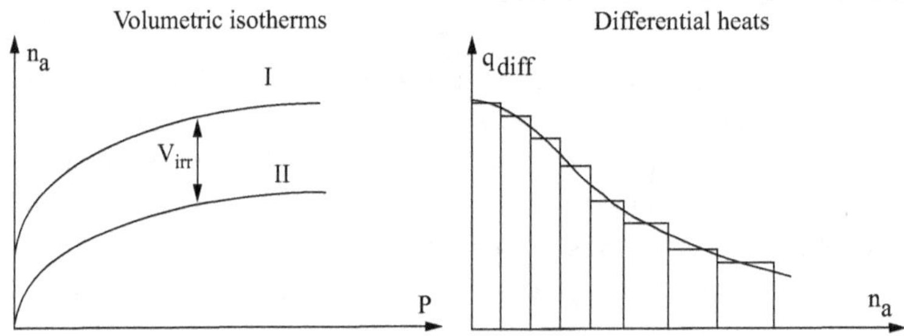

Fig. 4.1: Isotherms of adsorption (left) and differential heats of adsorption (right) obtained during a calorimetric experiment (adapted with permission from Springer Nature from [33], copyright 1997).

4.2.2 Liquid-phase calorimetry and "effective" acidity and basicity

Adsorption microcalorimetry in the liquid phase is suited to the determination of the "effective" acidity. Isothermal titration calorimetry makes it possible to directly study liquid-solid or liquid-liquid interactions. In this setup, a heat-flow calorimeter is usually employed, such as a Tian-Calvet calorimeter, fitted with a stirring system. The calorimeter is linked to a programmable syringe pump apparatus by capillary tubes. Phenylethylamine (PEA) is commonly used as a basic probe to titrate the acidity, while benzoic acid is used to probe the basicity. The procedure is quite similar to gas-phase adsorption microcalorimetry: the syringe pump system is programmed to inject successive doses of probe molecules with a known concentration. For each injection dose, an exothermic or endothermic signal is measured by the calorimeter. This signal, which is the sum of adsorption and displacement of adsorption contributions, can be used to plot the differential heat of adsorption versus time and consequently determine the strength of surface active sites. The contributions of dilution and mixing phenomena can be canceled out thanks to the presence of two cells (reference and sample) and an auxiliary furnace dedicated to the preheating of the probe molecule solution. The scheme of the apparatus is shown in Fig. 4.2.

In addition to that, the apparatus can be coupled to a UV-Vis spectrometer, making it possible to determine the adsorbed amount of probe molecule and thus to plot

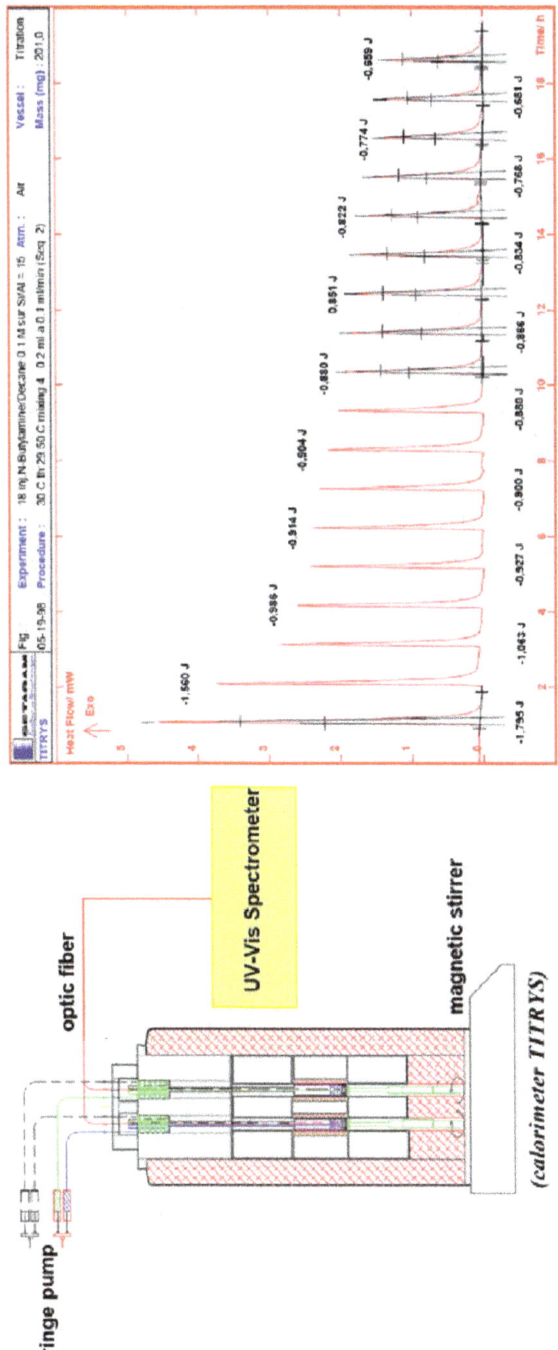

Fig. 4.2: Scheme of TITRYS apparatus and heats of adsorption recorded during liquid-phase calorimetry experiments (adapted with permission from Springer Nature from [31], copyright 2013).

adsorption isotherms. The combined information provided by calorimetric measurements and adsorption isotherms gives access to the number, strength, and strength distribution of the surface active sites.

4.3 Production of 5-HMF from biomass: case studies

Sugars can be used as feedstocks for the production of various value-added chemicals. To achieve this goal, carbohydrates are first converted to 5-HMF, which is an important intermediate for the production of some biofuels, for the synthesis of polymers, or in the pharmaceutical industry [34]. One of the most interesting carbohydrates for this process is fructose, which can be directly dehydrated to form 5-HMF. Glucose, which can be synthesized by hydrolysis of cellobiose, can be isomerized to fructose and thus can also be used to produce 5-HMF.

Fructose dehydration and hydrolysis of cellobiose are acid-catalyzed reactions. It is thus important to characterize the acid-base properties of catalysts for these reactions to improve their conversion and selectivity to the desired products. Adsorption microcalorimetry is a useful technique to achieve this goal. This section reports on different case studies concerning the characterization of the acid-base properties of various families of catalysts [15, 35–37] to optimize carbohydrates conversion and 5-HMF selectivity.

4.3.1 Fructose dehydration

Dehydration of fructose to produce 5-HMF (Scheme 4.1) has been extensively studied over various catalysts [6–12, 36]. Since this reaction is known to be acid-catalyzed, the determination of the acidic properties of the catalysts is important to control the reaction. For example, the acid-base properties of various catalysts, such as zeolites (H-MFI), ion exchange resins (Nafion), silica-alumina (Si-Al), niobic oxide (NbO), and niobium phosphate (NbP) have been investigated and then correlated to fructose conversion and 5-HMF selectivity.

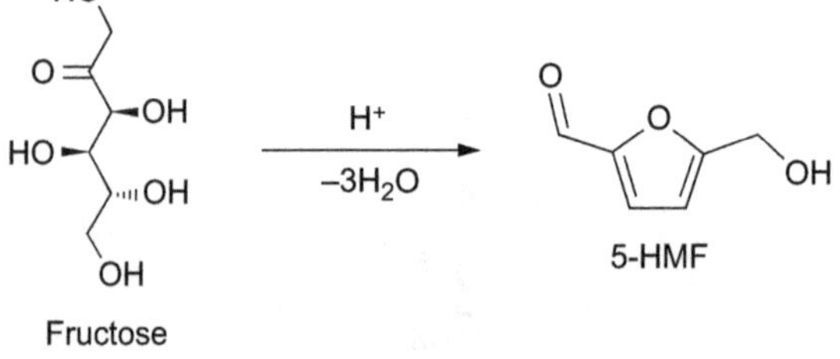

Scheme 4.1: Acid-catalyzed dehydration of fructose to HMF.

Figure 4.3 displays the differential heats of adsorption of ammonia in gas phase (top) and phenylethylamine (PEA) in liquid phase (bottom) versus adsorbed volume of probe molecule over H-MFI zeolite, Nafion ion exchange resin, silica-alumina, niobic oxide, and niobium phosphate.

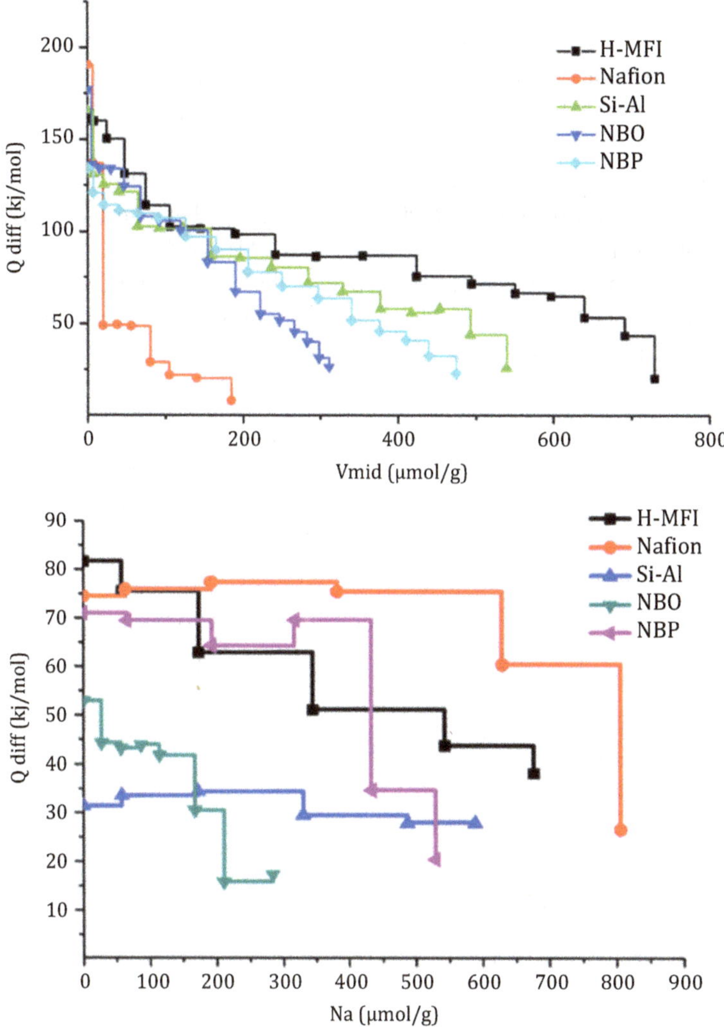

Fig. 4.3: Differential heats of adsorption of ammonia in gas phase (top) and PEA in liquid-phase (bottom) zeolites (H-MFI), ion exchange resins (Nafion), silica-alumina (Si-Al), niobic oxide (NbO), and niobium phosphate (NbP).

In the gas phase, H-MFI displays the highest acidity, with a large plateau in the domain of medium strength sites (100 kJ mol^{-1}), and the highest adsorbed volume (700 µmol g^{-1}). Silica-alumina, niobic oxide, and niobium phosphate display weaker, heterogeneous acidity, whereas ion exchange resin Nafion displays the lowest acidity with almost only weak sites (<50 kJ mol^{-1}). In the liquid phase, H-MFI zeolite still displays a large number of acidic sites; however, it is not the most acidic catalyst in the liquid phase. Surprisingly, Nafion resin, despite having the weakest acidity in the gas phase, displays a large plateau with a high amount of acidic sites. So, its acidity seems to be revealed by water, probably because it only manifests Brønsted sites.

Overall, the amounts of acid sites determined by microcalorimetry in the gas phase and liquid phase are close to each other, except in the case of Nafion, which possesses only Brønsted sites, whereas the other samples display both Brønsted and Lewis acidity. Catalytic tests have been performed on the same samples, and Fig. 4.4 displays the fructose conversion after 6 h reaction in water at 130 °C in an autoclave as a function of the number of acid sites. The results show that fructose conversion increases with the increase of the number of acid sites determined in the liquid phase.

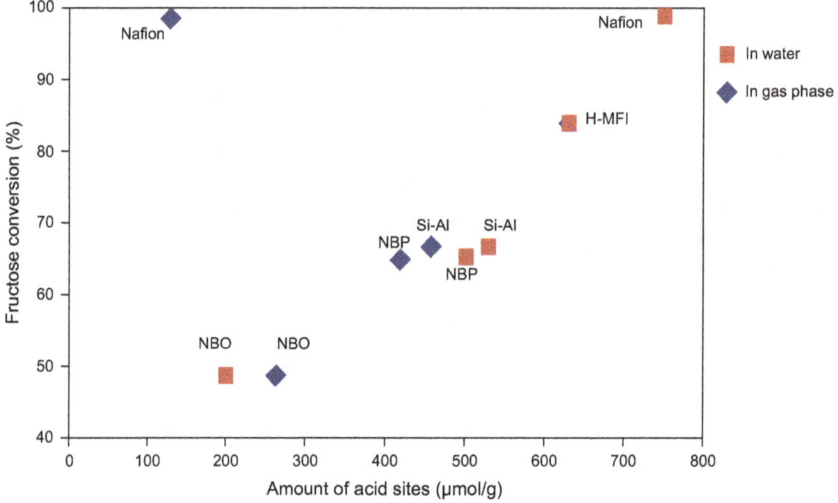

Fig. 4.4: Fructose conversion after 6 h reaction in water at 130 °C in an autoclave as a function of the number of acid sites.

In another study, fructose dehydration has been performed over tungsten oxide supported on zirconia [35]. In this work, catalytic dehydration of fructose and its conversion to 5-hydroxymethylfurfural was studied using tungstated zirconium oxides with various tungsten oxide loadings (1.2–20.9 wt%). To prepare catalysts, $Zr(OH)_4$ was impregnated with an ammonium metatungstate hydrate solution, to get a WO_3 loading ranging from 1 to 20 wt%. The synthesized catalysts were then air dried overnight at

85 °C followed by calcination at 700 °C under air for 4 h. The pure zirconia was calcined at the same temperature. The resulting catalysts were labeled m-WO_3/ZrO_2, where m indicates the percentage of WO_3 (in wt%).

Acid-base characterization was performed thanks to adsorption microcalorimetry in gas phase at 80 °C using sulfur dioxide (SO_2) and ammonia (NH_3) as probe molecules in a heat flow calorimeter (C80 from Setaram) linked to a conventional manometric apparatus equipped with a Barocel capacitance manometer for pressure measurements. Before adsorption, the samples were pretreated under vacuum at 400 °C overnight. Figure 4.5 represents differential heats of adsorption of SO_2 and NH_3 versus coverage.

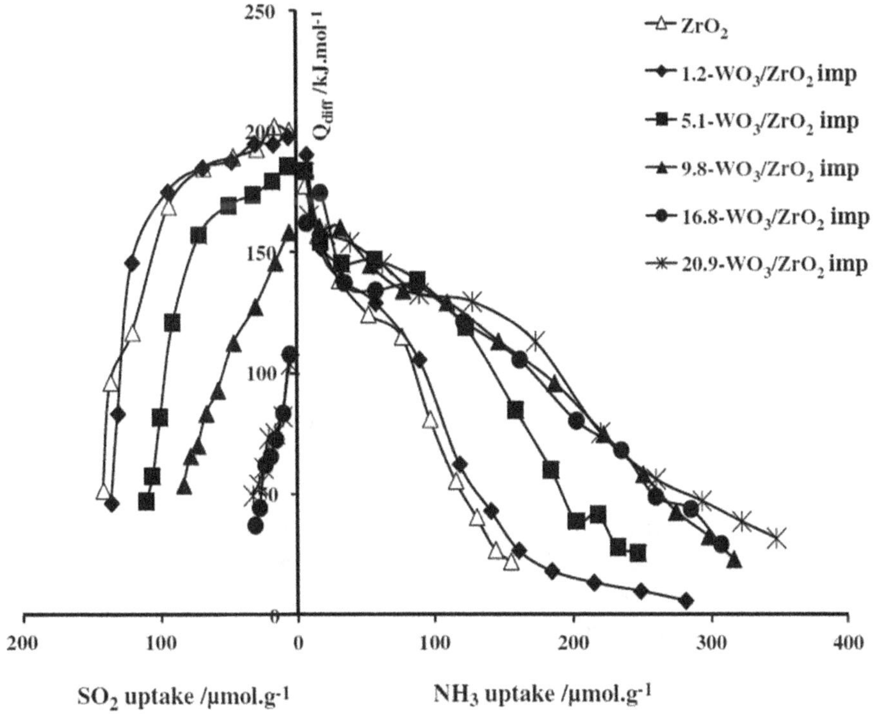

Fig. 4.5: Differential heats of adsorption of sulfur dioxide (left) and ammonia (right) versus probe molecule uptake at 80 °C on ZrO_2-impregnated WO_3 catalysts (reprinted by permission from Elsevier from [35], copyright 2013).

The differential heats of adsorption of SO_2 emphasize the presence of strong basic sites on the surface of ZrO_2. Besides, the quantity of basic sites decreases with increasing WO_3 content in the m-WO_3/ZrO_2 samples. These results are logical since tungsten oxide is considered acidic. Moreover, it appears that a higher tungsten oxide content leads to an increasing population of medium strength sites ($100 < Q_{diff} < 150$ kJ mol^{-1}).

Catalytic activity has been correlated with calorimetric results and Fig. 4.6 plots the 5-HMF yield versus the number of medium strength sites (left) and selectivity versus the ratio of basic to acidic sites (right).

It can be observed that the 5-HMF yield increased with the number of medium-strength sites. Besides, by plotting the selectivity as a function of the ratio of the numbers of basic to acidic sites, it can be seen that selectivity reaches a maximum (40.1%) for the catalyst containing 9.8% of WO_3 before a decrease with increasing WO_3 loading. This shape of the curve was expected by authors considering the reactivity of 5-HMF, which transforms to levulinic and formic acids over very strong acidic sites [35, 38]. The best selectivity is being attained for the higher dispersion and a high number of strong acid sites.

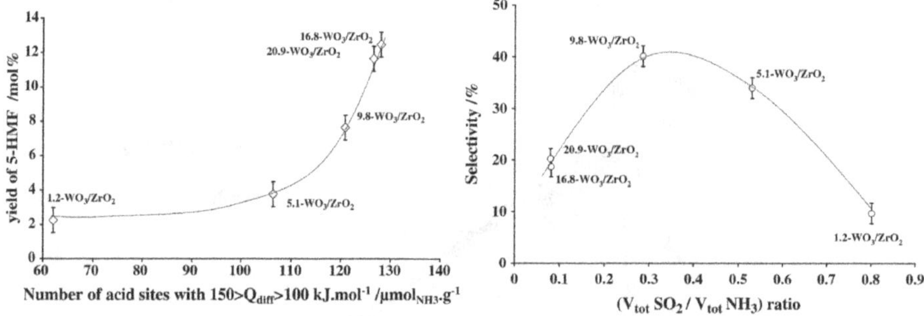

Fig. 4.6: 5-HMF yields versus the number of medium strength acid sites (left) and selectivity versus the ratio of basic to acidic sites (right) (adapted with permission from Elsevier from [35], copyright 2013).

In another work [36], fructose dehydration has also been studied over ceria/niobia catalysts. CeO_2–Nb_2O_5 mixed oxide catalysts were prepared by coprecipitation from ammonium–niobium oxalate complex (99.99%, Aldrich) and cerium nitrate (99.9%, Strem Chemicals) aqueous solutions. Ammonia solution (16 wt%) was gradually added dropwise to the mixture of two solutions with strong stirring until precipitation was finished (pH = 9). Cerium nitrate followed the same treatment to give rise to pure ceria. The precipitates were then filtrated, washed with hot distilled water, dried overnight at 110–120 °C, and then calcined at 400 °C for 10 h in air. The acid-base properties were investigated by adsorption microcalorimetry of NH_3 and SO_2 at 150 °C using the same procedure as described above. Samples were pretreated overnight at 300 °C under vacuum. The catalytic activity was investigated using a 100 mL stainless steel autoclave, at 130 °C. The procedure was to dissolve 600 mg of fructose in 60 ml of water and then add 80 mg of solid catalyst. Fructose, 5-HMF but also levulinic acid, and formic acid concentrations were monitored thanks to H^1 liquid NMR.

Figure 4.7 shows the differential heats of adsorption of NH_3 and SO_2 in gas phase on the investigated samples. The dominant acidic character of pure niobia can be easily seen as it adsorbs only ammonia. All the other catalysts adsorb both NH_3 and SO_2, evidencing their amphoteric character. It is interesting to note that the addition of ceria to niobia increases the amount of basic sites without too much alteration of the acidity. On the other hand, even though niobia does not exhibit basicity, its addition in a small quantity to ceria leads to the creation of strong basic sites whose content decreases with a further increase of niobia content. The authors suggested that it could indicate an interaction between ceria and niobia, or a change in the surface morphology of ceria as a result of mixing with niobia [36]. The correlation with catalytic activity has been studied: Fig. 4.8 represents the fructose conversion (in blue) and selectivity to 5-HMF (in red) versus the number of strong acidic sites per surface area (μmol m^{-2}). It can be seen that an increased density of acid sites enhances conversion and selectivity to 5-HMF. This drives to the conclusion, in agreement with a previous study [30], that niobium oxide is the main active component in the dehydration of fructose over these materials [36].

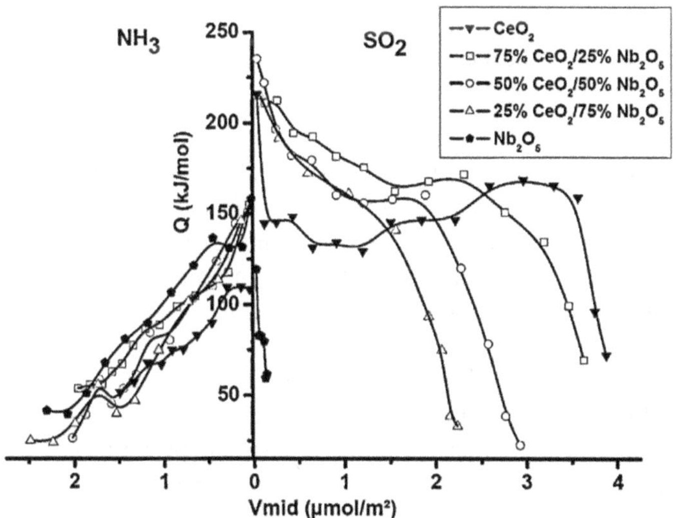

Fig. 4.7: Differential heats of adsorption of NH_3 and SO_2 versus probe molecule uptakes (reprinted by permission from Elsevier from [36], copyright 2012).

4.3.2 Hydrolysis of cellobiose

Producing 5-HMF is also possible starting directly from cellobiose, which is the disaccharide formed by two molecules of d-glucose. In this case, cellobiose is hydrolyzed to glucose which can then be isomerized to fructose to produce 5-HMF by dehydration. As for dehydration of fructose, hydrolysis of cellobiose is a reaction that requires acid catalysts such as WO_3 [15], zeolites, silica-alumina, resins [39], TiO_2-based catalysts [17],

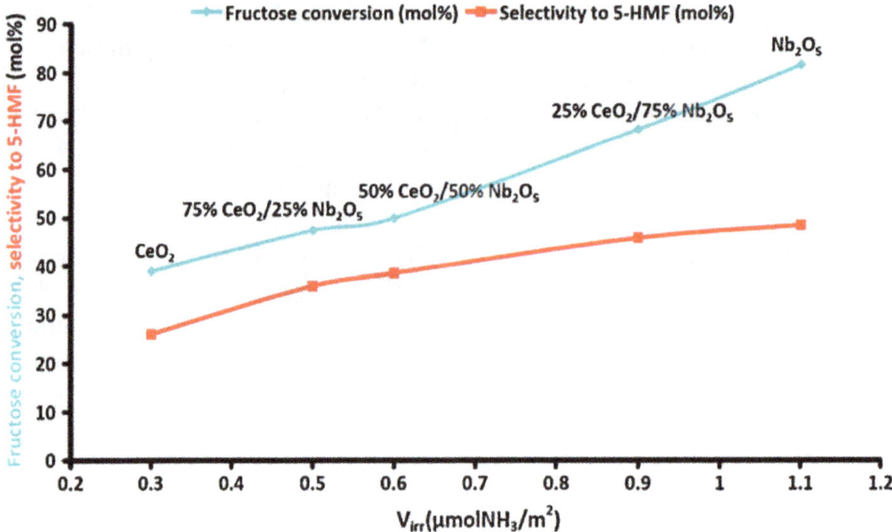

Fig. 4.8: Fructose conversion (in blue) and selectivity to 5-HMF (in red) versus the number of strong acidic sites (reprinted by permission from Elsevier from [36], copyright 2012).

heteropolyacids-based catalysts [40], or niobium phosphates [13]. This part presents the results of a microcalorimetry study of coprecipitated tungsta-zirconia and the correlation between their acidity and catalytic activity for cellobiose hydrolysis [15].

The catalysts were prepared by dissolving $ZrOCl_2 \cdot 8H_2O$ in water under stirring. Then, a solution of concentrated ammonium hydroxide (NH_4OH), distilled H_2O, and ammonium metatungstate hydrate [$(NH_4)_6H_2W_{12}O_{40} \cdot nH_2O$, Fluka, ≥99.0% WO_3] was added drop by drop to obtain a WO_3 loading ranging from 1 to 20 wt%. The final pH of the solution was adjusted to approximately 9 by the addition of concentrated NH_4OH. The suspension was then put in polypropylene bottles and placed in a steam box (100 °C) for 72 h. After filtration of the suspension, the product has been washed, dried overnight at 85 °C, and then calcined at 700 °C in flowing air for 4 h. For the sake of readability, catalysts were labeled as m-WO_3/ZrO_2-Cop where m indicates the percentage of WO_3 (in wt%) and cop the coprecipitation method.

To perform the kinetic study of cellobiose catalytic dehydration to glucose in water, a weighted amount of calcined catalyst sample of about 100 mg was crushed, sieved in 300 μm particle size and put into a glass thermostated reactor working at atmospheric pressure and constant temperature of 97 °C (reaction temperature ±1 °C) under vigorous magnetic stirring. The reactions were followed for 32 h. For each run, a new aqueous cellobiose solution (total volume of 10 mL) of 0.1 g mL^{-1} (corresponding to 0.3 M) was prepared.

The acid-base properties were determined by adsorption microcalorimetry of NH_3 and SO_2 at 80 °C using the same procedure described in Section 4.3.1. Before analysis, the samples were pretreated at 400 °C under vacuum for the night.

Figure 4.9 displays the differential heats of adsorption of NH_3 as a function of coverage on the left and the strength distribution of acidic sites on the right for WO_3/ZrO_2 coprecipitated catalysts. The sloping profile of the differential heat curves is indicative of the heterogeneous character of the acidity for all the catalysts which present acidic sites of different natures (Lewis acid sites from zirconia and both Brønsted and Lewis sites from WO_x) with various strengths. The adsorbed amounts increase with increasing tungsten oxide content until 15.2 wt% of WO_3 which corresponds to the maximum surface area [15]. The differential heat measurements make it possible to calculate the strength distribution of acidic sites, i.e., the number of sites within a given range of acid strength. In this case, it is possible to observe that the amount of medium strength sites ($90 < Q < 130$ kJ mol^{-1}), which could be assigned mainly to Brønsted acidity, follows the same trend as the differential heats and increases up to 15.2 wt% content of WO_3 while the number of strong sites (<130 kJ mol^{-1}), assigned to Lewis acidity, reaches its maximum at 9.9 wt% WO_3.

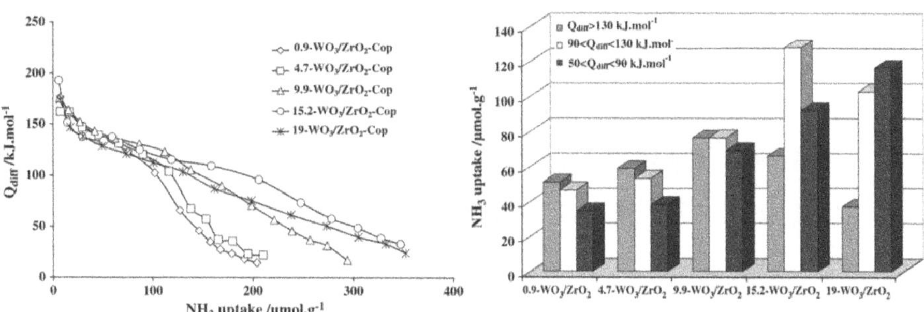

Fig. 4.9: Differential heats of adsorption of NH_3 a function of coverage (left) and strength distribution of acidic sites (right) for WO_3/ZrO_2 catalysts (reprinted by permission from Elsevier from [15], copyright 2012).

The kinetic activity of these catalysts has been studied, and the reaction rates for the hydrolysis of cellobiose are reported in Fig. 4.10 as a function of the WO_3 content. It can be seen that catalytic activity decreases from 0.9 to 4.7% of WO_3 content and then increases to reach its maximum at 19%. Kourieh *et al.* [15] suggested that the presence of monomeric WO_x species partly covering active zirconia surface was linked to the loss of activity. Indeed, the amounts of tungsten oxide were insufficient to create a WO_3 structure with Brønsted acid sites. Increasing the content of tungsten oxide generated interconnected WO_x clusters which possess strong acid sites [41]. These results were correlated with those of a Fourier-transform infrared (FTIR) study of pyridine adsorption, which showed the absence of Brønsted acid sites on the catalyst with

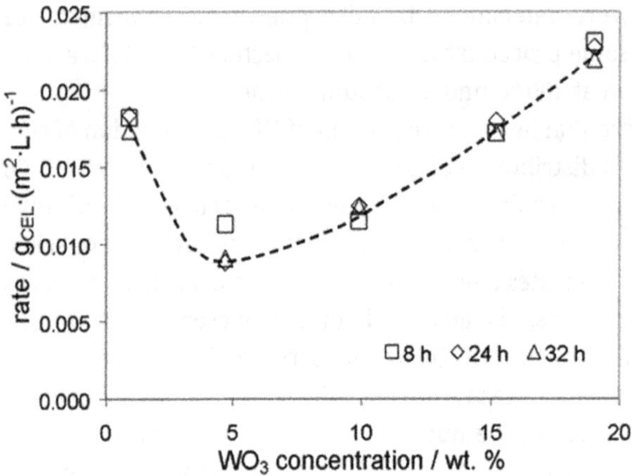

Fig. 4.10: Reaction rates expressed as gCELL/(m^2cat · L · h) for the hydrolysis of cellobiose over WO_3/ZrO_2 catalysts as a function of the WO_3 concentration determined at different reaction times (reprinted by permission from Elsevier from [15], copyright 2012).

4.7 wt% tungsten oxide content while the solids with 15.5 and 19 wt% tungsta exhibited Brønsted acid sites [15].

4.4 Guerbet reaction of methanol and ethanol to higher R-OH: a case study

Ethanol is probably one of the best known and most abundant bio-based chemicals. OECD-FAO Agricultural Outlook 2021–2030 report estimated the global production of ethanol at around 124 billion liters in 2021, with the United States and Brazil being the top two producers, and it should increase to 132 billion liters by 2030 [42].

Currently, as biofuel, bioethanol is used as a substitute or blended with gasoline [43]. Nonetheless, its hydrophilicity, low energy content, and differences in fuel properties compared to modern gasoline remain barriers to the further development of its applications as a replacement for fossil fuels [44, 45]. This has motivated research into other added-value applications for bio-ethanol.

A possible alternative to the direct use of ethanol as a biofuel is the Guerbet reaction (Scheme 4.2) which allows the synthesis of heavier alcohols from light alcohols with the loss of a water molecule. Thanks to this process, some chemicals such as 1-propanol or 1-butanol can be accessible by the reaction of ethanol and methanol. 1-Butanol is very interesting because of its similarity to gasoline, which makes it an attractive potential replacement product or additive to gasoline [45]. Besides, 1-butanol is also used directly as a solvent for paints or in the production of butyl acrylate and methacrylate [46].

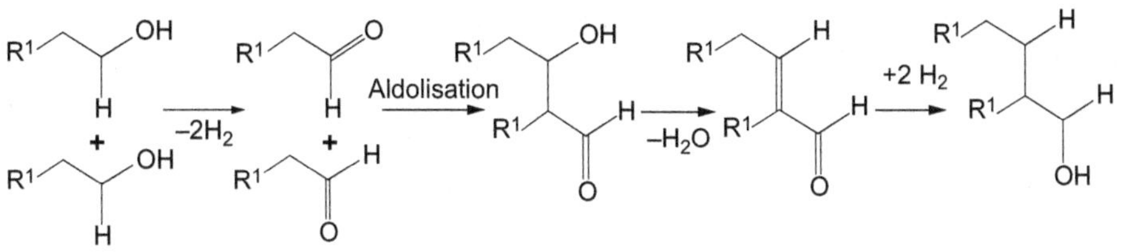

Scheme 4.2: General four-step sequence for Guerbet reaction.

Multiple catalysts have been tested for the Guerbet reaction, both in homogeneous phase with catalysts containing transition metals of the group VIII–X such as Ru, Rh, or Ir to convert ethanol to n-butanol [47, 48] and in heterogeneous phase with solid catalysts such as MgO [49], hydroxyapatites [50], supported salts on zeolites [51] or hydrotalcites [44]. It is widely known that surface properties, and especially basicity and acidity, can impact the mechanism and thus affect the selectivity toward the targeted products [52–54]. This part reports on a case study of the Guerbet reaction to convert methanol and ethanol to a mixture of 1-propanol and 1-butanol over hydrotalcites and the influence of the acid-base properties of catalysts on the selectivity to the products [44]. In this work, six different hydrotalcites with Mg/Al ratios varying from 2 to 7 were prepared and tested.

Hydrotalcites were prepared by co-precipitation of Mg and Al nitrates (0.15 M $Mg(NO_3)_2 \cdot 6H_2O$ and 0.05 M $Al(NO_3)_3 \cdot 9H_2O$ both from Sigma-Aldrich) in a 1 M sodium carbonate (Na_2CO_3) solution (99%, Sigma-Aldrich). The desired stoichiometry was obtained by adaptation of metallic salt quantities. The solution was drop-wisely added under strong stirring and pH was kept between 9 and 10 at room temperature thanks to the addition of a 2 M sodium hydroxide (NaOH) solution (98%, Sigma-Aldrich). The final solution was stirred for 1 h and the obtained precipitate was filtered, washed with deionized water (50 °C), and then dried at 63 °C for 12 h. The final solid was crushed and calcined for 4 h under air at 400 °C [44].

Their acidic and basic properties were measured using adsorption microcalorimetry of ammonia (for acidic sites titration) and SO_2 (for basic sites titration) probe molecules in gas phase using a Tian-Calvet heat-flow microcalorimeter (C80, Setaram) coupled to a manometric line equipped with a Barocel capacitance manometer. The catalytic test was composed of a fixed-bed glass reactor which was filled with 200 mg of catalyst mixed with 200 mg of silicon carbide (SiC). The reactor feed was a mixture of alcohols (10 vol.% methanol and 10 vol.% ethanol) in He gas with a total flow rate of 50 ml min^{-1}. The reaction was performed at 400 °C for 2 h, after which the products were analyzed online by chromatography. In addition to 1-propanol and 1-butanol, numerous products were synthesized during the catalytic tests, such as ethylene, acetaldehyde, dimethyl ether (DME), and methoxyethane, with very different yields depending on the samples.

First, it is interesting to see that surface areas markedly increased after calcination from around 27–32 $m^2\ g^{-1}$ to 72–145 $m^2\ g^{-1}$, consequence of the elimination of carbonates anions leading to formation of mesopores and thus the destruction of layered structure which has been observed by the authors on XRD results [44].

Moreover, the surface areas after calcination are more heterogeneous with three samples (Mg/Al ratio = 3, 4, 5) exhibiting surface between 128 and 145 $m^2\ g^{-1}$ and three others with surface below 100 $m^2\ g^{-1}$.

Study of acid-properties of theses samples is very interesting; Fig. 4.11 displays isotherms of adsorption of ammonia in the left and SO_2 in the right. By looking the isotherms of adsorption of SO_2 which quantify the number of basic sites, we can see that samples with Mg/Al = 3, 4, and 5 are more basic than the three others even if we consider the difference of specific surface areas.

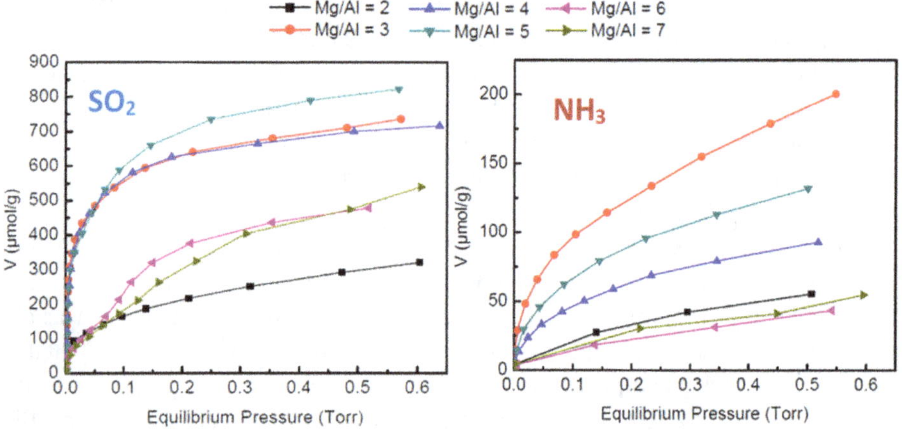

Fig. 4.11: Isotherms of adsorption of SO_2 (left) and NH_3 (right) for the studied hydrotalcites (adapted with permission from Elsevier from [44], copyright 2017).

Di Cosimo *et al.* [55] stated that addition of small quantities of aluminum to magnesia (MgO) drastically decreases the density of surface basic sites due to Al enrichment of the surface. Indeed, authors describes that with low Al contents (where Mg/Al > 5) the formation of surface amorphous AlO_y structures partially covers the Mg-O pairs and then decreases the concentration of surface O^{2-} anions. When Al contents increases (5 > Mg/Al > 1), Al^{3+} cations in MgO lattice generates a defect to compensate the positive charge produced and the adjacent oxygen anions become coordinatively unsaturated leading to a higher basicity. This can explain the fact that we can discriminate two different groups among basicity results.

Besides, it is interesting to notice the shapes of SO_2 adsorption isotherms. For the CHT with Mg/Al = 6 and 7, we can see two steps during adsorption. This is characteristic of sulfate formation during SO_2 adsorption at the surface. This behavior agrees with the literature who already reported that MgO, which is in higher quantity for CHT with Mg/Al ratio of 6 and 7, provide favorable sites for sulfate formation [56, 57].

Concerning acidity measures, CHT with Mg/Al = 3 exhibit the highest adsorbed ammonia amount. Increasing or decreasing this Mg/Al ratio leads to adsorb lower quantities of ammonia. As for basicity, samples with Mg/Al = 2, 6, and 7 show lower adsorption capacities that samples with Mg/Al = 3, 4, or 5. It is worth noticing that amounts of basic sites are particularly higher than amounts of acidic sites for all the studied hydrotalcites.

Methanol and ethanol conversion have been correlated to the number of strong basic sites (i.e., those on which SO_2 is irreversibly adsorbed). The results are displayed in Fig. 4.12.

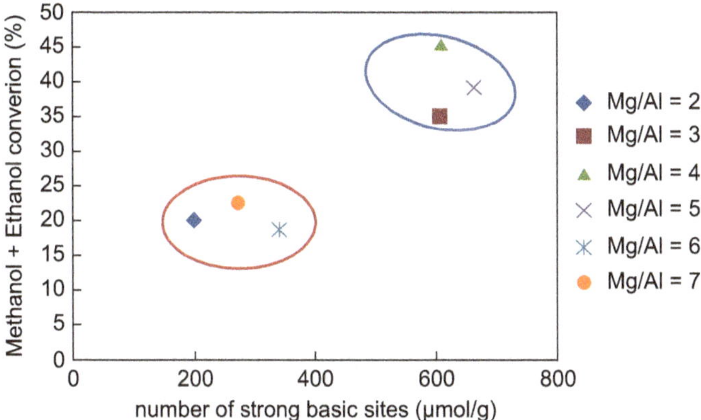

Fig. 4.12: Methanol and ethanol conversion versus the number of strong basic sites (reprinted by permission from Elsevier from [44], copyright 2017).

Two groups of three samples can be seen in the figure. The first group is constituted by hydrotalcites with Mg/Al ratios of 2, 6, and 7, while the second one is composed of solids with Mg/Al ratios equal to 3, 4, and 5. These results can be correlated to surface areas. Indeed higher conversions are displayed by samples showing higher surface areas. Moreover, it can be also observed that the three hydrotalcites which exhibit the highest conversions are also the ones with the highest amounts of basic sites.

Among the different products, acetaldehyde was synthesized over all samples; selectivity was not correlated to the acidic or basic character of the catalysts, probably because the aldolization reaction can take place over both basic or acidic sites and acetaldehyde activated in different modes.

Ethylene is the product of ethanol dehydration and thus, its synthesis is in competition with acetaldehyde production (dehydrogenation of ethanol), impacting the final yield of the desired products. It is therefore important to understand the mechanism of ethylene formation over catalysts for this reaction. Stošić *et al.* [44] observed a correlation between the ethylene selectivity for each sample and the ratio of the numbers of strong basic to the number of strong acidic sites. The resulting curve is plotted in Fig. 4.13.

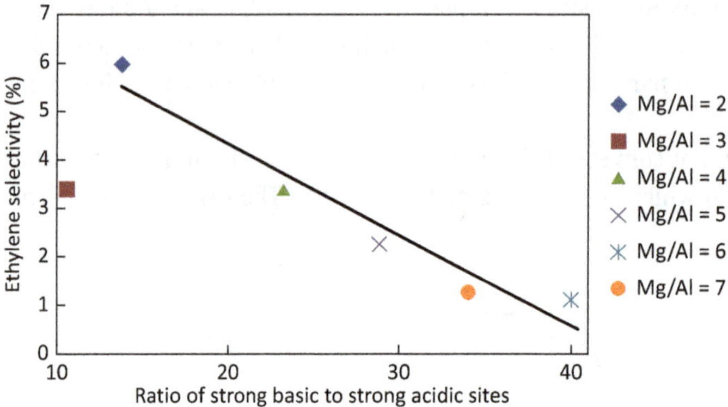

Fig. 4.13: Ethylene selectivity versus the ratio of the numbers of strong basic to strong acidic sites (reprinted by permission from Elsevier from [44], copyright 2017).

From this figure, it can be seen that when the ratio of strong basic to strong acidic sites increases (meaning that the basic character of the sample increases), the selectivity to ethylene decreases.

The expected reaction products, namely 1-propanol and 1 butanol, were observed on all the catalysts and the highest selectivities were displayed by hydrotalcites with Mg/Al equal to 3, 4, and 5. Figure 4.14 presents the selectivity to 1-propanol and 1-butanol versus the number of strong acidic sites (left-hand side) or strong basic sites (right-hand side). It can be observed that selectivity to alcohols increases with the basic character, as anticipated by the authors [44]. Indeed, basic sites have two possible roles during reactions: dehydrogenation of alcohols and aldol formation. Interestingly, the presence of acidic sites also has a positive influence, as the selectivity to alcohols increases with the number of acidic sites until a maximum at $V_{irr} = 26\ \mu mol\ g^{-1}$, after which point a further increase in acidity did not affect the selectivity. This influence of acidic sites has been explained by the fact that aldolization and aldol dehydration are also catalyzed by acidic sites [44]. These correlations underline the necessity to have both basic and acidic sites to permit the Guerbet reaction of ethanol and methanol to get heavier alcohols.

The results obtained in this study demonstrate the importance of acid-base properties, their characterization, and their fine-tuning to optimize the properties of the catalyst for Guerbet reaction.

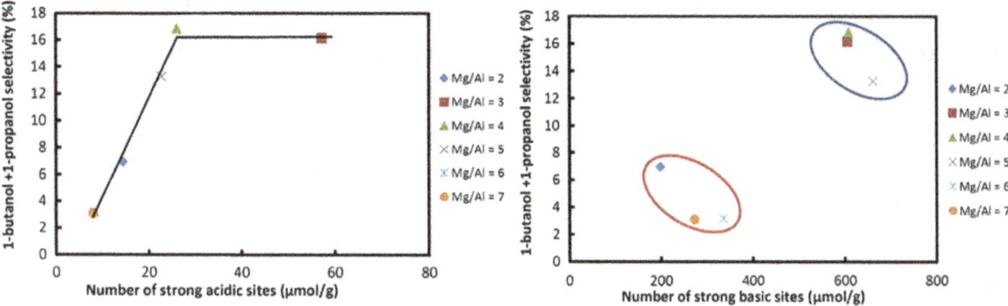

Fig. 4.14: 1-Propanol + 1-butanol selectivity versus the number of strong acidic sites (left) and strong basic sites (right) (adapted with permission from Elsevier from [44], copyright 2017).

4.5 Acrolein production from biomass: case studies

Acrolein is a widely used intermediate in industrial chemistry to produce various added-value chemicals such as acrylic acid, superabsorbent polymers, biocides, or methionine [18]. Whereas acrolein is currently produced by propylene oxidation over multicomponent catalysts [19], the depletion of fossil fuel resources has motivated research into alternative processes using bio-based feedstocks. Different alternatives have been studied. The best known of these is probably glycerol dehydration, but other alternatives such as oxidative coupling of alcohols (OCA) have also been investigated. These two methods of production of acrolein both require an acid-base catalyst. For this purpose, a characterization of acid-base properties is inevitable to optimize the catalyst properties and thus improve the acrolein yield. This section presents two case studies involving the characterization of the acid-base properties of different catalysts and their correlation to catalytic activity for glycerol dehydration in one instance and oxidative coupling of a mixture of alcohols in the other.

4.5.1 Glycerol dehydration

Glycerol dehydration is a very promising alternative to produce acrolein. Generated during the production of biodiesel, glycerol is a rather cheap source of carbon. Besides, the possibility of manufacturing acrolein directly on-site could reduce costs and increase safety (major accidents with acrolein have been caused by storage and transportation issues) [58].

It is known that for glycerol dehydration, the acidity of the active catalytic phase has a very important influence on acrolein production [37, 59–62]. Glycerol dehydration has been widely studied over heteropolyacids [63], phosphates [60], zeolites [60, 64, 65] or mixed oxides [23, 66]. Mixed oxides such as niobium oxides or tungsten supported on oxides have shown particular promise, with some studies achieving an interesting selectivity to acrolein at almost complete glycerol conversion [66].

We report on a couple of studies [37, 59] focusing on the determination of the acid-base properties of mixed oxide catalysts by microcalorimetry of adsorption of NH_3 and SO_2 and their correlation with catalytic performance for gas-phase glycerol dehydration. The materials used in these studies were zirconia and titania-based catalysts whose acid-base properties were adjusted by the addition of other oxides such CeO_2, La_2O_3, WO_3, or supported phosphotungstic acid (HPW) [37, 59]. Zirconia- and titania-supported phosphotungstic acid were prepared by incipient wetness impregnation with an aqueous solution of phosphotungstic acid, dried at 110 °C followed by calcination at 400 or 500 °C under air.

WO_3–ZrO_2 (10 wt% of WO_3), La_2O_3–ZrO_2 (30 wt% of La_2O_3), and CeO_2–ZrO_2 (25 wt% of CeO_2) were commercial (purchased from Dai Ichi Kigenso Kabushiki Kaisha) just like ZrO_2, TiO_2 anatase, and TiO_2 rutile (purchased form Saint-Gobain Norpro). These materials were all calcined under air at 500 °C for 5 h before use.

Adsorption microcalorimetry experiments using NH_3 and SO_2 probes were performed at 150 °C using the same procedure described in Section 4.3.1. Before adsorption, the samples were pretreated for 2 h under vacuum at 300 °C. The catalytic performance was evaluated at 280 °C with a gas hourly space velocity (GSHV) of 4,400 h^{-1}. First of all, the catalysts were crushed and sieved (300–500 μm), then they were mixed with quartz sand (50–80 mesh) and loaded in a fixed-bed Pyrex reactor (in the middle) with quartz wool packed in both ends. Initial pretreatment of the catalysts has been performed at 300 °C in a nitrogen flow for 2 h. An aqueous solution of glycerol (30 wt%) was introduced in the flow of nitrogen thanks to a syringe pump with a flow of 0.48 g h^{-1}. Feed gas composition was N_2: H_2O: glycerol (18.7:75:6.3 in vol%). The mixture at the reactor outlet has been analyzed by gas chromatography with flame ionization detector (FID) [37, 59].

Figure 4.15 displays the isotherms of adsorption, i.e., the adsorbed amounts of probe molecule plotted as a function of the equilibrium pressure, while Fig. 4.16 displays the differential heats of adsorption of both NH_3 and SO_2 probe molecules as a function of the adsorbed amounts at 150 °C for the zirconia-based catalysts.

The profiles of the differential heat curves of adsorption of NH_3 indicate that all the catalysts exhibit a heterogeneous acidity. As expected, samples containing WO_3 and HPW show a strong acidic character. ZrO_2, CeO_2/ZrO_2, and La_2O_3/ZrO_2 display both acidic and basic sites confirming their amphoteric character.

The addition of WO_3 and/or HPW to zirconia does not seem to affect the strength and strength distribution of acidity, unlike the case of CeO_2 and La_2O_3, whose addition provokes a decline in both the number and the strength of the acidic sites. Concerning basicity, the addition of WO_3 and HPW leads to a drastic decrease in both number and strength, indicating the predominant acidic character of these solids which is not surprising, the strong acidic character of WO_3 was already stated in prior microcalorimetric,

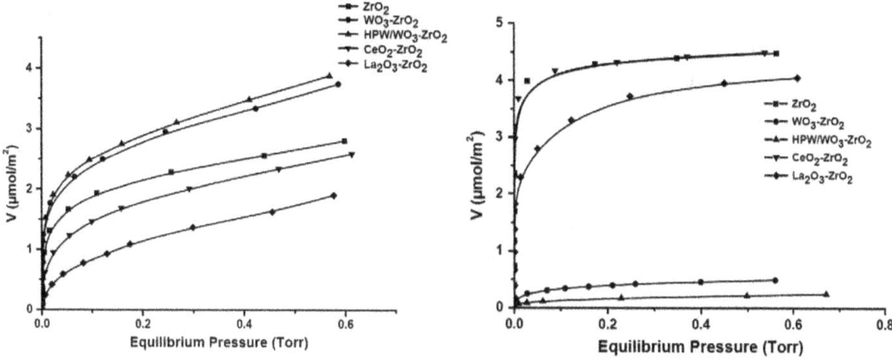

Fig. 4.15: Isotherms of adsorption as a function of surface coverage for adsorption of NH_3 (left) and SO_2 (right) at 150 °C on zirconia-based catalysts (adapted with permission from Elsevier from [59], copyright 2012).

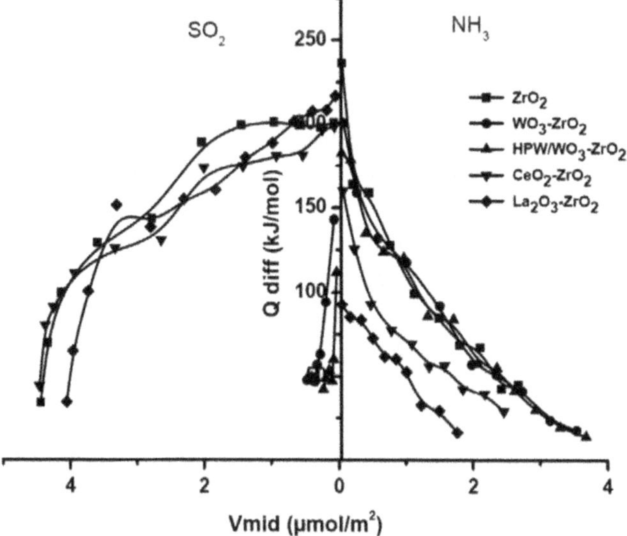

Fig. 4.16: Differential heats of adsorption of NH_3 and SO_2 as a function of surface coverage at 150 °C on zirconia-based catalysts (reprinted by permission from Elsevier from [59], copyright 2012).

spectroscopic, or TPD studies over WO_x/ZrO_2 [15, 35, 67, 68] but also on tungsta-titania catalysts [69] or WO_3/Al_2O_3 composite oxides [70].

For the other samples, the differential heats of SO_2 adsorption remain similar to those of zirconia; the latter three zirconia-based catalysts present an amphoteric character. The addition of HPW to titania catalysts, which express a predominantly acidic character, has also been studied by microcalorimetry of adsorption of NH_3 and SO_2 [59].

Complementary data about the acid-base properties of these materials have been obtained in an FTIR spectroscopic study of pyridine and CO_2 adsorption [37]. Indeed, whereas adsorption microcalorimetry informs on the number, the strength, and the strength distribution, FTIR spectra after adsorption of a probe molecule such as pyridine or CO_2 can provide information about the nature of the acidic and basic sites respectively [71].

FTIR analyses were performed over self-supported small pellets. Prior to probe molecule adsorption, each pellet was pretreated under vacuum in a quartz infrared cell (with KBr windows) at 400 °C for 2 h. After pretreatment, the cell was cooled, and an already known probe molecule amount (133.3 Pa) was added to the cell. Then, spectra were recorded after 30 min of evacuation at different temperatures (rt, 50, 100, 150, 200, 250, and 300 °C) thanks to a Nicolet Nexus IR spectrometer equipped with an extended KBr beam splitter and a mercury cadmium telluride (MCT) detector. The used spectral resolution was 4 cm^{-1}. All the displayed spectra were obtained after subtraction of fresh sample spectrum (without probe molecule adsorption) [37].

Figure 4.17 displays the FTIR spectra for pyridine desorption on zirconia and titania-based catalysts. After adsorption of pyridine on zirconia, four bands can be evidenced between 1,400 and 1,700 cm^{-1} with two major ones at 1,442 and 1,603 cm^{-1} attributed to ν_{19b} and ν_{8a} modes, respectively. The authors assigned it to pyridine adsorbed on Lewis acidic sites already observed in the literature [72, 73]. After increasing desorption temperature, the authors observed a decrease of these bands' intensities coupled to a shift toward higher wavenumbers. They attributed this behavior to lateral interactions between adsorbed species resulting from the decrease of electron enrichment of the surface with desorption of pyridine (acting as an electron donor) [37].

After pyridine adsorption over ZrO_2-CeO_2, two bands (ν_{8a}) at 1,602 and 1,596 cm^{-1}, which have been attributed to pyridine adsorbed on Zr^{4+} and Ce^{4+} centers respectively, appeared. Besides, the authors observed a decrease in the intensity of both bands after an increase in desorption temperature. They noticed that band at 1,596 cm^{-1} assigned to ν_{8a} Py-Ce^{4+} was particularly affected by this intensity decrease. This observation allowed to indicate that the acid strength of Ce^{4+} centers was much lower than the acid strength of Zr^{4+} centers. Interestingly, the authors reported, at 300 °C, the apparition of new bands at 1,459, 1,560, and 1,389 cm^{-1} resulting from a surface reaction corresponding probably to oxidation of pyridine by ceria (residual surface products corresponding to carbonate species), which has already been observed elsewhere [74].

Concerning the adsorption of pyridine on La_2O_3-ZrO_2, it leads to the formation of two major bands at 1,597 and 1,444 cm^{-1}, characteristic of pyridine adsorbed on Lewis acid sites. Upon increasing the temperature, these peaks decrease in intensity, letting appear a shoulder at 1,605 cm^{-1} characteristic of pyridine adsorbed on ZrO_2, underlining the heterogeneity of the surface sites and their strength, Zr^{4+} sites being stronger than La^{3+} sites [37]. WO_3–ZrO_2 and HPW/WO_3–ZrO_2 spectra were relatively close to each other, in addition to bands at 1,610 and 1,450 cm^{-1} characteristic of Lewis acidic sites, new peaks appear at 1,640 and 1,540 cm^{-1}. These new bands are attributed to the

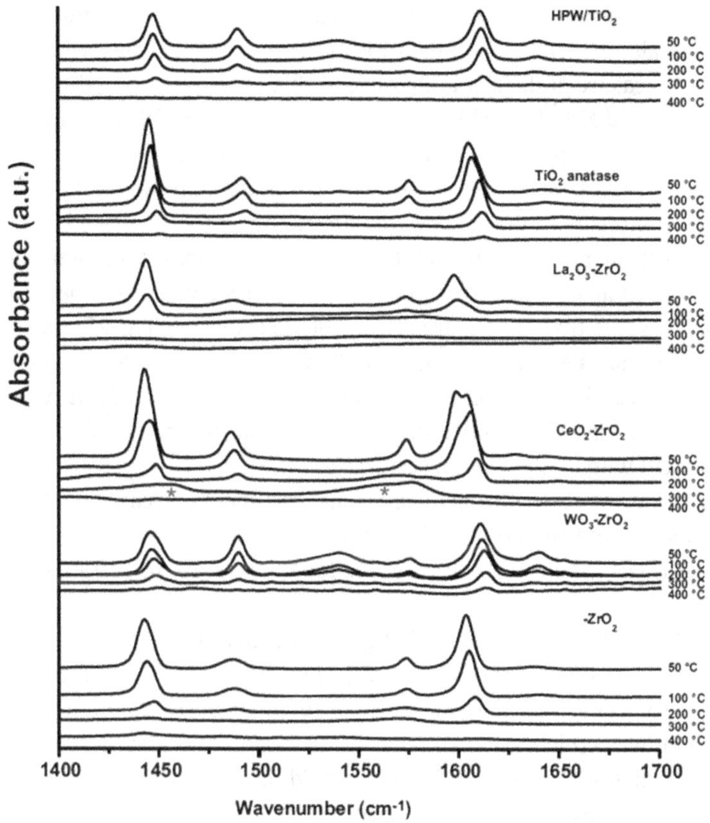

Fig. 4.17: FTIR spectra for pyridine desorption on zirconia- and titania-based materials at different temperatures (new bands that appear due to the surface reaction of pyridine on CeO_2–ZrO_2 sample are marked by *) (reprinted by permission from Elsevier from [37], copyright 2014).

presence of pyridinium species and thus of Brønsted acid sites [73]. By comparing the evolution of the peaks at 1,540 and 1,450 cm^{-1} upon increasing the temperature, it can be observed that pyridine desorbs preferentially from Brønsted acid sites [37]. This FTIR study shows the complementarity of the data given by spectroscopy and calorimetry.

A CO_2 adsorption study by FTIR was also performed over the different catalysts. On zirconia, after CO_2 adsorption, several bands appear at 3,610, 1,620, 1,600, 1,444, 1,442, 1,223, and 1,063 cm^{-1}. The authors attributed these results to the formation of hydrogen carbonate species [37].

Bands at 1,620 and 1,600 cm^{-1} were assigned to ν3′(C = O), those at 1,444 and 1,422 cm^{-1} to ν3′(C = O), bands at 1,063 and 1,223 cm^{-1} were assigned to ν(C-O) and δ (OH) respectively. Finally, the band at 3,610 cm^{-1} has been attributed to ν(OH) vibration of hydrogen carbonate species. The attribution to hydrogen carbonate species is in agreement with more recent work performed on zirconia [75].

Figure 4.18 plots the selectivity toward acrolein as a function of the total number of basic sites (left), and also as a function of the ratio of the number of strong acidic to strong basic sites (as a measure of the acidic character) which corresponds to the irreversibly amount of adsorbed NH_3 divided by the irreversibly amount of SO_2 (right)

From Fig. 4.18 left, it is obvious that the selectivity to acrolein decreases when the number of basic sites increases. Moreover, as seen in Fig. 4.18 right, the acrolein selectivity is enhanced by an increase in the acidic dominant character.

However, this improvement is limited, and the acrolein selectivity reaches a maximum when the ratio of strong acidic to strong basic sites reaches a value of 6; beyond that point, further increases in the acidic character do not affect the acrolein selectivity. Besides, the authors also indicated that the addition of La_2O_3 to zirconia was responsible for the decrease of glycerol conversion and increases the yield of by-products without enhancing acetol selectivity. The addition of CeO_2 to zirconia led also to a decrease of glycerol conversion but with an increase in acetol selectivity.

They explained this lower glycerol conversion compared to pure zirconia by their lower acidity. Moreover, from the results of FTIR measurements of adsorbed CO_2, they assumed that basic OH groups were responsible for the increase of acetol selectivity [37].

Previous studies have already suggested that glycerol dehydration to acrolein is controlled by strong Brønsted acidic sites [20, 76, 77]. However, glycerol can react in several manners over catalysts, especially those which simultaneously exhibit acidity and basicity, to form either acrolein on acidic sites or acetol on basic sites. Given this fact, it becomes evident that an increase in basicity would favor the reaction of glycerol over basic sites and thus lead to a decrease in selectivity to acrolein.

The obtained results allowed to suggest that the good catalytic performances observed over WO_3–ZrO_2, HPW/WO_3–ZrO_2, and HPW–TiO_2 were not only due to their high Brønsted acidity but also because of the hindering of the numbers/strength/action of the basic sites. Finally, thanks to these studies, it can be easily understood that to enhance performances of a targeted reaction it is important to control not only the amount of desired sites (acidic sites in this case) but also to try to manage and hinder as much as possible the number/strength/action of the undesired ones [37, 59].

4.5.2 Oxidative coupling of alcohols

Glycerol offers numerous advantages for acrolein production, but also drawbacks such as the quick deactivation of the catalysts, requiring frequent regeneration [62, 78]. Besides, due to multiple changes in the EU biofuels regulation, the market of biodiesel did not develop as expected and nowadays, the glycerol available volumes in Europe are not sufficient to consider a large-scale plant [79]. This has motivated further research into other alternative processes which could satisfy technical but also environmental and economic criteria. Among alternatives, "oxidative coupling of alcohols" has emerged as a

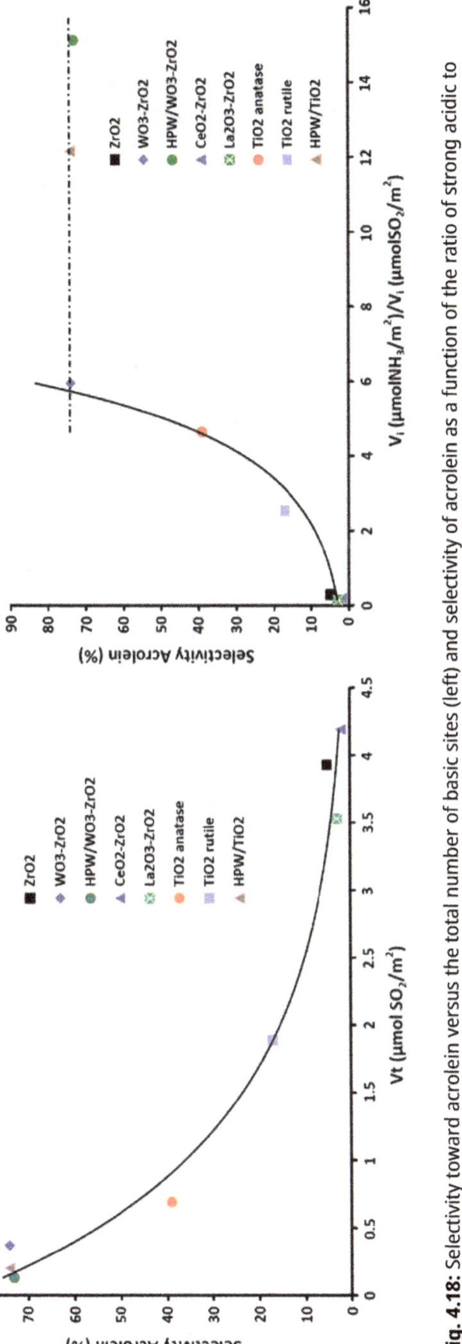

Fig. 4.18: Selectivity toward acrolein versus the total number of basic sites (left) and selectivity of acrolein as a function of the ratio of strong acidic to strong basic sites (right) (adapted with permission from Elsevier from [59], copyright 2012).

promising option. This process, patented by Dubois *et al.*, is performed in two steps [80]. First, a mixture of methanol and ethanol is oxidized to formaldehyde and acetaldehyde, respectively. Then, acrolein is synthesized in the second step by cross-aldolization of the aldehydes. Acrolein production through cross-aldolization of formaldehyde and acetaldehyde has been already widely studied under various conditions [81–89]. The first industrial production of acrolein by Degussa in 1942 used this method [19]. In the oxidative coupling process, the reaction could be performed in a single reactor starting directly from alcohols, thus reducing not only costs but also risks because the only reagents would be methanol and ethanol [25–29, 90, 91]. Aldolization reactions are well-known to proceed along both basic and acidic reaction pathways and thus to be driven by acid-base properties [81, 82, 84, 85, 87, 89, 92, 93]. Therefore, the acid-base properties of aldolization catalysts need to be controlled and, if needed, modulated, to improve the acrolein production. This part reports on obtained results concerning acrolein production by oxidative coupling of alcohols over numerous catalysts [25, 26] and particularly the correlation between the acid-base properties and the catalytic results for acrolein production. The catalysts used in these studies included acidic materials such as heteropolyacids, basic catalysts such as pure magnesia, but also amphoteric solids, for instance, alumina or mixed oxides deposited on silica. Supported mixed oxides on silica (Mg/SiO_2, K/SiO_2, Ca/SiO_2, Na/SiO_2) were prepared by wet impregnation using an aqueous solution of NaOH (VWR Chemicals), KOH (VWR Chemicals), $Ca(NO_3)_2 \cdot 4H_2O$ (Fluka), and $Mg(NO_3)_2 \cdot 6H_2O$ (Sigma–Aldrich) mixed with colloidal silica Ludox TMA (Sigma–Aldrich) and then evaporated for 3 h (50 °C, 1 mbar). The salt solutions were prepared to have a theoretical bulk metal/silicon molar ratio equal to 0.1.

The powders were calcined at 500 °C under inert atmosphere (N_2) for 4 h to give Na/SiO_2 (3.8 wt% Na_2O), K/SiO_2 (5.2 wt% K_2O), Ca/SiO_2 (5.6 wt% CaO), and Mg/SiO_2 (3.7 wt% MgO). For the sake of readability, the samples were then labeled Na/Si, K/Si, Mg/Si, and SiO_2 (which represent the support)

The acid-base properties of the various catalysts were determined by adsorption microcalorimetry in the gas phase at 150 °C, following the same typical procedure described previously. Figure 4.19 represents the differential heats of adsorption of ammonia and SO_2 versus coverage for several amphoteric and basic samples.

By looking at the microcalorimetry results in Fig. 4.18, it can be seen that magnesia exhibits a large number of basic sites with high differential heats of adsorption ($Q > 150$ kJ mol^{-1}) which characterize the presence of strong basic sites. Besides, MgO displays a very weak acidity confirming the strong basic dominant character of magnesia as already mentioned by previous studies [94, 95]. All the other magnesium-containing mixed oxides considered in this study display a clear amphoteric character, exhibiting both heterogeneous acidity and basicity, with the presence of strong sites as evidenced by the high initial heats of adsorption (varying from 125 to 240 kJ mol^{-1}). Silica-based catalysts also show an amphoteric character exhibiting both heterogeneous acidity and basicity, except for pure SiO_2 which displays only weak acidity. As expected, the authors observed that deposition of basic oxides on the silica surface enhances the surface

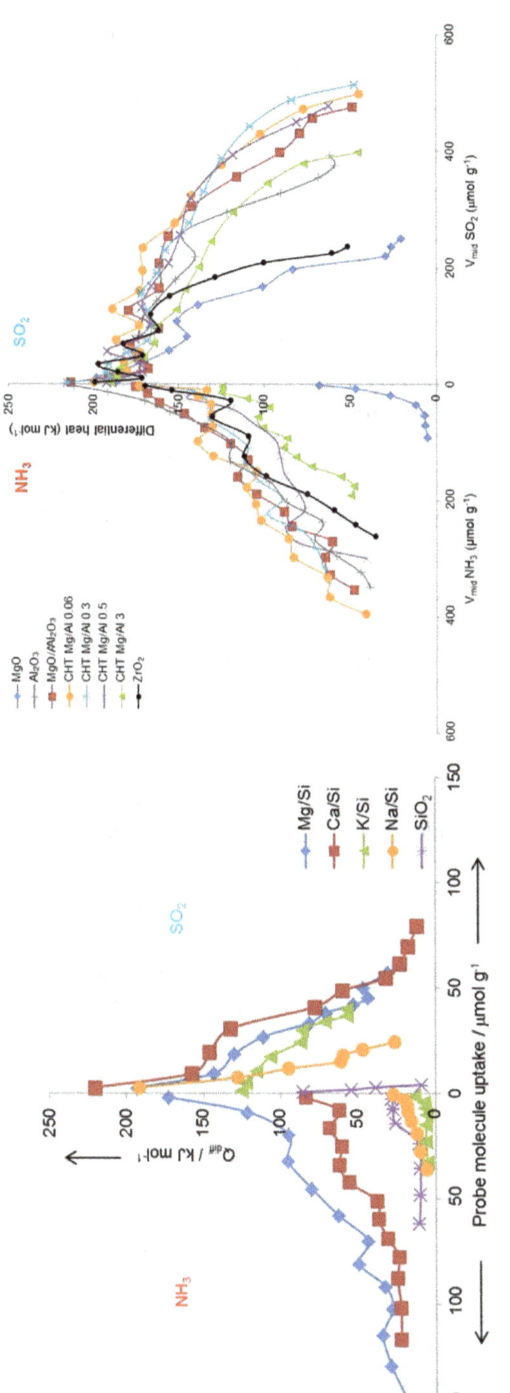

Fig. 4.19: Differential heats of adsorption of NH_3 and SO_2 as a function of surface coverage for various amphoteric and basic samples (adapted with permission from John Wiley and Sons from [25, 26], copyright 2017).

basicity. The highest amount of basic sites was displayed by CaO/SiO_2 followed by MgO/SiO_2, KO/SiO_2 while Na_2O/SiO_2 showed the lowest amount of basic sites. The authors also noticed that, for elements in the same group of the periodic table, basicity was increasing with decreasing electronegativity.

They also observed that the higher the electronegativity of the added element, the higher the initial heats of adsorption of NH_3. Moreover, the authors noticed that the addition of K_2O or Na_2O to SiO_2 decreases the acidity compared to bare SiO_2, whereas adding MgO or CaO increases the acidity, in agreement with a previous study performed by Gervasini *et al.* [96].

To study the influence of their acidic and basic properties on acrolein production, catalytic tests were performed over two sequential continuous-flow reactors working close to the atmospheric pressure and independently heated.

Methanol, ethanol, O_2, and N_2 ($MeOH/EtOH/O_2/N_2$ molar ratio = 4:2:8:86) were first injected in the first reactor (R1) filled with iron molybdate catalyst (FeMoOx) at 260 °C with a GHSV of 5,000 or 10,000 h^{-1}. The second reactor (R2) was filled with mixed oxide to perform the aldolization reaction. The temperature in R2 was varying between 250 and 340 °C with GHSV of 5,000 and 10,000 h^{-1}[25]. It is worth noting that acrolein is already produced during the first step of oxidation of alcohols to aldehydes over FeMoOx catalyst, as also reported by Borowiec *et al.* [90, 91]. Among all the studied catalysts, the best result was displayed by MgO/SiO_2, with a 35% yield of acrolein and 10% of COx at 320 °C with a GHSV of 5,000 h^{-1} [25]. MgO and Na_2O/SiO_2 also revealed themselves to be relatively good catalysts, displaying 28% and 25% acrolein yield, respectively, under their optimal conditions. The other catalysts showed less reactivity [25, 26]. These results make it possible to establish the desired acidity/basicity profile of a good catalyst for this reaction. Plotting the acrolein yield as a function of the ratio of strong basic to strong acidic sites, representing the basic character of the catalyst (Fig. 4.20) makes it apparent that the more pronounced the basic character of the catalyst, the lower the acrolein yield. This underlines the fact that excessive basicity is not favorable to acrolein production. A recent study performed over magnesium aluminate spinel catalysts confirmed the detrimental effect of a too high strong basicity [27]. In addition to that, it is interesting to underline that thanks to the use of adsorption microcalorimetry, this study, performed over spinel catalyst, permitted to show the preferential adsorption of formaldehyde compared to acetaldehyde thus explaining the absence of crotonaldehyde due to the isolation of acetaldehyde at the surface of the catalyst [27]. In another recent study performed over perovskites, the catalytic activities for the aldolization reaction were promoted by coordination between both basic and acidic sites confirming the importance to get a good balance of acidic and basic properties for the production of acrolein [97].

Lilić et al. also noted that some catalysts such as some hydrotalcites (CHT) produced high amounts of CO_x, contrary to the purely acidic catalysts which exhibited very little CO_x production. They connected the increase of COx production over CHT catalysts with the increase of the amount of strong basicity ($Q > 150$ kJ mol^{-1}). To

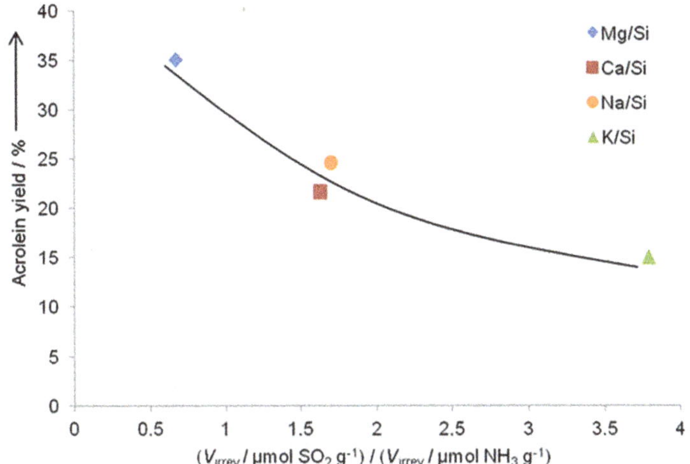

Fig. 4.20: Acrolein yield versus the ratio of strong basic to strong acidic sites (reprinted by permission from John Wiley and Sons from [25], copyright 2017).

evidence this fact, the authors proposed Fig. 4.21 which shows the correlation between the acid-base properties of catalysts as determined by adsorption calorimetry and CO_x yield. It can be seen that a larger number of basic sites, represented by blue histograms, leads to an increase in carbon oxide production. On the other hand, the data suggest that acidity does not have a direct influence on carbon oxide production.

In conclusion, the characterization of aldolization catalysts by adsorption microcalorimetry provided information that made it possible to correlate the catalytic activity of these solids to their acid-base properties. The results indicate that aldol condensation of formaldehyde and acetaldehyde is driven by both acidic and basic sites acting together cooperatively. However, the number of strong basic sites needs to be kept below a certain threshold due to their detrimental effect on acrolein yield and carbon oxides production. The optimal amount of strong basic sites appears to be approximately similar to the number of strong acidic sites; the amount of the latter does not seem to have any influence on acrolein yield nor carbon oxides production. Nonetheless, those acid-base properties are not the only parameter to control acrolein production. Indeed, in a more recent study performed over magnesium aluminate spinels where some transition metals partly or totally substitute magnesium, the acrolein production was correlated to the ionic radius of transition metals leading to the conclusion that not only the acid-base properties but also other parameters such as redox or electronic properties can influence the acrolein production [29].

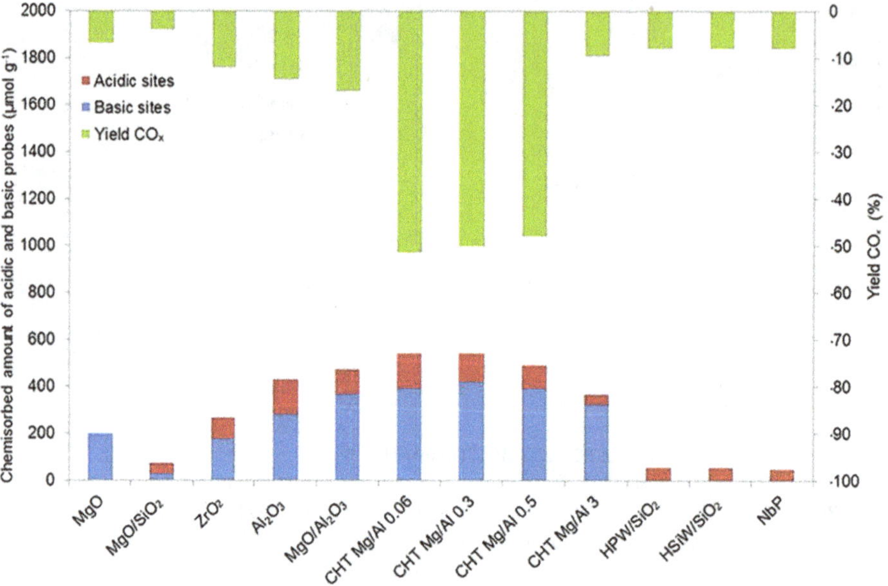

Fig. 4.21: Correlation between acid-base properties of catalysts as determined by adsorption calorimetry and CO_x yield (300 °C, GHSV=10000 h^{-1}) (reprinted by permission from John Wiley and Sons from [26], copyright 2017).

4.6 Conclusions

Catalytic biomass conversion reactions are one of the hottest topics in modern environmental catalysis. Many of these reactions are driven by the acid-base properties of the catalysts, which makes it necessary to characterize their acidity and basicity to enhance their selectivity and also to understand the reaction mechanisms involved. This chapter presented adsorption microcalorimetry of probe molecules in the gas or the liquid phase and its application to various biomass conversion case studies. This technique is a very powerful tool to characterize precisely the number of active sites on a catalytic surface, as well as the strength and the strength distribution of the titrated surface sites, particularly for acidity and basicity. The data obtained by this technique are of great help to understand the surface properties of catalytic solids in various media (liquid or gas) and explain their catalytic behavior. Moreover, when adsorption calorimetry is coupled to other techniques such as FTIR spectroscopy, supplementary information can be provided on the nature of the active sites (e.g., Brønsted or Lewis sites), making it possible to further investigate the correlations between the nature, strength, and number of the sites and the catalytic activity. Due to the valuable information it provides, adsorption microcalorimetry ranks among the indispensable tools for the study of catalytic surfaces.

References

[1] Liu, B.; Zhang, Z. Catalytic conversion of biomass into chemicals and fuels over magnetic catalysts. *ACS Catal.* **2016**, *6*, 326–338, DOI:10.1021/acscatal.5b02094.

[2] Werpy, T.; Petersen, G. *Top Value Added Chemicals from Biomass: Volume I – Results of Screening for Potential Candidates from Sugars and Synthesis Gas*; United States Department of Energy, 2004, doi.org/10.2172/15008859.

[3] Román-Leshkov, Y.; Barrett, C. J.; Liu, Z. Y. ; Dumesic, J. A. Production of dimethylfuran for liquid fuels from biomass-derived carbohydrates. *Nature* **2007**, *447*, 982–985, DOI:10.1038/nature05923.

[4] Zhang, Z.; Deng, K.; Recent advances in the catalytic synthesis of 2,5-FurandicarboxylicAcid and its derivatives. *ACS Catal.* **2015**, *5*, 6529–6544, DOI:10.1021/acscatal.5b01491.

[5] Dutta, S.; De, S.; Saha, B. Advances in biomass transformation to 5-hydroxymethylfurfural and mechanistic aspects. *Biomass Bioenergy* **2013**, *55*, 355–369, DOI:10.1016/j.biombioe.2013.02.008.

[6] Kuster, B. F. M. 5-hydroxymethylfurfural (HMF). A review focussing on its manufacture. *Starch/Stärke* **1990**, *42*, 314–321, DOI:10.1002/star.19900420808.

[7] Roman-Leshkov, Y. Phase modifiers promote efficient production of hydroxymethylfurfural from fructose. *Science* **2006**, *312*, 1933–1937, DOI:10.1126/science.1126337.

[8] Carniti, P.; Gervasini, A.; Biella, S.; Auroux, A. Niobic acid and niobium phosphate as highly acidic viable catalysts in aqueous medium: Fructose dehydration reaction. *Catal. Today* **2006**, *118*, 373–378, DOI:10.1016/j.cattod.2006.07.024.

[9] Benvenuti, F.; Carlini, C.; Patrono, P.; Raspolli Galletti, A. M.; Sbrana, G.; Massucci, M. A.; Galli, P. Heterogeneous zirconium and titanium catalysts for the selective synthesis of 5-Hydroxymethyl-2-Furaldehyde from carbohydrates. *Appl. Catal. A: Gen.* **2000**, *193*, 147–153, DOI:10.1016/S0926-860X(99)00424-X.

[10] Rac, V.; Rakić, V.; Stošić, D.; Otman, O.; Auroux, A. Hierarchical ZSM-5, beta and USY zeolites: Acidity assessment by gas and aqueous phase calorimetry and catalytic activity in fructose dehydration reaction. *Microporous Mesoporous Mater* **2014**, *194*, 126–134, DOI:10.1016/j.micromeso.2014.04.003.

[11] Sonsiam, C.; Kaewchada, A.; Pumrod, S.; Jaree, A. Synthesis of 5-hydroxymethylfurfural (5-HMF) from fructose over cation exchange resin in a continuous flow reactor. *Chem. Eng. Process. Process Intensif.* **2019**, *138*, 65–72, DOI:10.1016/j.cep.2019.03.001.

[12] Qu, Y.; Li, L.; Wei, Q.; Huang, C.; Oleskowicz-Popiel, P.; Xu, J. One-pot conversion of disaccharide into 5-hydroxymethylfurfural catalyzed by imidazole ionic liquid. *Sci. Rep.* **2016**, *6*, 26067, DOI:10.1038/srep26067.

[13] Carniti, P.; Gervasini, A.; Bossola, F.; Dal Santo, V. Cooperative action of Brønsted and lewis acid sites of niobium phosphate catalysts for cellobiose conversion in water. *Appl. Catal. B Environ.* **2016**, *193*, 93–102, DOI:10.1016/j.apcatb.2016.04.012.

[14] Marzo, M.; Gervasini, A.; Carniti, P. Hydrolysis of disaccharides over solid acid catalysts under green conditions. *Carbohydr. Res.* **2012**, *347*, 23–31, DOI:10.1016/j.carres.2011.10.018.

[15] Kourieh, R.; Bennici, S.; Marzo, M.; Gervasini, A.; Auroux, A. Investigation of the WO_3/ZrO_2 surface acidic properties for the aqueous hydrolysis of cellobiose. *Catal. Commun.* **2012**, *19*, 119–126, doi:10.1016/j.catcom.2011.12.030.

[16] Deng, W.; Lobo, R.; Setthapun, W.; Christensen, S. T.; Elam, J. W.; Marshall, C. L. Oxidative hydrolysis of cellobiose to glucose. *Catal. Lett.* **2011**, *141*, 498–506, DOI:10.1007/s10562-010-0532-8.

[17] Vilcocq, L.; Rebmann, É.; Cheah, Y. W.; Fongarland, P. Hydrolysis of cellobiose and xylan over TiO_2 -based catalysts. *ACS Sustainable Chem. Eng.* **2018**, *6*, 5555–5565, DOI:10.1021/acssuschemeng.8b00486.

[18] Liu, L.; Ye, X. P.; Bozell, J. J. A Comparative Review of petroleum-based and bio-based acrolein production. *Chem. Sus. Chem.* **2012**, *5*, 1162–1180, DOI:10.1002/cssc.201100447.

[19] Arntz, D.; Fischer, A.; Höpp, M.; Jacobi, S.; Sauer, J.; Ohara, T.; Sato, T.; Shimizu, N.; Schwind, H. *Acrolein and Methacrolein. In Ullmann's Encyclopedia of Industrial Chemistry*; Wiley-VCH Verlag GmbH & Co. KGaA, Ed.; Wiley-VCH Verlag GmbH & Co. KGaA: Weinheim, Germany, 2007, DOI:10.1002/14356007.a01_149.pub2.
[20] Katryniok, B.; Paul, S.; Bellière-Baca, V.; Rey, P.; Dumeignil, F. Glycerol dehydration to acrolein in the context of new uses of glycerol. *Green Chem.* **2010**, *12*, 2079, DOI:10.1039/c0gc00307g.
[21] Cespi, D.; Passarini, F.; Mastragostino, G.; Vassura, I.; Larocca, S.; Iaconi, A.; Chieregato, A.; Dubois, J.-L.; Cavani, F. Glycerol as feedstock in the synthesis of chemicals: A life cycle analysis for acrolein production. *Green Chem.* **2015**, *17*, 343–355, DOI:10.1039/C4GC01497A.
[22] Neher, A.; Haas, T.; Arntz, D.; Klenk, H.; Girke, W. Process for the Production of Acrolein, US5387720A, **1995**.
[23] Cavani, F.; Guidetti, S.; Marinelli, L.; Piccinini, M.; Ghedini, E.; Signoretto, M. The control of selectivity in gas-phase glycerol dehydration to acrolein catalysed by sulfated zirconia. *Appl. Catal. B Environ.* **2010**, *100*, 197–204, DOI:10.1016/j.apcatb.2010.07.031.
[24] Xie, Q.; Li, S.; Gong, R.; Zheng, G.; Wang, Y.; Xu, P.; Duan, Y.; Yu, S.; Lu, M.; Ji, W.; et al. Microwave-assisted catalytic dehydration of glycerol for sustainable production of acrolein over a microwave absorbing catalyst. *Appl. Catal. B Environ.* **2019**, *243*, 455–462, DOI:10.1016/j.apcatb.2018.10.058.
[25] Lilić, A.; Bennici, S.; Devaux, J. -F.; Dubois, J. -L.; Auroux, A. Influence of catalyst Acid/Base properties in acrolein production by oxidative coupling of ethanol and methanol. *Chem. Sus. Chem.* **2017**, *10*, 1916–1930, DOI:10.1002/cssc.201700230.
[26] Lilić, A.; Wei, T.; Bennici, S.; Devaux, J. -F.; Dubois, J. -L.; Auroux, A. A comparative study of basic, amphoteric, and acidic catalysts in the oxidative coupling of methanol and ethanol for acrolein production. *Chem. Sus. Chem.* **2017**, *10*, 3459–3472, DOI:10.1002/cssc.201701040.
[27] Folliard, V.; Postole, G.; Devaux, J.-F.; Dubois, J.-L.; Marra, L.; Auroux, A. Oxidative coupling of a mixture of bio-alcohols to produce a more sustainable acrolein: An in depth look in the mechanism implying aldehydes co-adsorption and acid/base sites. *Appl. Catal. B Environ.* **2020**, *268*, 118421, DOI:10.1016/j.apcatb.2019.118421.
[28] Folliard, V.; Postole, G.; Marra, L.; Dubois, J.-L.; Auroux, A. Synthesis of acrolein by oxidative coupling of alcohols over spinel catalysts: Microcalorimetric and spectroscopic approaches. *Catal. Sci. Technol.* **2020**, *10*, 1889–1901, DOI:10.1039/D0CY00094A.
[29] Folliard, V.; Postole, G.; Marra, L.; Dubois, J.-L.; Auroux, A. Sustainable acrolein production from bio-alcohols on spinel catalysts: influence of magnesium substitution by various transition metals (Fe, Zn, Co, Cu, Mn). *Appl. Catal. A: Gen.* **2020**, *608*, 117871, DOI:10.1016/j.apcata.2020.117871.
[30] Carniti, P.; Gervasini, A.; Biella, S.; Auroux, A. Intrinsic and effective acidity study of niobic acid and niobium phosphate by a multitechnique approach. *Chem. Mater.* **2005**, *17*, 6128–6136, DOI:10.1021/cm0512070.
[31] Stošić, D.; Auroux, A. Couplings. In *Calorimetry and Thermal Methods in Catalysis*; Springer-Verlag Berlin Heidelberg, 2013, DOI:10.1007/978-3-642-11954-5_3.
[32] Mekki-Berrada, A.; Auroux, A. Thermal Methods. In *Characterization of Solid Materials and Heterogeneous Catalysts*; Che, M., Védrine, J.C., Eds.; Wiley-VCH Verlag GmbH & Co. KGaA: Weinheim, Germany, 2012, DOI:10.1002/9783527645329.ch18.
[33] Auroux, A. Acidity characterization by microcalorimetry and relationship with reactivity. *Topics Catal.* **1997**, *4*, 71–89, DOI:10.1023/A:1019127919907.
[34] van Putten, R.-J.; van der Waal, J. C.; de Jong, E.; Rasrendra, C. B.; Heeres, H. J.; de Vries, J. G. Hydroxymethylfurfural, A versatile platform chemical made from renewable resources. *Chem. Rev.* **2013**, *113*, 1499–1597, DOI:10.1021/cr300182k.
[35] Kourieh, R.; Rakic, V.; Bennici, S.; Auroux, A. Relation between surface acidity and reactivity in fructose conversion into 5-HMF using tungstated zirconia catalysts. *Catal. Commun.* **2013**, *30*, 5–13, DOI:10.1016/j.catcom.2012.10.005.

[36] Stošić, D.; Bennici, S.; Rakić, V.; Auroux, A. CeO_2–Nb_2O_5 mixed oxide catalysts: preparation, characterization and catalytic activity in fructose dehydration reaction. *Catal. Today* **2012**, *192*, 160–168, DOI:10.1016/j.cattod.2011.10.040.
[37] Stošić, D.; Bennici, S.; Sirotin, S.; Stelmachowski, P.; Couturier, J.-L.; Dubois, J.-L.; Travert, A.; Auroux, A. Examination of acid–base properties of solid catalysts for gas phase dehydration of glycerol: FTIR and adsorption microcalorimetry studies. *Catal. Today* **2014**, *226*, 167–175, DOI:10.1016/j.cattod.2013.11.047.
[38] Qi, X.; Watanabe, M.; Aida, T. M.; Smith, R. L. Sulfated zirconia as a solid acid catalyst for the dehydration of fructose to 5-hydroxymethylfurfural. *Catal. Commun.* **2009**, *10*, 1771–1775, DOI:10.1016/j.catcom.2009.05.029.
[39] Vilcocq, L.; Castilho, P. C.; Carvalheiro, F.; Duarte, L. C. Hydrolysis of oligosaccharides over solid acid catalysts: A review. *Chem. Sus. Chem.* **2014**, *7*, 1010–1019, DOI:10.1002/cssc.201300720.
[40] Shimizu, K.; Furukawa, H.; Kobayashi, N.; Itaya, Y.; Satsuma, A. Effects of Brønsted and Lewis acidities on activity and selectivity of heteropolyacid-based catalysts for hydrolysis of cellobiose and cellulose. *Green Chem.* **2009**, *11*, 1627, DOI:10.1039/b913737h.
[41] Zhou, W.; Ross-Medgaarden, E. I.; Knowles, W. V.; Wong, M. S.; Wachs, I. E.; Kiely, C. J. Identification of active Zr–WOx clusters on a ZrO_2 support for solid acid catalysts. *Nat. Chem.* **2009**, *1*, 722–728, DOI:10.1038/nchem.433.
[42] OECD-FAO Agricultural Outlook 2021-2030. *OECD-FAO* 2021, DOI:10.1787/19428846-en.
[43] Aditiya, H. B.; Mahlia, T. M. I.; Chong, W. T.; Nur, H.; Sebayang, A. H. Second generation bioethanol production: A critical review. *Renew. Sustainable Energy Rev.* **2016**, *66*, 631–653, DOI:10.1016/j.rser.2016.07.015.
[44] Stošić, D.; Hosoglu, F.; Bennici, S.; Travert, A.; Capron, M.; Dumeignil, F.; Couturier, J.-L.; Dubois, J.-L.; Auroux, A. Methanol and ethanol reactivity in the presence of hydrotalcites with Mg/Al ratios varying from 2 to 7. *Catal. Commun.* **2017**, *89*, 14–18, DOI:10.1016/j.catcom.2016.10.013.
[45] Riittonen, T.; Toukoniitty, E.; Madnani, D. K.; Leino, A.-R.; Kordas, K.; Szabo, M.; Sapi, A.; Arve, K.; Wärnå, J.; Mikkola, J.-P. One-pot liquid-phase catalytic conversion of ethanol to 1-butanol over aluminium oxide – the effect of the active metal on the selectivity. *Catalysts* **2012**, *2*, 68–84, DOI:10.3390/catal2010068.
[46] Billig, E. Butyl Alcohols. In *Kirk-Othmer Encyclopedia of Chemical Technology*; Wiley-VCH Verlag GmbH, 2001. DOI:10.1002/0471238961.0221202502091212.a01.pub2.
[47] Tanaka, Y.; Utsunomiya, M. Process of Producing Alcohol, US8318990B2, **2012**.
[48] Aitchison, H.; Wingad, R. L.; Wass, D. F. Homogeneous ethanol to butanol catalysis – guerbet renewed. *ACS Catal.* **2016**, *6*, 7125–7132, DOI:10.1021/acscatal.6b01883.
[49] Hanspal, S.; Young, Z. D.; Shou, H.; Davis, R. J., Multiproduct steady-state isotopic transient kinetic analysis of the ethanol coupling reaction over hydroxyapatite and magnesia. *ACS Catal.* **2015**, 1737–1746, DOI:10.1021/cs502023g.
[50] Silvester, L.; Lamonier, J.-F.; Lamonier, C.; Capron, M.; Vannier, R.-N.; Mamede, A.-S.; Dumeignil, F. Guerbet reaction over strontium-substituted hydroxyapatite catalysts prepared at various (Ca+Sr)/P ratios. *Chem. Cat. Chem* **2017**, *9*, 2250–2261, DOI:10.1002/cctc.201601480.
[51] Gotoh, K.; Nakamura, S.; Mori, T.; Morikawa, Y. Supported alkali salt catalysts active for the guerbet reaction between methanol and ethanol. *Stud. Surf. Sci. Catal.* **2000**, *130*, 2669–2674, DOI:10.1016/S0167-2991(00)80873-3.
[52] León, M.; Díaz, E.; Vega, A.; Ordóñez, S.; Auroux, A. Consequences of the iron–aluminium exchange on the performance of hydrotalcite-derived mixed oxides for ethanol condensation. *Appl. Catal. B Environ.* **2011**, *102*, 590–599, DOI:10.1016/j.apcatb.2010.12.044.
[53] Chieregato, A.; Velasquez Ochoa, J.; Bandinelli, C.; Fornasari, G.; Cavani, F.; Mella, M. On the chemistry of ethanol on basic oxides: Revising mechanisms and intermediates in the lebedev and guerbet reactions. *Chem. Sus. Chem.* **2015**, *8*, 377–388, DOI:10.1002/cssc.201402632.

[54] Hanspal, S.; Young, Z. D.; Prillaman, J. T.; Davis, R. J. Influence of surface acid and base sites on the guerbet coupling of ethanol to butanol over metal phosphate catalysts. *J. Catal.* **2017**, *352*, 182–190, DOI:10.1016/j.jcat.2017.04.036.
[55] Di Cosimo J. I.; Díez V. K.; Xu M.; Iglesia E.; Apestuguía C. R. Structure and surface and catalytic properties of Mg-Al basic oxides. *J. Catal.* **1998**, *178*, 499–510,.
[56] Pacchioni G., Clotet A., Josep M. R. A theoretical study of the adsorption and reaction of SO, at surface and step sites of the MgO(100) surface, *Surf. Sci.* **1994**, *315*, 337–350, doi.org/10.1016/0039-6028(94)90137-6.
[57] Corma A., Palomares A. E., Rey F. Optimization of SOx additives of FCC catalysts based on MgO-Al_2O_3 mixed oxides produced from hydrotalcites, *Appl. Catal. B: Environ.* **1994**, *4* 29–43. DOI:10.1016/0926-3373(94)00007-7.
[58] ARIA (Analysis, Research and Information on Accidents) Database Available online: https://www.aria.developpement-durable.gouv.fr/?s=acroleine (accessed on 2 April 2021).
[59] Stošić, D.; Bennici, S.; Couturier, J.-L.; Dubois, J.-L.; Auroux, A. Influence of surface acid–base properties of zirconia and titania based catalysts on the product selectivity in gas phase dehydration of glycerol. *Catal. Commun.* **2012**, *17*, 23–28, DOI:10.1016/j.catcom.2011.10.004.
[60] Ren, X.; Zhang, F.; Sudhakar, M.; Wang, N.; Dai, J.; Liu, L. Gas-phase dehydration of glycerol to acrolein catalyzed by hybrid acid sites derived from transition metal hydrogen phosphate and meso-HZSM-5. *Catal. Today* **2019**, *332*, 20–27, DOI:10.1016/j.cattod.2018.08.012.
[61] Chai, S.-H.; Wang, H.-P.; Liang, Y.; Xu, B.-Q. Sustainable production of acrolein: investigation of solid acid–base catalysts for gas-phase dehydration of glycerol. *Green Chem.* **2007**, *9*, 1130–1136, DOI:10.1039/B702200J.
[62] Jiang, X. C.; Zhou, C. H.; Tesser, R.; Di Serio, M.; Tong, D. S.; Zhang, J. R. Coking of catalysts in catalytic glycerol dehydration to acrolein. *Ind. Eng. Chem. Res.* **2018**, *57*, 10736–10753, DOI:10.1021/acs.iecr.8b01776.
[63] Tsukuda, E.; Sato, S.; Takahashi, R.; Sodesawa, T. Production of acrolein from glycerol over silica-supported heteropoly Acids. *Catal. Commun.* **2007**, *8*, 1349–1353, DOI:10.1016/j.catcom.2006.12.006.
[64] Huang, L.; Qin, F.; Huang, Z.; Zhuang, Y.; Ma, J.; Xu, H.; Shen, W. Hierarchical ZSM-5 zeolite synthesized by an ultrasound-assisted method as a long-life catalyst for dehydration of glycerol to acrolein. *Ind. Eng. Chem. Res*. **2016**, *55*, 7318–7327, DOI:10.1021/acs.iecr.6b01140.
[65] Shan, J.; Li, Z.; Zhu, S.; Liu, H.; Li, J.; Wang, J.; Fan, W. Nanosheet MFI zeolites for gas phase glycerol dehydration to acrolein. *Catalysts* **2019**, *9*, 121, DOI:10.3390/catal9020121.
[66] Lauriol-Garbey, P.; Postole, G.; Loridant, S.; Auroux, A.; Belliere-Baca, V.; Rey, P.; Millet, J. M. M. Acid–base properties of niobium-zirconium mixed oxide catalysts for glycerol dehydration by calorimetric and catalytic investigation. *Appl. Catal. B Environ.* **2011**, 94–112 DOI:10.1016/j.apcatb.2011.05.011.
[67] Kourieh, R.; Bennici, S.; Auroux, A. Study of acidic commercial WOx /ZrO_2 catalysts by adsorption microcalorimetry and thermal analysis techniques. *J. Therm. Anal. Calorim.* **2010**, *99*, 849–853, DOI:10.1007/s10973-009-0407-7.
[68] Ulgen, A.; Hoelderich, W. Conversion of glycerol to acrolein in the presence of WO3/ZrO2 catalysts. *Catal. Lett.* **2009**, *131*, 122–128, DOI:10.1007/s10562-009-9923-0.
[69] Ramis, G.; Busca, G.; Cristiani, C.; Lietti, L.; Forzatti, P.; Bregani, F. Characterization of tungsta-titania catalysts. *Langmuir* **1992**, *8*, 1744–1749, DOI:10.1021/la00043a010.
[70] Zhang, R.; Jagiello, J.; Hu, J. F.; Huang, Z.-Q.; Schwarz, J. A.; Datye, A. Effect of WO3 loading on the surface acidity of WO_3/Al_2O_3 composite oxides. *Appl. Catal. A: Gen.* **1992**, *84*, 123–139, DOI:10.1016/0926-860X(92)80111-O.
[71] Lavalley, J. C. Infrared spectrometric studies of the surface basicity of metal oxides and zeolites using adsorbed probe molecules. *Catal. Today* **1996**, *27*, 377–401, DOI:10.1016/0920-5861(95)00161-1.
[72] Parry, E. An infrared study of pyridine adsorbed on acidic solids. Characterization of surface acidity. *J. Catal.* **1963**, *2*, 371–379, DOI:10.1016/0021-9517(63)90102-7.

[73] Travert, A.; Vimont, A.; Sahibed-Dine, A.; Daturi, M.; Lavalley, J.-C. Use of pyridine CH(D) vibrations for the study of lewis acidity of metal oxides. *Appl. Catal. A: Gen.* **2006**, *307*, 98–107, DOI:10.1016/j.apcata.2006.03.011.

[74] Zaki, M. I.; Hasan, M. A.; Al-Sagheer, F. A.; Pasupulety, L. In situ FTIR spectra of pyridine adsorbed on SiO_2–Al_2O_3, TiO_2, ZrO_2 and CeO_2: general considerations for the identification of acid sites on surfaces of finely divided metal oxides. *Colloids Surf. A Physicochem. Eng. Asp.* **2001**, *190*, 261–274, DOI:10.1016/S0927-7757(01)00690-2.

[75] Aboulayt, A.; Onfroy, T.; Travert, A.; Clet, G.; Maugé, F. Relationship between phosphate structure and acid-base properties of phosphate-modified zirconia – application to alcohol dehydration. *Appl. Catal. A: Gen.* **2017**, *530*, 193–202, DOI:10.1016/j.apcata.2016.10.030.

[76] Katryniok, B.; Paul, S.; Dumeignil, F. Recent developments in the field of catalytic dehydration of glycerol to acrolein. *ACS Catal.* **2013**, *3*, 1819–1834, DOI:10.1021/cs400354p.

[77] Alhanash, A.; Kozhevnikova, E. F.; Kozhevnikov, I. V. Gas-phase dehydration of glycerol to acrolein catalysed by caesium heteropoly salt. *Appl. Catal. A: Gen.* **2010**, *378*, 11–18, DOI:10.1016/j.apcata.2010.01.043.

[78] Katryniok, B.; Paul, S.; Capron, M.; Dumeignil, F. Towards the sustainable production of acrolein by glycerol dehydration. *Chem. Sus. Chem.* **2009**, *2*, 719–730, DOI:10.1002/cssc.200900134.

[79] Folliard, V.; de Tommaso, J.; Dubois, J.-L. Review on alternative route to acrolein through oxidative coupling of alcohols. *Catalysts* **2021**, *11*, 229, DOI:10.3390/catal11020229.

[80] Dubois, J. L.; Capron, M.; Dumeignil, F. Method for directly synthesizing unsaturated aldehydes from alcohol mixtures, US9365478B2, **2012**.

[81] Dumitriu, E.; Hulea, V.; Bilba, N.; Carja, G.; Azzouz, A. Synthesis of acrolein by vapor phase condensation of formaldehyde and acetaldehyde over oxides loaded zeolites. *J. Mol. Catal.* **1993**, *79*, 175–185, DOI:10.1016/0304-5102(93)85100-8.

[82] Dumitriu, E.; Hulea, V.; Chelaru, C.; Catrinescu, C.; Tichit, D.; Durand, R. Influence of the acid–base properties of solid catalysts derived from hydrotalcite-like compounds on the condensation of formaldehyde and acetaldehyde. *Appl. Catal. A: Gen.* **1999**, *178*, 145–157, DOI:10.1016/S0926-860X(98)00282-8.

[83] Dumitriu, E.; Hulea, V.; Fechete, I.; Auroux, A.; Lacaze, J.-F.; Guimon, C. The aldol condensation of lower aldehydes over MFI zeolites with different acidic properties. *Microporous Mesoporous Mater* **2001**, *43*, 341–359, DOI:10.1016/S1387-1811(01)00265-7.

[84] Ai, M. Formation of acrylaldehyde by vapor-phase aldol condensation 1. Basic oxide catalysts. *Bull. Chem. Soc. Jpn.* **1991**, *64*, 1341–1345, DOI:10.1246/bcsj.64.1342.

[85] Ai, M. Formation of acryladehyde by vapor-phase aldol condensation II. Phosphate catalysts. *Bull. Chem. Soc. Jpn.* **1991**, *64*, 1346–1350 DOI:10.1246/bcsj.64.1346.

[86] Ungureanu, A.; Royer, S.; Hoang, T. V.; Trong On, D.; Dumitriu, E.; Kaliaguine, S. Aldol condensation of aldehydes over semicrystalline zeolitic-mesoporous UL-ZSM-5. *Microporous Mesoporous Mater* **2005**, *84*, 283–296, DOI:10.1016/j.micromeso.2005.05.038.

[87] Cobzaru, C.; Oprea, S.; Dumitriu, E.; Hulea, V. Gas phase aldol condensation of lower aldehydes over clinoptilolite rich natural zeolites. *Appl. Catal. A: Gen.* **2008**, *351*, 253–258, DOI:10.1016/j.apcata.2008.09.024.

[88] Azzouz, A.; Messad, D.; Nistor, D.; Catrinescu, C.; Zvolinschi, A.; Asaftei, S. Vapor phase aldol condensation over fully ion-exchanged montmorillonite-rich catalysts. *Appl. Catal. A: Gen.* **2003**, *241*, 1–13, DOI:10.1016/S0926-860X(02)00524-0.

[89] Dumitriu, E.; Bilba, N.; Lupascu, M.; Azzouz, A.; Hulea, V.; Cirje, G.; Nibou, D. Vapor-phase Condensation of formaldehyde and acetaldehyde into acrolein over zeolites. *J. Catal.* **1994**, *147*, 133–139 DOI:10.1006/jcat.1994.1123.

[90] Borowiec, A.; Devaux, J. F.; Dubois, J. L.; Jouenne, L.; Bigan, M.; Simon, P.; Trentesaux, M.; Faye, J.; Capron, M.; Dumeignil, F. An acrolein production route from ethanol and methanol mixtures over FeMo-based catalysts. *Green Chem.* **2017**, *19*, 2666–2674, DOI:10.1039/C7GC00341B.
[91] Borowiec, A.; Lilić, A.; Morin, J.-C.; Devaux, J.-F.; Dubois, J.-L.; Bennici, S.; Auroux, A.; Capron, M.; Dumeignil, F. Acrolein production from methanol and ethanol mixtures over la- and ce-doped femo catalysts. *Appl. Catal. B Environ.* **2018**, *237*, 149–157, DOI:10.1016/j.apcatb.2018.05.076.
[92] Palion, W. J.; Malinowski, S. Gas phase reactions of acetaldehyde and formaldehyde in the presence of mixed solid catlysts containing silica and alumina. *React. Kinet. Catal. Lett.* **1974**, *1*, 461–465, DOI:10.1007/BF02074480.
[93] Malinowski, S.; Basinski, S. Kinetics of aldolic reactions in the gaseous phase on solid catalysts of basic character. *J. Catal.* **1963**, *2*, 203–207, DOI:10.1016/0021-9517(63)90044-7.
[94] Auroux, A.; Vedrine Jacques, C. Microcalorimetric characterization of acidity and basicity of various metallic oxides. *Stud. Surf. Sci. Catal.* **1985**, *20*, 311–318, DOI:10.1016/S0167-2991(09)60180-4.
[95] Shen, J.; Kobe, J. M.; Chen, Y.; Dumesic, J. A. Synthesis and surface acid/base properties of magnesium-aluminum mixed oxides obtained from hydrotalcites. *Langmuir* **1994**, *10*, 3902–3908, DOI:10.1021/la00022a082.
[96] Gervasini, A.; Bellussi, G.; Fenyvesi, J.; Auroux, A. Acidity generation of binary metal oxide catalysts. *Stud. Surf. Sci. Catal.* **1993**, *75*, 2047–2050, DOI:10.1016/S0167-2991(08)64222-6.
[97] Essehaity, A.-S. M.; Abd Elhafiz, D. R.; Aman, D.; Mikhail, S.; Abdel-Monem, Y. K. Oxidative coupling of bio-alcohols mixture over hierarchically porous perovskite catalysts for sustainable acrolein production. *RSC Adv.* **2021**; *11*, 28961–28972, DOI:10.1039/D1RA05627A.

María-Guadalupe Cárdenas-Galindo and Brent E. Handy

Chapter 5 The correspondence of calorimetric studies with DFT simulations in heterogeneous catalysis

Abstract: Molecular modeling, being computationally intensive, started with simple molecules and several atom clusters. Given that surface sites in heterogeneous catalysts involve complex structure, multielement mixtures, the effects of coverage, coadsorbates, and support interaction, simulations must continue to evolve in sophistication. An essential aspect is to build and validate these models with direct experimental evidence from calorimetric and spectroscopic results. In this regard, experimental heats of adsorption from single-crystal adsorption calorimetry and Tian-Calvet microcalorimetry on porous catalysts serves to benchmark the energies predicted by density functional theory (DFT) simulations, which are dependent upon the particular exchange functionals. The development of energy of adsorption databases is a key to improve the sophistication of DFT simulations. Adsorption databases to-date have focused on single crystal metal surfaces. Experience with modeling the energetics of even simple molecule adsorption on single crystal and supported metal catalyst surfaces has motivated the development of increasingly sophisticated DFT models. The particular case study of CO adsorption documents the challenges associated with developing DFT models that agree with experimental energetics. Studies with zeolites have shown that the van der Waals term associated with adsorbate–wall interactions must be incorporated into the DFT model for better agreement between prediction and experimental calorimetric data. This interaction is specific to the micropore structure and quantifying its contribution serves with heat of adsorption data to quantify the energetics of reactive intermediates in microporous structures.

Keywords: microcalorimetry, SCAC, Density Functional Theory, adsorption energy, benchmarking, exchange-correlation, functional

The construction of microkinetic models of catalytic reactions involves adsorption-desorption steps and surface reactions. In the calculation of the energetics of these elementary steps, electronic structure calculations have become more accessible with the advances in computers and software availability. The electronic structure methods are based on the solution of the time-independent Schrödinger equation:

María-Guadalupe Cárdenas-Galindo and Brent E. Handy, Facultad de Ciencias Químicas, Universidad Autónoma de San Luis Potosí, Av, Manuel Nava No. 6, Zona Universitaria, C.P. 78210 San Luis Potosí, S.L.P., México

https://doi.org/10.1515/9783110590449-005

$$\hat{H}\psi = E\psi$$

where $\hat{H}$ is the Hamiltonian operator, E is the energy of the system (eigenvalue), and ψ is a wave function (eigenfunction). A simple way to solve this eigenvalue problem is with the Hartree-Fock (HF) approach. The eigenfunctions are described by Slater determinants and have dimensionality $3N$, where N is the number of electrons in the system [1]. The calculation of the wave function is a complex problem, especially in heterogeneous catalysis where the number of atoms that describe the solid catalyst and the species involved in the reaction is very large if all the interactions that occur during the catalytic cycle are reasonably described. Density functional theory (DFT) offers an alternative to the use of a wave function in the context of the HF approach, since the dimensionality of the system is reduced with the use of electron density, a three-dimensional variable. A key aspect of the DFT methodology calculations relies on selecting and exchange-correlation functional. The development of new functionals over the past 20 years has led to more accurate descriptions of the catalyst and the surface species as they exist at any particular reaction condition.

The intention of this chapter is to highlight studies where experimental adsorption microcalorimetry has served up fundamental reference energetics for validating DFT functionals and evolving them into more refined ones. This is an ongoing area of analysis to continually improve the accuracy of DFT predictions for adsorption, catalytic activity, to capture the essential surface chemistry will ultimately serves the purpose of microkinetic analysis and catalytic reaction synthesis.

5.1 DFT Studies for prediction of adsorption energetics and activated complexes

The fundamentals of DFT theory were set by the work of Hohenberg and Kohn [2] and Kohn and Sham [3]. The Kohn-Sham implementation of DFT considers exchange-correlation energies ($\boldsymbol{E_{xc}}[r]$), in addition to kinetic energy ($\boldsymbol{T_s}[r]$), nuclear-electron potential energy ($\boldsymbol{V_{Ne}}[r]$), and electron-electron repulsion energy ($\boldsymbol{J}[r]$). The addition of these components gives the DFT ground state energy, $\boldsymbol{E}[r]$:

$$\boldsymbol{E}[r] = \boldsymbol{T_s}[r] + \boldsymbol{V_{Ne}}[r] + \boldsymbol{J}[r] + \boldsymbol{E_{xc}}[r]$$

The forms of all the functionals are known with exception of the $\boldsymbol{E_{xc}}[r]$, which describes the many body electron interactions. The development of exchange-correlation functionals (XC) is an active field of research. The approaches used to define the density functional approximations improve in chemical accuracy according to the Jacob's ladder (Fig. 5.1) defined by Perdew and Schmidt [4]. At the bottom rung of the ladder is the Hartree world where no exchange-correlation exists; at the top is the perfect XC. Computational cost increases as one ascends the Jacob's ladder.

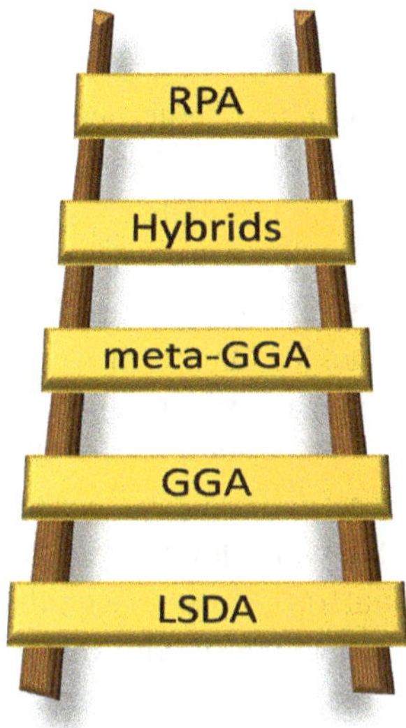

Fig. 5.1: Jacob's ladder of density functional approximations for the exchange-correlation energy (reproduced from [4] with the permission of AIP Publishing).

5.1.1 Exchange correlation functionals

Hundreds of nonempirical and semiempirical density functionals have been developed. A detailed description of the different types of functionals can be found in reviews by Mardirossian and Head-Gordon [5] and Chen [6]. The performance of the functional has been the focus of several publications. Noteworthy are the works by Cohen et al. [7], Wellendorff et al. [8], Mardirossian and Head-Gordon [5], and Göltl et al. [9]. Brief descriptions of relevant functionals are discussed further.

5.1.1.1 Local spin-density approximation (LSDA)

The local spin-density approximation (LSDA), a simple extension of the LDA functional, is based on the assumption of a homogeneous electron gas with a density that depends on the position (local density). The exchange-correlation energy $\mathbf{E_{xc}}$ involves the exchange energy $\mathbf{E_x}$ and the correlation energy $\mathbf{E_c}$:

$$E_{XC} = E_X + E_C$$

The correlation energy $\mathbf{E_c}$ does not have an analytical part and different parameterizations have been developed for its calculation. Some examples of parameterization are: Vosko–Wilk–Nusair (VWN5) [10], Perdew-Zunger (PZ81) [11], and Perdew-Wang (PW92) [12]. However, it should be realized that in heterogeneous catalysis the density distribution is in fact inhomogeneous and the prediction of molecular properties is highly inaccurate with LSDA.

5.1.1.2 Generalized gradient approximation (GGA)

Generalized gradient approximation introduces a density gradient to account for inhomogeneities in the electron gas density. Compared with LSDA, the predictions are improved with GGA XC functionals, leading to its frequent use in calculations of the exchange and correlation energy terms in DFT studies applied to heterogeneous catalysis. The XC functionals that have been widely used in this area are the Perdew-Wang (PW91) [12] and the Perdew-Burke-Ernzerhof (PBE) [13]. Some XC functionals share the same correlation functional, but different exchange functional (exchange + correlation). Examples of these combinations are revPBE (revPBE + PBE) and RPBE (RPBE + PBE). The accuracy of DFT-GGA is comparable to that of the second-order MP perturbation theory (MP2), which is the simplest post-HF method, but with a lower computational cost. This is due to the fact that their respective computational costs are on the order of $O(N^3)$ and $O(N^5)$, respectively. The chemical accuracy obtained with GGA is low, but better than with LSDA. For instance, in studies of the adsorption of several molecules on catalytic materials, the energy can be overestimated by as much as 30 kJ/mol for strong adsorption, and 70 kJ/mol when the adsorbate-surface dispersion interactions are significant [6]. Often, the adsorption site predicted is not in agreement with experimental studies [14].

5.1.1.3 Meta-GGA

The errors in DFT-GGA predictions stem from the self-interaction error, resulting from the interaction between an electron and its own density. To compensate for this problem the functionals in the third rung of the Jacob's ladder, the meta-GGA, emerged. As in GGA, these functionals depend on density and its gradient, but consider also the second derivative of the electron density or the kinetic energy density. Examples of nonempirical meta-GGA XC functionals are Tao, Perdew, Staroverov, and Scuseria (TPSS) [15], MS2 [16], and the correlation functional strongly constrained and appropriately normed (SCAN) [17]. Examples of semiempirical functionals are the Minnesota 6 (M06L) [18], and the Minnesota 15 (MN15-L) [19]. One of the problems with these types of functionals is that in hydrogen-bonded systems there is the tendency to underestimate binding energies; however, in the prediction of atomization energies and barrier heights they are better than GGA functionals [5].

5.1.1.4 Hybrid GGA/meta-GGA

The fourth rung of Jacob's ladder has the hybrid GGA/meta-GGA functionals. The hybrid functionals combine a fraction of the exact exchange evaluated with HF/post-HF methods and the exchange correlation functional to reduce the self-interaction error:

$$E^{\mathrm{HF/DFT}} = c_X E_X^{\mathrm{HF}} + (1 - c_X) E_X{}^{\mathrm{DFT}} + E_c{}^{\mathrm{DFT}}$$

where $\mathbf{E}^{\mathrm{HF/DFT}}$ corresponds to the hybrid functional, $\mathbf{E_x}^{\mathbf{HF}}$ corresponds to HF or post-HF method, $\mathbf{E_x}^{\mathbf{DFT}}$ refers to the DFT exchange and $\mathbf{E_c}^{\mathbf{DFT}}$ to the DFT correlation functional, and $\mathbf{c_x}$ is the fraction that represents the contribution of $\mathbf{E_x}^{\mathbf{DFT}}$ to the exchange functional. The most popular functional of this type is the B3LYP [20] where $\mathbf{c_x}$ has a value of 0.20 [5].

5.1.1.5 Van der Waals interactions

A major shortcoming of DFT calculations is the lack of precision in systems where van der Waals interactions play an important role. This is the circumstance, for example, in the adsorption of molecules in the micropore channels of a zeolite, where wall effects strongly influence adsorption energetics. This was noted early on as a contributing factor for the lack of accuracy in DFT-GGA calculations and prompted the development of new functionals by modifying them according to one of two approaches: the inclusion of nonlocal correlation effects, or with empirical van der Waals corrections [21]. Examples of nonlocal density functionals are optPBE-vdW, optB86b-vdW the Rutgers–Chalmers–van der Waals density functional (vdW-DF) [22] and the more recently developed XC functional BEEF-vdW, a Bayesian error estimation semiempirical functional parametrized to experimental data [23]. The methods developed by Grimme (D1, D2, D3, D3bj and D4) [24–28], and by Tkatchenko and Scheffler (TS) [29] are examples of empirical corrections that in combination with other functionals account for van der Waals interactions. This is the case of PBE-D3, a GGA functional with the D3 correction for treatment of London-dispersion interactions, and PBEsol-D3bj, which is the PBEsol GGA functional with the D3bj dispersion correction.

5.1.1.6 Random phase approximation (RPA)

Rung 5 functionals include the random phase approximation. In RPA the self-interaction error is eliminated with the calculation of the exact exchange energy with the incorporation of the wave function exchange, additionally, van der Waals interactions are included because of its nonlocal character. The semilocal functionals overestimate the energy of adsorption, but with RPA it is possible to obtain energies closer to the actual experimental energies, as well as the observed adsite complex configuration, making possible to have a better description of a catalytic surface and the adsorbed species on it, but at high computational cost [6, 30].

5.1.2 Databases for benchmarking

The ongoing development of more functionals makes more important and difficult the selection process of the one that best describes the energetics and structure of the reacting species and the catalytic material being investigated. It is thus important to rely on experimental information to benchmark the calculation results. Information typically used as benchmarking are related to structure, bonding, and energetics. Several databases are available for benchmarking the functionals. For chemistry applications, the Main-Group Chemistry DataBase (MGCDB84) is a compilation of 84 data-sets with a total of 4,986 data points that refer to noncovalent interactions, isomerization energies, thermochemistry, and barrier heights [5], but it does not contain adsorption energies, which is of upmost importance in heterogeneous catalysis. Another database used in benchmarking XC functionals is GMTKN55, a collection of 55 different test sets with 1,505 relative energies related to thermochemistry, kinetics, and noncovalent interactions, but adsorption energies are not included [31]. Wellendorff et al. [8] prepared a database for adsorption bond energies of CO, NO, H_2, O_2, NH_3, I, CH_3I, CH_3OH, CH_4, CH_2I_2, D_2O, and several hydrocarbons on transition metal surfaces (see Tab. 5.1) to benchmark DFT functionals. Most of this information is for coverages below 25% on single crystals and the experimental data are a compilation of the results with diverse techniques that include single-crystal adsorption calorimetry (SCAC). ADS41 is a database that includes similar information for chemisorption and physisorption on single crystals [32] and SBH10 has experimentally observed reference barrier heights for dissociation reactions on single crystals of transition metal surfaces [33]. A similar database does not exist with information of adsorption or reaction energy barriers on real catalysts. The case studies documented below describe the work of several authors who benchmarked their calculations with experimentally measured heats of adsorption on catalytic surfaces, in some cases using the heats of adsorption from Tian-Calvet measurements of porous catalysts.

Tab. 5.1: Database of adsorption energies compiled from experimental thermal methods and isotherms compiled on single crystals (from Wellendorff et al. [8]).

	Adsorbate	Surfaces	Coverage (ML)[a]	Exp. method[b]
1	CO	Ni(111)	1/4	SCAC, EQLB, TPD
2	CO	Pt(111)	1/4	SCAC, EQLB, TPD
3	CO	Pd(111)	1/4	EQLB, TPD, MMB
4	CO	Pd(100)	1/4	EQLB
5	CO	Rh(111)	1/4	MMB-TREELS, MMB-He scattering, EQLB-LITD
6	CO	Ir(111)	1/4	TPD
7	CO	Cu(111)	1/4	EQLB
8	CO	Ru(001)	1/4	EQLB, TPD
9	CO	Co(001)	1/4	TPD

Tab. 5.1 (continued)

	Adsorbate	Surfaces	Coverage (ML)[a]	Exp. method[b]
10	NO	Ni(100)	1/8	SCAC
11	NO	Pt(111)	1/4	SCAC
12	NO	Pd(111)	1/4	TPD
13	NO	Pd(100)	1/4	SCAC
14	O_2	Ni(111)	1/4	SCAC
15	O_2	Ni(100)	1/4	SCAC
16	O_2	Pt(111)	1/9	SCAC, TPD
17	O_2	Rh(100)	1/4	SCAC
18	H_2	Pt(111)	1/4	TPD, EQLB, MMB
19	H_2	Ni(111)	1/4	EQLB, TPD
20	H_2	Ni(100)	1/4	EQLB, TPD
21	H_2	Rh(111)	1/4	TPD
22	H_2	Pd(111)	1/4	TPD
23	I	Pt(111)	1/4	TPD
24	NH_3	Cu(100)	1/4	MMB- TREELS
25	CH_3I	Pt(111)	1/4	SCAC
26	CH_3OH	Pt(111)	1/4	SCAC
27	CH_3I	Pt(111)	1/25	SCAC
28	CH_4	Pt(111)	1/2	TPD
29	C_2H_6	Pt(111)	1/3	TPD
30	C_3H_8	Pt(111)	1/4	TPD
31	C_4H_{10}	Pt(111)	1/5	TPD
32	C_6H_6	Pt(111)	1/9	SCAC
33	C_6H_6	Cu(111)	1/9	TPD
34	C_6H_6	Ag(111)	1/9	TPD
35	C_6H_6	Au(111)	1/9	TPD
36	C_6H_{10}	Pt(111)	1/9	SCAC
37	CH_2I_2	Pt(111)	1/12	SCAC

[a]ML, monolayer coverage.
[b]SCAC, single-crystal adsorption calorimetry; TPD, temperature programmed desorption; EQLB, equilibrium measurements of coverage versus temperature and pressure; MMB, modulated molecular beam measurements of surface residence times versus temperature; TREELS, time-resolved electron energy loss spectroscopy; EQLB-LITD, equilibrium measurements using laser induced thermal desorption.

5.1.3 Other aspects to consider

In addition to the selection of exchange and correlation functionals, DFT calculations involve the selection of a basis set. A basis set is a set of equations that represent the electronic wave function. In heterogeneous catalysis, DFT calculations commonly use a plane-wave basis set, which is well suited to solving systems involving three-dimensional periodic boundary conditions. Another important aspect is the selection of the unit cell that represents the catalytic surface. As the number of atoms in the surface model increases, it becomes possible to explore in more detail the effect of adsorbate coverage, as well as the interactions between different surface species and their mobility through surface or bulk diffusion. An interesting review that covers all these aspects was prepared by Kratzer and Neugebauer [34].

5.2 Adsorption microcalorimetry in catalysis

Direct heat of adsorption measurements focused on catalytic systems have existed since the late 1950s with the introduction of sensitive microcalorimeters, notably the heat conduction Tian-Calvet microcalorimeter, which is uniquely suited for studies with porous catalysts and adsorbents, where the thermal response to the adsorption process is small and slow [35–37]. In contrast, the SCAC technique [38] is a more recent development and centered on studies of well-defined surfaces of an exposed crystal face that represents the active catalytic phase under ultra-high vacuum conditions. It is instructive to consider the strengths and limits of these two experiments when relating the results to the DFT modeling studies.

The SCAC technique can be applied to both well-defined metal and metal oxide surfaces, and generates well-resolved profiles of the variation of differential heat with coverage up to and beyond one monolayer [38, 39]. The features of this high-vacuum instrument have been described in detail [40]. A 0.2 μm thin, oriented foil target is exposed to a pulsed molecular beam of adsorbate molecules. The fraction of impinging molecules that adsorb locally heat the target, which is radiated on the opposite side and detected by a sensitive IR detector, or in the case of studies at sub-ambient temperatures by a pyroelectric sensor in thermal contact with the backside of the target. The slight rise in gas pressure from nonadsorbed molecules is detected by mass spectrometry to quantify the gas amount adsorbed in each pulse. Molecular pulsing is rapid (50 ms), with pulses small enough to represent dosing increments down to ca. 1% saturation coverage, and the sticking probability for the coverage increment can be determined for each gas pulse. Heat calibration is performed by use of pulses of known duration and power emitted from a He-Ne laser beam passing along the same trajectory toward the target. The uncertainty in enthalpy of adsorption values is reported to be 5–7% [41] and a detection limit (with high sticking coefficient) of ca. 10 kJ/mol. The variation of the sticking probability of adsorbate with coverage is also determined, and there is consistency with desorption energies reported from thermal desorption spectra (TDS) data obtained with the same well-defined surface/adsorbate structure. Schießer [41] reports that measurements with SCAC are limited to ambient or sub-ambient temperatures if molecular beam pulsing is employed and high sticking coefficients are required of the adsorbate molecule. While SCAC measurements are typically conducted on surfaces at ambient or sub-ambient temperatures, experiments on porous catalysts with Tian-Calvet calorimeters are typically conducted at elevated temperatures in order to achieve thermodynamic site-filling (strongest sites occupied first during initial gas dosing) when intraparticle diffusion may be limiting. For instance, ammonia adsorption on microporous MFI on 150 ± 5 kJ/mol sites [42–44] are performed at 480 K to allow adequate probe molecule pore mobility necessary to titrate all available adsites within the time scale of each gas dose.

While Tian-Calvet adsorption microcalorimetry may not provide a direct correspondence between a particular faceting of supported metal particles, adsorption heat, and coverage, the inherent complexity of supported metal particles in real catalysts provides a more pragmatic picture of variations of site enthalpy with coverage. It also has been shown that calorimetric data can be effectively combined with TPD studies of porous catalyst [45] to provide better estimates of desorption energies, pre-exponential factors, and entropies of adsorption that are necessary for microkinetic analysis of real catalyst systems.

In spite of differences in methodology, sample form, and adsorption conditions, SCAC and Tian–Calvet microcalorimetry yield results of adsorption heats within a narrower range than the variation in values predicted by the various functionals used in DFT calculations. In fact, the stablest species predicted by DFT are typically much higher (overbinding) in energy than are shown by the experimental calorimetric data.

5.3 Benchmarking DFT methods through comparisons with heat of adsorption data

To date, DFT and microcalorimetry have appeared in countless studies that address important themes of adsorption and surface reaction related to acidity, basicity, reactions with hydrocarbons, CO, NO, and alcohols. This is not an extensive review of all literature where DFT simulations and microcalorimetry have been used. It is only to focus on several studies where microcalorimetry has been useful for benchmarking the XC functionals used in DFT calculations. As was described above, the selection of the exchange and correlation functionals is an important step in setting up the electronic structure calculation. In a heterogeneous reaction one of the major problems is to identify a functional that is well suited for a solid material and the chemical species that participate in the reaction, both with very different properties. This task is a difficult one as it is evident from the following cases.

The development of the microkinetic model of the reaction system requires a reliable prediction of the energetics for reaction steps for which it is not possible to obtain direct experimental measurements. In the development of new catalytic materials, it is important to know how the composition and structure of the catalytic material affects the microkinetic model. Therefore, the accuracy of the predictions of the DFT calculations depend on how good the description is of the system under study. As the following cases demonstrate, in heterogeneous catalysis the adsorption studies are an ideal source of information for benchmarking energetics. Differences observed between experimental results and DFT calculations of catalytic systems have driven intensive research to test new XC functionals and to modify the existing ones. A classic example is the adsorption of CO on Pt surfaces, a source of intensive interest since the 1980s [46], and which continues to motivate investigators in computational chemistry to this day [47, 48].

5.3.1 CO adsorption on metal surfaces

The benchmarking of the DFT functional for applications in heterogeneous catalysis relies on experimental information to assure that the following aspects are accurately predicted by the calculations: lattice constants, surface formation energies, adsorption site configuration, and adsorption energies. The experimental results generated by the techniques in Tab. 5.2 have been used in benchmarking CO adsorption in DFT calculations.

Tab. 5.2: Experimental information used in benchmarking of DFT calculations of CO adsorption as in selected works.

Technique	Adsorption system	References
Tian-Calvet microcalorimetry	CO on 1% Pt/BaK-L	[49]
Tian-Calvet microcalorimetry	CO on 1.7% Pt/SiO_2	[50]
Tian-Calvet microcalorimetry	CO on H-FER zeolite, Si-FER and silicalite	[51]
Tian-Calvet microcalorimetry	CO on Ca-FER	[52]
Tian-Calvet microcalorimetry	CO adsorption on low silica FAU zeolites	[53]
Adiabatic calorimetry	CO on ZSM5	[54]
SCAC	CO on Pt(111)	[41]
SCAC	CO on Pt(111)	[55]
SCAC	CO on Pt(111)	[56]
SCAC	CO on Pt(111)	[14]
TPD	CO adsorption on low silica FAU zeolites	[53]
Variable temperature IR spectrometry	CO adsorption on FAU, MFI and CHA zeolites	[53]

The initial heat of CO adsorption on Pt/SiO_2 catalyst of 10% metal dispersion was reported to be 140 kJ/mol [57], corresponding chiefly to terminally bonded Pt–CO adsorbates as shown by numerous IR studies [58]. For comparison, considerably higher initial heat values of 175 kJ/mol were registered in Pt/Ba-K-L zeolite [49], characterized as having small Pt metal clusters in the zeolite framework. The heat value decreased to 90 kJ/mol at saturation coverage. These results could be simulated using a 10-atom DFT cluster model, configured as four exposed 6-atom faces in near (111) arrangement, though it was noted that the equilibrated structure contained slightly elongated Pt-Pt distances and bulging of each face.

The diminishing adsorption heat with increasing coverage could be modeled by adding up to six CO molecules to the metal cluster. This coincided with added CO bonding as complexes that bridge two adjacent Pt atoms. The simulations also cross-checked with shifts in the vibrational frequency of the ν(C–O) bond upon adsorption in metal cluster versus bulk Pt particles, noting that small Pt metal clusters are negatively charged, increasing the degree of π-backbonding to the CO anti-bonding orbital and strengthening the Pt-CO adsorbate bond. However, DFT calculated adsorption heats consistently overestimated the experimental values, by 19% when compared with the result reported on Pt/ BaK-L zeolite, of Watwe [49] and by 65% the value on Pt(111) by SCAC [41]. The DFT functional used by Watwe was B3LYP, a hybrid functional that is capable of very accurate predictions, but in this case inadequately predicted the energy, perhaps also due to the insufficient number of atoms involved in the model. An additional problem with DFT calculations based on the Pt cluster model was the wrong prediction of the predominant CO adsorption mode. From spectroscopic measurements it is known that CO adsorption on Pt occurs mostly in "on top" mode, yet the DFT calculations predicted the CO adsorbing in fcc or hcp "hollow" sites as being most stable. Subsequent attempts to address this particular discrepancy has come to be known as the "CO puzzle," and received much attention in the literature.

DFT calculations based upon periodic surface models (slabs several monolayers thick), to which the adsorbate molecules can be added in increasing number to represent increased coverage levels are commonly used in adsorption studies over metals. Feibelman et al. [46] followed this approach in the study of CO adsorption on Pt(111). The DFT calculations were performed with different software codes (VASP, Dacapo, and WIEN), the Pt(111) surface was represented by slabs of 4 to 6 layers, and the orbital sets were plane-wave or linearized augmented plane-wave depending on the software. The XC functionals tested were the Ceperley-Alder version of the LDA functional [59] and the GGA functionals PW91, PBE, and RPBE. In all cases, the calculations erroneously predicted the preference of CO for the fcc hollow site. The difference in binding energies between the CO on fcc hollow and on top configurations depended on the type of XC functional and were larger with LDA than with GGA functionals. It has become evident that the choice of XC functional is very important to calculate accurate adsorption energies. However, the accurate prediction of the on-top configuration remains elusive. Relativistic effects and van der Waals interactions were addressed by several researchers trying to solve the CO puzzle but with inconclusive results. Olsen et al. [60] incorporated scalar relativistic corrections in the DFT calculations and were able to predict the proper adsorption site, but with a very small difference in the energies between the on-top and the fcc adsorption sites. The calculated adsorption energies were clearly overestimated when compared with the experimental measurements. XC functionals being the next rung on Jacob's ladder, meta-GGA, were later tested. The M06-L functional falls in this category, and its use in DFT calculations predicts the correct adsorption site with an adsorption energy very close to the experimental measurements if dispersion effects are not considered. However, the difference between the energies of adsorption on the hollow and the top site are insignificant [14, 61]. The dispersion effects (van der Waals interactions) play an important role in the stabilization of the on top position and if they are considered in M06-L the adsorption energy is clearly overestimated [14].

CO adsorption has been studied on other metallic systems. The DFT prediction of CO adsorption on Cu(111) and Rh(111) with GGA functionals also failed to predict the experimentally observed on-top adsorption, favoring instead a highly coordinated site [46]. This problem was solved with the use of the hybrid XC functionals PBE0 and HSE03. The reduction of the self-interaction error with these functionals made possible the prediction of the on top adsorption of CO on Cu and Rh, but failed with Pt(111) [62, 63]. Significantly, the energies of adsorption were overestimated in all the cases. More recently developed functionals that include the van der Waals nonlocal correlation functionals optPBE-vdW, optB86b-vdW, and the BEEF-vdW, were tested with similar results. BEEF-vdW gives the best results for adsorption energies, but they are still 15% above the experimental value [55]. The most recent benchmarking study of CO adsorption compared functionals of different types: GGA, meta-GGA, hybrid and including van der Waals interactions [14]. None of these functionals were able to predict the stability of the on top site for CO on Pt(111), but this study concluded that dispersion effects must be included even when the outcome is a consistent overestimation of the adsorption energy. It was emphasized that the accuracy of existing density functionals needs to be improved.

As the experience with CO adsorption has shown, adsorption energy and mode of adsorption have been difficult to predict on Pt(111), and the results are mixed on other transition metal surfaces. Wellendorff [8] used 39 surface reactions on transition metal surfaces to benchmark LDA, PBEsol, PW91, PBE, RPBE, and BEEF-vdW functionals. The surface reactions included the adsorption of CO, NO, O_2, H_2, C_1 to C_4 paraffins, CH_3I, CH_3OH, C_6H_6, and C_6H_{10}. An ideal would be to have a global functional that could accurately predict the thermochemical properties of all the transition metals, but this is not the case. The RPBE and the BEEF-vdW functionals were the best for adsorption processes where van der Waals interactions do not play an important role in the stabilization of the adsorbate on the surface. Systems where van der Waals interactions are important are best simulated with either PBEsol or BEEF-vdW [8].

The importance of having experimental information for validating the DFT methodology is of upmost importance not only for CO adsorption on metal surfaces, but also on other surfaces, such as zeolites. To have a reliable DFT methodology that accurately describes the adsorption of CO on metal surfaces is of special interest in the study of electrocatalytic reactions for CO_2 reduction, water gas shift reaction, etc.

5.3.2 Benzene adsorption on metal surfaces

The exploration of reaction pathways and the effect of changes in the catalytic formulation for a given reaction requires a good description of the catalytic surface and the species on the surface. Ideally, the same XC functional should allow an accurate prediction of the thermochemical properties of all of them. Since a large number of reactions are catalyzed by various transition metals there is great interest in the evaluation of the performance of existing XC functionals with the surfaces of Pt, Pd, Cu, Rh, and Ru. GGA XC functionals describe accurately metal properties such as lattice constants, but the surface formation energies require meta-GGA XC functionals for accurate predictions, as were demonstrated in studies on the (111) surfaces of Rh, Pt, Cu, Ag, and Pd. A more complex story arises when the system includes an adsorbate, and here CO is not the only challenging example. In the adsorption of π-bound unsaturated hydrocarbons on Pt(111), DFT calculations with functionals that included vdW terms (optPBE-vdW or PBE-dDsC) give a good prediction of adsorption energies [55]. As mentioned earlier for CO, these functionals overestimate the CO adsorption energy; therefore, computational studies of coadsorption and reactivity between unsaturated hydrocarbons and CO remain a challenging task.

The study of benzene is of special interest in reactions catalyzed by transition metals (hydrogenations, oxidations, aromatization, etc.) [64–66]. On Pt(111) surfaces benzene binds with an orientation parallel to the surface forming four di-σ and two π bonds [67], resulting in a bonding where the van der Waals component is comparable to the covalent component. The heats of molecular adsorption of benzene onto Ni(111) and Pt(111) measured by SCAC have been used to validate theoretical calculations for the adsorption of benzene [68, 69]. The measured energetics were used to assess the DFT calculated energies of adsorption, showing that if the XC functionals used do not contain corrections for vdW interactions the bond energies of benzene to both Ni(111) and Pt(111) are underestimated. Functionals that include vdW interactions are much more accurate (Fig. 5.2), and additionally, the adsorbate configuration agrees with the experimental observations [68, 69].

It could be expected that the same functionals that perform well with benzene could perform equally well with other aromatic compounds; however, this is not always the case. SCAC heats of benzonitrile adsorption on Pt(111) were used to validate the adsorption energy predicted with DFT calculations similar to those used for benzene adsorption. Using PBE-vdW-DF, the same functional that gave good predictions with benzene, the adsorption energy (163.2 kJ/mol) is highly underestimated when compared with the experimental value (250 kJ/mol) for the same coverage ($\theta = 0.06$ ML). This may be due to the polar nature of the nitrile group that imparts different functionality. The DFT calculations predict that benzonitrile adsorbs with the phenyl group parallel to the surface and the cyano group is tilted upward [69].

Acrylonitrile has also a cyano group, but unlike with benzonitrile, the surface bonding occurs primarily via a di-σ-mode, sp-hybridized interaction with the CN group, the methyl group being inclined away from the surface. Here, the energies calculated with

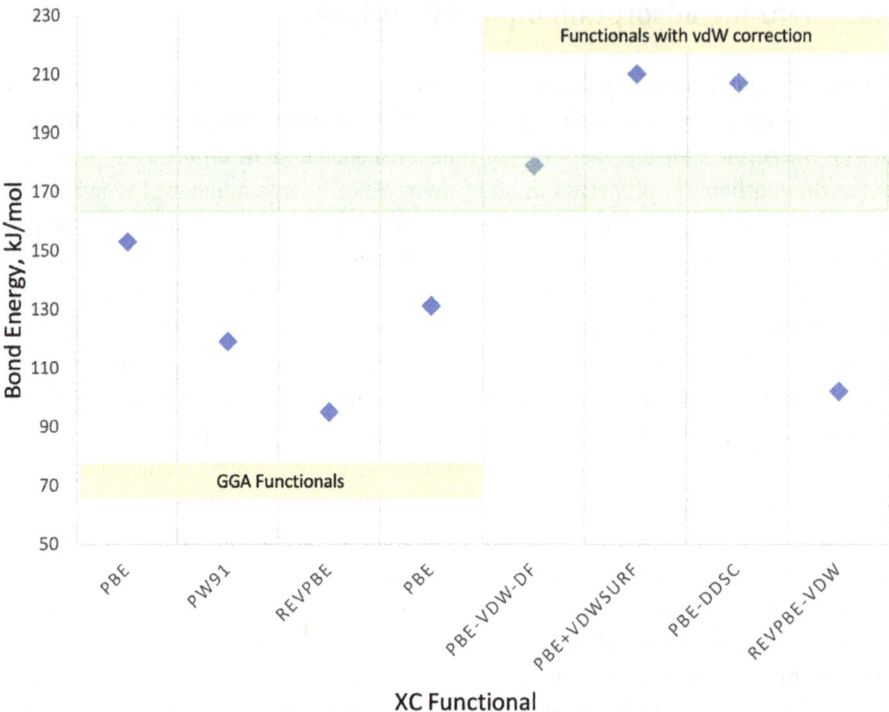

Fig. 5.2: Adsorption energies calculated with GGA functionals with and without van der Waals interactions [70–75] for benzene. The green band represents the range of experimental adsorption heats measured with SCAC studies [68, 76].

vdW corrected XC functionals are 131.3 kJ/mol, highly overestimating the SCAC measured value (74 kJ/mol, at 0.06 ML) [69]. In the work of Pašti, XC functionals without vdW (PBE) predicted a lower adsorption energy of 77.2 kJ/mol that is much closer to the experimental value [77]. These results show how vdW-corrected DFT functionals have their limitations and are not easily applicable to adsorbates with different modes of adsorption. The inclusion of van der Waals interactions in the functionals gives accurate energetics and configuration for the π-bonding of benzene and phenyl groups, but the behavior of the CN group is better treated by functionals that don't include van der Waals interactions.

5.3.3 Acid site characterization in zeolites

The study of reactivity in zeolites with DFT is an extensive research area in heterogeneous catalysis where microcalorimetry data can be of special interest in validating the calculation results. Information than can be useful in this regard can be found in Tab. 5.3.

Tab. 5.3: Database of adsorption enthalpies compiled from adsorption microcalorimetry on zeolite catalysts.

Probe molecule	Catalyst	Qads[a] (kJ/mol)	Comments[b]	Ref.
Pyridine	HZSM5	200–160; 160 avg	≤10% coverage; avg to 1 molec/Al (Si/Al = 34)	[78]
Pyridine	HZSM5	195–205	Average value ≤1 molec/Al (Si/Al = 35)	[79]
Pyridine	HM	200	(Si/Al = 13)	[78]
Pyridine	HY	240–95	(Si/Al = 2.4)	[78]
Pyridine	HY	185–175	Average value ≤1 molec/Al (Si/Al = 30)	[42, 79]
Pyridine	HM	200–180	Average value ≤1 molec/Al (Si/Al = 15)	[42, 79]
Pyridine	fumed SiO_2	95	Calcined 723 K	[80]
Pyridine	SiO_2-Al_2O_3	219 (LA) 174 (BA, LA)	Ads at 473 K	[80]
Ammonia	HZSM5	150;145 avg	Average value ≤1 molec/Al (Si/Al = 35)	[42]
Ammonia	HY	150–135	(Si/Al = 30)	[42]
Ammonia	HM	160	Average value ≤1 molec/Al(Si/Al = 15)	[42]
Methylamine	HZSM5	185	Average value ≤1 molec/Al (Si/Al = 35)	[79]
Ethylamine	HZSM5	195	Average value ≤1 molec/Al (Si/Al = 35)	[79]
Isopropylamine	HZSM5	205	Uniform with coverage (Si/Al = 35)	[42]
Isopropylamine	HY	150–135	Uniform with coverage(Si/Al = 30)	[42]
Isopropylamine	HM	160	Uniform with coverage(Si/Al = 15)	[42]
n-Butylamine	HZSM5	220	Average value ≤1 molec/Al	[79]
Dimethylamine	HZSM5	205	Average value ≤1 molec/Al	[79]
Trimethylamine	HZSM5	205	Average value ≤1 molec/Al	[79]
Diethyl ether	HZSM5	135	Average value ≤1 molec/Al	[81]

[a]Values are initial heats of adsorption, unless a range is shown. High-to-low values indicate a decrease with increasing coverage.

The application of zeolites in oil refining and, more recently, their potential in the production of valuable chemicals from biomasses and in the upcycling of plastics motivates further study with DFT to understand the relationship between structure, thermochemical properties, and reactivity in these catalysts.

Here, the DFT approach requires a reasonable description of zeolite frameworks, the electrostatic field in the channels and cavities, and the van der Waals interactions between the surface of the zeolite and the adsorbed molecules [82, 83]. An intrinsic property of zeolites is their acidity, and knowledge about the location of Brønsted and Lewis sites is of utmost importance to determine the reactants accessibility to them. To characterize the acidity and basicity of zeolites, probe molecules are used in experimental studies. NH_3 adsorption calorimetry is successfully used to determine acid site density and strength, however the type and location of the sites are out of its scope. Other probe molecules used in acidity studies are CO and pyridine. Comparisons among these probe molecules often show differences, due to molecular size and type of interaction functionals. To elucidate the locations of Lewis and Brønsted acid sites within the zeolite pores, IR spectroscopy is used in combination with microcalorimetry. DFT offers a tool to connect the results provided by both methodologies as it is described in Bucko's work with CO as a probe molecule [84]. The choice of probe molecule is not trivial, and the relationship between acidity and reactivity can be lost if the wrong choice is made, as the work by Liu et al. [53] shows. Both of these studies made evident the importance of dispersion forces in the system. Bucko's work explored the nature of Lewis and Brønsted acidity in mordenite, considering locations in the large channel, side pockets, or small cages. Additionally, the interaction between CO and other hydroxyl groups neighboring an acid site was considered. CO bonding is based upon the concept of donation from the 5σ MO of CO to the acidic hydroxyl and back-donation into the 2π^* unoccupied orbitals, which can be easily seen with IR spectroscopy by shifts in the ν(OH) and ν(CO) bands. On Brønsted sites, IR evidence shows the association between the carbon end of CO molecules and acidic hydroxyl groups located in the main channel of mordenite induces a stronger red shift of the fundamental C-O vibration than in CO molecules adsorbed in side-pockets. However, DFT shows red shift to occur in the opposite sense. DFT predicts vanishing or even endothermic CO adsorption, such that the large red shift should not be observable. Bucko et al. [84] created a slab model of the (001) surface of mordenite for periodic DFT calculations containing only T-O_4 to represent the bulk structure. This periodic model is representative of the zeolite channel structure, and Al cations were located at different positions in the channels. The authors discovered that there was little variation in the energetics of protons located at these sites. The GGA XC functionals PW91 and RPBE were used in the DFT calculations and did not include a van der Waals correction.

Microcalorimetric measurements performed by others with CO on mordenite have shown that CO interaction energies were in fact similar regardless of acid site location [85]; thus, all BA sites accessible to CO were energetically homogeneous. The experimental adsorption energy was 26.5 kJ/mol, considerably different from the DFT result, calculated with RPBE functionals, of 9.0 kJ/mol. RPBE also predicts that the side-pockets are not accessible. When applying the PW91 functional, however, both main channel and side pockets are accessible, and the calculated energy difference was smaller (2.3 kJ/mol) for the main channel, and about the same calculated energy for the side-pocket. If both PW91 and RPBE functionals had included van der Waals interaction energy, there would be better agreement with experimental values. The importance of including van der Waals interactions in zeolites was later corroborated in DFT calculations with the PBE XC functional, where the dispersion correction was added using the D3 Grimme's method [85]. The results showed that CO adsorbed in the smaller channels had a component of dispersion energy of 25 kJ/mol, and in the larger channel of 8 kJ/mol.

The protonation of isobutylene as an acid-mediated reaction serves as a test case for developing correlations between adsorbate bond strength on the acid site and reactivity. Using PBE functionals with a dispersion-corrected method (DFT-D2), van Santen and co-workers [53] did DFT calculations for a number of probe molecules (CO, CH_3CN, NH_3, and $(CH_3)_3$ N, and pyridine) in different zeolites of FAU, CHA, and

MFI topologies, using different Si/Al ratios. Their objective was to develop an acidity scale based on the adsorption of these bases. Some models involved Fe or Ga substitution of the Al framework structure within the FAU topology. The DFT energies were in agreement with the existing microcalorimetric data for the heats of adsorption of the alkylamine probe molecules [87]. It was concluded that the heat of adsorption measured is the sum of the chemical bonding interaction between probe molecule and the zeolite acid site, and the van der Waals interactions (dispersion forces). The effect of the dispersion forces is larger with molecules of larger size. It was found that within the same zeolite topology, there is a strong correlation between NH_3 heat of adsorption and the activation energy of isobutylene protonation. With a different topology, however, the dispersion forces contribution is different, and the same scaling trend for acidity does not apply.

5.3.4 Energetics of the active site in real catalysts at two operating conditions

A study by Wang et al. [88] shows how the enthalpy of adsorption measurements can be conducted on zeolite catalyst where previous studies highlighted different SCR kinetics depending upon the temperature of activation for the reaction:

$$4\,NO + 4\,NH_3 + O_2 = 4\,N_2 + 6\,H_2O$$

which is used in diesel fuel NO_x abatement technology. Under reaction conditions, oxygen activation has been found to be kinetically significant, with notable differences in activity between catalyst operation at temperatures below 250 °C and above this temperature. Previous studies had shown that for temperatures below 250 °C, the oxygen activation is second order in Cu, postulated to involve the interaction between molecular oxygen and a pair of $NH_3–Cu^+–NH_3$ complexes. On the other hand, at higher temperatures the Cu-ammonia complex is unstable, with the Cu^+ cations reverting to established Al locations within the zeolite framework, with reaction kinetics manifesting a different activation energy. DFT periodic calculations were based on the use of a GGA functional (PBE) and the D3 approach to describe van der Waals interactions. The presence of copper in the zeolite requires the use of a Hubbard term (PBE + U) to account for the highly localized 3d-electrons in the oxidized copper. This correction allows a better description of strongly correlated electronic states, such as those in *d* and *f* orbitals, while the rest of the valence electrons are treated by the normal DFT methodology [89]. The zeolite model consisted of a hexagonal unit cell with 36 tetrahedral positions containing two Al cations to represent a Si/Al ratio of 17.

Using dynamic adsorption with a flow differential calorimeter, the procedure used was to activate the catalyst in situ (2000 NO, 2,400 ppm NH_3 flowstream) at 250 °C before exposing the catalyst at 200 °C to O_2/inert flowstream of varying O_2 concentration. In a separate experiment, following activation, the catalyst was heated in inert flow to

500 °C, thus decomposing the Cu-ammonia complex, the Cu^+ cations migrating to framework locations within the chabazite.

The calorimetric result showed that O_2 adsorption on the Cu-ammonia complex was 79 kJ/mol, whereas adsorption on framework Cu sites was 120 kJ/mol. With the DSC-type calorimeter used, the variation of O_2 adsorption heats were investigated over the temperature interval 75–275 °C. This was done with catalyst having been pretreated to assume either the Cu-ammonia complex or the Cu-framework states. These results show increasing adsorption heat with temperature, documenting also the destruction (heat drop off for $T \geq 200$ °C) of the Cu-ammonia complexes formed under low temperature SCR conditions, and additionally, apparent activation energies were found to be different for each site. The activation energy on Cu-ammonia complexes was significantly lower than for Cu-framework sites (15 vs 39 kJ/mol). This direct enthalpy information both qualitatively and quantitatively serves as benchmarks for the activation energies being modeled with DFT. Finally, by performing O_2 exposure at increasing O_2 concentration, a Langmuir isotherm could be generated and analyzed with the calorimetric data to estimate the entropy of adsorption of −142 J/mol-K. Thus, the adsorption could be interpreted as an O_2 molecule losing gas-phase freedom by entering the micropores, with smaller degrees of entropy losses upon association with the Cu-ammonia pair sites, which also loose additional entropy by reconfiguring upon coordination with the O_2 molecule.

In summary, to best describe a zeolite with guest species within its micropores requires a model that accurately represents the structure and the location of the active sites, the inclusion of van der Waals interactions in the XC functional, and the modification of the DFT calculation with the inclusion of terms such as the Hubbard term to account for the presence of transition metals with electrons in *d* or *f* orbitals. The use of microcalorimetry to validate the adsorption energies calculated with DFT is evident. Considering that the Hubbard parameter (U) is not necessarily known, and its calculation can be empirical, machine learning is a new field of study that is being explored to determine this contribution [90]. Again, benchmarking the new methodologies with microcalorimetry offers new applications of this methodology.

5.4 Concluding remarks

While uncertainties of 5–10% exist in experimental heat of adsorption measurements on porous catalysts and in SCAC, even larger errors can exist in calculated energies unless DFT methods are benchmarked by experimental energetic data. The energies predicted by DFT depend upon the particular method and functionals applied, and experience with DFT is necessary to match the type of DFT functional used to the catalytic surface and probe molecule in a particular catalyst system. Benchmarking is most effectively done with adsorption microcalorimetry, often in conjunction with vibrational spectroscopy to confirm the adsorbate-active site configuration, and with TPD for cross-checking with desorption energetics and for providing desorption pre-exponentials in the ads-des mechanistic steps of the microkinetic mechanism. It is advantageous to perform calorimetry on real catalysts, since the energy values realistically involve surfaces that are more complex and often contaminated with other molecules. As the experience to-date with CO adsorption (hollow vs on-top) has shown, often DFT predicts that the most stable complex is not what reality shows, as the experimental calorimetric data and other evidence infer a less stable configuration. Better agreement can be achieved, but at a higher computational cost or with improved XC functionals.

There is already a database with simple molecules that have been studied with adsorption microcalorimetry for benchmarking DFT functionals, and this will continuously be added to. Given the current interest in biomass processing and the transformation of waste plastics to chemical products, new catalyst development can be guided by DFT studies, but it is important to identify which DFT type and functional(s) are appropriate for modeling the new biomass reactions. This will require benchmarking through use of experimental adsorption microcalorimetry of probe molecules on new catalytic materials.

References

[1] Slater, J. C.;. *Quantum Theory of Molecules and Solids, Vol. 4, the Self-Consistent Field for Molecules and Solids*; McGraw–Hill: New York; 1974, ISBN-10:0070580383; ISBN-13:978-0070580381.

[2] Hohenbert, P.; Kohn, W. Inhomogeneous electron gas. *Phys. Rev.* **1964**, *136(3B)*, 864–871, DOI:10.1103/PhysRev.136.B864.

[3] Kohn, W.; Sham, L. J. Self-consistent equations including exchange and correlation effects. *Phys. Rev.* **1965**, *140(4A)*, A1133, DOI:10.1103/PhysRev.140.A1133.

[4] Perdew, J. P.; Schmidt, K. Jacob's ladder of density functional approximations for the exchange-correlation energy. *AIP Conference Proceedings; AIP*. **2001**, *577*, 1–20, DOI:10.1063/1.1390175.

[5] Mardirossian, N.; Head-Gordon, M. Thirty years of density functional theory in computational chemistry: An overview and extensive assessment of 200 density functionals. *Molec. Phys*. **2017**, *115(19)*, 2315–2372, DOI:10.1080/00268976.2017.1333644.

[6] Chen, B. W.; Xu, L.; Mavrikakis, M. Computational methods in heterogeneous catalysis. *Chem. Rev.* **2021**, *121(2)*, 1007–1048, DOI:10.1021/acs.chemrev.0c01060.

[7] Cohen, A. J.; Mori-Sánchez, P.; Yang, W. Challenges for density functional theory. *Chem. Rev.* **2012**, *112(1)*, 289–320, DOI:10.1021/cr200107z.

[8] Wellendorff, J.; Silbaugh, T. L.; Garcia-Pintos, D.; Nørskov, J. K.; Bligaard, T.; Studt, F.; Campbell, C. T. A benchmark database for adsorption bond energies to transition metal surfaces and comparison to selected DFT functionals. *Surf. Sci.* **2015**, *640*, 36–44, DOI:10.1016/j.susc.2015.03.023.

[9] Göltl, F.; Murray, E. A.; Tacey, S. A.; Rangarajan, S.; Mavrikakis, M. Comparing the performance of density functionals in describing the adsorption of atoms and small molecules on Ni(111). *Surf. Sci.* **2020**, *700*, 121675, DOI:10.1016/j.susc.2020.121675.

[10] Vosko, S. H.; Wilk, L.; Nusair, M. Accurate spin-dependent electron liquid correlation energies for local spin density calculations: A critical analysis. *Can. J. Phys.* **1980**, *58(8)*, 1200–1211, DOI:10.1139/p80-159.

[11] Perdew, J. P.; McMullen, E. R.; Zunger, A. Density-functional theory of the correlation energy in atoms and ions: A simple analytic model and a challenge. *Phys. Rev. A.* **1981**, *23(6)*, 2785, DOI:10.1103/PhysRevA.23.2785.

[12] Perdew, J. P.; Wang, Y. Accurate and simple analytic representation of the electron-gas correlation energy. *Phys. Rev. B.* **1992**, *45(23)*, 13244, DOI:10.1103/physrevb.45.13244.

[13] Perdew, J. P.; Burke, K.; Ernzerhof, M. Generalized gradient approximation made simple. *Phys. Rev. Lett.* **1996**, *77*, 3865–3868, DOI:10.1103/PhysRevLett.77.3865.

[14] Janthon, P.; Vines, F.; Sirijaraensre, J.; Limtrakul, J.; Illas, F. Adding pieces to the CO/Pt (111) puzzle: The role of dispersion. *J. Phys. Chem. C.* **2017**, *121(7)*, 3970–3977, DOI:10.1021/acs.jpcc.7b00365.

[15] Tao, J.; Perdew, J. P.; Staroverov, V. N.; Scuseria, G. E. Climbing the density functional ladder: Nonempirical meta-generalized gradient approximation designed for molecules and solids. *Phys. Rev. Lett.* **2003**, *91(14)*, 146401, DOI:10.1103/PhysRevLett.91.146401.

[16] Sun, J.; Haunschild, R.; Xiao, B.; Bulik, I. W.; Scuseria, G. E.; Perdew, J. P. Semilocal and hybrid meta-generalized gradient approximations based on the understanding of the kinetic-energy-density dependence. *J. Chem. Phys.* **2013**, *138(4)*, 044113, DOI:10.1063/1.4789414.

[17] Sun, J.; Ruzsinszky, A.; Perdew, J. P. Strongly constrained and appropriately normed semilocal density functional. *Phys. Rev. Lett.* **2015**, *115(3)*, 036402, DOI:10.1103/PhysRevLett.115.036402.

[18] Zhao, Y.; Schultz, N. E.; Truhlar, D. G. Design of density functionals by combining the method of constraint satisfaction with parametrization for thermochemistry, thermochemical kinetics, and noncovalent interactions. *J. Chem. Theory Comput.* **2006**, *2(2)*, 364–382, DOI:10.1021/ct0502763.

[19] Yu, H. S.; He, X.; Truhlar, D. G. MN15-L: A new local exchange-correlation functional for Kohn–Sham density functional theory with broad accuracy for atoms, molecules, and solids. Journal of chemical theory and computation. *J. Chem. Theory Comput.* **2016**, *12(3)*, 1280–1293, DOI:10.1021/acs.jctc.5b01082.

[20] Becke, A. D.;. A new mixing of Hartree–Fock and local density-functional theories. *J. Chem. Phys.* **1993**, *98(2)*, 1372–1377, DOI:10.1063/1.464304.

[21] Berland, K.; Cooper, V. R.; Lee, K.; Schröder, E.; Thonhauser, T.; Hyldgaard, P.; Lundqvist, B. I. Van der Waals forces in density functional theory: A review of the vdW-DF method. *Rep. Prog. Phys.* **2015**, *78(6)*, 066501, DOI:10.1088/0034-4885/78/6/066501.

[22] Dion, M.; Rydberg, H.; Schröder, E.; Langreth, D. C.; Lundqvist, B. I. Van der Waals density functional for general geometries. *Phys. Rev. Lett.* **2004**, *92(24)*, 246401, DOI:10.1103/PhysRevLett.92.246401.

[23] Wellendorff, J.; Lundgaard, K. T.; Møgelhøj, A.; Petzold, V.; Landis, D. D.; Nørskov, J. K.; Bligaard, T.; Jacobsen, K. W. Density functionals for surface science: Exchange-correlation model development with Bayesian error estimation. *Phys. Rev. B.* **2012**, *85(23)*, 235149, DOI:10.1103/PhysRevB.85.235149.

[24] Grimme, S. Accurate description of van der Waals complexes by density functional theory including empirical corrections. *J. Comput Chem.* **2004**, *25(12)*, 1463–1473, DOI:10.1002/jcc.20078.

[25] Grimme, S. Semiempirical GGA-type density functional constructed with a long-range dispersion correction. *J. Comput Chem.* **2006**, *27(15)*, 1787–1799, DOI:10.1002/jcc.20495.

[26] Grimme, S.; Antony, J.; Ehrlich, S.; Krieg, H. A consistent and accurate ab initio parametrization of density functional dispersion correction (DFT-D) for the 94 elements H-Pu. *J. Chem. Phys.* **2010**, *132(15)*, 154104, DOI:10.1063/1.3382344.

[27] Grimme, S.; Ehrlich, S.; Goerigk, L. Effect of the damping function in dispersion corrected density functional theory. *J. Comput Chem.* **2011**, *32(7)*, 1456–1465, DOI:10.1002/jcc.21759.

[28] Caldeweyher, E.; Bannwarth, C.; Grimme, S. Extension of the D3 dispersion coefficient model. *J. Chem. Phys.* **2017**, *147(3)*, 034112, DOI:10.1063/1.4993215.

[29] Tkatchenko, A.; Scheffler, M. Accurate molecular van der Waals interactions from ground-state electron density and free-atom reference data. *Phys. Rev. Lett.* **2009**, *102(7)*, 073005, DOI:10.1103/PhysRevLett.102.073005.

[30] Schimka, L.; Harl, J.; Stroppa, A.; Grüneis, A.; Marsman, M.; Mittendorfer, F.; Kresse, G. Accurate surface and adsorption energies from many-body perturbation theory. *Nat. Mater.* **2010**, *9(9)*, 741–744, DOI:10.1038/nmat2806.
[31] Goerigk, L.; Hansen, A.; Bauer, C.; Ehrlich, S.; Najibi, A.; Grimme, S. A look at the density functional theory zoo with the advanced GMTKN55 database for general main group thermochemistry, kinetics and noncovalent interactions. *Phys. Chem. Chem. Phys.* **2017**, *19(48)*, 32184–32215, DOI:10.1039/C7CP04913G.
[32] Sharada, S. M.; Karlsson, R. K.; Maimaiti, Y.; Voss, J.; Bligaard, T. Adsorption on transition metal surfaces: Transferability and accuracy of DFT using the ADS41 dataset. *Phys. Rev. B.* **2019**, *100(3)*, 035439, DOI:10.1103/PhysRevB.100.035439.
[33] Sharada, S. M.; Bligaard, T.; Luntz, A. C.; Kroes, G. J.; Nørskov, J. K. SBH10: A benchmark database of barrier heights on transition metal surfaces. *J. Phys. Chem. C.* **2017**, *121(36)*, 19807–19815, DOI:10.1021/acs.jpcc.7b05677.
[34] Kratzer, P.; Neugebauer, J. The basics of electronic structure theory for periodic systems. *Front. Chem.* **2019**, *7*, 106, DOI:10.3389/fchem.2019.00106.
[35] Cardona-Martinez, N.; Dumesic, J. A. Applications of adsorption microcalorimetry to the study of heterogeneous catalysis. *Adv. Catal.* **1992**, *38*, 149–244, DOI:10.1016/S0360-0564(08)60007-3.
[36] Gravelle, P. C. Heat-flow microcalorimetry and its application to heterogeneous catalysis. *Adv. Catal.* **1972**, *22*, 91–263, DOI:10.1016/S0360-0564(08)60248-5.
[37] Gravelle, P. C. Application of adsorption calorimetry to the study of heterogeneous catalysis reactions. *Thermochim. Acta.* **1985**, *96*, 365–376, DOI:10.1016/0040-6031(85)80075-7.
[38] Brown, W. A.; Kose, R.; King, D. A. Femtomole adsorption calorimetry on single-crystal surfaces. *Chem. Rev.* **1998**, *98*, 797–831, DOI:10.1021/cr9700890.
[39] Campbell, C. T.; Sellers, J. R. V. Enthalpies and entropies of adsorption on well-defined oxide surfaces: Experimental measurements. *Chem. Rev.* **2013**, *113*, 4106–4135, DOI:10.1021/cr300329s.
[40] Stuck, A.; Wartnaby, C. E.; Yeo, Y. Y.; Stuckless, J. T.; Al-Sarraf, N.; King, D. A. An improved single crystal adsorption calorimeter. *Surf. Sci.* **1996**, *349*, 229–240, DOI:10.1016/0039-6028(95)01070-X.
[41] Schießer, A.; Hörtz, P.; Schäfer, R. Thermodynamics and kinetics of CO and benzene adsorption on Pt (111) studied with pulsed molecular beams and microcalorimetry. *Surf. Sci.* **2010**, *604(23-24)*, 2098–2105, DOI:10.1016/j.susc.2010.09.001.
[42] Parrillo, D. J.; Gorte, R. J. Characterization of acidity in H-ZSM-5, H-ZSM-12, H-Mordenite, and H-Y using microcalorimetry. *J. Phys. Chem.* **1993**, *97*, 8786–8792, DOI:10.1021/j100136a023.
[43] Parrillo, D. J.; Lee, C.; Gorte, R. J. Heats of adsorption for ammonia and pyridine in H-ZSM-5: Evidence for identical Brønsted-acid sites. *Appl. Catal. A.-Gen.* **1994**, *110*, 67–74, DOI:10.1016/0926-860X(94)80106-1.
[44] Parrillo, D. J.; Gorte, R. J. Design parameters for the construction and operation of heat-flow calorimeters. *Thermochim Acta.* **1998**, *312*, 125–13, DOI:10.1016/S0040-6031(97)00446-2.
[45] Sharma, S. B.; Meyers, B. L.; Chen, D. T.; Miller, J.; Dumesic, J. A. Characterization of catalyst acidity by microcalorimetry and temperature-programmed desorption. *Appl. Catal. A.-Gen.* **1993**, *102*, 253–265, DOI:10.1016/0926-860X(93)80232-F.
[46] Feibelman, P. J.; Hammer, B.; Nørskov, J. K.; Wagner, F.; Scheffler, M.; Stumpf, R.; Dumesic, J. A.; Watwe, R. The CO/Pt(111) Puzzle. *J. Phys. Chem. B.* **2001**, *105(18)*, 4018–4025, DOI:10.1021/jp002302t.
[47] Maiti, S.; Maiti, K.; Curnan, M. T.; Kim, K.; Noh, K. J.; Han, J. W. Engineering electrocatalyst nanosurfaces to enrich the activity by inducing lattice strain. *Energy Environ. Sci.* **2021**, *14(7)*, 3717–3756, DOI:10.1039/D1EE00074H.
[48] Liao, X.; Lu, R.; Xia, L.; Liu, Q.; Wang, H.; Zhao, K.; Wang, Z.; Zhao, Y. Density functional theory for electrocatalysis. *Energy Environ. Mater.* **2022**, *5(1)*, 157–185, DOI:10.1002/eem2.12204.

[49] Watwe, R. M.; Spiewak, B. E.; Cortright, R. D.; Dumesic, J. A. Density functional theory (DFT) and microcalorimetric investigations of CO adsorption on Pt clusters. *Catal. Lett.* **1998**, *51(3)*, 139–147, DOI:10.1023/A:1019038512945.

[50] Alcala, R.; Shabaker, J. W.; Huber, G. W.; Sanchez-Castillo, M. A.; Dumesic, J. A. Experimental and DFT studies of the conversion of ethanol and acetic acid on PtSn-based catalysts. *J. Phys. Chem. B.* **2005**, *109(6)*, 2074–2085, DOI:10.1021/jp049354t.

[51] Rubeš, M.; Trachta, M.; Koudelková, E.; Bulánek, R.; Klimes, J.; Nachtigall, P.; Bludský, O. Temperature dependence of carbon monoxide adsorption on a high-silica H-FER zeolite. *J. Phys. Chem. C.* **2018**, *122(45)*, 26088–26095, DOI:10.1021/acs.jpcc.8b08935.

[52] Voleská, I.; Nachtigall, P.; Ivanova, E.; Hadjiivanov, K.; Bulánek, R. Theoretical and experimental study of CO adsorption on Ca-FER zeolite. *Catal. Today.* **2015**, *243*, 53–61, DOI:10.1016/j.cattod.2014.07.029.

[53] Liu, C.; Tranca, I.; van Santen, R. A.; Hensen, E. J.; Pidko, E. A. Scaling relations for acidity and reactivity of zeolites. *J. Phys. Chem. C.* **2017**, *121(42)*, 23520–23530, DOI:g/10.1021/acs.jpcc.7b08176.

[54] Kumashiro, R.; Fujie, K.; Kondo, A.; Mori, T.; Nagao, M.; Kobayashi, H.; Kuroda, Y. Development of a new analysis method evaluating adsorption energies for the respective ion-exchanged sites on alkali-metal ion-exchanged ZSM-5 utilizing CO as a probe molecule: IR-spectroscopic and calorimetric studies combined with a DFT method. *Phys. Chem. Chem. Phys.* **2009**, *11(25)*, 5041–5051, DOI:10.1039/B818323F.

[55] Gautier, S.; Steinmann, S. N.; Michel, C.; Fleurat-Lessard, P.; Sautet, P. Molecular adsorption at Pt (111). How accurate are DFT functionals? *Phys. Chem. Chem. Phys.* **2015**, *17(43)*, 28921–28930, DOI:10.1039/C5CP04534G.

[56] Karp, E. M.; Campbell, C. T.; Studt, F.; Abild-Pedersen, F.; Nørskov, J. K. Energetics of oxygen adatoms, hydroxyl species and water dissociation on Pt (111). *J. Phys. Chem. C.* **2012**, *116(49)*, 25772–25776, DOI:10.1021/jp3066794.

[57] Sharma, S. B.; Miller, M. T.; Dumesic, J. A. Microcalorimetric study of silica- and zeolite-supported platinum catalysts. *J. Catal.* **1994**, *148*, 198–204, DOI:10.1006/jcat.1994.1201.

[58] de Ménorval, L.-C.; Chaqroune, A.; Coq, B.; Figueras, F. Characterization of mono- and bi-metallic platinum catalysts using CO FTIR spectroscopy. *J. Chem. Soc. Faraday Trans.* **1997**, *93(20)*, 3715–3720, DOI:10.1039/a702174g.

[59] Perdew, J. P.; Zunger, A. Self-interaction correction to density-functional approximations for many-electron systems. *Phys. Rev. B.* **1981**, *23(10)*, 5048, DOI:10.1103/PhysRevB.23.5048.

[60] Olsen, R. A.; Philipsen, P. H. T.; Baerends, E. J. CO on Pt (111): A puzzle revisited. *J Chem. Phys.* **2003**, *119(8)*, 4522–4528, DOI:10.1063/1.1593629.

[61] Luo, S.; Zhao, Y.; Truhlar, D. G. Improved CO Adsorption energies, site preferences, and surface formation energies from a meta-generalized gradient approximation exchange–correlation functional, M06-L. *J. Phys. Chem. Lett.* **2012**, *3(20)*, 2975–2979, DOI:10.1021/jz301182a.

[62] Stroppa, A.; Termentzidis, K.; Paier, J.; Kresse, G.; Hafner, J. CO adsorption on metal surfaces: A hybrid functional study with plane-wave basis set. *Phys. Rev. B.* **2007**, *76(19)*, 195440, DOI:10.1103/PhysRevB.76.195440.

[63] Stroppa, A.; Kresse, G. The shortcomings of semi-local and hybrid functionals: What we can learn from surface science studies. *New J. Phys:.* **2008**, *10(6)*, 063020, DOI:10.1088/1367-2630/10/6/063020.

[64] Chen, Z.; Li, J.; Yang, P.; Cheng, Z.; Li, J.; Zuo, S. Ce-modified mesoporous γ-Al_2O_3 supported Pd-Pt nanoparticle catalysts and their structure-function relationship in complete benzene oxidation. *Chem. Eng. J.* **2019**, *356*, 255–261, DOI:10.1016/j.cej.2018.09.040.

[65] Yang, K.; Liu, Y.; Deng, J.; Zhao, X.; Yang, J.; Han, Z.; Dai, H. Three-dimensionally ordered mesoporous iron oxide-supported single-atom platinum: Highly active catalysts for benzene combustion. *Appl. Catal., B.* **2019**, *244*, 650–659, DOI:10.1016/j.apcatb.2018.11.077.

[66] Guo, Y.; Gao, Y.; Li, X.; Zhuang, G.; Wang, K.; Zheng, Y.; Li, Q. Catalytic benzene oxidation by biogenic Pd nanoparticles over 3D-ordered mesoporous CeO_2. *Chem. Eng. J.* **2019**, *362*, 41–52, DOI:10.1016/j.cej.2019.01.012.

[67] Gao, W.; Zheng, W. T.; Jiang, Q. Dehydrogenation of benzene on Pt (111) surface. *J. Chem. Phys.* **2008**, *129(16)*, 164705, DOI:10.1063/1.3001610.

[68] Carey, S. J.; Zhao, W.; Campbell, C. T. Energetics of adsorbed benzene on Ni (111) and Pt (111) by calorimetry. *Surf. Sci.* **2018**, *676*, 9–16, DOI:10.1016/j.susc.2018.02.014.

[69] Shayeghi, A.; Krähling, S.; Hörtz, P.; Johnston, R. L.; Heard, C. J.; Schäfer, R. Adsorption of acetonitrile, benzene, and benzonitrile on Pt (111): Single crystal adsorption calorimetry and density functional theory. *J. Phys. Chem. C.* **2017**, *121(39)*, 21354–21363, DOI:10.1021/acs.jpcc.7b05549.

[70] Yildirim, H.; Greber, T.; Kara, A. Trends in adsorption characteristics of benzene on transition metal surfaces: Role of surface chemistry and van der Waals interactions. *J. Phys. Chem. C.* **2013**, *117(40)*, 20572–20583, DOI:10.1021/jp404487z.

[71] Sabbe, M. K.; Lain, L.; Reyniers, M. F.; Marin, G. B. Benzene adsorption on binary Pt_3M alloys and surface alloys: A DFT study. *Phys. Chem. Chem. Phys.* **2013**, *15(29)*, 12197–12214, DOI:10.1039/C3CP50617G.

[72] Yildirim, H.; Greber, T.; Kara, A. Trends in adsorption characteristics of benzene on transition metal surfaces: Role of surface chemistry and van der Waals interactions. *J. Phys. Chem. C.* **2013**, *117(40)*, 20572–20583, DOI:10.1021/jp404487z.

[73] Peköz, R.; Donadio, D. Effect of van der Waals interactions on the chemisorption and physisorption of phenol and phenoxy on metal surfaces. *J. Chem. Phys.* **2016**, *145(10)*, 104701, DOI:10.1063/1.4962236.

[74] Zhang, R.; Hensley, A. J.; McEwen, J. S.; Wickert, S.; Darlatt, E.; Fischer, K.; Steinrück, H. P. Integrated X-ray photoelectron spectroscopy and DFT characterization of benzene adsorption on Pt (111), Pt (355) and Pt (322) surfaces. *Phys. Chem. Chem. Phys.* **2013**, *15(47)*, 20662–20671, DOI:10.1039/C3CP53127A.

[75] Réocreux, R.; Huynh, M.; Michel, C.; Sautet, P. Controlling the adsorption of aromatic compounds on Pt (111) with oxygenate substituents: From DFT to simple molecular descriptors. *J. Phys. Chem. Lett.* **2016**, *7(11)*, 2074–2079, DOI:10.1021/acs.jpclett.6b00612.

[76] Ihm, H.; Ajo, H. M.; Gottfried, J. M.; Bera, P.; Campbell, C. T. Calorimetric measurement of the heat of adsorption of benzene on Pt (111). *J. Phys. Chem. B.* **2004**, *108(38)*, 14627–14633. DOI:10.1021/jp040159o.

[77] Pašti, I. A.; Marković, A.; Gavrilov, N.; Mentus, S. V. Adsorption of acetonitrile on platinum and its effects on oxygen reduction reaction in acidic aqueous solutions – Combined theoretical and experimental study. *Electrocatalysis.* **2016**, *7(3)*, 235–248, DOI:10.1007/s12678-016-0301-6.

[78] Chen, D. T.; Sharma, S. B.; Filimonov, I.; Dumesic, J. A. Microcalorimetric studies of zeolite acidity. *Catal: Lett.* **1992**, *12*, 201–212, DOI:10.1007/BF00767202.

[79] Parrillo, D. J.; Gorte, R. J.; Farneth, W. E. A calorimetric study of simple bases in H-ZSM-5: A comparison with gas-phase and solution-phase acidities. *J. Am. Chem. Soc.* **1993**, *115*, 12441–12445, DOI:10.1021/ja00079a027.

[80] Cardona-Martínez, N.; Dumesic, J. A. Acid strength of silica-alumina and silica studied by microcalorimetric measurements of pyridine adsorption. *J. Catal.* **1990**, *125*, 427–444, DOI:10.1016/0021-9517(90)90316-C.

[81] Lee, -C.-C.; Gorte, R. J.; Farneth, W. E. Calorimetric study of alcohol and nitrile adsorption complexes in H-ZSM-5. *J. Phys. Chem. B.* **1997**, *101*, 3811–3817, DOI:10.1021/jp970711s.

[82] Chiu, C. C.; Vayssilov, G. N.; Genest, A.; Borgna, A.; Roesch, N. Predicting adsorption enthalpies on silicalite and HZSM-5: A benchmark study on DFT strategies addressing dispersion interactions. *Journal of Computational Chemistry.* **2014**, *35(10)*, 809–819, DOI:10.1002/jcc.23558.

[83] Ma, S.; Liu, Z. P. The role of zeolite framework in zeolite stability and catalysis from recent atomic simulation. *Top. Catal.* **2022**, *65(1)*, 59–68, DOI:10.1007/s11244-021-01473-6.

[84] Bucko, T.; Hafner, J.; Benco, L. Adsorption and vibrational spectroscopy of CO on mordenite: Ab initio density-functional study. *J. Phys. Chem. B.* **2005**, *109(15)*, 7345–7357, DOI:10.1021/jp050151u.

[85] Savitz, S.; Myers, A. L.; Gorte, R. J. Calorimetric Investigation of CO and N_2 for characterization of acidity in zeolite H– MFI. *J. Phys. Chem. B.* **1999**, *103(18)*, 3687–3690, DOI:10.1021/jp990157h.

[86] Boronat, M.; Corma, A. What is measured when measuring acidity in zeolites with probe molecules? *ACS Catal.* **2019**, *9(2)*, 1539–1548, DOI:10.1021/acscatal.8b04317.

[87] Parrillo, D. J.; Gorte, R. J. Characterization of acidity in H-ZSM-5, H-ZSM-12, H-Mordenite, and HY using microcalorimetry. *J. Phys. Chem.* **1993**, *97(34)*, 8786–8792, DOI:10.1021/j100136a023.

[88] Wang, X.; Chen, L.; Vennestrøm, P. N. R.; Janssens, T. V. W.; Jansson, J.; Grönbeck, H.; Skoglundh, M. Direct measurement of enthalpy and entropy changes in NH_3 promoted O_2 activation over Cu-CHA at low temperature. *Chem. Cat. Chem.* **2021**, *13*, 2577–2582, DOI:10.1002/cctc.202100253.

[89] Benrezgua, E.; Zoukel, A.; Deghfel, B.; Boukhari, A.; Amari, R.; Kheawhom, S.; Mohamad, A. A. A review on DFT+ U scheme for structural, electronic, optical and magnetic properties of copper doped ZnO wurtzite structure. *Mater. Today Commun.* **2022**, 103306, DOI:10.1016/j.mtcomm.2022.103306.

[90] Yu, M.; Yang, S.; Wu, C.; Marom, N. Machine learning the Hubbard U parameter in DFT+ U using Bayesian optimization. *Npj Comput. Mater.* **2020**, *6(1)*, 1–6, DOI:10.1038/s41524-020-00446-9.

Mohamed Zbair, Elliot Scuiller, Patrick Dutournié, and Simona Bennici*

Chapter 6 Major concern regarding thermophysical parameters' measurement techniques of thermochemical storage materials

Abstract: Thermochemical heat storage (thermochemical energy storage, TES) is a methodology that allows storing heat at ambient temperature and release it in a second time for heating or cooling applications. The main application of TES is to provide heat load temporal shifting to use this energy to compensate for periods of intense heat needs. It can be combined with renewable energies, such as solar thermal heating systems, for providing a consistent heat supply. The use of renewable energies can then gain in efficiency and competitiveness, facilitating their deployment. Optimization and implementation of innovative heat storage devices (TES applications) required a detailed and comprehensive study of the system's behavior. To this end, numerical modeling and simulation tools can be applied, but a prior assessment of the thermal properties of the heat storage material is required. The parameters that need to be measured are:

- thermal conductivity,
- heat of reaction, and
- heat capacity.

Researchers are looking for standards, methods, and procedures for estimating the thermal properties of TES materials. The thermal characterization of these materials is presented and discussed in this chapter which focuses on the calorimetry technique and thermal analysis dedicated methods to estimate the thermal conductivity, the thermal diffusivity, and the effusivity. This chapter was also driven by the need to fill a discrepancy in the literature while reviewing the thermophysical property assessments utilized in the TES material study. Researchers working in materials science for TES materials development and thermal engineers working on TES systems will find this chapter of interest. Overall, the TES community will confront numerous future obstacles to reach a common understanding that will result in similar outcomes across

***Corresponding author: Simona Bennici,** Université de Haute-alsace, CNRS, IS2M UMR 7361, F-68100 Mulhouse, France, Université de Strasbourg, France, e-mail: simona.bennici@uha.fr

Mohamed Zbair, Université de Haute-alsace, CNRS, IS2M UMR 7361, F-68100 Mulhouse, France, Université de Strasbourg, France, e-mail: mohamed.zbair@uha.fr

Elliot Scuiller, Université de Haute-alsace, CNRS, IS2M UMR 7361, F-68100 Mulhouse, France, Université de Strasbourg, France, e-mail: elliot.scuiller@uha.fr

Patrick Dutournié, Université de Haute-alsace, CNRS, IS2M UMR 7361, F-68100 Mulhouse, France, Université de Strasbourg, France, e-mail: patrick.dutournie@uha.fr

https://doi.org/10.1515/9783110590449-006

investigations. This entails optimizing sample preparation and method procedure across all approaches to achieve high accuracies for various TES materials.

Keywords: Heat storage material, thermophysical properties, thermal conductivity, specific heat capacity, hygroscopic materials

6.1 Introduction

Due to environmental issues (in particular global warming due to the unreasonable use of fossil fuels), the international community has taken steps to promote the use of renewable energies. Solar energy is a promising alternative to existing energy sources. Yet, its intermittent and seasonal nature is a major disadvantage, resulting in a supply-demand gap. Energy storage is an effective way to close the gap [1].

Thermochemical energy storage (TES) is a potential method for preserving energy and exploiting variable renewable energy sources and waste heat. Regarding TES technology, there are three types of systems:

- Sensible heat storage (SHS): stores energy proportionally correlated to a medium's temperature change
- Latent heat storage (LHS): related to the phase change enthalpies of different phase transitions
- Thermochemical heat storage (THS): connected to endothermic/exothermic transformations (e.g., chemical reactions, adsorption, or absorption) [2]

SHS is the most advanced method for storing thermal energy. The SHS system stores energy without phase change, and the energy storage density is determined by the mass and specific heat capacity of the material and the difference in temperature. The SHS system's benefits include low cost and thermal stability. At the same time, its disadvantages are a lower energy storage density than LHS and THS systems, requiring larger volume, significant thermal losses, and imbalanced discharge temperature.

The LHS system stores the heat of a phase change of a material (solid-solid or solid-liquid) [3, 4]. The energy stored is directly proportional to the latent heat of the phase change. The LHS system stores more energy than the SHS systems for the same volume and releases the heat at a consistent and almost constant discharge temperature. The main drawbacks of the most common materials used in LHS systems are the poor heat conductivity, certain organic chemicals' flammability, and in specific cases, the corrosiveness [5, 6].

Thermochemical heat storage systems use chemical or ad/absorption reactions to store heat by changing operation conditions and consequently reach the condition for reaction spontaneity. Two types of THS systems are discussed in the literature: sorption-based THS and reaction-based THS. A THS technique has several major advantages: high energy storage density, minor losses, and storage at ambient temperature for prolonged periods of time. Sorption-based THS has piqued the curiosity of scientific researchers all over the world for these reasons.

6.2 Principles of sorption-based THS

The term sorption refers to two distinct processes that are distinguished from the state of material, namely liquid sorption and solid sorption [1]. Despite the reactants used, the principle of a sorption thermochemical heat storage (STHS) can be schematically represented as shown in Fig. 6.1: during the charging phase (heat storage), the compound AB is separated into two reactants, A and B, by providing the heat of desorption, ΔH_{desorp}, at high temperature (endothermic reaction). The thermal energy can be stored as long as two reactants are separated. At medium temperature, A and B react during the discharging phase to form AB and to release the stored energy, ΔH_{sorp} (exothermic reaction).

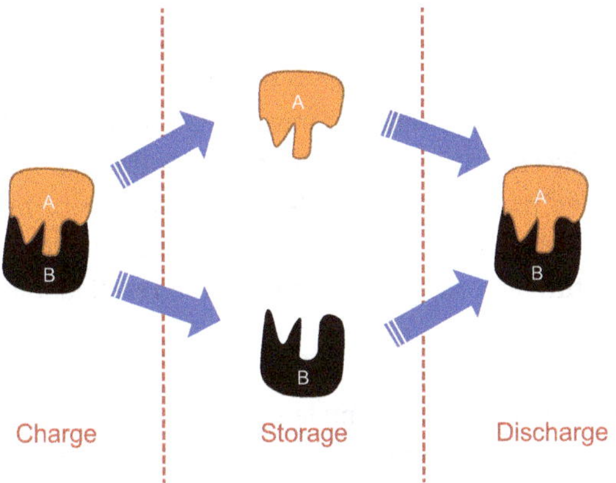

Fig. 6.1: Sorption heat storage: charging and discharging phases.

6.3 Closed and open system for sorption-based THS

Different systems and set-ups have been investigated and proposed to meet different energy needs, such as hot water and domestic heating, by deeply studying different sorbent/sorbate reactants. Depending on the targets, these systems can operate continuously or in batches (open or closed systems). On the basis of whether or not they exhibit mass and heat transfer to the surrounding environment, STHS processes have indeed been split into closed and open systems. Figure 6.2 depicts their distinct operation principles:

- STHS closed systems. The system consists of a reactor with the sorbent material and a reactor-exchanger containing the sorbate, both separated from the atmospheric environment. During the charging operation, the saturated material is regenerated (separation of the sorbent and sorbate) when heated up to the complete separation. After closing the separation valve, the pressure decreases in the reactor containing the unsaturated sorbent. When heat is needed, the separation valve is opened. The liquid sorbate is evaporated (cold generation in the reactor-exchanger), and after reactive contact with the sorbent material, it releases the heat. This process can be used to supply both heating and cooling, and it can be specially used to supply cooling systems via the latent heat exchanged.

The possibility to use certain toxic sorbates (e.g., NH_3), which cannot be released into the natural environment, and the ability of the system to provide cool in the summer and heat in the winter are two benefits of a closed system. The reactor and the exchanger are more technically challenging to design and produce because of the complicated structure operating at low pressure and requiring a good heat transfer of the sorbent material [7].

- STES open systems. In this system, the air is used both as a heat and sorbate carrier. It consists of a reactor containing the sorbent, where the air flows through. When heat is needed, moist air (from the surrounding atmosphere or humidified) is injected into the reactor. During the charging process, dry air flows through the sorbent material to dehydrate it. The heat can be provided by the hot air or by direct heat supply into the reactor. During the discharge phase, moist air flows through the sorbent material (unsaturated), supplying water for the reaction and carrying thermal energy.

An open system's key advantages are its comparatively simple structure and the ease of fabrication and maintenance. An open system provides better heat and mass transfer than a closed system. Its weaknesses include the risk of changing the moisture content of the room air during operation. This problem can be avoided using exhaust (stale) and fresh air from a heat recovery ventilation system. Furthermore, hazardous materials cannot be used as sorbent materials.

6.4 Sorption storage materials

In the development of sorption thermal storage systems, storage materials are critical. The following are some key guidelines for selecting appropriate materials [2, 8, 9]:
- Sorbate uptake is high (g sorbate/g sorbent)
- High energy density at system operating temperatures
- Regeneration should take place at low temperature

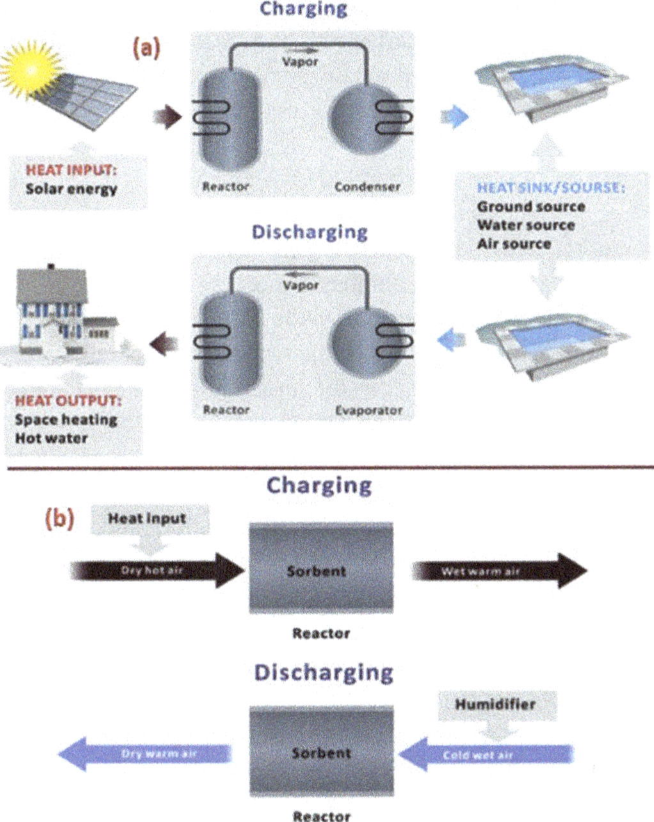

Fig. 6.2: (**a**) Principal operation of closed sorption system and (**b**) principal operation of open sorption system (reprinted with permission from ref [1]., Copyright 2013, Elsevier).

- Adsorption mass and heat transfer are good
- High thermal conductivity
- Nonpoisonous and easy to handle
- Low cost, nontoxic, nonflammable, noncorrosive
- There is no deterioration due to thermal stability

There is currently no material that can meet all of the specifications. As a result, the material selection should be evaluated and optimized. The basics for finding good storage materials are sorption characteristics. The full examination of sorption properties of a mass of accessible sorbents takes time because it necessitates precise measurements of a collection of sorption isobars, isosteres, and isotherms over a large temperature and pressure range. Aristov [10, 11] developed a novel method for finding acceptable adsorbents for adsorption heat transfer (AHT) cycles – adsorption cooling, heating, and storage cycles – called target-oriented design (or tailoring). The approach's core idea is that there is an ideal adsorbent for each AHT cycle with specific climatic and boundary conditions. Two steps comprise the target-oriented design:

- The first stage is to determine the temperature requirements for a specific application.
- The second step is to produce an adsorbent with characteristics that meet those needs.

The modification of composites with "salt inside the porous matrix" was seen as a suitable example of nanotailoring sorption capabilities to meet practical needs.

Sorbate selection criteria are limited in comparison to sorbent selection criteria. Due to its ecologically favorable features and high evaporation heat, several working pairs or materials evaluated primarily use water as the sorbate. Methanol and ammonia should be considered if the ambient temperature is below 0 °C.

Because part of the energy is discharged into or recovered from the environment, several factors affect the performance of a sorption system. The charging (temperature and pressure) and discharging operating conditions (temperature and pressure) affect the energy storage density. As a result, proper criteria should be specified while comparing sorbents' storage density values. The necessity of choosing an appropriate heat source is driven by the charging temperature, making it the most important parameter. Adsorbent materials with high energy storage densities and low charging temperatures are the most coveted sorption materials.

Figure 6.3 shows a possible classification of sorption materials, divided into four groups: liquid absorption, solid adsorption, chemical reaction, and composite materials. This classification is done regarding the involved reactions. Table 6.1 lists the most well-studied sorption storage materials, along with their key features under various conditions.

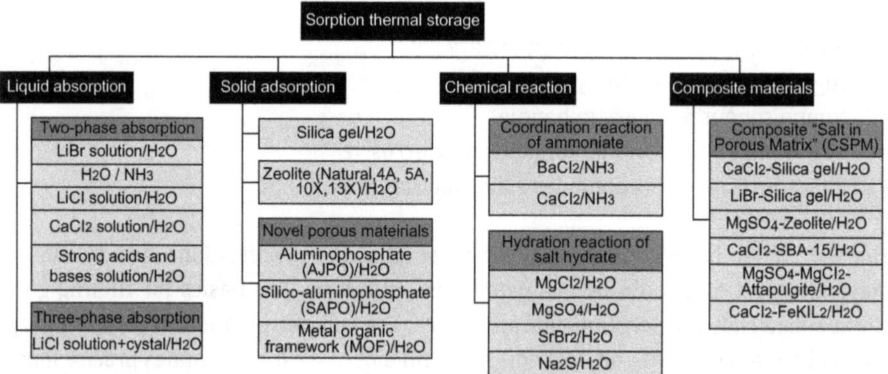

Fig. 6.3: Classification of sorption thermochemical heat storage materials (reprinted with permission from ref [1]., Copyright 2013, Elsevier).

Tab. 6.1: Theoretical energy storage density, thermal conductivity, and density of sorption storage materials.

Sorption storage material	Experimental conditions	Water uptake/ enthalpy reaction	Theoretical energy storage density (GJ/m^3)	Thermal conductivity (W/m K)	Density (kg/m^3)
Zeolites 4A	$T_{charging}$: 130 °C, $T_{discharging}$: 65 °C [12]	n.a.	0.58 (experimental) [12]	Solid grain: 0.18–0.4 [13, 14]	n.a.
	$T_{charging}$: 180 °C [1]		0.45 [1]		700 [1]
	$T_{charging}$: 150 °C [1]		0.353 [1]		700 [1]
Zeolites 13X	$T_{charging}$: 180 °C [1]		0.396 [1]		739 [1]
	$T_{charging}$: 130 °C, T_{cond}: 40 °C [1]		0.349 [1]	Granular bed: 0.03–0.15 [15] (modulated radial-heat flow technique) [16]	896 [1]
	$T_{discharging}$: 65 °C, T_{evap}: 5 °C [1]				
Zeolite MgNaX	$T_{charging}$: 180 °C [1]		0.461 [1]		732 [1]
Zeolite LiX			0.576 [1]		711 [1]
Zeolite CaNaA-60			0.418 [1]		671 [1]
Zeolite NaCa 5A	$T_{charging}$: 130 °C, T_{cond}: 40 °C [1]		0.205 [1]		801 [1]
Zeolite Y	$T_{discharging}$: 65 °C, T_{evap}: 5 °C [1]		128 Wh/kg [1]		n.a.
Microporus silica gel SG-127B	$T_{charging}$: 90 °C, T_{cond}: 40 °C [1]	n.a.	0.09 [1]	Solid grain: 0.37–0.8 [14] (steady state by vertical cylindrical apparatus) [17]	710 [1]
Macroporus silica gel SG-LE32	$T_{discharging}$: 40 °C, T_{evap}: 15 °C [1]		0.054 [1]	Granular bed: 0.05–0.20 [14]	620 [1]

(continued)

Tab. 6.1 (continued)

Sorption storage material	Experimental conditions	Water uptake/ enthalpy reaction	Theoretical energy storage density (GJ/m^3)	Thermal conductivity (W/m K)	Density (kg/m^3)
$CaCl_2 \cdot 2H_2O/CaCl_2$	95 °C	n.a.	1.1 [12]	Granular bed: 0.1–0.2 [14]	n.a.
$CaCl_2 \cdot 2H_2O/CaCl_2$	$T_{discharging}$: 174 °C [12] $T_{charging}$: 95 °C [12] $T_{discharging}$: 35 °C [12]	n.a.	0.60 [12] 0.72 [12]	n.a.	n.a.
$Al_2(SO_4)_3 \cdot 6H_2O/Al_2(SO4)_3$	150 °C [12]	n.a.	1.9 [12]	n.a.	n.a.
$MgSO_4 \cdot 7H_2O/MgSO_4$	Turnover temperature: 122 °C [14]	n.a.	2.8 [14] 1.5 [12]	n.a.	n.a.
$MgSO_4 \cdot 7H_2O/MgSO_4 \cdot H_2O$	$T_{charging}$: 150 °C [12] $T_{discharging}$: 105 °C [12]	n.a.	2.3 [12]	n.a.	n.a.
$MgSO_4 \cdot 7H_2O/$ $MgSO_4 \cdot 0.1H_2O$	$T_{charging}$: 150 °C [1]	n.a.	3.33 [1]	n.a.	1,330 [1]
$MgSO_4 \cdot 6H_2O/MgSO_4 \cdot H_2O$	$T_{charging}$: 72 °C [12]	n.a.	2.37 [12] 1.83 (experimental) [12]	n.a.	n.a.
$CaSO_4 \cdot 2H_2O/CaSO_4$	Turnover temperature: 89 °C [14]	n.a.	1.4 [14]	n.a.	n.a.
$MgCl_2 \cdot 6H_2O/MgCl_2 \cdot H_2O$	$T_{charging}$: 150 °C [12] $T_{discharging}$: 50–30 °C [12]	n.a.	2.5 [12] 0.71 (experimental) [12]	n.a.	n.a.

$MgCl_2 \cdot 4H_2O/MgCl_2 \cdot 2H_2O$	$T_{charging}$: 118 °C [12]	n.a.	1.27 [12] 1.10 (experimental) [12]	n.a.	n.a.
$MgCl_2 \cdot 6H_2O/MgCl_2 \cdot 2H_2O$	$T_{charging}$: 130 °C [1]	n.a.	2.00 [1]	n.a.	1,165 [1]
$Na_2S \cdot 5H_2O/Na_2S \cdot 1/2H_2O$	$T_{charging}$: 80 °C [1], [12] $T_{discharging}$: 65 °C [12]	n.a.	2.7 [12] 3.56 [1]	n.a.	928 [1]
$SrBr_2 \cdot 6H_2O/SrBr_2 \cdot H_2O$	$T_{discharging}$: 23.5 °C [12]	n.a.	2.3 [12] 2.08 (experimental) [12]	n.a.	n.a.
$Li_2SO_4 \cdot H_2O/LiSO_4$	$T_{charging}$: 103 °C [12]	n.a.	0.92 [12] 0.80 (experimental) [12]	n.a.	n.a.
$CuSO_4 \cdot 5H_2O/CuSO_4 \cdot H_2O$	$T_{charging}$: 92 °C [12]	n.a.	2.07 [12] 1.85 (experimental) [12]	n.a.	n.a.
$SrBr_2 \cdot 6H_2O/SrBr_2 \cdot H_2O$ + ENG	$T_{charging}$: 80 °C [1]		0.63 [1]		1165 [1]
$CaCl_2$ + ENGP-S-1	From 100 to 200 °C	n.a.	n.a.	1.64 [18]	85 [18]
$CaCl_2$ + ENGP-S-1		n.a.	n.a.	0.74 [18]	46 [18]
$CaCl_2$ + ACF VI		Dehydration: 0.10 g/g, 235 J/g Hydration: 0.12 g/g, 198 J/g	n.a.	1.03 [18]	649 [18]
Aluminosilicate + 30% $CaCl_2$	$T_{charging}$: 180 °C [1]	n.a.	0.62 [1]	n.a.	972 [1]

(continued)

Tab. 6.1 (continued)

Sorption storage material	Experimental conditions	Water uptake/ enthalpy reaction	Theoretical energy storage density (GJ/m^3)	Thermal conductivity (W/m K)	Density (kg/m^3)
Mesoporous silica gel + 33.7% $CaCl_2$	$T_{charging}$: 90 °C, T_{cond}: 40 °C [1] $T_{discharging}$: 40 °C, T_{evap}: 15 °C [1]	n.a.	0.31 [1]	n.a.	646 [1]
Zeolite 13X + 15% $MgSO_4$	$T_{charging}$: 150 °C [1] $T_{discharging}$: 30 °C [1]	n.a.	0.60 [1]	n.a.	922 [1]
Zeolite 4A + 10% $MgSO_4$	$T_{charging}$: 180 °C [1]	n.a.	0.64 [1]	n.a.	797 [1]
Bentonite + 40%$CaCl_2$	$T_{charging}$: 150 °C [1]	n.a.	0.49 [1]	n.a.	695 [1]
Attapulgite + $MgSO_4$ and $MgCl_2$ (mass ratio 20/80)	$T_{charging}$: 130 °C [1] $T_{discharging}$: 30 °C [1] 85% RH	n.a.	442 Wh/kg [1]	n.a.	n.a.

6.5 Thermophysical material characterization

Materials' thermophysical characteristics are critical parameters. They define the behavior of materials toward heat transfers (thermal conductivity, thermal diffusivity, and thermal effusivity), as well as toward their capacity to store or transform energy, which contributes to the material's temperature rise or transformation (specific heat capacity, enthalpy of reaction). These characteristics are influenced by the material's nature, as well as by the temperature. This section will overview the various strategies and existing methods for measuring these thermal properties. This outlook will not be exhaustive due to the large number of existing approaches published in the literature. However, to demonstrate the richness of this problem, we will refer to a few of them:

- DSC/calorimetry for C_p measurement
- Guarded hot plate method-GHP
- Method known as radial flow or coaxial cylinder cell
- Hot-wire method
- Transient plane source (TPS) method "hot disk"
- Laser flash method

In this section, we will discuss the most common methodologies utilized to determine the thermal properties of THS materials. The goal is to demonstrate the various classes of existing methods, as well as their application fields and potential restrictions.

6.5.1 Enthalpy of hydration reaction-ΔH_h

Differential scanning calorimetry (DSC) is the most appropriate and effective technique to estimate the heat exchanged by a STHS material [19–22]. It allows the measurement of the heat flow during the sorption/desorption phases. These experiments can be conducted under THS operating conditions (Fig. 6.4a). Frazzica and coworkers [19] suggested a technique using a modified DSC/TG equipment to measure sorption/desorption equilibrium curves and the associated heat stored or released (integration of heat flow peak) by performing experiments under saturated vapor conditions.

After determining the quantity of heat released or absorbed during the sorption or desorption process, ΔH_h can be calculated by dividing the measured heat by the amount of water exchanged ($\Delta H_h = -Q/n(H_2O)$). Lastly, to calculate the heat storage capacity, multiply the ΔH_h by the sorbate uptake observed during the sorption/desorption phase. Figure 6.5 shows an example of a heat of sorption evaluation in a TG/DSC apparatus for a given sorption step. The heat associated with the sorption process is obtained by integrating the heat flux (blue curve).

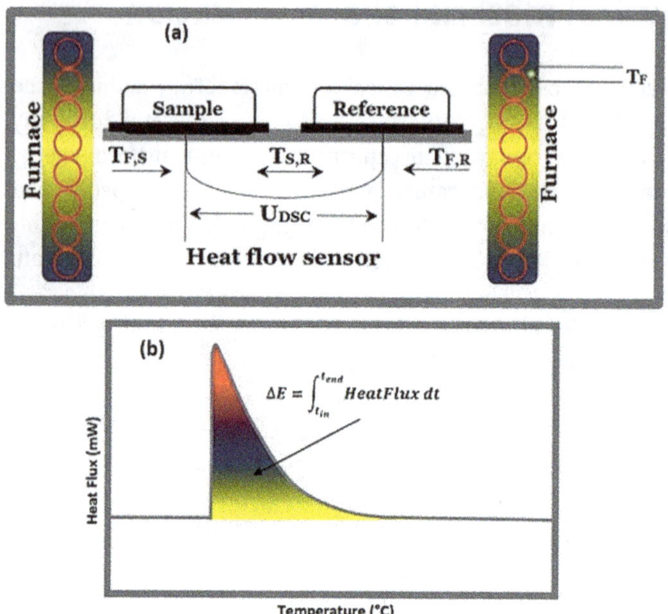

Fig. 6.4: (**a**) Graphics of a heat flow DSC with a disk-type sensor – heat flow paths. (**b**) Schematic procedure for the evaluation of enthalpy of adsorption.

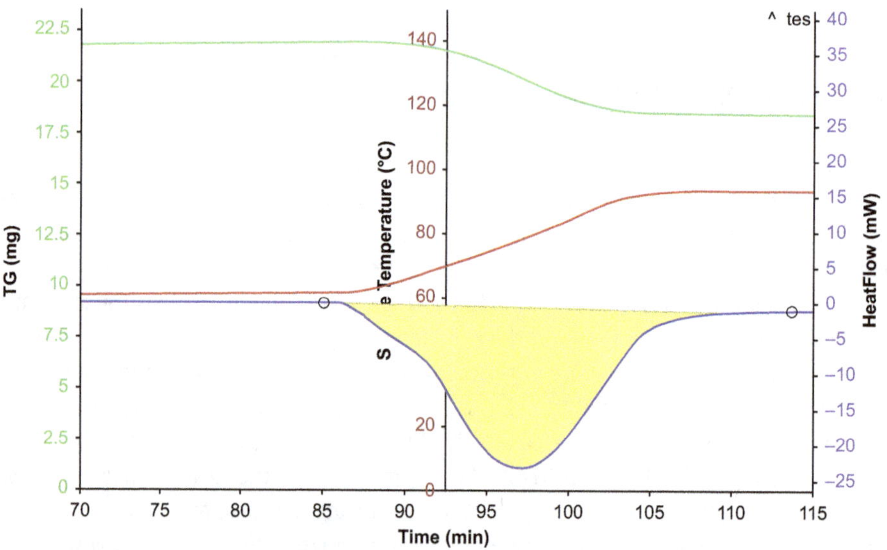

Fig. 6.5: Heat flux (blue), temperature (red), and weight (green) during the desorption stage at P = Cte (reprinted with permission from ref [19]., Copyright 2014, Elsevier).

Brancato et al. [23] conducted a study on the experimental characterization of newly shaped water sorbent, LiCl/vermiculite, for thermal energy storage applications. In a dedicated TG/DSC apparatus, the material's sorption ability and the thermal storage capacity (TSC) were examined under two significant boundary conditions: condition 1 (C1) and condition 2 (C2) storage applications. The heat released value for LiCl [52]/vermiculite in the C1 cycle was 1.88 ± 0.02 kJ/g at T_{ch} = 75–80 °C and was independent of the charging temperature (Tab. 6.2). The heat released ΔH_r was 2.8 ± 0.2 kJ/g_{wat} (Tab. 6.2), which is slightly higher than the enthalpy of water evaporation (2.42 kJ/g at T = 35 °C), indicating a strong interaction between the sorbed water and the LiCl [52]/vermiculite composite.

Tab. 6.2: TSC of the composite for temperature ranging between 75 and 85 °C [23].

Cycle	T_{ch} (°C)	Δw (g/g)	ΔH (kJ/g_{wat})	TSC (kJ/g_{mat})
C1	75	0.64 ± 0.01	2.9 ± 0.1	1.87 ± 0.01
	80	0.64 ± 0.01	2.9 ± 0.1	1.89 ± 0.01
C2	75	0.45 ± 0.03	2.8 ± 0.2	1.23 ± 0.04
	80	0.58 ± 0.04	2.8 ± 0.2	1.49 ± 0.07
	85	0.75 ± 0.03	2.8 ± 0.2	2.15 ± 0.01

mat, material (adsorbent); W, water (sorbate); T_{ch}, charging temperature.

6.5.2 Specific heat capacity, C_p

The specific heat capacity of sorbent materials is required to stimulate THS systems [24, 25]. The specific heat capacity (C_p) is expressed in J/kg/K. It reflects the ability of a material to absorb and release heat, and it is involved in unsteady thermal problems. The importance of knowing the specific heat capacity of materials for THS systems (specifically ρC_p), as well as the lack of a clear and consistent technique in the literature, motivated by the necessity to focus on this subject.

Material specific heat capacity (C_p), particularly (ρC_p) contributes to the thermal inertia and sensible heat storage. There are various techniques available to explore these properties, such as:

- Laser flash method [26]
- Transient experiments with hot disk [27]
- Differential scanning calorimetry [28]

These conventional approaches determine the C_p of material through transient measurements using a temperature gradient. In the measurement of C_p of adsorbent materials, especially hygroscopic ones, temperature variations occur, which result in variations in solvent contents and, as a result, in weight. Besides, if the heat of the reaction is high, the temperature is significantly altered, creating measurement inaccuracy. Materials employed in THS applications are an example of this comportment [2, 29]. These materials (zeolites and hydrated salts) are highly hygroscopic, and the adsorption reaction is extremely exothermic [30, 31].

Several studies reported in the literature deal with the estimation or measurement of specific heat capacity (C_p) or heat capacity (ρC_p), for example, the works of Blanco [32], Schick [33], and Vyazovkin [34]. However, this chapter section is not devoted to describe these significant contributions, but to simply provide some operational guidance.

Experiments using a transient heat source ("hot disk") can be used to measure the heat capacity (ρCp) of the materials. However, to prevent variations in the material's hydration level, these studies should only be carried out in a temperature/humidity-controlled chamber at a fixed temperature and consequently at hydric equilibrium.

Today, DSC instruments can be found in almost all academic and industrial laboratories, and they are used (sometimes incorrectly) not only by experts in the field of calorimetry and thermal analysis, but also by researchers and technical employees who have only a basic training derived from reading the user's manual. Thus, the major goal is to offer them some insight about how to use these instruments. Because instrumentations continuously evolve, researchers are brought to work with new devices and software. The gain in experience helps to select the most appropriate instrument and operating conditions, which will be discussed in this section of the chapter.

In the literature, there are only rare studies dealing with the measurement the C_p of THS materials, and they are not always correctly estimated. The thermal properties, on the other hand, are essential to identify the right operating conditions and are required to be accurate enough for sizing the heat storage systems. Even a 10% difference in C_p can have a negative impact on material inertia and cause delays in the heat production.

As sorbent materials are hygroscopic and reactive, experimental data on the specific heat or the heat capacity are not numerous in the literature, especially as a function of temperature and water content. The limited data available provide estimated C_p values at unrelated operating conditions (temperatures, water uptake, pressures . . .). Moreover, for hydrated salts, for example, some investigations and results can be questionable. This makes comparing different thermal characteristics incredibly challenging.

A selection of literature data for relevant materials is shown in Tab. 6.3. The heat capacity values presented in the literature show large variability, without uncertainty, and sometimes are inconsistent. According to Bird et al. [35] and the methods used for the experiment, the C_p values can show important variations of results with around 100% inaccuracy.

Tab. 6.3: Specific heat capacity of materials from the literature.

Material	Temperature	Specific heat capacity, C_p	Ref.
Zeolite 13X-H_2O	50	910 J/(kg K)	[36]
Zeolite 13X	>20 °C	880 J/(kg K)	[13]
Zeolite 13X	>20 °C	800–900 J/(kg K)	[37]
Zeolite A3	−50 °C to 50 °C	700–870 J/(kg K)	[36]
Zeolite A3-H_2O	−50 °C to 50 °C	3,000–5,000 J/(kg K)	[36]
Zeolite F9	−50 °C to 50 °C	620–820 J/(kg K)	[36]
Zeolite F9-H_2O	−50 °C to 50 °C	3000–4,500 J/(kg K)	[36]
Zeolite 13X	−50 °C to 50 °C	790–910 J/(kg K)	[36]
Zeolite 13X-H_2O	−50 °C to 50 °C	3,000–4500 J/(kg K)	[36]
K_2CO_3	25 °C	114.4 J/(mol K)	[35, 38]
$LiAlH_4$	25 °C	83.2 J/(mol K)	[35, 38]
$LiBH_4$	25 °C	82.6 J/(mol K)	[35, 38]
$MgCO_3$	25 °C	75.5 J/(mol K)	[35, 38]
MgO	25 °C	37.2 J/(mol K)	[35, 38]
Na_2CO_3	25 °C	160.0 J/(mol K)	[35, 38]
NaCl	–	850.2 J/(kg K)	[35, 39]
		770 J/(kg K)	[35]
		870 J/(kg K)	[35]
		50.5 J/(mol K)	[35, 38]
$SrCO_3$	25 °C	81.4 J/(mol K)	[35, 38]

The C_p of dry zeolite 4A was measured using DSC in three different studies [24, 40, 41]. Even though zeolite 4A is dry and reactive, we can observe that the temperature ranges examined were different in the various studies.

The C_p determination is linked to the temperature and water content [42–44]. Therefore, the determination of C_p of hygroscopic and reactive adsorbent materials is a challenge [24, 35, 36, 45, 46]. In fact, the majority of traditional methods for studying C_p are based on a temperature differential in the material [35]. A change in temperature modifies the water content (sorption equilibrium) in this scenario, resulting in the generation of heat [47–49].

A computational method, as reported by Aristov [24], does not provide an accurate determination of thermodynamic parameters. Therefore, the direct measurement is the most appropriate approach. Aristov presented an empirical model [24] for the estimation of the specific heat capacity of the hydrated zeolite 4A. The contributions of the dry material and the water, however, are not discussed in depth.

Simonot-Grange et al. [50] evaluated the apparent specific heat capacity of the zeolite 13X – water pair at different temperatures from 16 to 350 °C and water content combinations. The authors reported that a distinction should be made between adsorbed water at high fill rates and adsorbed water at low fill rates. In the area of high fill rates, the apparent specific heat capacity of adsorbed water is of the order of magnitude of the specific heat capacity of pure water. The adsorbed water can be assimilated to pure water with weak adsorbent-adsorbate interactions (dispersive forces). The thermal capacity of the hydrated zeolite is then practically the sum of the thermal capacities of the two pure constituents, the anhydrous zeolite and water.

Hirasawa and Urakami [36] used a differential scanning calorimeter to investigate the C_p of various dry zeolites (A3, F9, and 13X) in the temperature range of −50 to 50 °C. The results reveal a gradual and linear increase in temperature for all dry zeolites with a connection between C_p and temperature. The C_p of the adsorbed water was also measured by DSC (water mass ratio: 5%, 10%, 15%, and 20%). In comparison to other ratios, the C_p of A3 for low uptake ratios is quite high (at −10 °C), and the C_p decreases as the temperature rises. These occurrences might be explained by the fact that the A3 micropore size is 3, which is on the same scale as a water molecule, and the physical properties of adsorbed water are different from the nature of a water cluster. The C_p increased progressively with the temperature in the case of F9 and 13X. The adsorbed water in the relatively large adsorption site is found to be in the condition of frozen water as well. The C_p of 5% water uptake in F9 and 13X is similar to other ratios that are not the same as in A3. This was explained by the fact that the adsorbed water at the relatively large adsorption site had a low water cluster property.

As conclusion, it is difficult to approve that the C_p of the adsorbed phase is intermediate between that of the gas and liquid states. The only way to obtain a correct C_p value by DSC is to arrive at fixing the water uptake during the measurement, especially for hygroscopic and reactive materials.

6.5.3 Thermal conductivity, λ

The thermal conductivity λ is expressed in W/m.K and corresponds to the heat flow passing through a material (the ability to transfer heat) subjected to a temperature differential between the entering and exit faces. If the thermal conductivity is high, the material will be a good conductor of heat. There are several methods to measure the thermal conductivity, but this section will focus just on the most used methods. The thermal conductivity depends also on the shape and nature of particles (spheres, slites . . .)

Before starting the description of methods used for λ measurement. We should not that the λ and C_p can be measured indirectly from thermal effusivity and thermal diffusivity.

- **Thermal effusivity**

Noted *E*, it is expressed in J/m^2 s$^{1/2}$ K^1. It reflects the ability of a material to absorb heat from the surrounding environment. If the thermal conductivity is higher, the heat contribution of the surrounding environment toward the material will be important. In addition, the greater the heat capacity at constant pressure, the less the stored heat will contribute to the rise in temperature and leave the possibility of accumulating more. The thermal effusivity is then expressed as follows:

$$E = \sqrt{\lambda \rho C_p} \tag{6.1}$$

This parameter is especially important when two materials at different temperatures come into contact. The "contact temperature" is directly proportional to the temperature of the two bodies and their relative thermal effusivities.

– **Thermal diffusivity**

The thermal diffusivity noted as D is, as its name indicates, a coefficient of thermal diffusion that intervenes directly in the equation of the heat. It is expressed in m^2/s and describes the rate at which heat is diffused in the material. The thermal diffusivity is expressed as follows:

$$D = \frac{\lambda}{\rho C_p} \tag{6.2}$$

6.5.3.1 Steady-state methods

Overall, the steady-state methods are based on a unidirectional and stationary model. We get rid of the time dependence present in the heat equation by assuming that the system has reached a situation of equilibrium. The thermal flow (constant) and the temperature difference generated in the system are measured to calculate the thermal resistance of the sample. Particular geometries are chosen, making it possible to consider the problem as unidirectional. In this way, it is possible to arrive at an analytical expression for thermal resistance, an expression involving thermal conductivity and geometric factors. The thermal conductivity is then deduced from the latter.

6.5.3.1.1 Guarded hot plate method-GHP

The GHP technique is the most commonly employed stationary method for calculating the apparent thermal conductivity of different materials (also insulating materials) [51–53]. This technique is the most used and precise method for solid materials [54, 55].

As illustrated in Fig. 6.6, the sample (solid, powder, granular, etc.) is sandwiched between two parallel plates. The temperature of the two surfaces is controlled to have a constant temperature gradient in the material. When the steady state is reached, the power supplied keeps the system thermally balanced for the measurement. In practice, the thickness of the studied material is small regarding the other directions, and the edges are insulated in order to assume a mono-dimensional heat transfer [51, 56]. In a steady state, for a homogeneous material, the heat flow is related to the temperature difference via Fourier's equation [57]:

$$\Phi = -\lambda A\frac{\Delta T}{\Delta x} \tag{6.3}$$

The thermal conductivity can be calculated from this equation, where ΔT is the difference of temperature of the plates, Δx is the sample thickness, Φ is the supplied power, λ is the apparent thermal conductivity (W/(m K)), and A is the surface (in m^2) of the material.

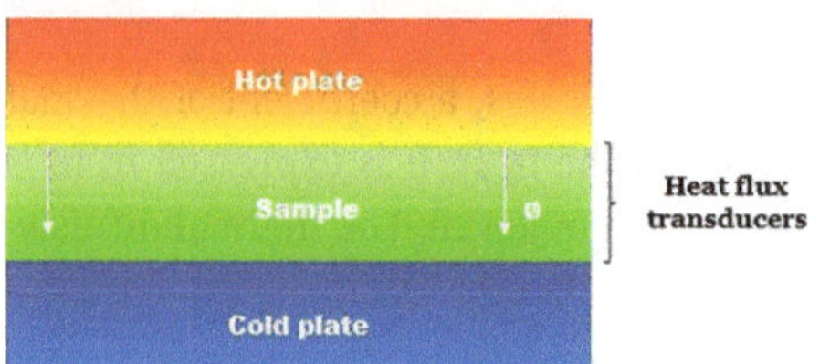

Fig. 6.6: Illustration of guarded hot plate method.

When examining the measurement methods for the thermal conductivity of material for TES applications reported in the literature [58–66], the GHP technique is the most used one. Commercial equipment operates at temperatures ranging from room temperature to 700 °C to accurately measure low thermal conductivity values (up to 2 W/(m K)) with an uncertainty of 2–5% [66]. Researchers estimated that the inaccuracy is roughly 4–6%, but few studies discussed the repeatability and the number of experiments needed. Three repetitions were estimated to be necessary for reliable results.

In comparison to nonsteady-state approaches, the GHP experiments are very time-consuming. The sample size is important, ranging from 42 × 42 to 500 × 500 mm for squared samples and a material thickness of 5 to 113 mm, according to the literature [67]. The diameter of round samples is 50.8 mm, and their thickness ranges from 0.5 to 38 mm. Furthermore, measurement errors increase at high temperatures due to heat losses through the lateral walls [68]; this problem worsens for materials with high thermal conductivity [68]. The preparation of samples in this technique necessitates highly flat and parallel surfaces (beads and powder materials can also be measured). Roughness can significantly increase the contact thermal resistance (material/plate interfaces). Furthermore, when characterizing storage materials, it is necessary to ensure that the material does not leak; in order to prevent equipment deterioration due to contact with chemicals. Likewise, the material expansion during the phase shift must be considered, as it might expand by more than 20%, producing equipment issues and erroneous data due to the modification of surface contact with the storage material.

6.5.3.1.2 Method known as radial flow or coaxial cylinder cell

The principle of the radial heat flow method is in certain aspects similar to the one previously described. The main difference is in the model's geometry, which is in this case, cylindrical [68]. A radial flow cell, as illustrated in Fig. 6.7, is made up of two coaxial cylinders: the first cylindrical tube encompasses all the other elements and acts as a furnace for performing thermal conductivity measurements depending on the temperature. A second cylinder is placed inside, surrounded by a metal conductor that serves as a heat source. The thermal flow is determined by the electrical power dissipated by the Joule effect in the conductor. The sample to be characterized is placed in the intermediate part.

Temperature sensors, such as thermocouples, are positioned along the sample and on the interior and external surfaces of the cell to measure the temperature gradient and to verify that the temperature remains constant throughout the central axis. These sensors are either introduced through holes drilled in the various cylinders, or they are placed simultaneously during the insertion of the sample. A protective sheath could also be used in this case. However, the difficulty in positioning the temperature sensors undoubtedly makes the method of the guarded hot plate preferred. In addition, the main drawback encountered when using this method is, as with the majority of the stationary methods, the relatively long waiting time before being able to perform a measurement. In fact, in principle, it is necessary to reach the equilibrium condition (steady state) to be able to proceed with the measurement.

According to Presley and coworkers [69], basic systemic errors are due to the longitudinal heat dissipation, thermal expansion of the material, and the position and the way to place the thermocouples [69]. As for the axial heat flow, Gurgel and coworkers [70] confirmed that the 1-D heat transfer assumption is not always satisfied. The authors found that for materials with low thermal conductivity, thermal losses in the axial direction would have a minor (less than 1%) impact on the apparent thermal conductivity for a length-to-diameter ratio of the sample bed higher than 2.5 [70].

According to the literature, the apparent thermal conductivity of hygroscopic salts is variable. Uncertain values due to different measurement methods and operating conditions have been published for the same material. The temperature dependency of the salt thermal conductivity is reported by Lele et al. [54] as shown in Fig. 6.8a for low thermal conductivity materials (below 1 W/(m K)). The apparent thermal conductivity of salt hydrates tends to decrease in the 20–70 °C temperature range, while for anhydrous salts, it tends to increase. In addition, the hydrated salts exhibited a higher thermal conductivity than anhydrous salt simply because of the presence of water. Indeed, water enhances the conductive heat transfer by reducing the void fraction (by increasing apparent density) and consequently facilitates internal conduction. Tanashev and coworkers [45] state that the water film improves thermal contact between salt grains and, therefore, heat transfer among adjacent particles.

In the same work of Lele et al. [54], a comparison of the two approaches (DSC and radial heat flow) is made showing similar results (Fig. 6.8b). The thermal conductivity seems to be overestimated for some of the salts (the first four), and underestimated for others. The authors found that the confidence interval between the two methods is between 15 and 20%.

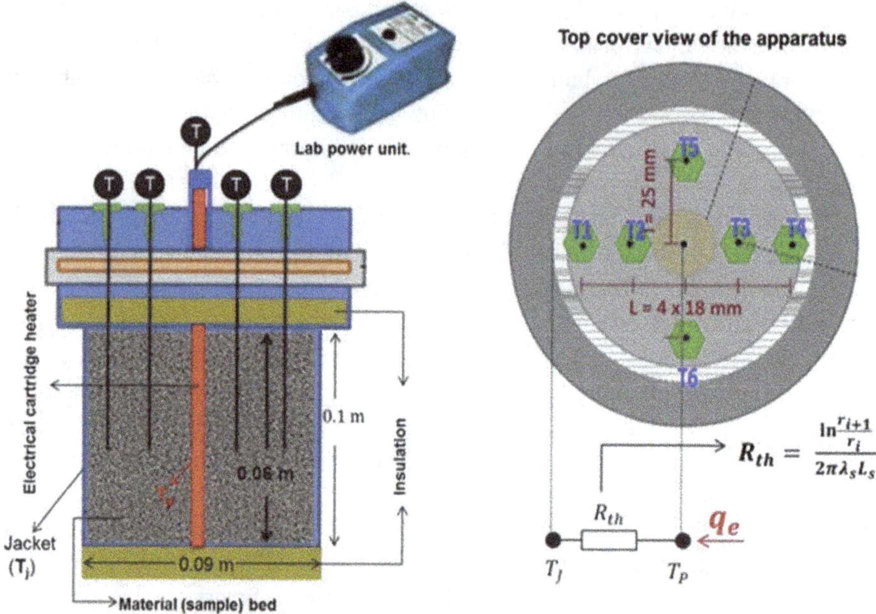

Fig. 6.7: Radial heat flow equipment for assessing thermal conductivity (reprinted with permission from ref [54]., Copyright 2015, Elsevier).

In steady-state measurements, the apparent thermal conductivity of a sample is calculated by measuring the temperature difference between the plates for a given supplied power or vice versa [71]. The apparent thermal conductivity of the material is deduced by plotting results from experiments performed at different operating conditions in a power/difference of temperature diagram (slope of the curve) [72]. These techniques allow the successful estimation of the apparent thermal conductivity of composite materials. However, these experiments are very time-consuming and require a large amount of material compared to unsteady investigations. They also require to carefully considering the lateral heat losses and contact resistances [72]. The major advantage of steady-state methods is that the unidirectional temperature gradient, perpendicular to a larger surface, allows conducting experiments on thick and anisotropic samples (composite materials, for example).

6.5.3.2 Transient methods

The transient methods are time-dependent. In addition, unlike steady-state methods, they make it possible to evaluate parameters such as the thermal diffusivity and heat capacity. For a transient measurement, the system is initially at the thermal equilibrium before being destabilized by a constant heat flow. A characteristic equation

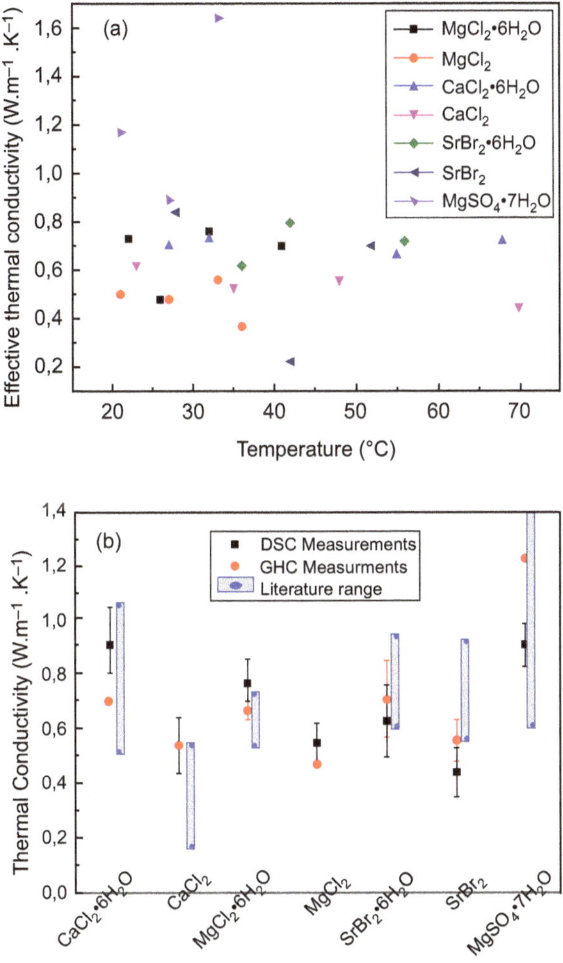

Fig. 6.8: (**a**) Effect of temperature dependence on thermal conductivity measurement carried out with radial heat flow apparatus (reprinted with permission from ref [54]., Copyright 2015, Elsevier).
(**b**) Comparison of apparent thermal conductivity of salt hydrates measured on DSC and radial heat flow apparatus (reprinted with permission from ref [54]., Copyright 2015, Elsevier).

(obtained from the heat equation) describes the temperature variations as a function of time (thermogram) for the system moving toward a new state of equilibrium. The thermophysical properties can be deduced thanks to this temporal evolution of the temperature.

Transient methods record a change of temperature in the sensor or in the material due to a pulse or periodic heat input [71]. During the heating process, these techniques allow the estimation of the thermal effusivity (*E*) or diffusivity (*D*).

The unsteady experiments are carried out under a small temperature difference and with short-duration experiments (only a few minutes) in order to minimize radiative and convective heat losses. The major advantage is the estimation of all thermal properties (thermal diffusivity, conductivity, effusivity, and heat capacity), simultaneously. It can also work at high temperatures and pressures and needs minimal sample amounts. Though, the relevance and the precision of the results are lower than previously described approaches. Indeed, with this technique, it is necessary to perform additional experiments to obtain the specific heat capacity and density [73] that are necessary to deduce the thermal conductivity values, which increases the measurement error.

6.5.3.2.1 Hot-wire method

Today, the hot-wire method is the reference method for measuring the thermal conductivity of liquids and gases [74, 75]. This method was initially introduced for measurements on non-conductive liquids, but has gradually been adapted for the characterization of liquids in general, gases, and even solids [76]. For the hot-wire method, the heat source is ideally a linear source, an infinitely long source with no thermal capacity and infinite thermal conductivity [77]. This ideal source is immersed in the medium to be characterized (liquid or solid) initially at a stabilized temperature T. As in the previous example, symmetry considerations bring to the definition of a one-dimensional system, and the use of cylindrical coordinates is particularly suitable. The thermal perturbation obtained by heating the source imposes a constant linear heat flow.

The transient hot-wire technique is presented in Fig. 6.9. It consists in placing the linear heat source, which also acts as a temperature sensor, in the material along the main axis to provide a uniform heat flow. The wire serves as both an output heater and a temperature sensor. The heat is transferred from the wire to the sample. Assuming a uniform radial heat transfer, the thermal conductivity is directly calculated. The temperature rise is dependent on the logarithm of the experimental time and the thermal conductivity, and it is determined by using the following equation:

$$\lambda = \frac{Q}{4\pi a} \quad \text{and} \quad a = \frac{\Delta T}{\ln(t)} \tag{6.4}$$

where Q is the linear heating power W/m; a is the slope of the line of the temperature increase over the logarithm of time, using eq. (6.4); t is the experimental time in seconds; and ΔT is the increase of temperature.

In terms of operating conditions, the transient hot-wire approach has some restrictions; the material shall be a liquid, a powder, a bed of pellets, or a bulk solid (after drilling a hole to permit the insertion of the probe). The radial thickness must be large enough to assume a half-space heat transfer. In thin films, the thickness of the sensor is comparable to the thickness of the sample. Heat losses are not negligible when samples are too thin [75]. Large measurement errors are caused by conductive (electric) and

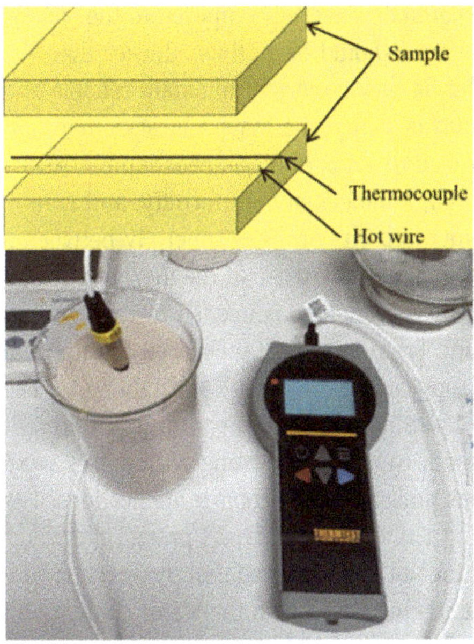

Fig. 6.9: Illustration and apparatus of the hot-wire method.

radiative heat transfer for semi-transparent materials [78, 79]. It is critical to ensure that the sample is sufficiently insulated to eliminate the uncertainty related to contact resistance. In order to obtain reliable results, thermal resistance due to connection leads should be minimized, by using thermal paste, for example. The transient hot strip method is a modified technique that uses a metallic hot strip as a probe sensor instead of a hot wire. It can also be applied to electrical insulating materials such as powders, granules, pellets of minerals, of wood, rocks, . . . [80, 81]. As for steady-state methods, unsteady techniques are effective for measuring homogeneous and isotropic materials. In the case of granules or pellets, the size must be very small compared to the radial depth of the temperature gradient. For thermochemical heat storage applications, to ensure a good mass transfer, the used materials are generally in the form of pellets. For these materials, the hot-wire technique is the most widely utilized approach. Because of the adaptability of this technology, fast and accurate measurements, and experimental investigations can be performed with a wide range of materials at different operating temperatures. Despite the benefits of this technology, only a few commercially viable devices exist, probably due to the complexity of the setup and the weakness of the very thin wire, which is easily broken [82].

As previously described, this technique can be used for a variety of materials, including zeolite, salts, activated carbon, and bio-char with no change in the uncertainty. In general, the error is higher in measurements for liquid at high temperatures (3–5%) than in measurements taken at ambient temperature or in the solid state (1–3%). According to the literature, between 3 and 10 repetitions need to be recorded to acquire a consistent and precise measurement of the thermal conductivity ([83–90]).

This technique also requires a smaller amount of material compared to the steady-state methodologies. At the minimum, the material should be 10 times thicker than the wire diameter (<2 mm). The sample preparation is very important to obtain reliable and consistent results. For example, with nanofluids or composite materials, agglomeration and poor mixing may occur and consequently lead to a non-homogeneous sample. As the heat transfer is assumed to be one-dimensional; homogeneity, uniformity, and consistency are the key parameters for a good estimation. The probe is generally between 5 and 10 cm long and requires, for good measurements, no change of the thermo-physical properties in the length direction. For this reason, vigilance is needed toward gravity stratification (mixing or nanoparticles in fluid). Freni et al. [43] measured the thermal conductivity of composite materials for TES applications ($CaCl_2/SiO_2$ and $LiBr/SiO_2$). The experimental measurements were conducted by using a hot-wire apparatus at different operating conditions (P_{H2O}, T, and water content (w)). The operating parameters were chosen to be in accordance with the conditions present during cooling applications. The range of temperature studied is between 40 and 130 °C, and the partial pressure of water is between 10 and 70 mbar. As a major result, the authors observed that the thermal conductivity of their composite materials significantly increased with the water content and slightly with the temperature and pressure conditions.

One of the advantages of transient methods compared to stationary methods is that they offer the possibility to make rapid measurements and follow the evolution of thermo-physical properties over time. For example, if we are interested in hygroscopic and reactive samples such as hydrated salts, their thermal conductivity is often correlated to their water content. The latter evolves strongly and more or less rapidly with time. The use of a stationary method implies obtaining a thermal equilibrium situation before starting the measurement, waiting during which the properties of the sample will have evolved, removing any error in the measurement.

6.5.3.2.2 Transient plane source (TPS) method "hot disk"

The "hot disk" technique can measure the thermal properties of bulk and powder solid samples [67]. The hot disk approach (developed by Gustafsson [91]) consists of a planar heater constituted of concentric electrical wires isolated from each other (as shown in Fig. 6.10). The modified transient plan source approach is employed in bulk thermal conductivity testing [91] to limit the effect of contact resistance.

This technique is comparable to the transient hot wire. The heater (disk) is placed between two portions of the same material to obtain a symmetrical thermal plane. Constant electric power is applied for few minutes and the temperature of the sensor is recorded. As the electrical resistance of the sensor changes with temperature, the voltage and current are continuously adjusted to maintain the constant energy input.

This technique was developed to estimate all thermal properties (heat capacity, thermal diffusivity, effusivity, and conductivity) at the same time [92]. It is tempting to be able to measure multiple thermophysical parameters at once using a simple approach.

Fig. 6.10: Transient plane source technique "hot disk" for thermal conductivity measurement.

Acem et al. [93] used a transient hot plate to evaluate the thermal properties of graphite/salt. Authors assumed that the hot disk and the sample were in perfect contact to consider a one-dimensional heat transfer. On silica gel, accurate measurements of steady and transient conductivity were taken [17, 70]. The results, which were found to be depending on the water uptake, were used for modeling the heat and mass transfer in solar sorption systems.

The same technique was used by Pinheiro et al. [94] to determine the thermal conductivity of zeolite 13X in the 25–115 °C temperature range. The calculated value was lower than 0.3 W/(m K). When used for low-conductivity materials, the "hot disk" approach presents certain limitations [95]. This is particularly true in the case of a bed of zeolite granules that presents a high percentage of void (please see Fig. 6.10). As a result, the bed's packing density must be considered. To the best of the author's knowledge, the works dealing with conductivity measurement of hygroscopic or reactive materials are few in number. For example, the thermal conductivity reported in the literature for hydrated zeolite 13X ranges from 76 to 400 mW/(m K) [46, 94, 96, 97]. The estimation of thermal conductivity is related to the shape and size of the materials, so these values must be taken with caution.

According to the theory, the thermal conductivity and the thermal diffusivity are estimated by fitting the experimental temperature of the sensor as follows:

$$\Delta T = \frac{P_0}{\pi^{\frac{3}{2}} \lambda r} \int_0^{\tau} \frac{d\sigma}{\sigma} \int_0^1 v dv \int_0^1 \exp\left(\frac{-(u^2+v)}{4\sigma^2}\right) I_0\left(\frac{uv}{2\sigma^2}\right) u du \tag{6.5}$$

P_0 is the power input, I_0 is the zero-order modified Bessel functions, and $\tau = \sqrt{\frac{Dt}{r^2}}$ is the dimensionless time with t as the experimental time and r the radius of the sensor.

In practice, due to its complexity, this equation is not used for calculation [98, 99]. As verified by comparing the computations made by the software of a hot-disk device from ThermoConcept and the calculations performed applying eq. (6.5), the commercial software carries out the calculation by applying a simplified equation. To solve the problem, simplifications are made. It is then assumed that the measurements are carried out in a half-space one-dimensional plane system (material), that the hot disk provides a constant heat flow, and that the contact between the material and the disk is perfect. The analytical solution of this unsteady problem is then the following equation:

$$T(x,t) = T(t=0) + \frac{P_0}{\lambda S}\sqrt{Dt}\left[\frac{1}{\sqrt{\pi}} e^{-\frac{x^2}{4Dt}} - \frac{x}{2\sqrt{Dt}} erfc\left(\frac{x}{2\sqrt{at}}\right)\right] \tag{6.6}$$

The temperature of the sensor is (for $x = 0$):

$$T(x=0,t) = T(t=0) + \frac{P_0}{\lambda S}\sqrt{Dt}\left[\frac{1}{\sqrt{\pi}}\right] \tag{6.7}$$

The estimation of the thermophysical characteristics is then possible by fitting the experimental temperature by a square root of a function of the time. The material effusivity $E = \sqrt{\rho C_p \lambda}$ is exactly proportional to the slope of the linear $T(t)$ versus $\sqrt{t}$ curve.

The fundamental disadvantage of this technique is that the tested materials and the sensor need to have a perfect and plane contact, difficult to obtain for powders or particles [100]. Furthermore, some studies reveal other sources of errors: the thermal contact resistances, the inertia of the hot disk, and some problems inherent to the apparatus (slight electrical power change and the difference between the measured temperature and the real temperature of the heating coil used for the calculation/correction of the electrical resistance) [71]. In addition, at high temperature, temperature calibration errors and lateral heat losses can take place. Starting from eq. (6.7), the software can then estimate the thermal effusivity. Then, by using a polynomial approximation of eq. (6.6), it calculates the thermal conductivity. The other thermal properties are calculated starting from thermal effusivity and conductivity.

This approach is primarily used to evaluate the thermal conductivity of phase change materials and materials for thermal energy storage applications. The majority of the results available in the literature have been obtained by experiments performed at room temperature, with a maximum reported temperature of 250 °C (see Tab. 6.4). In terms of uncertainty of the experiments, the authors reported very good data accuracy and repeated the measurements more than three times in most cases (see Tab. 6.4). Many studies, however, do not describe each test repetition (Tab. 6.4).

Tab. 6.4: Thermal conductivity measurements by transient plane source (TPS) (literature sources) for materials used in thermal energy storage.

Composition	Type of apparatus	Uncertainty	Repetitions	Property reported	Temperature measurement	Ref.
$LiCl/H_2O$ and activated carbon	Hot disk TPS2,500, Sweden AB	±5%	N/A	Thermal conductivity	Room temperature	[101]
$LiCl/H_2O$ and expanded natural graphite matrix	Hot disk TPS2500, Sweden AB	±5%	N/A	Thermal conductivity	Room temperature	[102]
$LiNO_3/KCl$ and expanded graphite	Hot disk TPS2500, Sweden AB	±2%	N/A	Thermal conductivity	Room temperature	[103]
$CaCl_2 \cdot 6H_2O$/ expanded graphite	Hot disk TPS2500, Sweden AB	0.035–0.169	Seven times	Thermal conductivity	Room temperature	[104]
TCM/silica gel/ vermiculite/ activated carbon/ Zeolite 13×	Setaram TCi modified transient plane source	N/A	N/A	Dry-state thermal conductivity	Room temperature	[105]
Strontium bromide/ graphite	NETZSCH LFA 447 (NanoFlash)	N/A (error bars in the graph)	Three times	Thermal conductivity	50–250 °C	[106]
Magnesium sulfate and magnesium chloride/zeolite	Model LFA 427 Netzsch	N/A	Three laser shots (just one measurement)	Thermal conductivity	30/80/150 °C	[107]

Small amounts of sample are needed to apply this method, significantly lower sample size than that required for steady-state techniques. Depending on the material properties and the sensor size, the sample thickness must be higher than the heat penetration layer, and the diameter is important enough to limit the lateral heat loss. For TES materials, various sample preparation issues can contribute to erroneous measurements. For example, to limit the thermal resistance (bad contact), the sample preparation must ensure a flat surface.

6.5.3.2.3 Laser flash method

The flash or laser flash method measures thermal diffusivity. If the specific heat capacity and density are known, this technique allows calculating the thermal conductivity. This method was initially introduced in 1961 by Parker et al. [108]. The laser flash achieves great accuracy by using non-intrusive and non-destructive temperature sensors. The laser flash approach is shown in Fig. 6.11. The energy input provided during some seconds by a laser beam impact on the sample's surface. Temperature variations on the other side of the sample are measured using infrared sensor or thermocouples.

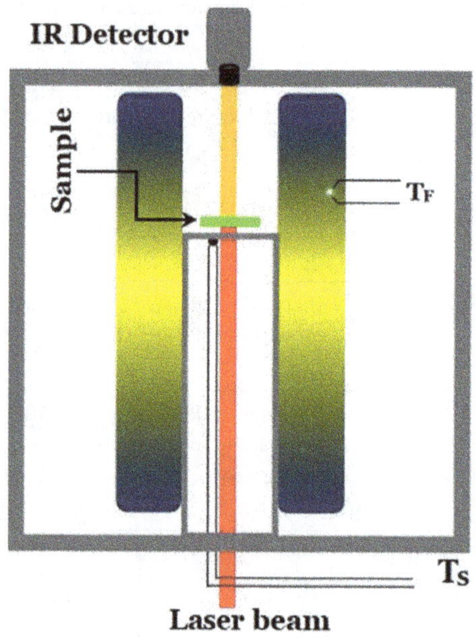

Fig. 6.11: Principle of an LFA for thermal diffusivity measurements.

The temperature increase on the material's surface is used to calculate the thermal diffusivity throughout the thickness of the sample (disk form). A two-dimensional heat conduction is assumed [71]. Equation (6.8) [71, 109] is used to calculate the thermal diffusivity, where d is the thickness of the sample (in m) and $t_{1/2}$ is the time at 50% of the maximum temperature increase (s):

$$D = 0.1388\ \frac{d^2}{t_{1/2}} \tag{6.8}$$

The sample's thermal conductivity (λ) may then be determined knowing the heat capacity (ρC_p):

$$\lambda = D.\rho.C_p \tag{6.9}$$

The laser flash method is fast, easy to use, and can estimate the thermal conductivity of tiny samples. Furthermore, the laser flash approach enables it to operate in a large temperature range [110]. However, this technique is not suitable for semi-transparent materials, presenting a low thermal conductivity, and for non-isotropic solid materials.

This technique can be used to estimate the thermal properties of materials used in TES applications [55]. It can be applied up to 700 °C, allowing the estimation of the thermal conductivity at high temperatures that is impossible with any other transient approach. For this reason, even if the laser flash approach is not fully adapted for measuring the thermal properties of liquids, semi-transparent and non-homogeneous materials attempts to enlarge the application field of the laser flash method are ongoing. The requirements for applying this experimental method are specified in the ASTM E1461 [111] and the ISO 22007–4 [112] regulations. Nevertheless, no criterion for sample preparation and how to proceed in the presence of hygroscopic and reactive materials are given, making it difficult to reach a consensus in the scientific community on the operating conditions required to perform such experiments.

6.5.3.3 Selection of a method

A large number of methods available for characterizing the thermophysical properties demonstrate that there is no one-size-fits-all answer to all problems. The method used must address the specific needs of the work being done. This necessitates the consideration of a variety of criteria. The first factor to consider is the identification of the property that needs to be measured (specific heat capacity, conductivity, diffusivity, or effusivity). Some methods directly provide the measurement of thermal conductivity (stationary method). Others focus on correlated properties such as thermal diffusivity or thermal effusivity. The second criterion is the state of the material to characterize (solid, liquid, gas, powder, and fiber) and the conditions of the measurement (temperature, pressure, and humidity). Some methods are adapted for measuring the thermophysical properties of solids, but they are difficult to use in the case of liquids and hygroscopic and reactive materials (such as guarded hot plate, DSC, and so on). The study of the thermophysical properties as a function of the temperature, the water uptake (hydration state) or pressure has to be considered. The type of sample and property to be measured determines the type of technique and equipment to be used. For example, for C_p measurements, the DCS is the best approach, especially for reactive and hygroscopic materials used in thermal storage systems. It is difficult to confirm that the C_p of the hydrated materials is an average between the C_p of the gas/liquid molecules and that of the dehydrated solid material. The only way to obtain a correct C_p value by DSC is to arrive at maintaining constant water uptake during the measurement.

It is always preferable to compare multiple measures of the same parameter using two different methods to ensure that the results are consistent.

6.6 General overview and summary

Many years, energy storage materials are of great interest in material science research [113]. Even if the heat storage techniques are numerous, the selection of suitable and efficient storage materials needs to be put in place. The scientific community is looking for standards, approaches, and techniques to appropriately assess the thermal properties of TES materials, as they are the critical parameters for the development of storage materials and systems. The issue is far to be solved, and the experts did not yet agree on a single, standardized methodology for measuring most of the thermal properties. Lazaro and coworkers [114] conducted an inter-comparative test to determine the properties of phase change materials (PCMs); the differences among the obtained results not only dependent on the applied method, but also on the type of equipment. An example is the over-performing (in terms of accuracy) method proposed by Cheng et al. [115] to determine the specific heat of PCM-concrete brick. This method is based on the resolution of the inverse problem that involves the measurements of temperatures, thermal conductivity, and density of PCM-concrete brick throughout the phase change process. Nevertheless, because it only applies to this material, this cannot be extended to other solids used in the same application. Mazzeo et al. [116] also attempted to develop a novel method for determining the thermo-physical properties of pure PCM's. The analytical model provides results close to those of the experimental model in terms of accuracy and predictability. The authors developed a model that can be used to calculate the thermal properties (thermal conductivity and specific heat capacity) in liquid and solid phase. Despite a strong activity to develop adequate techniques and methodologies for measuring the thermophysical properties, the research provided are too narrowly focused on a single type of material and operating conditions. Moreover, they are not representative of real applications.

The IEA working group reported on the major requirements to investigate TES material's properties for energy storage applications [117]. Two types of standards were identified for thermal conductivity measurements, one constituted of specific for polymers [112, 118], and the other of some thermal insulating materials [112]. Nevertheless, there is no requirement specification for assessing TES materials' thermal properties. In order to accurately conduct thermal conductivity tests, the most suitable technique must be chosen considering the material, operating conditions, amount and size, and targeted application (sensible, latent, or thermochemical heat storage). As previously described, many techniques (steady or unsteady) have been applied to obtain thermal conductivity values (hot-guarded plates, hot wire, hot disk, and laser flash). They all have some advantages and weaknesses [72]. The selection of TES materials is the first step in the design of a TES system, and it is based on the following criteria, as discussed by numerous researchers [119, 120]: high heat storage capacity, inexpensive, environmentally friendly, chemically stable, useable for industrial scaling-up, and good heat and mass transfer properties. In a TES operation, thermal conductivity plays an important role in the heat transport, promoting or no the efficiency of the process during the charging/discharging phases [121]. Unsuitably, the thermal conductivity of most TES materials is often poor. As a result, researchers have focused their activities to improve the thermal properties of TES materials [122–127], but the thermal conductivity value is strongly influenced by the testing conditions and technique used. Consequently, before comparing thermal conductivity investigation techniques, the operating conditions for performing experimental tests must be defined and normalized.

Several articles reporting on the estimation of the thermophysical characteristics of TES materials have been published in the literature, but there is no consensus on the most efficient technique neither on the most appropriate operating conditions to be used. In the future, TES materials thermophysical characteristics measurement will have to overcome numerous issues to reach a general agreement and get comparable results among the different research works. This entails optimizing sample preparation and selecting the appropriate methodology among the various available approaches, with a focus on a high accuracy of the measurements. When the measurement campaign is appropriately performed, the researchers can compare the respective results. At present, it is tough to compare results from different authors.

To conclude, this chapter emphasized the importance of establishing standards and normalized procedures for measuring the thermophysical parameters of TES medium, regardless of the end use. The fundamental challenge is to reach an international scientific consensus on analyzing the thermal properties of heat storage materials. The ultimate objective is to reduce the dispersion of data values caused by differences in the sample preparation, the sampling, and the technique used.

As a result, future challenges are to finally identify and set the experimental techniques to measure these properties and to develop new dedicated methodologies for determining the thermophysical properties of specific materials (nonisotropic and composites) to be used in pilot TES systems.

References

[1] Yu, N.; Wang, R. Z.; Wang, L. W. Sorption thermal storage for solar energy. *Prog. Energy Combust. Sci [Internet]*. **2013**, *39*(*5*), 489–514. Available from: https://linkinghub.elsevier.com/retrieve/pii/S0360128513000270.

[2] Zbair, M.; Bennici, S. Survey summary on salts hydrates and composites used in thermochemical sorption heat storage. *Rev. Energies [Internet]*. **2021**, *14*(*11*), 3105. Available from: https://www.mdpi.com/1996-1073/14/11/3105.

[3] Shabgard, H.; Bergman, T. L.; Sharifi, N.; Faghri, A. High temperature latent heat thermal energy storage using heat pipes. *Int. J. Heat Mass Transf [Internet]*. **2010**, *53*(*15–16*), 2979–2988. Available from: https://linkinghub.elsevier.com/retrieve/pii/S0017931010001766.

[4] Ermis, K.; Erek, A.; Dincer, I. Heat transfer analysis of phase change process in a finned-tube thermal energy storage system using artificial neural network. *Int. J. Heat Mass Transf [Internet]*. **2007**, *50*(*15–16*), 3163–3175. Available from: https://linkinghub.elsevier.com/retrieve/pii/S001793100700052X.

[5] Desai, F.; Sunku Prasad, J.; Muthukumar, P.; Rahman, M. M. Thermochemical energy storage system for cooling and process heating applications: A review. *Energy Convers. Manag [Internet]*. **2021** Feb;*229*, 113617. Available from: https://linkinghub.elsevier.com/retrieve/pii/S0196890420311456.

[6] Iten, M.; Liu, S. A work procedure of utilising PCMs as thermal storage systems based on air-TES systems. *Energy Convers. Manag [Internet]*. **2014**, *77*, 608–627. Available from: https://linkinghub.elsevier.com/retrieve/pii/S0196890413006328.

[7] Zhang, Y.; Wang, R. Sorption thermal energy storage: Concept, process, applications and perspectives. *Energy Storage Mater [Internet]*. **2020**, *27*, 352–369. Available from: https://linkinghub.elsevier.com/retrieve/pii/S2405829720300726.

[8] N'Tsoukpoe, K. E.; Liu, H.; Le Pierrès, N.; Luo, L. A review on long-term sorption solar energy storage. *Renew. Sustain. Energy Rev [Internet]*. **2009**, *13(9)*, 2385–2396. Available from: https://linkinghub.elsevier.com/retrieve/pii/S1364032109001129.

[9] Wongsuwan, W.; Kumar, S.; Neveu, P.; Meunier, F. A review of chemical heat pump technology and applications. *Appl. Therm. Eng [Internet]*. **2001**, *21(15)*, 1489–1519. Available from: https://linkinghub.elsevier.com/retrieve/pii/S1359431101000229.

[10] Aristov, Y. I.; Challenging offers of material science for adsorption heat transformation: A review. *Appl. Therm. Eng [Internet]*. **2013**, *50(2)*, 1610–1618. Available from: https://linkinghub.elsevier.com/retrieve/pii/S1359431111004832.

[11] Aristov, Y. I.; Novel materials for adsorptive heat pumping and storage: Screening and nanotailoring of sorption properties. *J. Chem. Eng. Jpn [Internet]*. **2007**, *40(13)*, 1242–1251. Available from: http://www.jstage.jst.go.jp/article/jcej/40/13/40_07WE228/_article.

[12] Solé, A.; Martorell, I.; Cabeza, L. F. State of the art on gas–solid thermochemical energy storage systems and reactors for building applications. *Renew. Sustain. Energy Rev [Internet]*. **2015**, *47*, 386–398. Available from: https://linkinghub.elsevier.com/retrieve/pii/S1364032115002300.

[13] Mette, B.; Kerskes, H.; Drück, H.; Müller-Steinhagen, H. Experimental and numerical investigations on the water vapor adsorption isotherms and kinetics of binderless zeolite 13X. *Int. J. Heat Mass Transf [Internet]*. **2014**, *71*, 555–561. Available from: https://linkinghub.elsevier.com/retrieve/pii/S001793101301106X.

[14] N'Tsoukpoe, K. E.; Restuccia, G.; Schmidt, T.; Py, X. The size of sorbents in low pressure sorption or thermochemical energy storage processes. *Energy [Internet]*. **2014**, *77*, 983–998. Available from: https://linkinghub.elsevier.com/retrieve/pii/S0360544214011621.

[15] Murashov, V. V.; White, M. A. Thermal properties of zeolites: Effective thermal conductivity of dehydrated powdered zeolite 4A. *Mater. Chem. Phys [Internet]*. **2002**, *75(1–3)*, 178–180. Available from: https://linkinghub.elsevier.com/retrieve/pii/S0254058402000512.

[16] Murashov, V. V.; White, M. A. Apparatus for dynamical thermal measurements of low-thermal diffusivity particulate materials at subambient temperatures. *Rev. Sci. Instrum [Internet]*. **1998**, *69(12)*, 4198–4204. Available from: http://aip.scitation.org/doi/10.1063/1.1149231.

[17] Gurgel, J. M.; Klüppel, R. P. Thermal conductivity of hydrated silica-gel. *Chem. Eng. J. Biochem. Eng. J. [Internet]*. **1996**, *61(2)*:133–138. Available from: https://linkinghub.elsevier.com/retrieve/pii/0923046796800200.

[18] Korhammer, K.; Druske, -M.-M.; Fopah-Lele, A.; Rammelberg, H. U.; Wegscheider, N.; Opel, O., et al. Sorption and thermal characterization of composite materials based on chlorides for thermal energy storage. *Appl. Energy [Internet]*. **2016**, *162*, 1462–1472. Available from: https://linkinghub.elsevier.com/retrieve/pii/S0306261915009708.

[19] Frazzica, A.; Sapienza, A.; Freni, A. Novel experimental methodology for the characterization of thermodynamic performance of advanced working pairs for adsorptive heat transformers. *Appl. Therm. Eng [Internet]*. **2014**, *72(2)*, 229–236. Available from: https://linkinghub.elsevier.com/retrieve/pii/S1359431114005535.

[20] Whiting, G. T.; Grondin, D.; Stosic, D.; Bennici, S.; Auroux, A. Zeolite–MgCl2 composites as potential long-term heat storage materials: Influence of zeolite properties on heats of water sorption. *Sol. Energy Mater. Sol. Cells [Internet]*. **2014**, *128*, 289–295. Available from: https://linkinghub.elsevier.com/retrieve/pii/S0927024814002669.

[21] Whiting, G.; Grondin, D.; Bennici, S.; Auroux, A. Heats of water sorption studies on zeolite–MgSO4 composites as potential thermochemical heat storage materials. *Sol. Energy Mater. Sol. Cells [Internet]*. **2013**, *112*, 112–119. Available from: https://linkinghub.elsevier.com/retrieve/pii/S0927024813000354.

[22] Bennici, S.; Polimann, T.; Ondarts, M.; Gonze, E.; Vaulot, C.; Le Pierrès, N. Long-term impact of air pollutants on thermochemical heat storage materials. *Renew. Sustain. Energy Rev [Internet]*. **2020** Jan;*117*:109473. Available from: https://linkinghub.elsevier.com/retrieve/pii/S1364032119306811.

[23] Brancato, V.; Gordeeva, L. G.; Sapienza, A.; Palomba, V.; Vasta, S.; Grekova, A. D., et al. Experimental characterization of the LiCl/vermiculite composite for sorption heat storage applications. *Int. J. Refrig [Internet]*. **2019**, *105*, 92–100. Available from: https://linkinghub.elsevier.com/retrieve/pii/S0140700718303025.

[24] Aristov, Y. I.; Adsorptive transformation of heat: Principles of construction of adsorbents database. *Appl. Therm. Eng [Internet]*. **2012**, *42*, 18–24. Available from: https://linkinghub.elsevier.com/retrieve/pii/S1359431111001049.

[25] Askalany, A. A.; Henninger, S. K.; Ghazy, M.; Saha, B. B. Effect of improving thermal conductivity of the adsorbent on performance of adsorption cooling system. *Appl. Therm. Eng [Internet]*. **2017**, *110*, 695–702. Available from: https://linkinghub.elsevier.com/retrieve/pii/S135943111631434X.

[26] Shinzato, K.; Baba, T. A laser flash apparatus for thermal diffusivity and specific heat capacity measurements. *J. Therm. Anal. Calorim [Internet]*. **2001**, *64*(*1*), 413–422. Available from: https://link.springer.com/article/10.1023/A:1011594609521.

[27] He, Y.; Rapid thermal conductivity measurement with a hot disk sensor. *Thermochim. Acta [Internet]*. **2005**, *436*(*1–2*), 122–129. Available from: https://linkinghub.elsevier.com/retrieve/pii/S004060310500345X.

[28] Rudtsch, S.; Uncertainty of heat capacity measurements with differential scanning calorimeters. *Thermochim. Acta [Internet]*. **2002**, *382*(*1–2*), 17–25. Available from: https://linkinghub.elsevier.com/retrieve/pii/S0040603101007304.

[29] Aydin, D.; Casey, S. P.; Riffat, S. The latest advancements on thermochemical heat storage systems. *Renew. Sustain. Energy Rev [Internet]*. **2015**, *41*, 356–367. Available from: https://linkinghub.elsevier.com/retrieve/pii/S1364032114007308.

[30] Palomba, V.; Sapienza, A.; Aristov, Y. Dynamics and useful heat of the discharge stage of adsorptive cycles for long term thermal storage. *Appl. Energy [Internet]*. **2019**, *248*, 299–309. Available from: https://linkinghub.elsevier.com/retrieve/pii/S0306261919307986.

[31] Gordeeva, L. G.; Aristov, Y. I. Adsorptive heat storage and amplification: New cycles and adsorbents. *Energy [Internet]*. **2019**, *167*, 440–453. Available from: https://linkinghub.elsevier.com/retrieve/pii/S0360544218321285.

[32] Blanco, I. The correctness of Cp measurements by DSC, actions to do and not to do. *Thermochim. Acta [Internet]*. **2020**, *685*, 178512. Available from: https://linkinghub.elsevier.com/retrieve/pii/S0040603119310408.

[33] Schick, C.Differential scanning calorimetry (DSC) of semicrystalline polymers. *Anal. Bioanal. Chem [Internet]*. **2009**, *395*(*6*), 1589–1611. Available from: http://link.springer.com/10.1007/s00216-009-3169-y.

[34] Vyazovkin, S.; Thermal Analysis. *Anal. Chem [Internet]*. **2010**, *82*(*12*), 4936–4949. Available from: https://pubs.acs.org/doi/10.1021/ac100859s.

[35] Bird, J. E. Humphries, T. D.; Paskevicius, M.; Poupin, L.; Buckley, C. E. Thermal properties of thermochemical heat storage materials. *Phys. Chem. Chem. Phys [Internet]*. **2020**;*22*(*8*):4617–4625. Available from: http://xlink.rsc.org/?DOI=C9CP05940G.

[36] Hirasawa, Y.; Urakami, W. Study on specific heat of water adsorbed in zeolite using DSC. *Int. J. Thermophys [Internet]*. **2010**, *31*(*10*):2004–2009. Available from: http://link.springer.com/10.1007/s10765-010-0841-6.

[37] Hauer, A. Sorption Theory for Thermal Energy Storage. In: *Thermal Energy Storage for Sustainable Energy Consumption [Internet]*. Dordrecht: Springer Netherlands; 2007. pp. 393–408. Available from: http://link.springer.com/10.1007/978-1-4020-5290-3_24.

[38] Haynes, W. M. *CRC Handbook of Chemistry and Physics [Internet]*. In Haynes, W. M.; Eds.;CRC Press; 2014. p. 3020. Available from: https://www.taylorfrancis.com/books/9781482208689.

[39] Clusius, K.; Goldmann, J.; Perlick, A. Ergebnisse der Tieftemperaturforschung VII. Die Molwärmen der Alkalihalogenide LiF, NaCl, KCl, KBr, KJ, RbBr und RbJ von 10° bis 273° abs.2. *Zeitschrift Für Naturforsch A [Internet]*. **1949**, *4*(*6*), 424–432. Available from: https://www.degruyter.com/document/doi/10.1515/zna-1949-0603/html.

[40] Tatlıer, M.; Tantekin-Ersolmaz, B.; Erdem-Şenatalar, A. A novel approach to enhance heat and mass transfer in adsorption heat pumps using the zeolite–water pair. *Microporous Mesoporous Mater [Internet]*. **1999**, *27*(*1*), 1–10. Available from: https://linkinghub.elsevier.com/retrieve/pii/S1387181198001747.

[41] Qiu, L.; Murashov, V.; White, M. A. Zeolite 4A: Heat capacity and thermodynamic properties. *Solid State Sci [Internet]*. **2000**, *2*(*8*), 841–846. Available from: https://linkinghub.elsevier.com/retrieve/pii/S129325580001102X.

[42] Dawoud, B.; Sohel, M. I.; Freni, A.; Vasta, S.; Restuccia, G. On the effective thermal conductivity of wetted zeolite under the working conditions of an adsorption chiller. *Appl. Therm. Eng [Internet]*. **2011**, *31*(*14–15*), 2241–2246. Available from: https://linkinghub.elsevier.com/retrieve/pii/S1359431111001487.

[43] Freni, A.; Tokarev, M.; Restuccia, G.; Okunev, A.; Aristov, Y. Thermal conductivity of selective water sorbents under the working conditions of a sorption chiller. *Appl. Therm. Eng [Internet]*. **2002**, *22*(*14*), 1631–1642. Available from: https://linkinghub.elsevier.com/retrieve/pii/S1359431102000765.

[44] Rocky, K. A.; Islam, M. A.; Pal, A.; Ghosh, S.; Thu, K.; Nasruddin, et al. Experimental investigation of the specific heat capacity of parent materials and composite adsorbents for adsorption heat pumps. *Appl. Therm. Eng [Internet]*. **2020**, *164*, 114431. Available from: https://linkinghub.elsevier.com/retrieve/pii/S1359431119337135.

[45] Tanashev, Y. Y.; Krainov, A. V.; Aristov, Y. I. Thermal conductivity of composite sorbents "salt in porous matrix" for heat storage and transformation. *Appl. Therm. Eng [Internet]*. **2013**, *61*(*2*), 401–407. Available from: https://linkinghub.elsevier.com/retrieve/pii/S1359431113006030.

[46] Liu, Z. Y.; Cacciola, G.; Restuccia, G.; Giordano, N. Fast simple and accurate measurement of zeolite thermal conductivity. *Zeolites [Internet]*. **1990**, *10*(*6*), 565–570. Available from: https://linkinghub.elsevier.com/retrieve/pii/S0144244905803136.

[47] Vučelić, V.; Vučelić, D. The heat capacity of water near solid surfaces. *Chem. Phys. Lett [Internet]*. **1983**, *102*(*4*), 371–374. Available from: https://linkinghub.elsevier.com/retrieve/pii/0009261483870584.

[48] Schwamberger, V.; Schmidt, F. P. Estimating the heat capacity of the adsorbate–adsorbent system from adsorption equilibria regarding thermodynamic consistency. *Ind. Eng. Chem. Res [Internet]*. **2013**, *52*(*47*), 16958–16965. Available from: https://pubs.acs.org/doi/10.1021/ie4011832.

[49] Berezin, G. I.; Kiselev, A. V.; Sinitsyn, V. A. Heat capacity of the H2O/KNaX zeolite adsorption system. *J. Chem. Soc. Faraday Trans. 1 Phys. Chem. Condens. Phases [Internet]*. **1973**, *69*, 614.Available from: http://xlink.rsc.org/?DOI=f19736900614.

[50] Simonot-Grange, M.-H.; Belhamidi-El Hannouni, F.; Bracieux-Bouillot, O. Proprietes physico-chimiques de l'eau adsorbee dans les zeolithes 13X et 4A. II. Capacites thermiques de l'eau adsorbee, du systeme zeolithe – Eau et de la zeolithe anhydre. *Thermochim Acta [Internet]*. **1986**, *101*, 217–230. Available from: https://linkinghub.elsevier.com/retrieve/pii/0040603186800569.

[51] Xamán, J.; Lira, L.; Arce, J. Analysis of the temperature distribution in a guarded hot plate apparatus for measuring thermal conductivity. *Appl. Therm. Eng [Internet]*. **2009**, *29(4)*, 617–623. Available from: https://linkinghub.elsevier.com/retrieve/pii/S1359431108001646.

[52] Siu, M. C. I.; Bulik, C. National Bureau of Standards line-heat-source guarded-hot-plate apparatus. *Rev. Sci. Instrum [Internet]*. **1981**, *52(11)*, 1709–1716. Available from: http://aip.scitation.org/doi/10.1063/1.1136518.

[53] ASTM C177-19, *Standard Test Method for Steady-State Heat Flux Measurements and Thermal Transmission Properties by Means of the Guarded-Hot-Plate Apparatus*, ASTM International, West Conshohocken, PA, 2019. In.

[54] Fopah Lele, A.; N'Tsoukpoe, K. E.; Osterland, T.; Kuznik, F.; Ruck, W. K. L. Thermal conductivity measurement of thermochemical storage materials. *Appl. Therm. Eng [Internet]*. **2015**, *89*, 916–926. Available from: https://linkinghub.elsevier.com/retrieve/pii/S1359431115006377.

[55] Palacios, A.; Cong, L.; Navarro, M. E.; Ding, Y.; Barreneche, C. Thermal conductivity measurement techniques for characterizing thermal energy storage materials – A review. *Renew Sustain Energy Rev [Internet]*. **2019**, *108*, 32–52. Available from: https://linkinghub.elsevier.com/retrieve/pii/S1364032119301625.

[56] Dubois, S.; Lebeau, F. Design, construction and validation of a guarded hot plate apparatus for thermal conductivity measurement of high thickness crop-based specimens. *Mater. Struct [Internet]*. **2015**, *48(1–2)*, 407–421. Available from: http://link.springer.com/10.1617/s11527-013-0192-4.

[57] Slifka, A. J.; Thermal-conductivity apparatus for steady-state, comparative measurement of ceramic coatings. *J. Res. Natl. Inst. Stand. Technol [Internet]*. **2000**, *105(4)*, 591.Available from: https://nvlpubs.nist.gov/nistpubs/jres/105/4/j54sli.pdf.

[58] Ricklefs, A.; Thiele, A. M.; Falzone, G.; Sant, G.; Pilon, L. Thermal conductivity of cementitious composites containing microencapsulated phase change materials. *Int. J. Heat Mass Transf [Internet]*. **2017**, *104*:71–82. Available from: https://linkinghub.elsevier.com/retrieve/pii/S0017931016306901.

[59] Kim, S.; Drzal, L. T. High latent heat storage and high thermal conductive phase change materials using exfoliated graphite nanoplatelets. *Sol. Energy Mater Sol Cells [Internet]*. **2009**, *93(1)*, 136–142. Available from: https://linkinghub.elsevier.com/retrieve/pii/S0927024808002857.

[60] Feldman, D.; Banu, D.; Hawes, D.; Ghanbari, E. Obtaining an energy storing building material by direct incorporation of an organic phase change material in gypsum wallboard. *Sol. Energy Mater [Internet]*. **1991**, *22(2–3)*, 231–242. Available from: https://linkinghub.elsevier.com/retrieve/pii/016516339190021C.

[61] Merlin, K.; Delaunay, D.; Soto, J.; Traonvouez, L. Heat transfer enhancement in latent heat thermal storage systems: Comparative study of different solutions and thermal contact investigation between the exchanger and the PCM. *Appl. Energy [Internet]*. **2016**, *166*, 107–116. Available from: https://linkinghub.elsevier.com/retrieve/pii/S0306261916000313.

[62] Sobolciak, P.; Karkri, M.; Al-Maadeed, M. A.; Krupa, I. Thermal characterization of phase change materials based on linear low-density polyethylene, paraffin wax and expanded graphite. *Renew. Energy [Internet]*. **2016**, *88*, 372–382. Available from: https://linkinghub.elsevier.com/retrieve/pii/S0960148115304730.

[63] Karkri, M.; Lachheb, M.; Albouchi, F.; Ben, N. S.; Krupa, I. Thermal properties of smart microencapsulated paraffin/plaster composites for the thermal regulation of buildings. *Energy Build [Internet]*. **2015**, *88*, 183–192. Available from: https://linkinghub.elsevier.com/retrieve/pii/S0378778814010354.

[64] Moulahi, C.; Trigui, A.; Karkri, M.; Boudaya, C. Thermal performance of latent heat storage: Phase change material melting in horizontal tube applied to lightweight building envelopes. *Compos. Struct [Internet]*. **2016**, *149*, 69–78. Available from: https://linkinghub.elsevier.com/retrieve/pii/S0263822316302501.

[65] Lachheb, M.; Younsi, Z.; Naji, H.; Karkri, M.; Ben Nasrallah, S. Thermal behavior of a hybrid PCM/plaster: A numerical and experimental investigation. *Appl. Therm. Eng [Internet]*. **2017**, *111*, 49–59. Available from: https://linkinghub.elsevier.com/retrieve/pii/S135943111631691X.

[66] Jia, S.; Zhu, Y.; Wang, Z.; Chen, L.; Fu, L. Improvement of shape stability and thermal properties of PCM using polyethylene glycol (PEG)/sisal fiber cellulose (SFC)/graphene oxide (GO). *Fibers Polym [Internet]*. **2017**, *18*(*6*), 1171–1179. Available from: http://link.springer.com/10.1007/s12221-017-7093-z.

[67] Mathis, N.; Transient thermal conductivity measurements: Comparison of destructive and nondestructive techniques. *High Temp. Press [Internet]*. **2000**, *32*(*3*), 321–327. Available from: http://www.hthpweb.com/abstract.cgi?id=htwu289.

[68] Venart, J. E. S.; A simple radial heat flow apparatus for fluid thermal conductivity measurements. *J. Sci. Instrum [Internet]*. **1964**, *41*(*12*), 727–731. Available from: https://iopscience.iop.org/article/10.1088/0950-7671/41/12/304.

[69] Presley, M. A.; Christensen, P. R. Thermal conductivity measurements of particulate materials 1. A review. *J. Geophys. Res. Planets [Internet]*. **1997**, *102*(*E3*), 6535–6549. Available from: http://doi.wiley.com/10.1029/96JE03302.

[70] Gurgel, J. M.; Filho, L. S. A.; Grenier, P.; Meunier, F. Thermal Diffusivity and Adsorption Kinetics of Silica-Gel/Water. *Adsorption [Internet]*. **2001**, *7*(*3*), 211–219. Available from: https://doi.org/10.1023/A:1012732817374.

[71] Zhao, D.; Qian, X.; Gu, X.; Jajja, S. A.; Yang, R. Measurement techniques for thermal conductivity and interfacial thermal conductance of bulk and thin film materials. *J. Electron. Packag [Internet]*. **2016**, *138*(*4*), 040802–19. Available from: https://asmedigitalcollection.asme.org/electronicpackaging/article/doi/10.1115/1.4034605/384410/Measurement-Techniques-for-Thermal-Conductivity.

[72] Terry, M. T. *Thermal Conductivity: Theory, Properties, and Applications (Physics of Solids and Liquids) [Internet]*. Tritt, T. M., Ed.; Springer US; 2004. p. 290. (Physics of Solids and Liquids). Available from: http://link.springer.com/10.1007/b136496.

[73] Li, Y.; Shi, C.; Liu, J.; Liu, E.; Shao, J.; Chen, Z., et al. Improving the accuracy of the transient plane source method by correcting probe heat capacity and resistance influences. *Meas Sci. Technol [Internet]*. **2014**, *25*(*1*), 015006. Available from: https://iopscience.iop.org/article/10.1088/0957-0233/25/1/015006.

[74] Roder, H. M.; A Transient Hot Wire Thermal Conductivity Apparatus for Fluids. *J. Res. Natl. Bur. Stand (1934) [Internet]*. **1981**, *86*(*5*), 457.Available from: https://nvlpubs.nist.gov/nistpubs/jres/086/jresv86n5p457_A1b.pdf.

[75] Assael, M. J.; Antoniadis, K. D.; Wakeham, W. A. Historical evolution of the transient hot-wire technique. *Int. J. Thermophys [Internet]*. **2010**, *31*(*6*), 1051–1072. Available from: http://link.springer.com/10.1007/s10765-010-0814-9.

[76] Richard, R. G.; Shankland, I. R. A transient hot-wire method for measuring the thermal conductivity of gases and liquids. *Int. J. Thermophys [Internet]*. **1989**, *10*(*3*), 673–686. Available from: http://link.springer.com/10.1007/BF00507988.

[77] Healy, J. J.; de Groot, J. J.; Kestin, J. The theory of the transient hot-wire method for measuring thermal conductivity. *Phys. B+C [Internet]*. **1976**, *82*(*2*), 392–408. Available from: https://linkinghub.elsevier.com/retrieve/pii/0378436376902035.

[78] Coquard, R.; Baillis, D.; Quenard, D. Experimental and theoretical study of the hot-wire method applied to low-density thermal insulators. *Int. J. Heat Mass Transf [Internet]*. **2006**, *49*(*23–24*), 4511–4524. Available from: https://linkinghub.elsevier.com/retrieve/pii/S0017931006003346.

[79] Santini, R.; Tadrist, L.; Pantaloni, J.; Cerisier, P. Measurement of thermal conductivity of molten salts in the range 100–500°C. *Int. J. Heat Mass Transf [Internet]*. **1984**, *27*(*4*), 623–626. Available from: https://linkinghub.elsevier.com/retrieve/pii/0017931084900346.

[80] Al-Ajlan, S. A.; Measurements of thermal properties of insulation materials by using transient plane source technique. *Appl. Therm. Eng [Internet]*. **2006**, *26*(*17–18*), 2184–2191. Available from: https://linkinghub.elsevier.com/retrieve/pii/S1359431106001256.

[81] Merckx, B.; Dudoignon, P.; Garnier, J. P.; Marchand, D. Simplified transient hot-wire method for effective thermal conductivity measurement in geo materials: microstructure and saturation effect. *Adv. Civ. Eng [Internet]*. 2012, *2012*, 1–10. Available from: http://www.hindawi.com/journals/ace/2012/625395/.

[82] Assael, M. J.; Antoniadis, K. D.; Metaxa, I. N.; Mylona, S. K.; Assael, J.-A. M.; Wu, J., et al. A novel portable absolute transient hot-wire instrument for the measurement of the thermal conductivity of solids *Int. J. Thermophys [Internet]*. **2015**, *36*(*10–11*), 3083–3105. Available from: http://link.springer.com/10.1007/s10765-015-1964-6.

[83] Sarı, A.; Karaipekli, A. Preparation, thermal properties and thermal reliability of palmitic acid/expanded graphite composite as form-stable PCM for thermal energy storage. *Sol. Energy Mater Sol Cells [Internet]*. **2009**, *93*(*5*), 571–576. Available from: https://linkinghub.elsevier.com/retrieve/pii/S0927024808004571.

[84] Karaipekli, A.; Sarı, A.; Kaygusuz, K. Thermal conductivity improvement of stearic acid using expanded graphite and carbon fiber for energy storage applications. *Renew. Energy [Internet]*. **2007**, *32*(*13*), 2201–2210. Available from: https://linkinghub.elsevier.com/retrieve/pii/S0960148106003351.

[85] Frusteri, F.; Leonardi, V.; Vasta, S.; Restuccia, G. Thermal conductivity measurement of a PCM based storage system containing carbon fibers. *Appl. Therm. Eng [Internet]*. **2005**, *25*(*11–12*), 1623–1633. Available from: https://linkinghub.elsevier.com/retrieve/pii/S1359431104002984.

[86] Babapoor, A.; Karimi, G. Thermal properties measurement and heat storage analysis of paraffinnanoparticles composites phase change material: Comparison and optimization. *Appl. Therm. Eng [Internet]*. **2015**, *90*, 945–951. Available from: https://linkinghub.elsevier.com/retrieve/pii/S1359431115007863.

[87] Cingarapu, S.; Singh, D.; Timofeeva, E. V.; Moravek, M. R. Nanofluids with encapsulated tin nanoparticles for advanced heat transfer and thermal energy storage. *Int. J. Energy Res [Internet]*. **2014**, *38*(*1*), 51–59. Available from: https://onlinelibrary.wiley.com/doi/10.1002/er.3041.

[88] Zhang, P.; Ma, Z. W.; Shi, X. J.; Xiao, X. Thermal conductivity measurements of a phase change material slurry under the influence of phase change. *Int. J. Therm. Sci [Internet]*. **2014**, *78*, 56–64. Available from: https://linkinghub.elsevier.com/retrieve/pii/S1290072913002810.

[89] Sarı, A.; Karaipekli, A. Thermal conductivity and latent heat thermal energy storage characteristics of paraffin/expanded graphite composite as phase change material. *Appl. Therm. Eng [Internet]*. **2007**, *27*(*8–9*), 1271–1277. Available from: https://linkinghub.elsevier.com/retrieve/pii/S1359431106004030.

[90] Fan, L.-W.; Fang, X.; Wang, X.; Zeng, Y.; Xiao, Y.-Q.; Yu, Z.-T., et al. Effects of various carbon nanofillers on the thermal conductivity and energy storage properties of paraffin-based nanocomposite phase change materials. *Appl. Energy [Internet]*. **2013**, *110*, 163–172. Available from: https://linkinghub.elsevier.com/retrieve/pii/S0306261913003383.

[91] Ahadi, M.; Andisheh-Tadbir, M.; Tam, M.; Bahrami, M. An improved transient plane source method for measuring thermal conductivity of thin films: Deconvoluting thermal contact resistance. *Int. J. Heat Mass Transf [Internet]*. **2016**, *96*, 371–380. Available from: https://linkinghub.elsevier.com/retrieve/pii/S0017931015314290.

[92] Log, T.; Gustafsson, S. E. Transient plane source (TPS) technique for measuring thermal transport properties of building materials. *Fire Mater [Internet]*. **1995**, *19*(*1*), 43–49. Available from: https://onlinelibrary.wiley.com/doi/10.1002/fam.810190107.

[93] Acem, Z.; Lopez, J.; Palomo Del Barrio, E. KNO3/NaNO3 – Graphite materials for thermal energy storage at high temperature: Part I. – Elaboration methods and thermal properties. *Appl. Therm. Eng [Internet]*. **2010**, *30*(*13*), 1580–1585. Available from: https://linkinghub.elsevier.com/retrieve/pii/S1359431110001237.

[94] Pinheiro, J. M.; Salústio, S.; Valente, A. A.; Silva, C. M. Adsorption heat pump optimization by experimental design and response surface methodology. *Appl. Therm. Eng [Internet]*. **2018**, *138*, 849–860. Available from: https://linkinghub.elsevier.com/retrieve/pii/S1359431117351219.

[95] Zheng, Q.; Kaur, S.; Dames, C.; Prasher, R. S. Analysis and improvement of the hot disk transient plane source method for low thermal conductivity materials. *Int. J. Heat Mass Transf [Internet]*. **2020**, *151*, 119331. Available from: https://linkinghub.elsevier.com/retrieve/pii/S0017931019362234.

[96] Wang, L.; Zhu, D.; Tan, Y. Heat transfer enhancement on the adsorber of adsorption heat pump. *Adsorption [Internet]*. **1999**, *5*(*3*), 279–286. Available from: https://doi.org/10.1023/A:1008964013879.

[97] Aittomäki, A.; Aula, A. Determination of effective thermal conductivity of adsorbent bed using measured temperature profiles. *Int. Commun. Heat Mass Transf [Internet]*. **1991**, *18*(*5*), 681–690. Available from: https://linkinghub.elsevier.com/retrieve/pii/073519339190080N.

[98] Malinarič, S.; Contribution to the transient plane source method for measuring thermophysical properties of solids. *Int. J. Thermophys [Internet]*. **2013**, *34*(*10*), 1953–1961. Available from: http://link.springer.com/10.1007/s10765-013-1502-3.

[99] Malinarič, S.; Dieška, P. Concentric circular strips model of the transient plane source-sensor. *Int. J. Thermophys [Internet]*. **2015**, *36*(*4*), 692–700. Available from: http://link.springer.com/10.1007/s10765-015-1848-9.

[100] Afriyie, E. T.; Karami, P.; Norberg, P.; Gudmundsson, K. Textural and thermal conductivity properties of a low density mesoporous silica material. *Energy Build [Internet]*. **2014**, *75*, 210–215. Available from: https://linkinghub.elsevier.com/retrieve/pii/S0378778814001236.

[101] Yu, N.; Wang, R. Z.; Lu, Z. S.; Wang, L. W. Study on consolidated composite sorbents impregnated with LiCl for thermal energy storage. *Int. J. Heat Mass Transf [Internet]*. **2015**, *84*, 660–670. Available from: https://linkinghub.elsevier.com/retrieve/pii/S0017931015000733.

[102] Yu, N.; Wang, R. Z.; Wang, L. W. Theoretical and experimental investigation of a closed sorption thermal storage prototype using LiCl/water. *Energy [Internet]*. **2015**, *93*, 1523–1534. Available from: https://linkinghub.elsevier.com/retrieve/pii/S0360544215013560.

[103] Huang, Z.; Gao, X.; Xu, T.; Fang, Y.; Zhang, Z. Thermal property measurement and heat storage analysis of LiNO3/KCl – Expanded graphite composite phase change material. *Appl. Energy [Internet]*. **2014**, *115*, 265–271. Available from: https://linkinghub.elsevier.com/retrieve/pii/S0306261913009124.

[104] Duan, Z.; Zhang, H.; Sun, L.; Cao, Z.; Xu, F.; Zou, Y., et al. CaCl2·6H2O/Expanded graphite composite as form-stable phase change materials for thermal energy storage. *J. Therm. Anal. Calorim [Internet]*. **2014**, *115*(*1*), 111–117. Available from: http://link.springer.com/10.1007/s10973-013-3311-0.

[105] Casey, S. P.; Elvins, J.; Riffat, S.; Robinson, A. Salt impregnated desiccant matrices for 'open' thermochemical energy storage – Selection, synthesis and characterisation of candidate materials. *Energy Build [Internet]*. **2014**, *84*, 412–425. Available from: https://linkinghub.elsevier.com/retrieve/pii/S0378778814006707.

[106] Cammarata, A.; Verda, V.; Sciacovelli, A.; Ding, Y. Hybrid strontium bromide-natural graphite composites for low to medium temperature thermochemical energy storage: Formulation, fabrication and performance investigation. *Energy Convers Manag [Internet]*. **2018**, *166*, 233–240. Available from: https://linkinghub.elsevier.com/retrieve/pii/S0196890418303662.

[107] Shere, L.; Trivedi, S.; Roberts, S.; Sciacovelli, A.; Ding, Y. Synthesis and characterization of thermochemical storage material combining porous zeolite and inorganic salts. *Heat Transf Eng [Internet]*. **2019**, *40*(*13–14*), 1176–1181. Available from: https://www.tandfonline.com/doi/full/10.1080/01457632.2018.1457266.

[108] Parker, W. J.; Jenkins, R. J.; Butler, C. P.; Abbott, G. L. Flash method of determining thermal diffusivity, heat capacity, and thermal conductivity. *J. Appl. Phys [Internet].* **1961**, *32(9)*, 1679–1684. Available from: http://aip.scitation.org/doi/10.1063/1.1728417.

[109] Lian, T.-W.; Kondo, A.; Akoshima, M.; Abe, H.; Ohmura, T.; Tuan, W.-H., et al. Rapid thermal conductivity measurement of porous thermal insulation material by laser flash method. *Adv. Powder Technol [Internet].* **2016**, *27(3)*, 882–885. Available from: https://linkinghub.elsevier.com/retrieve/pii/S0921883116000108.

[110] Storey, A. A.; Ramirez, J. M.; Quiroz, D.; Burley, D. V.; Addison, D. J.; Walter, R., et al. Radiocarbon and DNA evidence for a pre-Columbian introduction of Polynesian chickens to Chile. *Proc. Natl. Acad. Sci. [Internet].* **2007**;*104(25)*:10335–10339. Available from: http://www.pnas.org/cgi/doi/10.1073/pnas.0703993104.

[111] ASTM E1461. *Standard Test Method for Thermal Diffusivity by the Flash Method.* 2013.

[112] ISO 22007-4. Plastics – Determination of thermal conductivity and thermal diffusivity – Part 4: *Laser Flash Method [Internet].* **2017**. Available from: https://www.iso.org/standard/65085.html.

[113] Zalba, B.; Marín, J. M.; Cabeza, L. F.; Mehling, H. Review on thermal energy storage with phase change: Materials, heat transfer analysis and applications. *Appl. Therm. Eng [Internet].* **2003**, *23(3)*, 251–283. Available from: https://linkinghub.elsevier.com/retrieve/pii/S1359431102001928.

[114] Lazaro, A.; Peñalosa, C.; Solé, A.; Diarce, G.; Haussmann, T.; Fois, M., et al. Intercomparative tests on phase change materials characterisation with differential scanning calorimeter. *Appl. Energy [Internet].* **2013**, *109*, 415–420. Available from: https://linkinghub.elsevier.com/retrieve/pii/S0306261912008483.

[115] Cheng, R.; Pomianowski, M.; Wang, X.; Heiselberg, P.; Zhang, Y. A new method to determine thermophysical properties of PCM-concrete brick. *Appl. Energy [Internet].* **2013**, *112*, 988–998. Available from: https://linkinghub.elsevier.com/retrieve/pii/S030626191300055X.

[116] Mazzeo, D.; Oliveti, G.; de Gracia, A.; Coma, J.; Solé, A.; Cabeza, L. F. Experimental validation of the exact analytical solution to the steady periodic heat transfer problem in a PCM layer. *Energy [Internet].* **2017**, *140*, 1131–1147. Available from: https://linkinghub.elsevier.com/retrieve/pii/S0360544217314238.

[117] Gschwander, S.; Lazaro, A.; Cabeza, L. F.; Günther, E.; Fois, M.; Chui, J. Development of a test-standard for pcm and tcm characterization part 1: characterization of phase change materials compact thermal energy storage: Material development for system integration. *Int. Energy Agency.* **2011**;1–46.

[118] ISO 23993. Thermal insulation products for building equipment and industrial installations-Determination of design thermal conductivity (ISO 23993 : 2008, Corrected version 2009-10-01) [Internet]. 2010. Available from: https://www.iso.org/standard/37042.html.

[119] Pincemin, S.; Olives, R.; Py, X.; Christ, M. Highly conductive composites made of phase change materials and graphite for thermal storage. *Sol. Energy Mater. Sol. Cells [Internet].* **2008**, *92(6)*, 603–613. Available from: https://linkinghub.elsevier.com/retrieve/pii/S0927024807004291.

[120] Liu, L.; Su, D.; Tang, Y.; Fang, G. Thermal conductivity enhancement of phase change materials for thermal energy storage: A review. *Renew. Sustain. Energy Rev [Internet].* **2016** Sep;*62*:305–317. Available from: https://linkinghub.elsevier.com/retrieve/pii/S1364032116300909.

[121] Fan, L.; Khodadadi, J. M. Thermal conductivity enhancement of phase change materials for thermal energy storage: A review. *Renew Sustain Energy Rev [Internet].* **2011**, *15(1)*, 24–46. Available from: https://linkinghub.elsevier.com/retrieve/pii/S1364032110002595.

[122] Qureshi, Z. A.; Ali, H. M.; Khushnood, S. Recent advances on thermal conductivity enhancement of phase change materials for energy storage system. *Rev. Int. J. Heat Mass. Transf [Internet].* **2018** Dec;*127*:838–856. Available from: https://linkinghub.elsevier.com/retrieve/pii/S0017931018325511.

[123] Yang, C.; Navarro, M. E.; Zhao, B.; Leng, G.; Xu, G.; Wang, L., et al. Thermal conductivity enhancement of recycled high density polyethylene as a storage media for latent heat thermal

energy storage. *Sol. Energy Mater. Sol. Cells [Internet]*. **2016**, *152*, 103–110. Available from: https://linkinghub.elsevier.com/retrieve/pii/S0927024816000945.
[124] Palacios, A.; de Gracia, A.; Cabeza, L. F.; Julià, E.; Fernández, A. I.; Barreneche, C. New formulation and characterization of enhanced bulk-organic phase change materials. *Energy Build [Internet]*. **2018**, *167*, 38–48. Available from: https://linkinghub.elsevier.com/retrieve/pii/S037877881730230X.
[125] Alam, T. E.; Dhau, J. S.; Goswami, D. Y.; Stefanakos, E. Macroencapsulation and characterization of phase change materials for latent heat thermal energy storage systems. *Appl. Energy [Internet]*. **2015**, *154*, 92–101. Available from: https://linkinghub.elsevier.com/retrieve/pii/S0306261915005498.
[126] Alshaer, W. G.; Nada, S. A.; Rady, M. A.; Del Barrio, E. P.; Sommier, A. Thermal management of electronic devices using carbon foam and PCM/nano-composite. *Int. J. Therm. Sci [Internet]*. **2015**, *89*, 79–86. Available from: https://linkinghub.elsevier.com/retrieve/pii/S1290072914002968.
[127] Su, D.; Jia, Y.; Alva, G.; Tang, F.; Fang, G. Preparation and thermal properties of n–octadecane/stearic acid eutectic mixtures with hexagonal boron nitride as phase change materials for thermal energy storage. *Energy Build [Internet]*. **2016**, *131*, 35–41. Available from: https://linkinghub.elsevier.com/retrieve/pii/S0378778816308222.

Jean-Yves Coxam and Karine Ballerat-Busserolles

Chapter 7
Calorimetric methods for key properties in refrigeration cycles

Abstract: Refrigeration systems have at least one circuit containing a heat transfer fluid or working fluid. These systems cannot be totally hermetic. Because of potential leaks, new working fluids will have to be chosen according to their impacts on global warming. The physicochemical properties of these fluids will have to be determined in order to develop or optimize refrigeration processes. This chapter presents calorimetric techniques for the determination of heat capacities and heat of vaporization.

Keywords: adiabatic calorimeters, differential scanning calorimetry, electric heat pump, enthalpy, heat capacity, refrigerant, thermal heat pump

7.1 Cold production

Nowadays, the production of cold is a major challenge of our society. This is the case for chemical, pharmaceutical, or food industry where freezing and refrigeration remain the main methods for product preservation. With the domestic requirements such as individual food storage or air-conditioning, the global production of cold represents about 17% of the world electricity consumption.

One of the techniques for heat (heat pump) or cold (refrigerators) production consists of pumping heat energy from a cold source to a warm source. For heat production, the objectives will concern the "warm source" which receives the heat energy (Fig. 7.1) and, conversely, for cold production, it will concern the "cold source" from which the heat energy is removed.

The heat energy is pumped, or released, by an endothermic or an exothermic transformation of a working fluid. The transformations must be reversible to allow the machine to work on pumping/releasing calories cycles. The working fluid could be a pure fluid or a mixture of fluids, depending on the type of heat pump. We will here distinguish two heat pump types: the electric heat pump (EHP) and the thermal heat pump (THP).

Jean-Yves Coxam, Institut de Chimie de Clermont-Ferrand, Campus Universitaire des Cézeaux, TSA 60026 – CS 60026, 24, Avenue Blaise Pascal, 63178 Aubiere, France, e-mail: j-yves.coxam@uca.fr
Karine Ballerat-Busserolles, Institut de Chimie de Clermont-Ferrand, Campus Universitaire des Cézeaux, TSA 60026 – CS 60026, 24, Avenue Blaise Pascal, 63178 Aubiere, France, e-mail: Karine.BALLERAT@uca.fr

https://doi.org/10.1515/9783110590449-007

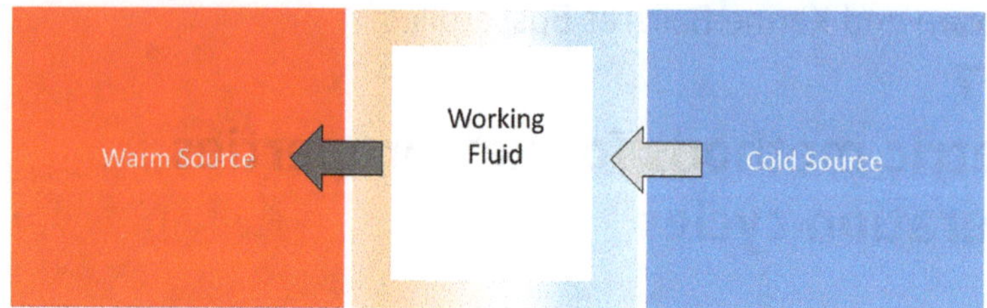

Fig. 7.1: Schematic diagram of principle for heat pump or refrigerator.

7.1.1 Electric heat pump (EHP)

For EHPs, the heat exchanges are caused by an endothermic evaporation of the working fluid at the cold source and, an exothermic condensation at the warm source. The principle is schematically represented in Fig. 7.2. The system necessitates external mechanical energy source at the compressor used between the evaporator and the condenser. The compressor is usually driven by an electric motor. This configuration, with a single cooling cycle, is adapted to residential cooling or refrigerator applications. For ultra-low temperatures, meaning cooling temperature below −40 °C, cascade refrigeration systems are usually adopted. It consists of a succession of two or more cooling cycles. In this configuration, single circuits are thermally coupled through a heat exchanger, as shown for a two cooling cycles in Fig. 7.3. The high temperature cycle will absorb the energy released by the low-temperature cycle at the condenser.

The reverse Rankine cycle (Fig. 7.4) is a fundamental operating cycle to represent such systems, where the working fluid undergoes evaporating and condensing cycles. An analysis of the thermodynamic cycle is necessary to select a working fluid and to

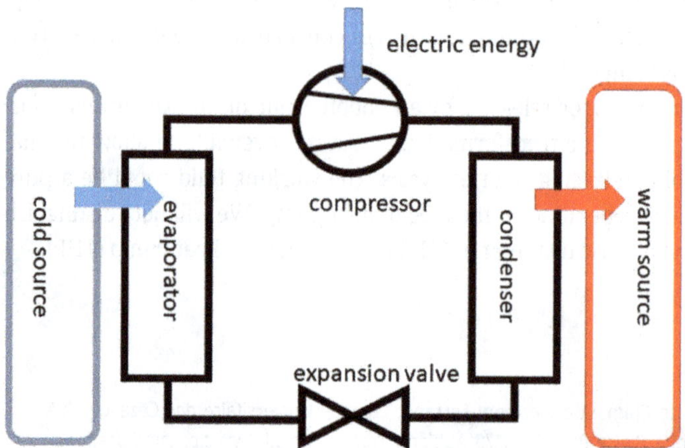

Fig. 7.2: Schematic representation of an electric heat pump working on evaporation/condensation cycles.

design and optimize the cooling machine. In order to establish these thermodynamic cycles, an important experimental work has to be carried out to collect vapor liquid equilibrium (VLE), calorimetric, and volume data.

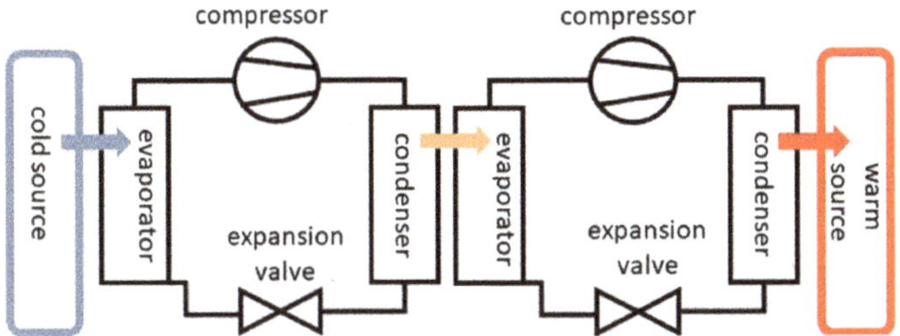

Fig. 7.3: Schematic representation of a cascade refrigeration with two cooling cycles.

7.1.2 Absorption heat pump (AHP)

The cooling principle for an absorption heat pump (AHP) is comparable to an EHP as the heat energy is transferred by vaporization or condensation of a working fluid. The difference lays in the fact that the compressor for EHP is replaced by an absorption/desorption unit. The working fluid flowing in the AEP is constituted of two fluids: a volatile one, called refrigerant, and a liquid one, called absorbent. The refrigerant and the absorbent mix in the absorber (Fig. 7.5) and separate (desorption) in the generator by heating. The absorbent flows back to the absorber while the refrigerant vapors are condensed into liquid in the condenser. During this step, the generator receives heat energy from an external source. The refrigerant is then vaporized (endothermic transformation) inside the evaporator before entering the absorber.

These AHPs present new interest because of environmental and sustainable energy issues. First, while EHPs operate mainly with fluorinated working fluids, AHP can work with more ecofriendly fluids. Secondly, as there is no compressor, AHPs do not require electric energy which is still mainly obtained from fossil energy. In this way, AHP can contribute to a reduction of carbon dioxide emissions.

7.1.3 Working fluids

The working fluid is chosen accordingly to the process condensation and vaporization temperatures. The constraints for condensation will be imposed by the “warm source,” i.e., the temperature range to be covered for heating. Conversely, the constraints of vaporization will be chosen as function of the domain of temperature required for cooling. Although there is no official classification for refrigeration temperature ranges, one can quote (Tab. 7.1) the one proposed by Linde gas industry [1]. The temperatures are

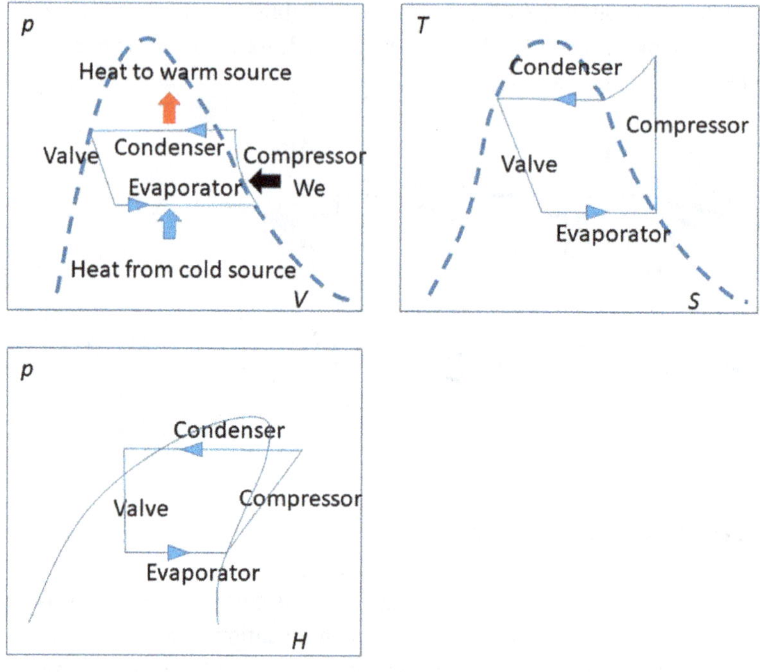

Fig. 7.4: Reverse Rankine cycles, pressure (p) versus volume (V); temperature (T) versus entropy (S) and pressure (p) versus enthalpy (H).

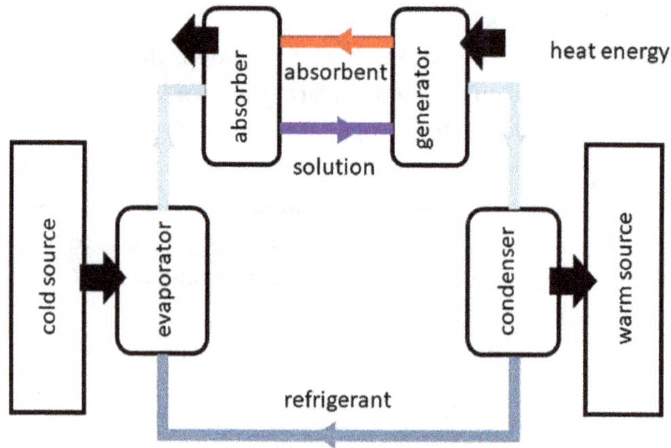

Fig. 7.5: Schematic representation of an absorption heat pump (AHP).

here classified in four groups, from high refrigeration temperature to ultra-low refrigeration temperatures, each group associated with specific applications, going from air conditioners to cryogenic freezers.

The working fluids can be pure refrigerants or a mixture of refrigerants. It must have specific physical and chemical properties, in addition to its thermodynamic properties, to insure a safe and environmentally friendly process. Many pure or mixed refrigerants have been already developed (Tab. 7.2) to meet the constraints imposed by process temperature range and specific application. The first refrigerants used in vapor compression systems were the carbon dioxide and ammonia. Sulfur dioxide and methyl chloride were later proposed for "high temperature" refrigeration with the development of domestic refrigeration. These so-called "natural" refrigerants were gradually replaced by chemical compounds from diverse halocarbon groups; the halocarbons are obtained from hydrocarbons by substitution of hydrogen(s) with chlorine and/or fluorine atom(s). However, the emissions of halocarbons have been found to be one of the major gas responsible for climate change. Their individual contribution is quantified by a high global warming potential (GWP) as reported in Tab. 7.2. To limit global warming, the nations have agreed to regulate or even ban the use of halocarbons. The Intergovernmental Panel on Climate Change (IPCC) stated in their fourth assessment report that the developed countries will have to reduce their greenhouse gas emissions by 80% to 95% below 1990 levels by 2050. This will force the replacement of chlorofluorocarbons (CFCs), hydrochlorofluorocarbons (HCFCs) or hydrofluorocarbons (HFCs). Consequently, many researches are currently carried out to propose climate-friendly alternatives to halocarbons. It is said that the refrigerant for future could be the carbon dioxide, hydrocarbons, ammonia, or hydrofloroolefins (HFOs).

Tab. 7.1: Refrigeration classification of temperature ranges and applications.

Refrigeration classification	Application	Evaporator temperature range
High-temperature refrigeration	Air conditioning	Above 0 °C
Medium-temperature refrigeration	Refrigerators-freezer	From 0 °C to −25 °C
Low-temperature refrigeration	Freezer	−25 °C down to −50 °C
Ultra-low temperature	Cryogenic engineering	Below −50 °C

7.2 Thermodynamic properties and calorimetric measurements for refrigeration systems

The thermodynamic and physicochemical properties of assumed alternative refrigerants are crucial data for the design, modeling, and optimization of new generations of cooling systems. The researches carried out in this field include both acquisition of experimental data and development of thermodynamic models or simulation tools. The experimental data must allow the characterization of the conditions of vaporization or condensation of the process. From a thermodynamic point of view and, to understand the interest of particular experimental data, we recall that the temperature and pressure conditions of VLE can be derived from the equality of fugacities:

$$f^{l}(T,p)=f^{v}(T,p) \tag{7.1}$$

where f^{φ} denotes the fugacity of a component in the φ phase (vap for vapor phase and liq for liquid phase). In case of blend refrigerant, the compositions of refrigerant i in the liquid phase (x_i) and in the gas phase (y_i) will be additional variables (eq. (7.2)); mass balance equations will complete the system of equations:

$$f_i^{l}(T,p,x_i)=f_i^{v}(T,p,y_i) \tag{7.2}$$

Using a γ-Φ approach, the equality of fugacity will be expressed as reported in eq. (7.3), where γ_i is the activity coefficient in the liquid phase and Φ_i^{vap}, the fugacity coefficients in the gas phase. Φ_i^{sat} is the fugacity coefficient of pure refrigerant i at saturation pressure:

$$x_i\gamma_i\Phi_i^{sat}p_i^{sat}\exp\left(\frac{v_i\left(p-p_i^{sat}\right)}{RT}\right)=y_i\Phi_i^{vap}p \tag{7.3}$$

The exponential term is the Poynting correction factor, including the molar volume of i (v_i). The fugacity coefficients are derived from an equation of state and the activity coefficients in the liquid phase, from activity coefficient models. Consequently, the thermodynamic analysis of the vaporization will necessitate various experimental data such as saturation pressure of pure refrigerant (p_i^{sat}), molar volume (v_i) and excess properties. The last ones, excess volume and excess enthalpy, are required to adjust the activity coefficient model.

Tab. 7.2: Example of chemical fluids used as refrigerant and their application.

CAS	Name	Formula	p_c (bar)	T_c (°C)	GWP	Application
75-69-4	R11	CCl_3F	45	198	4,750	Industrial process cooling
75-71-8	R12	CCl_2F_2	41	112	10,900	Refrigeration/air conditioning
75-72-9	R13	$CClF_3$	39	29	14,000	Low-temperature refrigerant
75-45-6	R22	$CHClF_2$	50	96	1,750	Air conditioning
75-10-5	R32	CH_2F_2	58	76	675	Industrial air conditioning
74-87-3	R40	CH_3Cl	67	143		Automotive refrigerants
76-13-1	R113	$C_2Cl_3F_3$		210	4,800	Industrial air conditioning and process cooling

Tab. 7.2 (continued)

CAS	Name	Formula	p_c (bar)	T_c (°C)	GWP	Application
76-14-2	R114	$C_2Cl_2F_4$	34	146	9,800	Industrial air conditioning and process cooling
76-15-3	R115	C_2F_5Cl	31	80	7,200	Used in blend refrigerant
354-33-6	R125	CHF_2CF_3	36	64	2,800	Air conditioning (replacement of R22)
811-97-2	R134a	CF_3CH_2F	41	101	1,430	Automotive air conditioning
420-46-2	R143a	$C_2H_3F_3$	38	73	3,800	Automotive air conditioning
75-37-6	R152a	$C_2H_4F_2$	45	113	140	Automotive air conditioning
754-12-1	R1234yf	$C_3H_2F_4$	34	93	4	Automotive air conditioning
7732-18-5	R718	H_2O	221	374		Air conditioning applications in absorption systems with brine
124-38-9	R744	CO_2	74	31	1	Long-term solution for refrigeration
7664-41-7	R717	NH_3	113	132	0	Refrigerant used with water in absorption systems for air conditioning
7446-09-5	R764	SO_2	78.9	157		Old-timey refrigerant
74-84-0	R170	CH_3CH_3	49	32		Very-low-temperature refrigeration applications
74-85-1	R1150	CH_2CH_2	51	9	3	Low-temperature refrigeration applications
75-28-5	R600a	$(CH_3)_2CHCH_3$	37	135	3	Domestic and small commercial refrigerator
Mixture	R404A	R125 + R134a + R143a	37	72	3,920	Automotive air conditioning, ice makers
Mixture	R410A	R32–R125	49	71	2,088	Industrial air conditioning
Mixture	R500	R12 + R152A				Residential air conditioning and domestic refrigerators
Mixture	R502	R22 + R115			4,657	Low-temperature freezers

Using a Φ–Φ approach, the fugacity will be expressed as reported in the following equations:

$$x_i \Phi_i^{\mathrm{Liq}} p = y_i \Phi_i^{\mathrm{vap}} p \tag{7.4}$$

where Φ_i^{φ} represents the fugacity coefficient of refrigerant i in the φ phase. This coefficient is derived from an equation of state. Using this approach, theoretically only *pVT* data will be necessary to adjust a model. However, additional experimental data are needed to test the model consistency.

The main difficulty in such modeling is the estimation of "interaction parameters" present in the equations of state or in the activity coefficient models. These are usually empirical terms and as a result, the robustness of the model will be strongly linked to the precision and the diversity of the experimental data. Many methods for collecting experimental data on pure or multicomponent systems are available in literature.

Both approaches, γ/Φ or Φ–Φ, make possible the calculation of the energy of vaporization. The molar Gibbs energy of vaporization of a refrigerant i is derived from fugacities in the liquid and vapor phases (eq. (7.5)). The enthalpy of vaporization is then obtained using a Gibbs-Helmholtz equation (eq. (7.6)). As the first temperature derivative of Gibbs energy, the accuracy on the enthalpy will be very dependent on the robustness of the models used to calculate fugacities. Direct measurements of enthalpies are thus always essential. This is even more important for calorific capacity, which is a second derivative of Gibbs energy.

$$\Delta_{\text{vap}} G_{(T,p)} = RT \ln \left(\frac{f_i^{\text{vap}}}{f_i^{\text{Liq}}} \right) \tag{7.5}$$

$$\left(\frac{\partial \Delta_{\text{vap}} G_{(T,p)} / T}{\partial T} \right)_p = - \frac{\Delta_{\text{vap}} H_{(T,p)}}{T^2} \tag{7.6}$$

In this chapter, we will focus on calorimetric measurements in temperature ranges from -50 °C to ambient and its application for determination of *VLE* diagrams and measurement of heat capacity and heat of reaction.

7.2.1 Measurement of heat of vaporization and phase diagrams

To study the thermodynamic cycles of cooling systems, first the phase diagrams of the working fluids have to be established. These phase diagrams are obtained from experiments which highlights temperature of phase change as function of pressure. The latent heats of vaporization (or enthalpies of vaporization) are key properties because they are directly related to the energy transfers between the working fluid and the external cold or warm sources. These properties provide the temperatures and the corresponding energy exchanged with the external sources. It is thus possible to define the operating conditions and the performances of a cooling process device.

The phase diagrams are usually obtained from calorimetric techniques using different equilibrium cells [2–4]. These equilibrium cells are basically constituted a specific vessel that can be temperature controlled and pressurized. The experimental setup generally makes it possible to extract small samples from each phase to analyze their compositions. As previously explained, the enthalpies of vaporization can be derived from the *VLE* experimental data. However, direct measurements of enthalpies remain essential. The calorimetric techniques also permit to determine temperatures of vaporization by analysis of a thermogram. Compared to classical technique using equilibrium cells, the calorimetric techniques present several advantages such as faster experiments and smaller sample quantities.

Concerning the calorimetric techniques, Tauqir H. Syed et al. [5] presented a method for measuring the enthalpies of vaporization at low temperature and high pressure. The measurements are performed with a low-temperature differential calorimeter equipped with a cylindrical cell, having a volume of 8.5 cm^3; the measuring cell is filled with a well-known mass of sample (between 0.5 and 3.5 g). To accurately fill the cell with volatile compound, a well-known volume of gas is condensed inside the cell, using a syringe pump and controlling both filling temperature and pressure. The syringe pump is then used in the experimental procedure to gradually withdraw fluid from the cell. Thus, increasing the volume of the pump results in a partial vaporization of the liquid. The authors point out that particular attention must be paid to the withdrawal rate to get consistent integration of the vaporization peaks in the thermogram. The full vaporization is performed in several steps. In the case of a mixture, the composition of the gas phase will change during vaporization and the technique can be adapted to make it possible to extract gas sample for a gas chromatography analysis. A reference cell, filled with nitrogen at atmospheric pressure, is used for differential measurements. The thermogram is basically constituted of an initial endothermic peak related to the sample condensation, following by several exothermic peaks of vaporization. An example of thermogram is represented in Fig. 7.6. The enthalpy of vaporization is the sum of the integrated peaks, divided by the initial mass or number of sample moles.

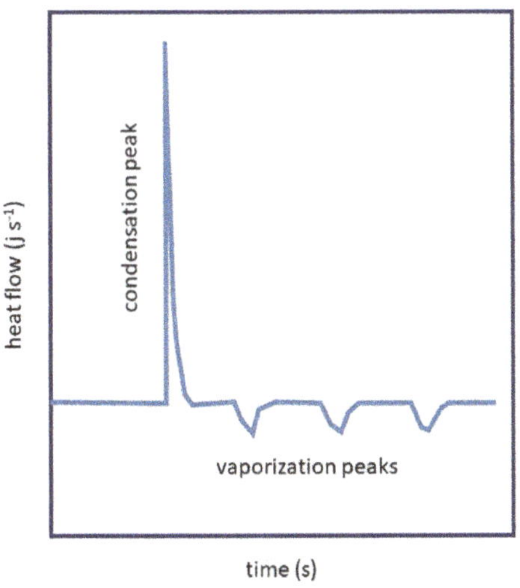

Fig. 7.6: Schematic representation of a thermogram using the method of Tauqir H. Syed et al. [5].

Sauerbrunn and Zemo [6] proposed a direct measurement of the heat of vaporization with a method based on thermogravimetry and differential calorimetry. The calorimetric cell is a small aluminum crucible, closed with a drilled lid adapted for measurement at ambient pressure. As reported by Fiorini et al. [7], the crucial point of this technique is the design of the lid, to allow a fluid evaporation without heat energy lost. The experiments are carried out at constant temperature and ambient pressure, with a nitrogen purge to aid in the vaporization of the sample. During an experiment, the rate of mass lost and the heat flux are recorded (Fig. 7.7) .The enthalpy of vaporization is calculated by dividing the heat flow signal by the rate of mass loss signal.

Apart from these particular techniques, the enthalpy of vaporization can also be directly obtained from calorimetric experiments, using classical micro differential scanning calorimeter. Such an experimental procedure using a thermal analysis calorimeter is described by Akisawa et al. [8]. For such measurements, a well-known mass of sample is introduced in a standard aluminum crucible presenting a pin hole on the cap. This pin hole allows vaporization at same pressure as the surrounding equipment. The crucible is placed inside a pressure DSC cell, (Fig. 7.8) and after temperature and pressure equilibrium, the system is heated at constant scanning rate. This method can be adapted for various commercial DSC calorimeters. For example, one can use a differential scanning Sensys Evo DSC calorimeter from Setaram equipped with a customized cell (Fig. 7.9). Here, the cell was designed to work at controlled pressures. The experiments are carried out at a constant scanning rate, on a range of temperature from ambient to a temperature 50 °C higher than the expected vaporization temperature. The measuring cell is filled with a well-known mass of sample, while the reference cell is kept empty. The sample cell is connected through stainless 1/16″ stainless steel tubes to a vacuum pump or a nitrogen cylinder to empty or pressurize the fluid in the cell.

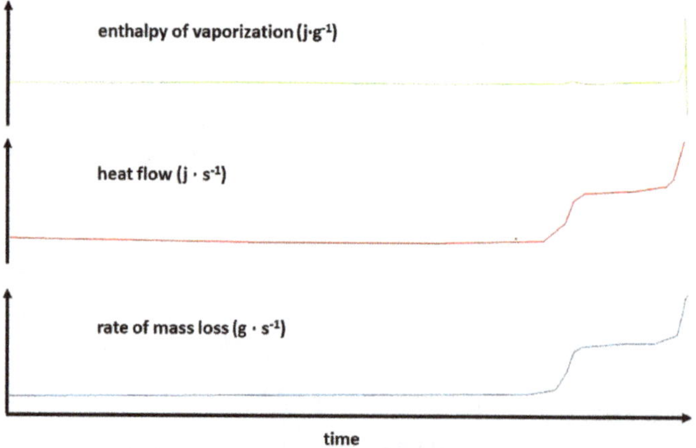

Fig. 7.7: Schematic representation of experimental data obtained by Sauerbrunn and Zemo [2] for water vaporization.

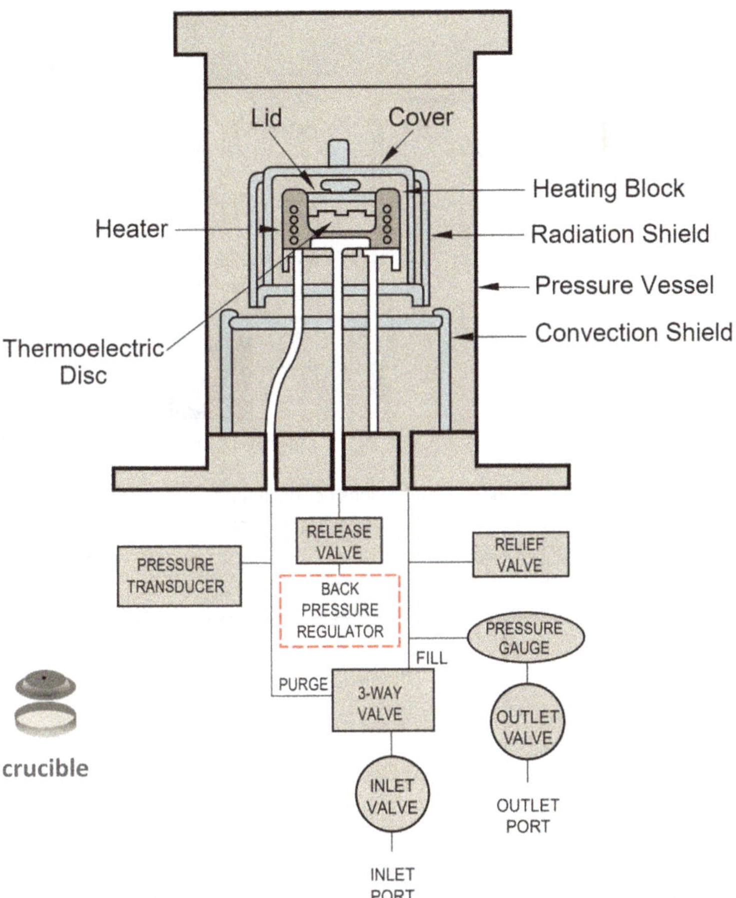

Fig. 7.8: Schematic representation of a DSC 2920 (thermal analysis) pressure cell equipped with a with standard aluminum crucible with pin hole on the cap (image courtesy of TA Instruments, available at http://www.tainstruments.com/pdf/literature/DSC_2920.pdf).

A schematic representation of the resulting thermogram is given in Fig. 7.10. The temperature of vaporization is determined by the intersection between the baseline before vaporization and the tangent of the peak at the inflection point (onset-*T* methods). The enthalpy of vaporization is then classically obtained by heat flux integration and mass sample.

7.2.2 Heat capacity measurements

The heat capacities are essential data for the development and optimization of cooling machines, particularly to dimension the heat exchangers. The historical technique for

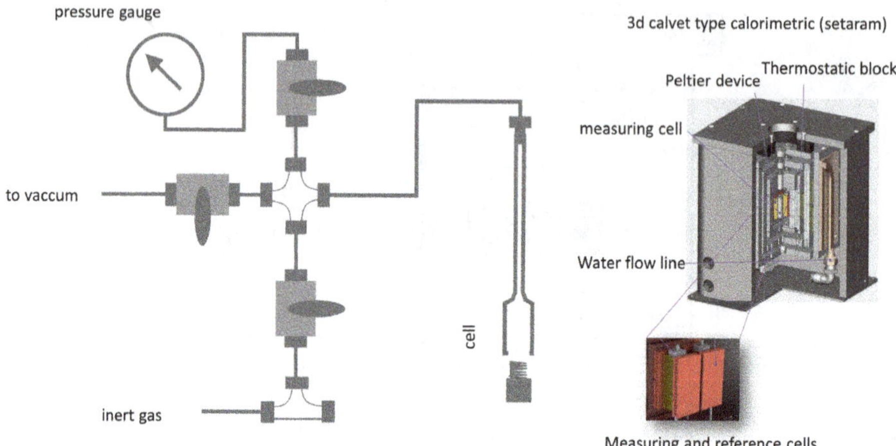

Fig. 7.9: Schematic representation of the cell and line connections designed for a Setaram 3D Calvet-type calorimetric sensor (image courtesy of KEP-Technology-Setaram, available at https://setaramsolutions.com/app/uploads/sites/2/2021/01/FR-MICROCALVET.pdf).

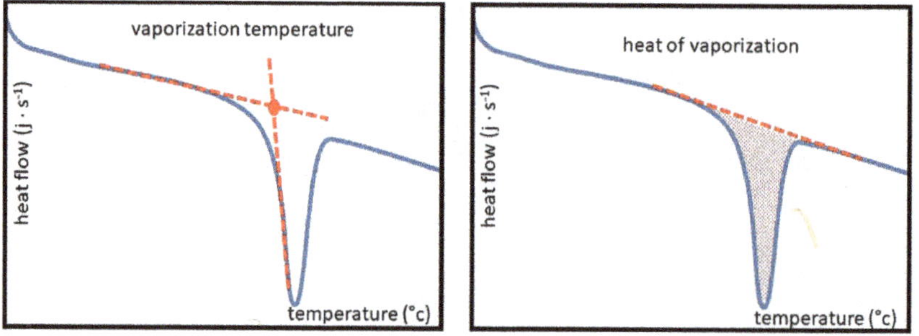

Fig. 7.10: Determination of temperature of vaporization and heat of vaporization from thermogram.

determining heat capacities at low and very low temperatures is adiabatic calorimetry. The principle of adiabatic calorimetry is based on the perfect knowledge of temperature variations of a sample caused by a given heat transfer. This method, using usually home-made calorimeters, is still used to determine heat capacities at temperatures below 120 K. However, for the last 50 years the calorimetric techniques have been improved in terms of instrumentation and automatization, and nowadays, commercial calorimeters, adiabatic or not, are available for heat capacity measurements down to 100 K.

7.2.2.1 Adiabatic calorimeters

The first development of adiabatic calorimetry at low temperature started in the 1910s with the works of Nernst. In the 1940s, Yost et al. [9] and Giauque et al. [10], developed an adiabatic calorimeter for measurement down to 60 K. The sample cell was cooled down by contact with a refrigerant tank filled with liquid nitrogen or helium. This first accurate calorimetry at equilibrium is nevertheless time-consuming and necessitates a relatively large amount of sample. The method remains still appropriated to determine specific heat capacities at very low temperatures [11].

An adiabatic calorimeter is essentially constituted of a sample cell equipped with a thermometer and an electric heater. The cell is located inside one or more concentric adiabatic shields. To reach quasi-adiabatic conditions, the cell surroundings must be maintained at the same temperature as the cell itself. The temperature is controlled using a set of differential thermocouples that link the cell and the adiabatic shields placed in a can connected to a vacuum pump. Such technique is described with more details by Southard and Andrews [12], together with a discussion on specific techniques that can be added to limit heat leaks between the cell and their surroundings. The calorimetric cell of Southard and Andrews [12] is made of gold-plated copper and has a volume of about 8 cm^3. The cell can be filled or emptied from a tube which is closed during experiments. A second tube permits the connection of the heater made of a lattice of platinum-iridium wires inserted in a mica cylinder maintained in thermal contact with the internal surface of the cell. The cell is suspended inside a temperature-controlled shield (Fig. 7.11). The shield is a cylinder, closed at both bottom and top, and temperature controlled by use of a heating wire covering all the cylinder surface. The temperature is adjusted thanks to a thermocouple which detects the temperature difference with the cell. The cell and the shield are finally suspended inside a large glass tube which can be connected to a vacuum pump. Such technique was then developed and used in the 1940s by Russell and coworkers [13], to determine the heat capacity of 1-1-1 trifluoroethane at very low temperatures, from 12 to 220 K [13]. The error on heat capacity was estimated to be 2% at temperature below 30 K and, to decrease down to 0.1% at for temperature above 30 K.

More recently, Tan et al. [14, 15] have developed similar adiabatic calorimeter for measuring heat capacities in temperature range from 60 to 350 K. The cylindrical cell E (Fig. 7.12), made of nickel or gold-plated copper, is 26 mm long, 0.4 mm thick, and has a diameter of 20 mm. The internal volume is about 6 cm^3. The sample is heated by a heating wire surrounded the cell, and the temperature is read by a platinum thermometer located at the bottom of the cell. Copper vanes were added inside the cell to shorten thermal equilibrium time. The cell is suspended in the inner adiabatic shield, suspended itself in an outer adiabatic shield. Each shield is entirely covered by heating wires to prevent temperature gradient inside the cell. The outer shield is hanged to a cylindrical vacuum can. The set is placed in a Dewar filled with liquid nitrogen or helium. The temperature of the outer adiabatic shield is kept slightly lower than that of the inner one. The cell is filled with a well-known quantity of sample and receives heat energy supplied by joule effect during a given temperature increment. The experimental procedure is repeated at different discrete temperatures. The initial and final temperatures of the increment are determined by recording the temperature as function of time before and after heating, up to reach the thermal equilibrium. The heat energy Q provided by the electric wire is given by the measurements of the potential difference U and the current I, during the heating time t:

$$Q = \Delta H = U \cdot I \cdot t \tag{7.7}$$

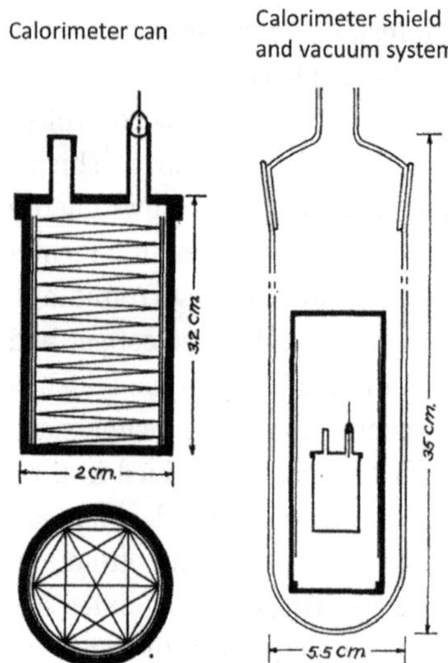

Fig. 7.11: Schematic representation of adiabatic calorimeter of Southard and Andrews [12]. (reprinted from Publication J. Frankl., 209(3), Southard JC and Andrews DH, "The heat capacities of organic compounds at low temperatures. iii. an adiabatic calorimeter for heat capacities at low temperatures.", Pages 349–360., Copyright (2023), with permission from Elsevier).

The same procedure is applied for a blank experiment to measure the heat energy consumed by only the cell. This quantity Q_{blank} will be subtracted in the calculation (eq. (7.8)) of the sample heat capacity. The precision on the heat capacity is estimated to be 0.28%, and the accuracy to be 0.40%.

$$\Delta H = Q - Q_{\text{blank}} = m\ cp\ \left(T_f - T_i\right) \tag{7.8}$$

In 1989, Saitoh et al. [16] proposed an automatic flow calorimeter for measuring isobaric heat capacity of liquid refrigerant. This technique necessitates precise flow control of the sample and adiabatic condition during measurement. The principle consists of measuring simultaneously the heat flux ($\dot{Q}$), the temperature increment ΔT and the mass flow rate $\dot{m}$ of the sample. The heat capacity is then obtained from eq. (7.9), if complete adiabatic conditions are achieved:

$$Cp = \dot{Q}/(\Delta T\ \dot{m}) \tag{7.9}$$

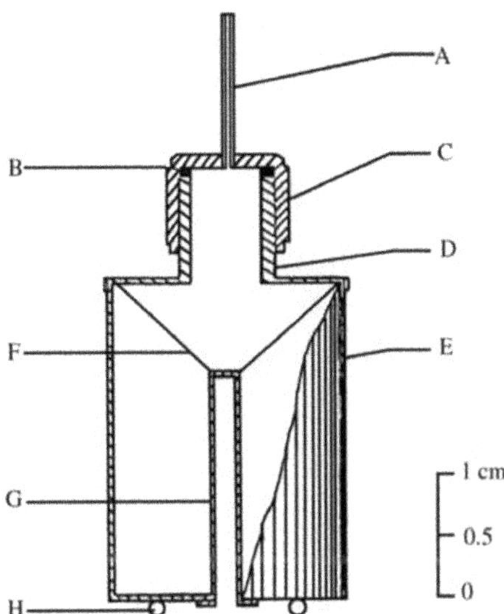

Fig. 7.12: Calorimetric cell for adiabatic calorimeter from Tan et al. [15]. A, copper capillary; B, Pb-Sn alloy gasket; C, copper screw cap; D, lid of the cell; E, main body of the cell; F, copper vane; G, re-entrant well for platinum resistance thermometer; H, sheath for differential thermocouples (reprinted from Publication J. Chem. Thermodyn., 34(9), Tan Z-C, Sun L-X, Meng S-H, Li L, Xu F, Yu P, Title of article "Heat capacities and thermodynamic functions of p-chlorobenzoic acid," 1417–29, Copyright (2023), with permission from Elsevier).

The calorimeter (Fig. 7.13) is immersed in a thermostatic bath. The sample fluid flows through a preheating zone to reach the bath temperature before entering the calorimeter. The fluid is heated by a DC electric microheater and the temperature before and after heating are read by two 100 ohms platinum resistance thermometers (inlet and outlet thermometers). The inside of the calorimeter is under vacuum to avoid heat loss. The fluid sample flows at low flow rates in a closed loop specially designed to reduce heat loss. The uncertainty of heat capacity is estimated to be less than 0.4%. Nakagawa et al. [17, 18] used similar technique for measuring heat capacity of 2,2-dichloro -1,1,1-trifluoroethane and 1-chloro-1,1difluoroethane and 1,1-difluoroethane at temperature from 276 to 440 K.

More recently, Zhong et al. [19] have proposed an adiabatic calorimeter, similar to the calorimeter developed by Kuroki et al. [20] for measuring isochoric specific heat capacity of compressed liquid at temperature ranges from 240 to 341 K and pressure up to 13 MPa. Isochoric heat capacity can be derived from isobaric heat capacity using thermodynamics relationships (eq. (7.10)) and an equation of state. Thus, direct measurements of isochoric heat capacity will be interesting data, complementary to isobaric heat capacity data.

$$C_p - C_v = T\left(\frac{\partial p}{\partial T}\right)_v \left(\frac{\partial V}{\partial T}\right)_p \tag{7.10}$$

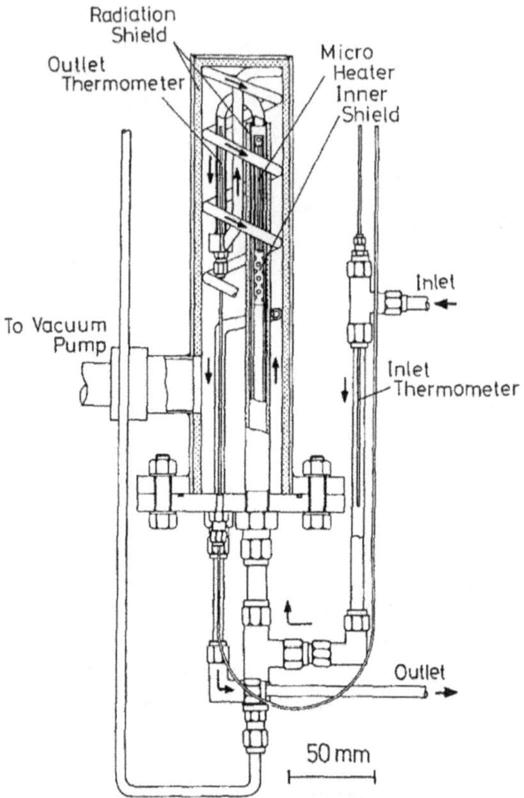

Fig. 7.13: Flow calorimeter developed by Saitoh et al. [16] (reprinted from Publication Int. J. Thermophys., 10(3), Saitoh A, Sato H, Watanabe K., "A new apparatus for the measurement of the isobaric heat capacity of liquid refrigerants," 649–59, Copyright (2023), with permission from Elsevier).

The experimental setup (Fig. 7.14) is constituted of a spherical bomb equipped with a filling tube, a platinum resistance thermometer, and a pressure transducer. The volume, including the bomb itself, filling tube, and pressure transducer cavity, is accurately determined by calibration. The bomb volume, about 70 mL, is expressed as function of temperature and pressure, with an uncertainty estimated to be 0.1 mL. The calorimetric setup is placed inside a container kept under vacuum. The bomb is heated by electrical wires coiled on its outer wall. Two adiabatic shields are used to limit radiation heat loss. The experimental procedure consists of heating the bomb during 4–5 min and recording the heat energy (Q) sent and the related temperature increase (ΔT). The experiment is carried out on the bomb kept empty (Q_0) and filled with m g of sample (Q). The isochoric heat capacity is given by eq. (7.11). The heat capacity calculation considers the pressure-volume work (W_{pv}) due to the volume change, calculated from eq. (7.12). This technique was applied to the refrigerant R1234yf, with an uncertainty of the isochoric specific heat capacity Cv estimated to be 2%.

$$C_v = (Q - Q_0 - W_{pv}) / (m\ \Delta T) \tag{7.11}$$

$$W_{pv} = \left(T \left(\frac{\partial p}{\partial T} \right)_v - \frac{1}{2} \Delta p \right) \Delta v \tag{7.12}$$

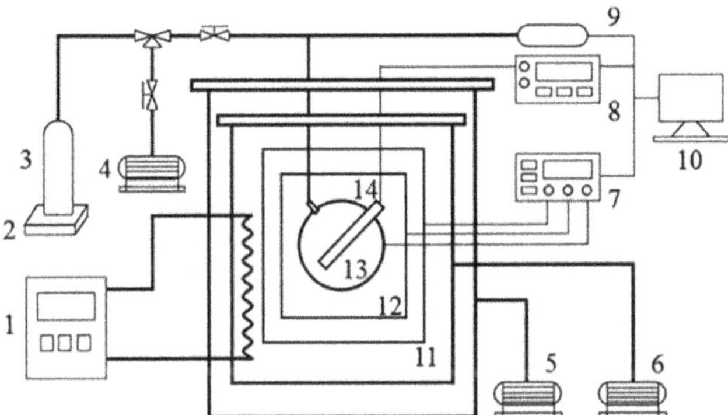

Fig. 7.14: Schematic representation of the adiabatic calorimeter developed by Zhong et al. [19]. 1, refrigeration machine; 2, scale; 3, gas cylinder; 4–6, vacuum pump; 7, DC power system; 8, multimeter; 9, pressure transducer; 10, computer; 11, outer adiabatic shield; 12, inner adiabatic shield; 13, bomb; 14, thermometer (reprinted from Publication J. Chem. Thermodyn., 125, Zhong Q, Dong X, Zhao Y, Wang J, Zhang H, Li H, "Adiabatic calorimeter for isochoric specific heat capacity measurements and experimental data of compressed liquid R1234yf," 86–92, Copyright (2023), with permission from Elsevier).

7.2.2.2 Differential scanning calorimetry

In the classical differential scanning calorimetry, the principle consists of scanning the temperature at a constant and controlled scanning rate, usually between 5 and 20 K/min. The measurements can be carried out at constant volume or pressure. The calorimetric signal is related to the difference between the thermal heat flux (mW) exchanged by the sample and the reference cells with the calorimetric block. During the temperature scanning, the heat flux exchanged by one cell is proportional to the heat capacity of the sample contained in the cell and the heat capacity of the cell itself. Three experiments at same scanning conditions are necessary: one with an empty cell (calorimetric signal expressed in eq. (7.13)), a second one with the sample (calorimetric signal expressed in eq. (7.14)) and a third one with a standard component (calorimetric signal expressed in eq. (7.15)). The reference cell must be filled with a same reference fluid or kept empty. The signals blank signals (S_{Blank}) are subtracted to the sample (S_{Sample}) and standard (S_{Standard}) signals to separate the signals related only to the sample (S_{Sample}) or the standard component (S_{Standard}). For each experiment, the initial and final isotherm signals are recorded to check and consider the signal zero value. The isobaric heat capacity of the sample is, for example, then given by the following equations:

$$S_{\text{Blank}} = K\left(cp_{\text{ Measuring cell}} - cp_{\text{ Reference cell}}\right) \tag{7.13}$$

$$S_{\text{Sample}} = K\left(cp_{\text{ Sample}} + cp_{\text{ Measuring cell}} - cp_{\text{Reference cell}}\right) \tag{7.14}$$

$$S_{\text{Standard}} = K\left(cp_{\text{ Standard}} + cp_{\text{ Measuring cell}} - cp_{\text{ Reference cell}}\right) \tag{7.15}$$

$$cp_{\text{ Sample}} = cp_{\text{ Sample}} \frac{S_{\text{Sample}} - S_{\text{Blank}}}{S_{\text{Standard}} - S_{\text{Blank}}} \tag{7.16}$$

The heat capacity of the sample (or of the standard compound) is the product of the sample mass contained inside the cell by its sample specific heat capacity. In the case of measurement at constant volume, the cell is closed and the sample mass is constant. For fluid sample measurement at constant pressure, the cell is designed to let the fluid expand with temperature. The cell is always fully filled and the sample mass inside the cell varies with temperature; the mass of sample is the product of the cell volume by the fluid density. This classical method was, for example, used by Kuroki et al.: [20] to measure the specific heat capacities of fluorinated propane and butane derivatives. The calorimeter was a Perkin Elmer DSC-4 differential scanning model. The sample is introduced in 75 µL stainless steel capsules. The main challenge is to fill out the crucible with pure component in its liquid state, avoiding any pollution. For this purpose, the capsules are filled in a cold and dry space. There procedure can be adapted to micro-DSC working at sub-ambient temperatures.

Hykdra et al. [21] and Gao et al. [22–24] used a similar experimental protocol adapted to a Calvet-type differential heat flux calorimeter, respectively, a BT2.15 and C80 Setaram microcalorimeter. Both models are differential fluxmetric calorimeters; the BT2.15 is specially designed for investigating sub-ambient temperatures. The cells are cylinders with internal volume about 10 cm^3. The measuring cell is connected to a buffer tank filled with the sample or standard fluid and connected to an inert tank to be pressurized (Fig. 7.15). Using this configuration (Fig. 7.16), the fluid inside the cell can be kept at constant pressure during the temperature scan. Gao et al carried out experiments on refrigerant such as R1234yf and R22, at temperatures range from 300 K to 360 K and pressure up to 5 MPa. The overall uncertainty of the isobaric heat capacity was estimated about 1.7%.

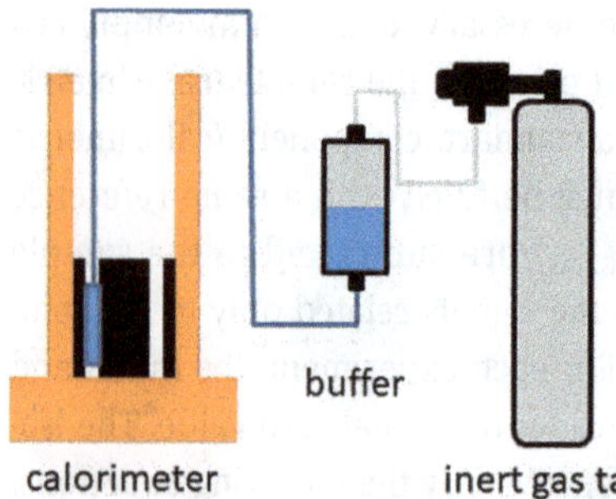

Fig. 7.15: Pressurized cell in differential Calvet-type calorimeter.

The low-temperature BT-215 Setaram calorimeter was slightly modified by Hykrda et al [21]. to improve temperature control during temperature scanning. The original cooling The calorimetric block (Fig. 7.16) is originally cooled down by vaporization of liquid nitrogen contained in a reservoir surrounding the cooling jacket. A modification was carried out by soldering a tubing to the liquid nitrogen inlet at the bottom of the cooling jacket. This tubing makes it possible to cool down the calorimeter by flowing a refrigerant from an external cryostat to the cooling jacket. This arrangement permits temperature scans from 200 K to 273 K with a better temperature control. Due to larger cells (10 mL) compared to classical DSC, the scanning rate is usually chosen between 1 and 0.5 K min[1]. Hykrda et al [21] have measured isobaric heat capacities of R227 (CF_3CHFCF_3) from 223 to 283 K at pressures up to 20 MPa with a total uncertainty estimated to be 0.5%.

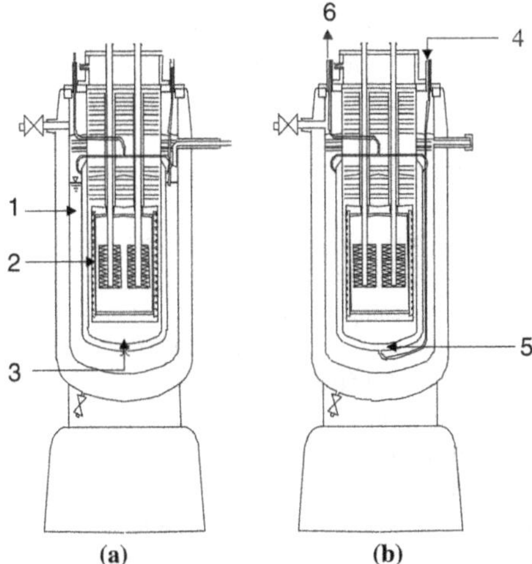

Fig. 7.16: Setaram BT-215 cooling systems using (a) vaporization of liquid nitrogen or (b) refrigerating liquid circulation. (1) Liquid nitrogen reservoir, (2) calorimetric block, (3) liquid nitrogen inlet for vaporization, (4) refrigerating liquid input, (5) cooling jacket, (6) cooling jacket output (reprinted from Publication Int. J. Thermophys., 25(6), Hykrda R, Coxam JY, Majer V., "Experimental Determination of Isobaric Heat Capacities of R227 (CF3CHFCF3) from 223 to 283 K at Pressures up to 20 MPa," 1677–94, Copyright (2023), with permission from Springer Nature).

References

[1] LINDE. Refrigeration: Processes & Temperatures [Internet]. Linde Gas. (Accessed April 23, 2020, at https://www.linde-gas.com/en/processes/refrigeration_and_air_conditioning/refrigeration_pro cesses_and_temperatures/index.html).

[2] Gui, X.; Wang, W.; Wang, C.; Zhang, L.; Yun, Z.; Tang, Z. Vapor–liquid phase equilibrium data of CO2 in some physical solvents from 285.19 K to 313.26 K. *J. Chem. Eng. Data*. **2014**, *59(3)*, 844–849.

[3] Wang, Q.; Xu, Y.-J.; Gao, Z.-J.; Qiu, Y.; Min, X.-W.; Han, X.-H. Isothermal vapor–liquid equilibrium data for the binary mixture ethyl fluoride (HFC-161)+1,1,1,2,3,3,3hepta fluoroproane (HFC-227ea) over a temperature range from 253.15K to 313.15K. *Fluid Phase Equilib.* **2010**, *297*(*1*), 67–71.

[4] Kandil, M. E.; May, E. F.; Graham, B. F.; Marsh, K. N.; Trebble, M. A.; Trengove, R. D., et al. Vapor–liquid equilibria measurements of methane +2-methylpropane (isobutane) at temperatures from (150 to 250) K and pressures to 9 MPa. *J. Chem. Eng. Data.* **2010**, *55*(*8*), 2725–2731.

[5] Syed, T. H.; Hughes, T. J.; May, E. F. Enthalpy of vaporization measurements of liquid methane, ethane, and methane + ethane by differential scanning calorimetry at low temperatures and high pressures. *J. Chem. Eng. Data.* **2017**, *62*(*8*), 2253–2260.

[6] Sauerbrunn, S.; Zemo, M. Measuring the Heat of Evaporation by TGA/DSC. (access April 2020 at http://www.americanlaboratory.com/914-ApplicationNotes/644-Measuring-the-Heat-of-Evaporation-by-TGA-DSC/).

[7] Fioroni, G. M.; Fouts, L.; Christensen, E.; Anderson, J. E.; McCormick, R. L. Measurement of heat of vaporization for research gasolines and ethanol blends by DSC/TGA. *Energy Fuels* **2018**, *32*(*12*), 12607–12616.

[8] Akisawa Silva, L. Y.; Matricarde Falleiro, R. M.; Meirelles, A. J. A.; Krähenbühl, M. A. Vapor–liquid equilibrium of fatty acid ethyl esters determined using DSC. *Thermochim. Acta.* **2011**, *512*(*1*), 178–182.

[9] Yost, D. M.; Garner, C. S.; Osborne, D. W.; Rubin, T. R.; Russell, H. A low temperature adiabatic calorimeter. The calibration of the platinum resistance thermometers. *J. Am. Chem. Soc.* **1941**, *63*(*12*), 3488–3492.

[10] Giauque, W. F.; Egan, C. J. The heat capacity and vapor pressure of the solid. The heat of sublimation. Thermodynamic and spectroscopic values of the entropy. *J. Chem. Phys.* **1937**, *5*(*1*), 45–54.

[11] Xu, F.; Sun, L.-X.; Tan, Z.-C.; Liang, J.-G.; Li, R.-L. Thermodynamic study of ibuprofen by adiabatic calorimetry and thermal analysis. *Thermochim. Acta.* **2004**, *412*(*1*), 33–57.

[12] Southard, J. C.; Andrews, D. H. The heat capacities of organic compounds at low temperatures. iii. an adiabatic calorimeter for heat capacities at low temperatures. *J. Frankl.* **1930**, *209*(*3*), 349–360.

[13] Russell, H.; Golding, D. R. V.; Yost, D. M. The heat capacity, heats of transition, fusion and vaporization, vapor pressure and entropy of 1,1,1-trifluoroethane. *J. Am. Chem. Soc.* **1944**, *66*(*1*), 16–20.

[14] Tan, Z.; Sun, G.; Sun, Y.; Yin, A.; Wang, W.; Ye, J. An adiabatic low-temperature calorimeter for heat capacity measurement of small samples. *J. Therm. Anal.* **1995**, *45*(*1*), 59–67.

[15] Tan, Z.-C.; Sun, L.-X.; Meng, S.-H.; Li, L.; Xu, F.; Yu, P. Heat capacities and thermodynamic functions of p-chlorobenzoic acid. *J. Chem. Thermodyn.* **2002**, *34*(*9*), 1417–1429.

[16] Saitoh, A.; Sato, H.; Watanabe, K. A new apparatus for the measurement of the isobaric heat capacity of liquid refrigerants. *Int. J. Thermophys.* **1989**, *10*(*3*), 649–659.

[17] Nakagawa, S.; Sato, H.; Watanabe, K. Isobaric heat capacity data for liquid HCFC-123 (CHCl2CF3, 2,2-dichloro-1,1,1-trifluoroethane). *J. Chem. Eng. Data.* **1991**, *36*(*2*), 156–159.

[18] Nakagawa, S.; Hori, T.; Sato, H.; Watanabe, K. Isobaric heat capacity for liquid 1-chloro-1,1difluoroethane and 1,1-difluoroethane. *J. Chem. Eng. Data.* **1993**, *38*(*1*), 70–74.

[19] Zhong, Q.; Dong, X.; Zhao, Y.; Wang, J.; Zhang, H.; Li, H. Adiabatic calorimeter for isochoric specific heat capacity measurements and experimental data of compressed liquid R1234yf. *J. Chem. Thermodyn.* **2018**, *125*, 86–92.

[20] Kuroki, T.; Kagawa, N.; Endo, H.; Tsuruno, S.; Magee, J. W. Specific heat capacity at constant volume for water, methanol, and their mixtures at temperatures from 300 K to 400 K and Pressures to 20 MPa. *J. Chem. Eng. Data.* **2001**, *46*(*5*), 1101–1116.

[21] Hykrda, R.; Coxam, J. Y.; Majer, V. Experimental determination of isobaric heat capacities of R227 (CF3CHFCF3) from 223 to 283 K at Pressures up to 20 MPa. *Int. J. Thermophys.* **2004**, *25*(*6*), 1677–1694.

[22] Gao, N.; Chen, G.; Li, R.; Wang, Y.; He, Y.; Yang, B. Measurements of the isobaric heat capacity of pressurized liquid trans-1,3,3,3-tetrafluoropropene [R1234ze(E)] by scanning calorimetry. *J. Therm. Anal. Calorim.* **2015**, *122*, 1469–1476.

[23] Gao, N.; Chen, G.; Li, R.; Lei, J.; He, Y.; He, H., et al. Isobaric heat capacities of R245fa and R236fa in liquid phase at temperatures from (315 to 365) K and pressures up to 5.5MPa. *J. Chem. Thermodyn.* **2015**, *90*, 46–50.
[24] He, Y.; Gao, N.; Jiang, Y.; Ren, B.; Chen, G. Isobaric heat capacity measurements for dimethyl ether and 1,1-difluoroethane in the liquid phase at temperatures from 305 K to 365 K and pressures up to 5 MPa. *J. Chem. Eng. Data*. **2014**, *59*(*9*), 2885–2890.

Rodica Chiriac, François Toche, and Olivier Boyron

Chapter 8 Calorimetry and thermal analysis for the study of polymer properties

Abstract: Nowadays, the production and intensive use of plastics having remarkable properties, such as transparency, lightness, resistance to water, solvents and corrosive agents, insulating character, and thermal stability, lead to the study of their physical and chemical properties and in particular to those that are strongly dependent on temperature. Among these properties, one can mention glass transition (T_g), enthalpy of melting and crystallization, decomposition temperature, thermal stability, and dimensional changes (such as softening, expansion, and shrinkage).

This chapter shows how polymers can be fully and very accurately characterized by using thermal analysis and calorimetry. Various examples are given in order to illustrate the potential of thermal analysis and calorimetry for measuring their physical properties, different types of transitions, aging, thermal stability, oxidation, degradation mechanism, effect of plasticizers, influence of synthesis, or shaping conditions. These examples will be presented using the following techniques: differential scanning calorimetry (DSC), pressurized-DSC, Calvet calorimetry, thermogravimetric analysis (TGA), thermomechanical analysis (TMA), TGA coupled to mass spectrometry (TGA-MS), TGA coupled to Fourier-transform infrared spectroscopy (TGA-FTIR), and TGA coupled to gas chromatography and mass spectrometry (TGA-GC-MS). The development of all these techniques and couplings between different instruments has greatly contributed to the understanding of the physical and chemical phenomena that occur within materials when they are synthesized, shaped, heated, burned, or degraded. These techniques are nowadays increasingly used in the field of plastic recyclability. They undeniably allow a better understanding of the mechanisms involved in the new recycling processes explored in research and development laboratories.

The term "polymer" comes from an old Greek word polus, meaning "many", "much", and meros, meaning "parts", and refers to a large molecule whose structure is composed of multiple repeating units.

Acknowledgments: We, the coauthors, would like to express our gratitude to the editors of this book, Dr. Aline Auroux and Dr. Ljiljana Damjanovic-Vasilic, who gave us the opportunity to write this chapter. We also wish to thank Dr. Franck Collas for the numerous and rich discussions about polymers.

Rodica Chiriac, François Toche, Laboratoire des Multimatériaux et Interfaces, UMR CNRS 5615, Univ Lyon, Université Claude Bernard Lyon 1, F-69622 Villeurbanne, France
Olivier Boyron, Univ Lyon, Université Claude Bernard Lyon 1, CPE Lyon, UMR CNRS 5128, Catalyse, Polymérisation, Procédés et Matériaux (CP2M), 43 Bd du 11 novembre 1918, 69616 Villeurbanne, France

https://doi.org/10.1515/9783110590449-008

Keywords: polymers, calorimetry, thermal analysis, chemical properties, physical properties

8.1 Introduction

A century ago, in 1920, H. Staudinger [1], *Nobel Prize in Chemistry in 1953*, demonstrated the existence of "long molecules," which he described as long chains made up of repeating molecular units linked by covalent bonds. Since then, polymeric materials, commonly called "plastics," are widely used for many purposes in electronics, construction, transportation, office automation, decoration, and so on.

In fact, natural plastics have been used since ancient times. The Egyptians used egg albumin or bone gelatin to glue various objects. Rubber and amber were heated and molded to obtain objects with multiple uses.

In the second part of the twentieth century, the considerable increase in the world production of plastics was due in particular to the development of petrochemistry, as many polymers were obtained from petroleum and its derivatives, such as thermoplastics (polyethylene, polypropylene, polyvinyl chloride, etc.).

Thanks to their properties such as density (*lightness* is one of the main criteria for the use of polymers), plasticity, elasticity, transparency, insulating character, and above all low production costs, plastics have rapidly replaced materials such as paper, glass, metal, and wood. Here are some examples of polymers with their main characteristics and properties:

- *Poly(vinyl chloride)* (PVC) is a thermoplastic polymer, non-flammable, and resistant to water, combustion and UV, chemical agents, and solvents; it is widely used in construction for pipes and window coverings, and as film in packaging.
- *Polyethylenes* (HDPE, LDPE, LLDPE), thermoplastic polymers, are excellent insulators (invaluable for the radio and electronic industry) and resist well to corrosive agents, solvents, and abrasion; they are suitable for the manufacture of containers, and with their barrier properties they are widely used in the packaging market.
- *Polystyrenes* (PS) are excellent electrical insulators, do not absorb humidity, and are resistant to chemical agents; they are also used as ultra-light foams; however, they are not resistant to high temperatures, nor to shocks or bending.
- *Poly(methyl methacrylate)* (PMMA) or *Plexiglas* is a thermoplastic polymer whose essential characteristics are a high transparency and a light transmission superior to that of glass, a lightness also superior to that of glass, and resistance to UV rays.
- *Polyamides* (PA), such as *Nylon*, *Rilsan*, or *Tergal*, are also thermoplastic polymers that have a high resistance not only to breakage and abrasion but also to chemical agents, solvents, and boiling water; some are excellent insulators (Rilsan).
- *Fluorinated polymers*, such as *polytetrafluoroethylene* (*Teflon*), are thermoplastic polymers that have exceptional qualities of resistance to solvents, acids, and even aqua

regia, which attacks gold and platinum; *Teflon* has a very high melting point compared to other plastics, giving it excellent thermal stability; it is a good electrical insulator; it is used as an antiadhesive coating, as a solid lubricant, or for sealing.

The stability over time of plastics, another of their qualities, is both an advantage when it is about to produce long-lasting equipment (buildings, vehicles, cables) and a disadvantage when disposable products (e.g. packaging) are left in nature. Studies regularly show that the manufacture and use of certain plastics are problematic, particularly because of the additives added to them (phthalates, bisphenol A, etc.), which are toxic to health and the environment.

In recent years, studies on biosourced plastics or bioplastics have intensified, particularly in the packaging sector. These bioplastics are synthesized from sources of totally or partially renewable origin (vegetable, animal, algal), but this does not mean a priori that they are biodegradable. Some polymers can be both biosourced and biodegradable. The use of biobased polymers is intended to address environmental issues such as the scarcity of fossil resources. Bioplastics represent only about 1% of the global production of polymeric materials [2], but their production is constantly increasing. Some examples of bio-based polymers are BioPE and BioPET (bio-polyethylene terephthalate), obtained from sugarcane; polylactic acid, obtained from corn starch; PHA (polyhydroxyalkanoates), obtained from corn starch, sugarcane, or beet.

8.2 Polymer classification

Polymers can be classified, according to their origin, as **natural** (cellulose, casein, collagen, proteins), **synthetic** (PE, PP, PVC, PET, PS, ABS, PC, etc.), or **semisynthetic** (gelatin, dextran, chitosan, etc.). The **synthetics**, according to their source, can be either of fossil origin or of totally or partially renewable origin as indicated previously, and they are considered as major in terms of production (above they are listed in decreasing order of tonnage).

In Tab. 8.1, several examples of polymers with their formula and the corresponding monomer are given. The monomer of polyethylene is ethylene (i.e. the repeating unit) as shown by its formula. Some polymers, such as polycarbonate, are obtained by reaction between two compounds, *for example* a polycondensation reaction. Proteins, long chains of amino acids formed from sequences of monomer amino acids, are also known by the name “polypeptides.” The amide linkage between the amino acids is called a peptide bond.

Another classification of polymers refers to their **macro-structure (chain organization)** that can be semicrystalline or amorphous. The chains of macromolecules are organized in a space way: compactness and order of “the packing” of chains, orientation or entanglement, as represented in Fig. 8.1.

Tab. 8.1: Examples of polymers with their formula and the corresponding monomer.

Monomer	Polymer	Formula
Ethene	Polyethylene, PE	$\left[CH_2-CH_2 \right]_n$
Styrene	Polystyrene, PS	$\left[CH(C_6H_5)-CH_2 \right]_n$
Chloroethene	Polychloroethylene, PVC	$\left[CH(Cl)-CH_2 \right]_n$
Tetrafluoroethene	Polytetrafluoroethylene, PTFE	$\left[CF_2-CF_2 \right]_n$
Polycondensation reaction (Bisphenol A + diphenyl carbonate)	Polycarbonate, PC	$\left[O-C_6H_4-C(CH_3)_2-C_6H_4-O-C(=O) \right]$
Sequence of amino acids (e.g., glycine + alanine)	Glycyl alanine (peptide, GlyAla)	$H_2N-CH_2-C(=O)-NH-CH(CH_3)-C(=O)-OH$

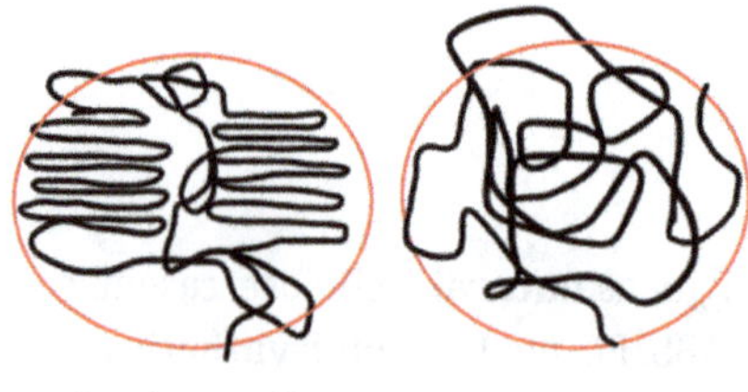

Fig. 8.1: Classification of polymers by function of their chain organization.

Thermoplastic polymers are either amorphous (PC, ABS, etc.) or semicrystalline (PE, PA, PP, PEEK, PET, etc.).

Polymers can also be classified by function of their thermal behavior into three groups as follows: thermoplastics, elastomers, and thermosets. Note that elastomers are usually thermosets (requiring vulcanization) but may also be thermoplastic. This classification is summarized in Tab. 8.2.

The thermal behavior of a polymer can be obtained by studying the polymer properties such as glass transition (T_g), the enthalpy of melting and crystallization, the decomposition temperature, thermal stability, dimensional changes (such as softening, expansion, and shrinkage), properties that are strongly dependent on temperature. Moreover, many factors can highly influence the thermal behavior of polymers and

Tab. 8.2: Classification of polymers by function of their thermal behavior.

Group of polymer	Thermoplastics	Elastomers	Thermosets
Examples	PE, PP, PS, PVC, PMMA, PET, PTFE, PA-6, PA-11, PVAc, PVA	Neoprene, polyurethanes, silicones, vulcanized natural rubber	Unsaturated polyesters, cured epoxy resins, silicone resins
Representation			
Properties	– Uncrossed linked polymer with high strength and low strain under load – Melt or flow on heating – Solidifies upon cooling – May be reshaped by heating	– Lightly cross-linked – Have viscoelasticity and very weak inter-molecular forces – Low creep resistance – Usually referred to rubber	– Heavily cross-linked – Do not melt but decompose – Do not reform upon cooling – Form irreversible chemical bonds during the curing process

thus their physical properties. Among these factors, we find not only the composition of polymers (i.e., the presence of fillers, plasticizers, antioxidants, and some other additives) but also the storage conditions, the mechanical stress during their synthesis, the nature of curing agents, the degree of curing or polymerization, the shaping of polymers (i.e., their transformation in pellets, films or fibers, etc.). For all these reasons, thermal analysis is a very important tool for the study of polymers, from their chemical composition to their physical organization.

This chapter describes various examples in order to illustrate the potential of thermal analysis and calorimetry for measuring their physical properties, different types of transitions (e.g., softening, melting, glass transition, and shrinkage), aging, thermal stability, oxidation, degradation mechanism, the effect of plasticizers, and the influence of synthesis or preparation conditions. These examples will be presented using the following techniques: differential scanning calorimetry (DSC), pressure-DSC, Calvet calorimetry, thermogravimetric analysis (TGA), Thermomechanical Analysis (TMA), TGA-coupled mass spectrometry (TGA-MS), TGA-coupled Fourier-transform infrared spectroscopy (TGA-FTIR), TGA-coupled gas-chromatography-mass spectrometry (TGA-GC-MS), and TGA-coupled micro-gas-chromatography-mass spectrometry (TGA-micro-GC-MS).

8.3 Calorimetry for the study of polymers

The vast majority of physical, chemical, or biological transformations involve the production or absorption of thermal energy. The measurement of this energy has become an almost universal way to follow these phenomena and to quantify them. This is done with the help of calorimetry. The **calorimeter**, a term proposed by Lavoisier in 1789, is the apparatus that allows the measurement of **heat** [3].

8.3.1 Differential scanning calorimetry

In this chapter part, several examples will highlight the **calorimetric** techniques like **DSC** (differential scanning calorimetry) through the determination of physicochemical parameters such as glass transition (T_g), purity, temperature and enthalpy of fusion or crystallization, oxidation, or polymerization. The **DSC** technique, which is now widespread not only in research laboratories but also in industrial control laboratories, allows rapid measurements and to obtain information on the kinetics of reactions as a function of temperature (aging of active ingredients in pharmaceuticals, food, polymers, chemical safety studies, etc.) or on the temperature of phase transitions or state. Most of the DSCs available on the market are equipped with a flat sensor (Fig. 8.2). In this case, only the bottom of the crucible measures the heat flux, so there is a risk of temperature gradient if the sample is too large. The quality of the contact between the sample and the crucible influences the result.

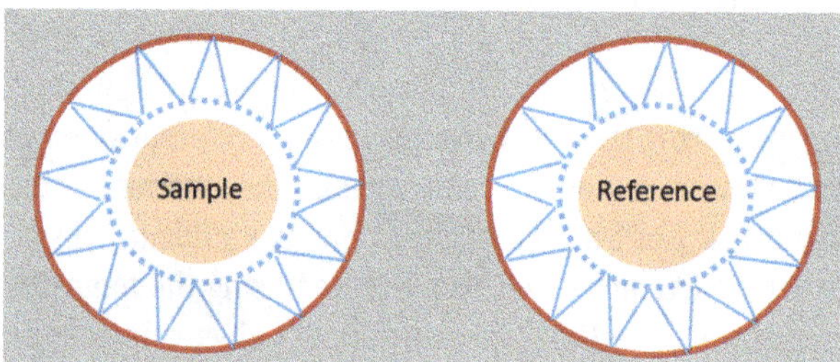

Fig. 8.2: Flat DSC sensor.

In the field of plastics, **cross-linking** is a very known phenomenon widely studied by DSC. This phenomenon allows by chemical or physical way to obtain from linear or branched polymers, a three-dimensional network of higher molecular mass and thus with physicochemical properties different from the initial polymer (e.g., insolubility in solvents, rigidity, and infusibility). The cross-linking is the outcome of the polymerization and is an irreversible process.

8.3.1.1 Examples

Study of the cross-linking of resins

The first example concerns the DSC study of an **epoxy resin**, which is a thermoset polymer. This material finds various and multiple applications such as adhesives, composites for automobile and aerospace industries, corrosion, and temperature-resistant coatings. The wide variety of curing compounds (amine, amide, mercaptans, phenolic) used to perform cross-linking of the epoxy resin lead to thermosets with tailored properties (strength, stiffness, temperature, and chemical resistance) and thus multiple choice of applications [4].

In order to construct the time–temperature transformation (TTT) diagram and determine the ideal curing cycle for an epoxy system, it is necessary to know the kinetic and thermal behavior of the resin. DSC is perfectly suited to analyze the polymerization and determine the conversion rate and glass transition, T_g, as a function of time. T_g is one of the most important properties of a thermosetting polymer and represents the passage from a rigid state to a rubber one. When this temperature is reached, the molecular chains acquire enough energy, allowing them to rearrange and relax, but since they are all tied together by cross-links, they will not flow.

Figure 8.3 shows the DSC curves at various heating rates of a thermoset polymer (epoxy resin). Analyses were run on the *METTLER TOLEDO* DSC1^{TM} by using Al crucibles and sample masses of around 7 mg. Two thermal phenomena appeared on each curve: a glass transition with a high endothermic relaxation peak and a large exothermic peak corresponding to the crosslinking (or the curing) of the resin. Both T_g and exothermic peak move to higher temperatures with an increasing heating rate.

Glass transition, characteristic of the amorphous part of the polymer, is a second-order transition. During the glass transition, there is a change of heat capacity, *Cp* (change of the baseline). The endothermal relaxation peak accompanying glass transition is linked to the thermal past of the sample. When relaxation is large, like in the Fig. 8.3, it means that the physical aging at temperatures below T_g was long. In addition, the presence of this relaxation peak gives information about the synthesis conditions, the stress exerted on the material, such as pressure, radiation, temperature, and shaping. In this example, the mid values of T_g varies from 63 to 72 °C with the increase of the heating rate from 2 to 20 °C min^{-1}.

At the maximum temperature of the large exothermal peak (T_{peak}), cross-linking is at its highest rate. Reaction kinetics explain the shift toward higher temperatures with the increasing heating rate. In order to obtain the enthalpy of the cross-linking reaction, it is possible to integrate the area under the surface of the curve and normalize it by the mass of the sample. The obtained enthalpy in J g^{-1} does not vary much with the heating rate: 27.6 ± 0.3 J g^{-1}.

From the heat flow curves, the degree of conversion as a function of temperature can be plotted (Fig. 8.4). The necessary condition to obtain this graph is the complete cross-linking of the polymer. To check this information, a second heating scan was run and no exothermal peak was observed. The conversion curves shown in Fig. 8.4 shift to higher temperatures with the increasing heating rate. For example, at a fixed temperature of 180 °C the degree of conversion is 95% at the lowest heating rate and decreases to 34% for the highest one.

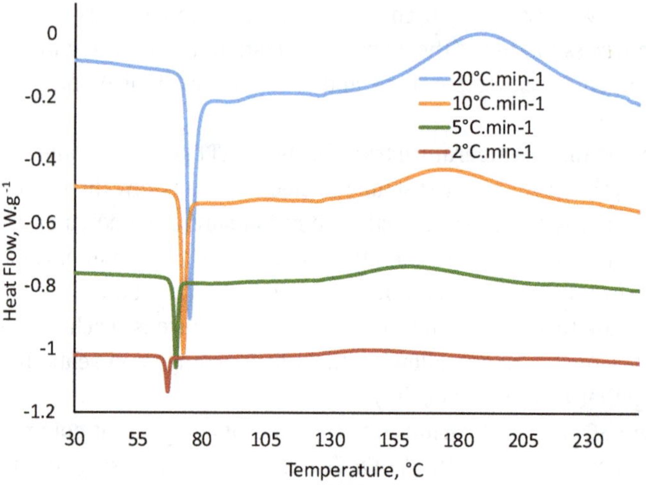

Fig. 8.3: DSC curves of an epoxy resin at 2, 5, 10, 20 °C min^{-1}.

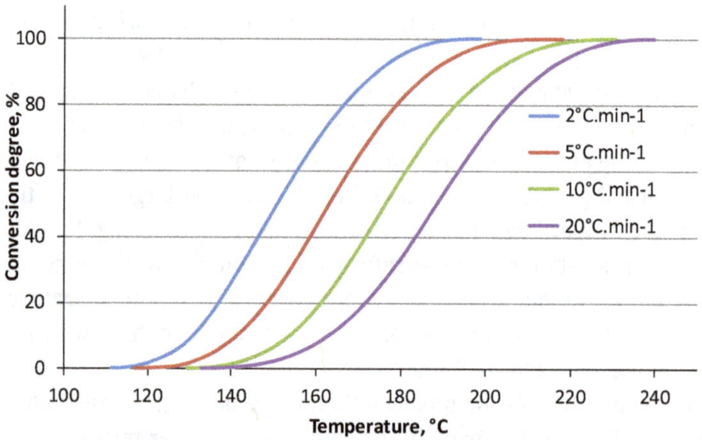

Fig. 8.4: Degree of conversion with temperature for an epoxy resin at four heating rates.

Effect of pressure on the cross-linking rate

The second example shows the influence of the pressure on the enthalpy of curing of an epoxy resin by using **pressurized DSC (P-DSC).** Tests were done with the *METTLER*

TOLEDO DSC827HP^{TM} and by using pierced aluminum crucibles of 40 μL with sample masses of about 13 mg.

The curing analyses of the epoxy resin and its reactivity were studied in isotherm (180 °C) by pressure-DSC at 10, 25, 50 and 80 bars of pure air. The cell containing the reference and sample crucibles was pressurized at room temperature just before starting the analysis. Pressure was maintained during the whole analysis with a flow rate of 50 mLmin^{-1}.

The enthalpy of curing increased with the increasing applied pressure (Fig. 8.5). This can be explained by the increasing reactivity at higher pressures. This leads to a denser macromolecular network. In order to check if the curing was complete, a heating run up to 300 °C was done on each sample tested at 180 °C. An exothermal peak corresponding to curing appeared and was more intense, especially for samples tested at lower pressures at 180 °C (10 and 25 bars). This means that the curing was not complete during the isotherm at 180 °C for lower pressures. This correlation between the degree of curing and the pressure has also been evidenced by Gracia-Fernàndez [5].

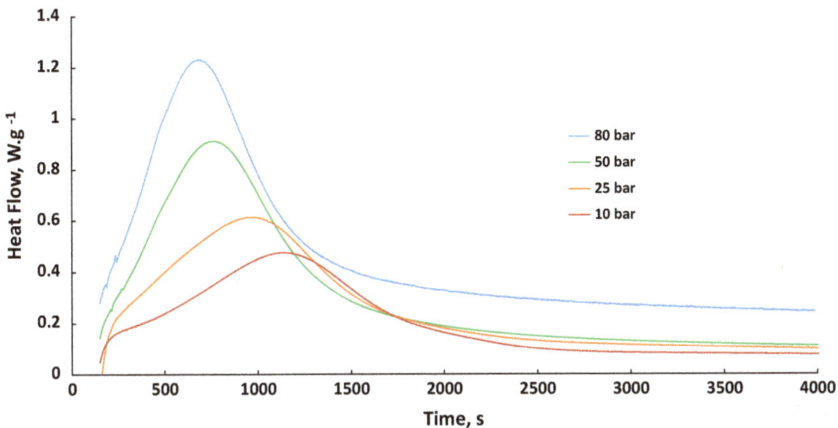

Fig. 8.5: Curing of an epoxy resin by pressure-DSC in isotherm at 180 °C (at 10, 25, 50, and 80 bars of pure air).

Effect of plasticizers on glass transition

Another example presents the influence of the plasticizer content on glass transition (T_g) for sample films of poly (vinyl acetate). Measurement was done from −100 to 50 °C at 10 °Cmin^{-1} with 40 μL aluminium pan and under pure nitrogen at a rate of 50 mL min^{-1}.

As indicated before, T_g is characteristic of the amorphous part of a polymer and depends on various parameters such as heating rate, thermomechanical history of the sample, plasticizer and moisture content, degree of curing, chemical compositions, and molar mass. [6]

Plasticizers are added in the composition of a polymer in order to change its thermoplastic properties. In Fig. 8.6, T_g moves to lower temperatures with the increasing percentage of plasticizer as the polymer becomes softer. The T_g becomes broader as the amount of plasticizer increases.

When publishing research data, the T_g value should always be indicated together with the measurement conditions (heating rate, T_g onset or T_g midpoint, etc.), otherwise the T_g value alone has no signification.

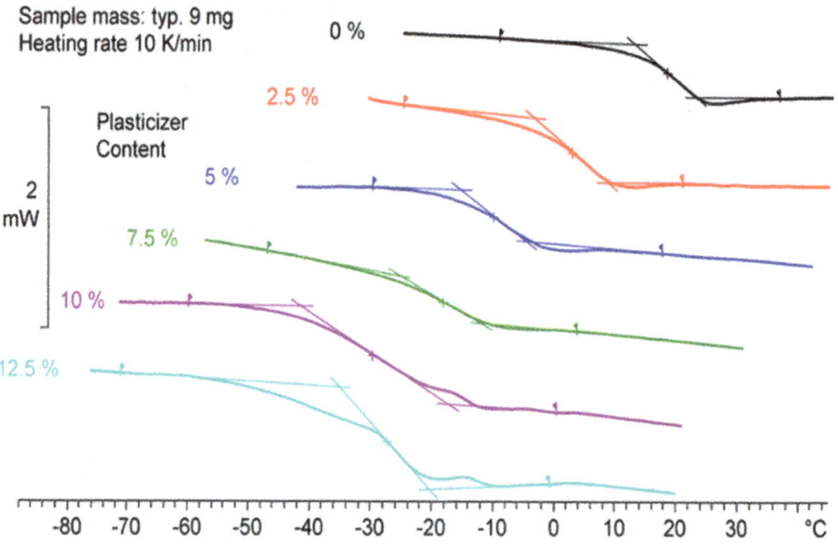

Fig. 8.6: DSC curves showing the impact of plasticizer content on glass transition for poly(vinyl acetate) films.

Determination of crystallinity by DSC

The example below shows how the polymer **crystallinity** can be determined **with DSC** by quantifying the heat associated with polymer's melting [7]. For the semicrystalline polymers, such as polyethylene (PE), DSC can characterize the crystalline phase by quantifying the heat associated with the melting of polymer.

Two types of samples have been used: a **high-density polyethylene (HDPE)** (i.e., unbranched or weakly branched PE) and **ethylene-hexene copolymers (LLDPE – linear low density PE)** with various hexene content, from 2 to 14 mol%, leading to different contents of short-chain branches. Incorporating branches into the backbone chain modifies the thermomechanical properties of the polymer. The more the polymer is branched, the softer it becomes.

As branching has a direct impact on crystallinity, DSC was used to quantify it. Figure 8.7 shows the DSC curve on a PE sample with two heating scans (from 25 to 180 °C) and one cooling (from 180 to −150 °C), all scans done at 10°C min^{-1}. To eliminate the influence of the synthesis and storage conditions, that is to erase the thermal past of the sample, a first heating scan is necessary. A second heating allows a proper exploitation of the results. Figure 8.8 presents the melting parameters of HDPE such as temperature and enthalpy.

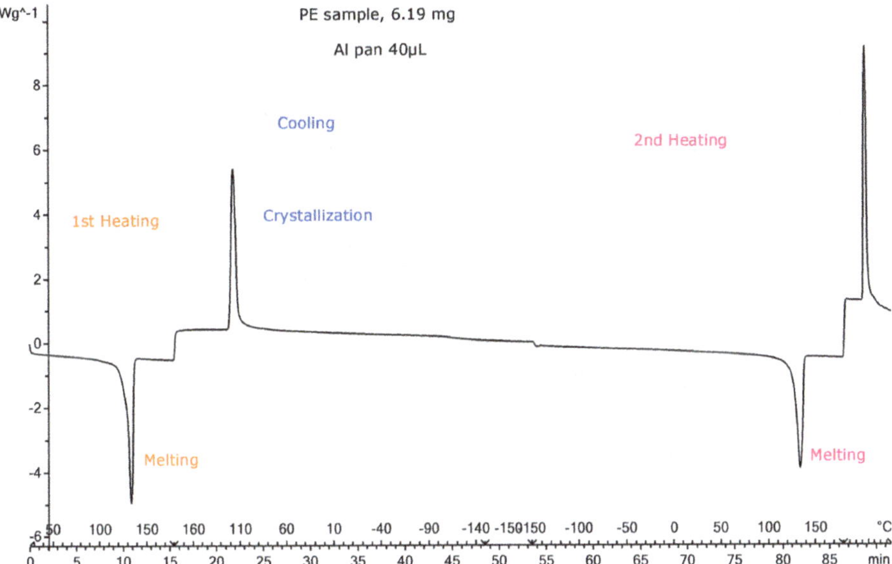

Fig. 8.7: DSC curve of a HDPE with 1 cooling and 2 heating scans.

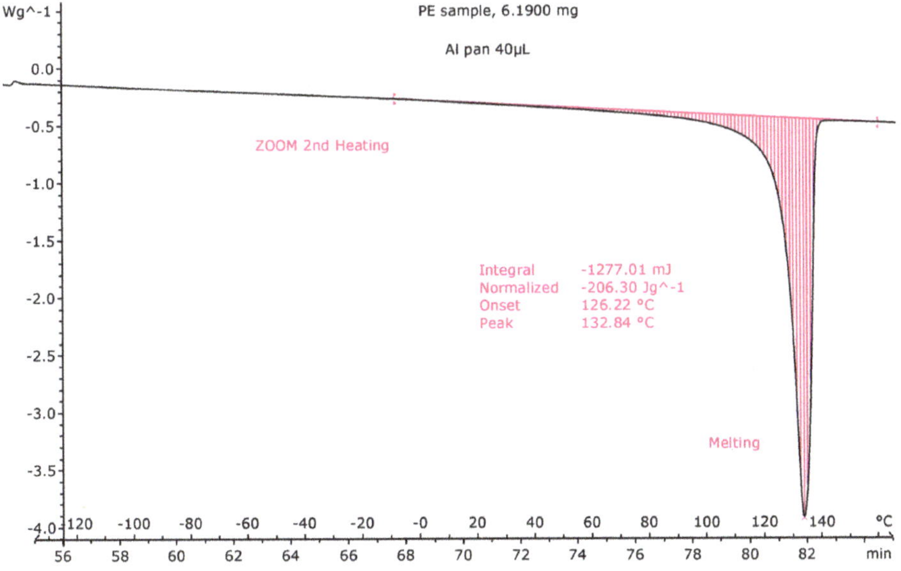

Fig. 8.8: DSC curve of a HDPE. Integration of melting peak during the second heating scan.

The endothermic peak corresponding to the fusion of the polymer is quite large and must be properly integrated by extending the baseline from the end of the melting to the intersection of the baseline before melting.

In the case of polymers presenting a large pic of melting, the **peak** value of the melting temperature (T_{peak}) is taken into consideration instead of the **onset** value as its determination is less accurate.

For very accurate analyses and an appropriate baseline, one can run a blank curve (i.e., a DSC analysis with empty crucibles and the same temperature program as for the sample), which will be then subtracted from the sample curve. The melting enthalpies of PE with and without subtraction of blank present quite similar values as shown below. The difference is low, of about 1%, but for a smaller value of the melting enthalpy, the difference could become important:

ΔH_{PE} = − 206.3 J g^{-1} (blank curve was subtracted from the PE curve)
ΔH_{PE} = − 204.2 J g^{-1} (no subtraction of blank curve from the PE curve)
$T_{m\ PE}$ = 133 °C (T_{peak} of melting)

As the melting enthalpy of the synthesized PE is known (ΔH_f = −206.3 J g^{-1}), the crystallinity (α) can be calculated with the equation below. For this, one needs to know the value of the melting enthalpy for 100% crystalline PE ($\Delta H_{f\ 100\%}$). This value was calculated and appears in the literature [8], as 100% crystalline PE does not exist or it cannot be synthesized. This theoretical enthalpy value is −293 J g^{-1} for 100% crystalline PE. For an experimental enthalpy value of −206.3 J g^{-1}, as obtained for the synthetized PE, the crystallinity is 70.4%. The formula for the calculation of crystallinity is as follows:

$$\alpha = \frac{\Delta H_f}{\Delta H_{f100\%}} \times 100 \tag{8.1}$$

For the copolymer **with 3.6 mol% of hexene**, the enthalpy obtained was −107.4 Jg^{-1} (the DSC thermogram is not shown here). Based on the 100% crystalline PE value and the equation (1), a crystallinity of 36.7% is calculated.

As expected, the branching has an impact on the polymer's crystallinity by highly decreasing it. One can also note that the peak melting temperature (T_{peak}) of the copolymer was decreased (105 °C) when we compare with the value obtained for the polymer alone (133 °C).

Figure 8.9 shows how the degree of branching of LLDPE (comonomer content) influences the melting peak measured by DSC. With the increasing molar percentage of hexene, the melting peak is smaller and appears at lower temperatures. Table 8.3 summarizes this influence. With the increasing of branching, the melting peak temperature and enthalpy decrease, as well as the crystallinity.

These results can be used to draw a DSC calibration curve with the mol% of hexene on y-axis and melting temperature values in abscise. In this way, for a newly

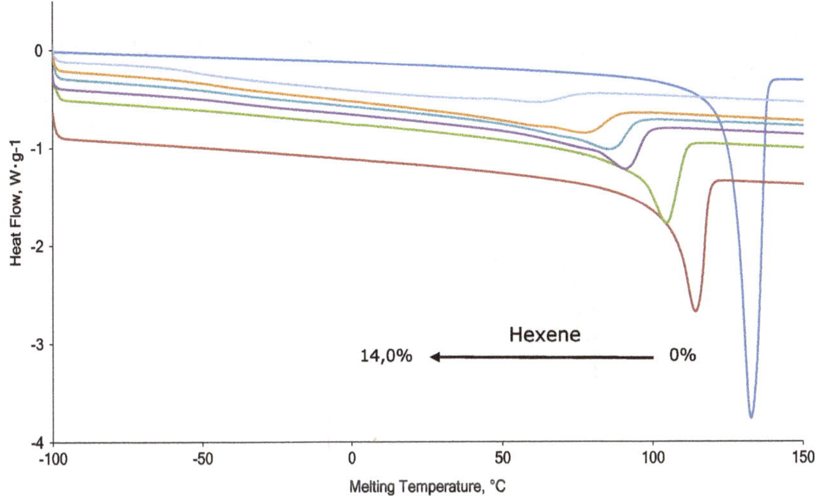

Fig. 8.9: Effect of the increasing mol% of hexene on the melting peak of LLDPE (*HDPE, without comonomer, in blue*).

Tab. 8.3: Evolution of the melting parameter values and crystallinities with the increasing of branching (i.e., the increasing of molar percentage of hexene).

1-Hexene (mol%)	Melting enthalpy (J g^{-1})	T_{peak} (°C)	Crystallinity (%)
2.2	123.6	114	42.2
3.6	107.4	105	36.7
6.3	91.1	91	31.8
7.6	83.3	86	28.4
9	74.8	77	25.5
13.6	64.1	60	22

synthesized copolymer, by running a DSC analysis and by using the DSC calibration curve, one can obtained the molar percentage of hexene. By replacing the hexene comonomer with another alkene like propylene, butene or octene, one can use the same protocol to determine the crystallinity.

8.3.2 Calvet calorimetry

To characterize polymers, it is also possible to use **Calvet calorimetry** *C80* from *SETARAM* (now commercialized as *CALVET*). With this type of *heat flow calorimeter* one can measure all heat evolved (or adsorbed) by radiation, convection, or conduction. This calorimeter has a 3D sensor with Tian-Calvet thermopiles. Thermopile is an assembly of thermocouples electrically mounted in series (to add the electrical voltages) but thermally in parallel (to drain efficiently the heat) [3]. Fig. 8.10 presents an image of the inside of calorimeter. Thermocouples are illustrated in red color. They are arranged all around the sample and reference cell, thus all heat evolved or adsorbed is measured. A wide range of sample vessels (standard cell, reversal-mixing cell, ampoule mixing cell, high-pressure cell, etc.) makes the calorimeter advantageous for various applications.

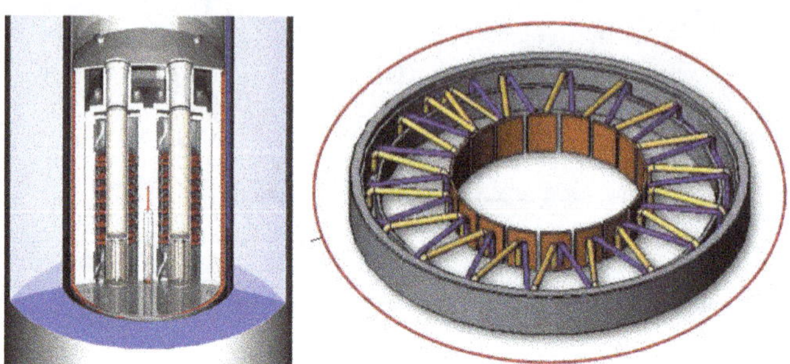

Fig. 8.10: Inside of calorimeter C80 (left side) and a ring with 38 thermocouples (right side). Reprinted with permission from SETARAM.

8.3.2.1 Example

The example below shows the use of 350 bar Hastelloy cells with pressure probe. In this case, the pressure signal is measured in the same time with DSC signal.

During the last two decades, a lot of research has been done on metal-organic frameworks (MOFs). They find applications in catalysis, electronic devices, photonic or sensors, and so on. The synthesis of MOF is generally done by **solvothermal method** into a stainless steel bomb, but in this type of reactor, the crystallization cannot be followed in terms of temperature, pressure, enthalpy, and so on.

An original approach to optimize MOF's crystallization conditions is to synthesize the coordination polymer *in situ*, that is into the cells of a Calvet calorimeter, and thus to better follow its crystallization.

The synthesis of H-KUST (MOF-199) was realized with the following program temperature: heating from ambient to 180 °C at a slow heating rate, followed by an isotherm of 12 h at 180 °C. Figure 8.11 shows the reaction scheme of the MOF synthesis.

$$+ \ Cu(NO_3)_2 \xrightarrow[180°C]{H2O/EtOH\ (1:1)}$$

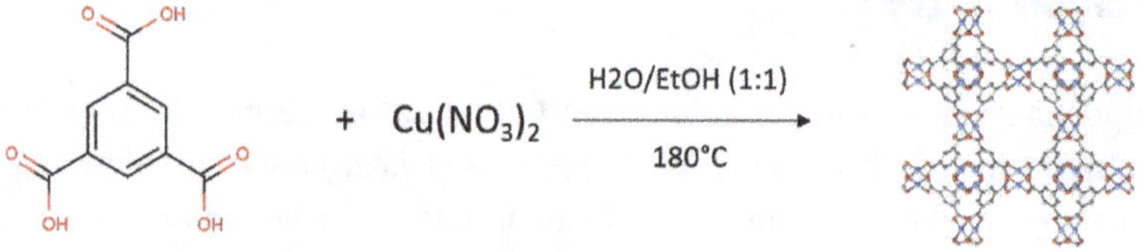

Fig. 8.11: Reaction scheme of the MOF synthesis.

The MOF polymerization reaction was studied by changing different parameters such as the ratio between the copper salt and ligand, the volume occupied in the calorimetric cell by reactants (ligand, copper salt, water, and ethanol), and the program temperature. Once the most appropriate analytical program was found, the reaction was studied at different heating rates as shown in Fig. 8.12. At higher heating rates, a large exothermic phenomenon appears, while for the lower ones, a second peak looking like a shoulder appears in the continuation of the first one. Two exothermal phenomena that seem to be very close to each other move to lower temperatures with the decreasing heating rate from 0.5 to 0.05 °C min^{-1}. The first phenomenon depends on the heating rate as it appears at lower temperatures with the decreasing heating rate; it is a kinetic phenomenon. The second exothermic phenomenon, in spite of the diminution of the heating rate, still remains closed and related to the first one; it probably corresponds to the crystallization of MOF.

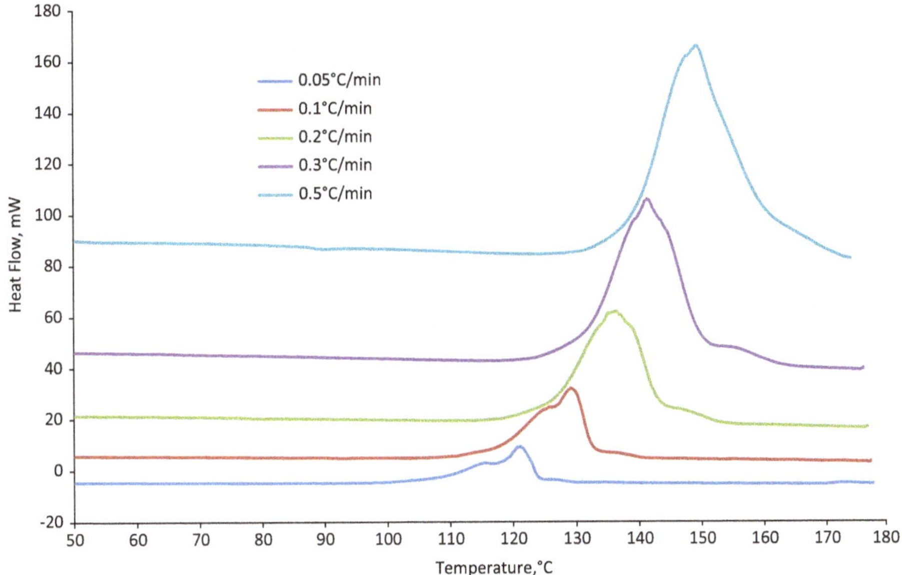

Fig. 8.12: Evolution of thermal phenomena with the increasing heating rate.

By considering the **onset**, $\boldsymbol{T_{max}}$ (or $\boldsymbol{T_{peak}}$), and **endset** values of the exothermic effect and by plotting their evolution (not shown here) with the heating rate, two observations were drawn:

- Regarding the **onset** and $\boldsymbol{T_{max}}$ evolution, no proportionality (non-linear behavior) with the heating rate was noted; there is a kinetic dependence; the system is out of thermodynamic equilibrium probably because the first peak, which is characterized by onset and T_{max}, corresponds to the formation of the complex.

– Concerning the second peak, which is related to **endset** values, the dependence with the heating rate shows a linear behavior; this corresponds to a thermodynamic phenomenon likely linked to the crystallization of MOF.

For the first exothermal phenomenon, one can determine $\boldsymbol{E_a}$ (activation energy) by Kissinger method either with **onset** or with $\boldsymbol{T_{max}}$ values. T_{max} value is generally used in this method, but the onset value was also used to check that the first peak corresponds to the formation of complex:

$$\log\ (\beta/T^2{}_{\max}) = -E_a/RT_{\max} + \log AR/E_a \tag{8.2}$$

where R is the gas constant, β is the heating rate, T_{max} is the maximum temperature of the peak, and A is the pre-exponential factor.

The slope ($-E_a/RT$) of the straight line gives $\boldsymbol{E_a}$. For **onset** and $\boldsymbol{T_{max}}$ values, the obtained equivalent **activation energies** were **38.3 kJ mol^{-1}** and **37.5 kJ mol^{-1}**, respectively, as the formation of Cu-ligand bonds take place between the two onset and T_{max} values.

Crystals obtained inside the calorimetric cells at different heating rates were observed by optical microscope technique as shown in Fig. 8.13.

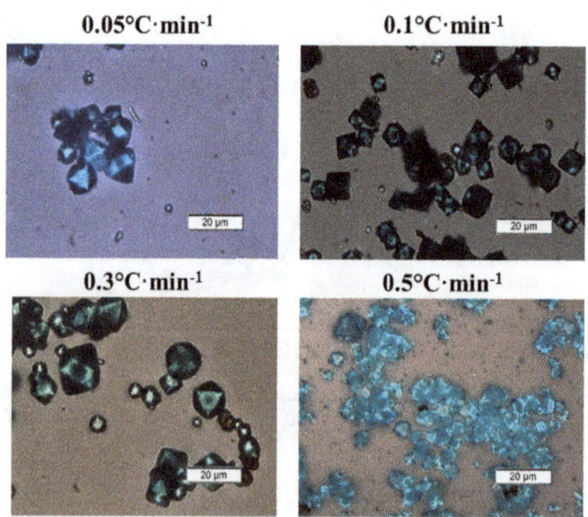

Fig. 8.13: Crystals obtained inside the calorimetric cells at various heating rates.

Crystals are more regular and well crystallized at a very low heating rate (0.05 °C min^{-1}), which corresponds to a duration analysis of 13 h. The morphology of crystals is less regular when the heating rate reaches 0.5 °C min^{-1}.

Important parameters on MOF synthesis such as the crystallization temperature and the time needed for the crystal growth were thus obtained by Calvet calorimetry.

8.4 TGA and polymers

Thermogravimetry (TG) became one of the most used technique in polymers characterization. Thermal stability, kinetics, and thermal decomposition are among the principal studies applied on polymers.

8.4.1 Principle

TGA alone gives information about the mass variation of a sample in function of time and temperature in a controlled program and atmosphere. By coupling TGA with DTA (differential thermal analysis), more information can be obtained about the thermal behavior of a sample. Today, all TGA manufacturers offer TGA-DSC sensors functioning over a wide range of temperature, from ambient temperature up to 1,600 °C. TGA coupled with DSC allows simultaneous acquisition of mass variation (TGA signal) and thermal phenomena (DSC signal).

8.4.1.1 Examples of TGA analyses

All analyses in the following examples were done on TGA/SDTA851, TGA/DSC2, and TGA/DSC3 + from *METTLER TOLEDO^{TM}*.

TGA-SDTA of Pluronic F127 copolymer

The example below (Fig. 8.14) concerns a copolymer of poly (ethylene glycol) and poly (propylene glycol), also known as Pluronic F127 copolymer. By superimposing the TGA signal (blue color) with SDTA (simulated differential thermal analysis) signal (orange color), one can see that the only mass loss step on TGA curve, corresponding to the large exothermal peak on SDTA signal, represents the polymer's decomposition. Another peak, thinner, an endothermic one, observed thanks to SDTA signal at about 50 °C, corresponds to the melting of the polymer. This can be important information when shaping of the polymer is envisaged.

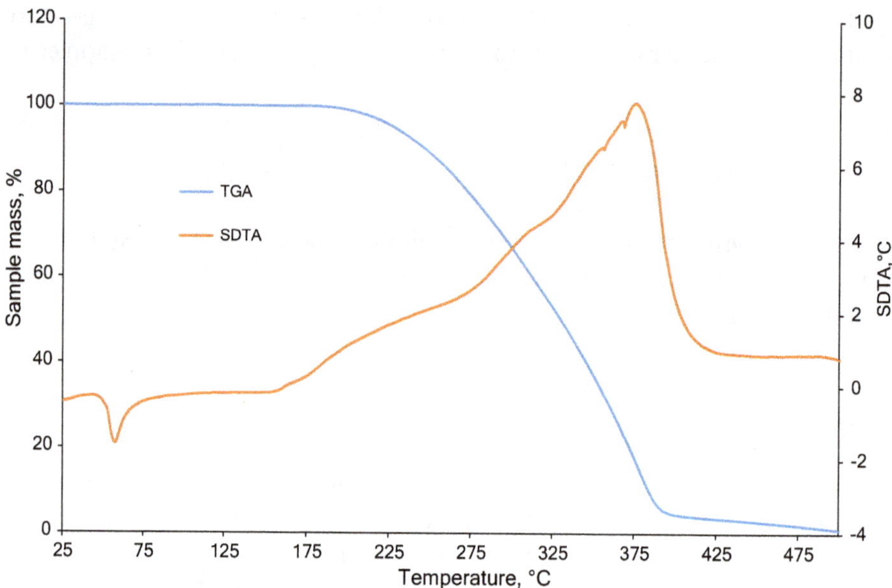

Fig. 8.14: TGA-SDTA of Pluronic F127 copolymer.

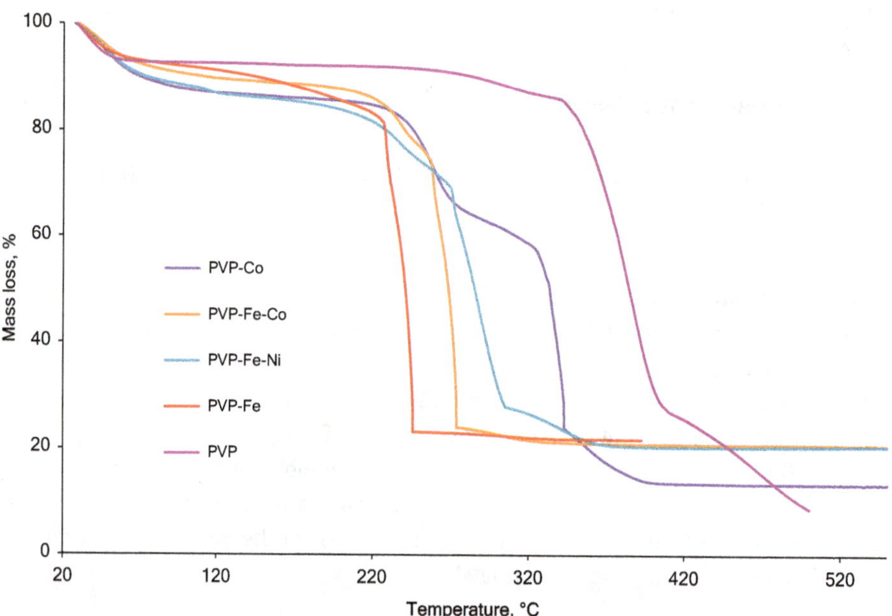

Fig. 8.15: TGA decomposition profiles under air, with a heating rate of 3 °C min^{-1}, for different electrospun PVP nanofilaments.

TGA decomposition profiles of PVP electrospun nanofilaments

In the next example, TGA has been used to compare the thermal decomposition profiles of different electrospun nanofilaments under air. Samples were synthesized with or without metal salt (Fig. 8.15). We notice that in the presence of metal salt, the decomposition profile of poly (vinyl pyrrolidone) (PVP) changes as the onset decomposition temperature decreases and the thermal profile becomes sharper than that of pristine PVP. The presence of iron salt strongly influences the decomposition profile of PVP by highly accelerating it. By TGA, the catalytic influence of iron salt was shown.

Depending on the application, the decomposition of polymers can be studied in an inert or oxidizing atmosphere. Attention should be paid to the nature of the reactive gas, which may greatly influence the thermal decomposition profile. Two examples will illustrate this influence.

TGA decomposition profile for a cyanate ester resin

The next example shows the decomposition profile of a cyanate ester resin. Concerning epoxies, cyanate esters have high temperature stability, low dielectric loss and low water absorption. In Fig. 8.16, the thermal profile changes with the nature of gas (air or nitrogen), especially at higher temperatures. Decomposition starts at

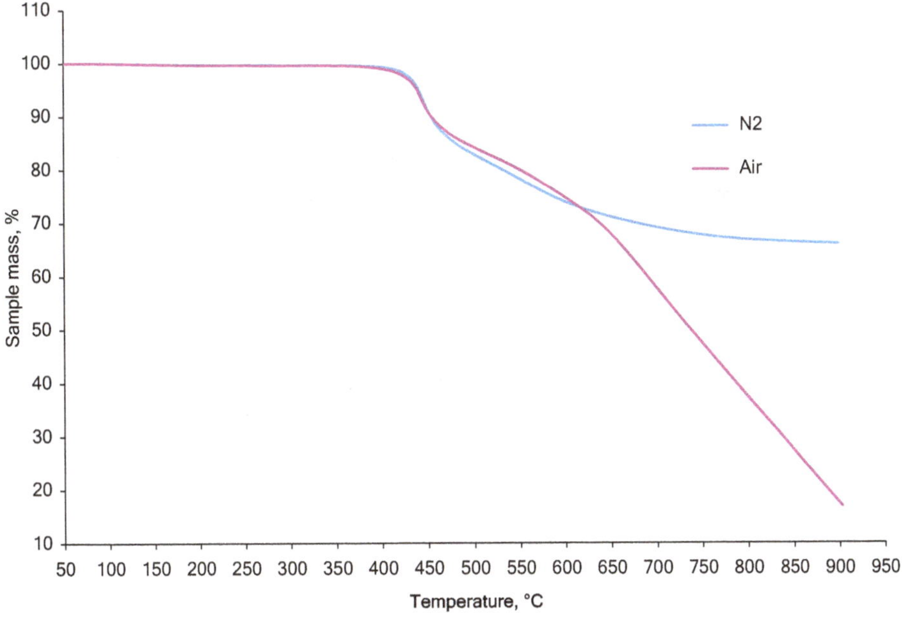

Fig. 8.16: TGA curves of a cyanate ester resin under N_2 and air (at 50 mL min^{-1}) at a heating rate of 10 °C min^{-1}.

425 °C, and it is exothermic either in air or in nitrogen. Under air, the total mass loss is about 83%, while it reaches only 35% under nitrogen. Even if the beginning of the decomposition is similar under air or under nitrogen, the decomposition is far to be finished under nitrogen, and differences in thermal profiles appear after 650 °C. This allows us to affirm that the only factor initiating decomposition is heat. For this example, the nature of the reactive gas does not influence the start of decomposition. The influence appears and becomes important when temperature exceeds the value of 650 °C.

TGA decomposition profiles for nanocomposites of PVA with iron salt

In other cases, the decomposition profile is highly influenced by the nature of gas. Nanocomposites of poly (vinyl alcohol) (PVA) with iron salt were analyzed under two different atmospheres: air and N_2, and the obtained profiles were quite different as shown in Fig. 8.17. As the thermal profile is sharper and finishes earlier under air rather than N_2, a third analysis was done under oxygen. The obtained profile is even sharper than that obtained under air and confirms the great effect of the gas nature on the decomposition profile. The superposition of DTA signals obtained under the three atmospheres shows that the exothermic peak corresponding to polymer decomposition was much more thin and intense when air or oxygen was used as reactive gas compared with the decomposition under nitrogen.

TGA to study the decomposition kinetics of a lignocellulosic biopolymer

In the kinetic studies, the polymers are tested at different heating rates. In this case, the thermal profile changes as can be seen in Fig. 8.18. The chosen example shows the study of the decomposition kinetics of a lignocellulosic biopolymer under air at heating rates ranging from 10 to 200 °C min^{-1}. It can be seen that the decomposition is complete towards the highest temperatures with the increase of the heating rate from 10 to 200 °C min^{-1}. At a fixed temperature of 550 °C, decomposition is complete (total mass loss of 100%) at 10 K min^{-1}, and then it falls at 74% for the heating rate of 200 °C min^{-1}. For the highest heating rate, decomposition becomes complete at 1,250 °C. The parameters obtained by TGA are very useful to complete different databases or modeling programs linked to the combustion of biomass [9].

In the literature, there are different methods of analysis of parameters coming from thermogravimetric experiments allowing the extraction of reliable kinetic data [9].

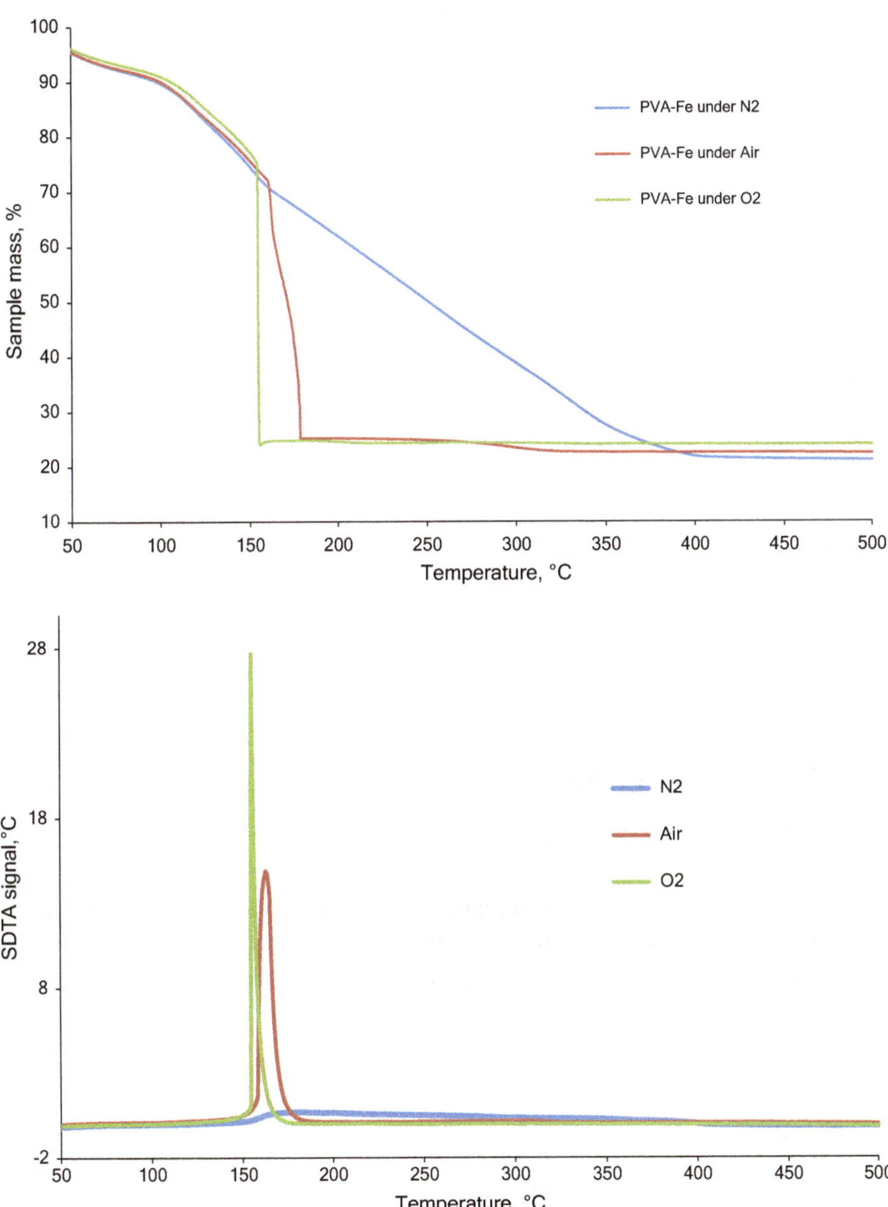

Fig. 8.17: TGA (a) and DTA curves (b) of electrospun PVA-iron salt nanofilaments under air, nitrogen, and oxygen at a heating rate of 3 °C min^{-1}.

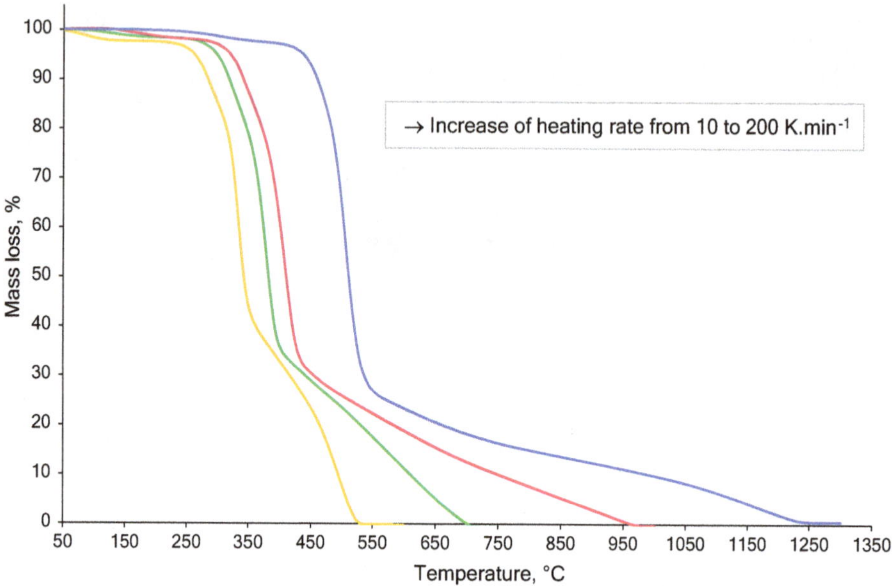

Fig. 8.18: Decomposition thermal profiles of a lignocellulosic biopolymer under air at four heating rates (10, 50, 100, and 200 °C min^{-1}).

8.5 Thermomechanical analyses (TMA) for the study of polymers

The stability of polymers in terms of dimensional change and mechanical behavior is a very important characteristic in their daily use, when they are subjected to strong temperature variations, either by accident or during their elaboration or use.

8.5.1 Principle

TMA measures the dimensional changes of a material (length, thickness or volume) as a function of temperature while submitted to a constant mechanical load or deformation. The sample is heated or cooled at a certain rate, or maintained at a fixed temperature (isothermal mode), in a controlled atmosphere. The stress or strain may be compression, tension, flexure, or torsion. Some materials will deform under the applied stress at a specific temperature that often relates to the material softening and/or melting. In function of the applied force, different modes are available as shown below.

Dilatometric mode (A): is the most commonly used mode in thermomechanical analysis. The expansion coefficient is determined as a function of temperature. To some extent, the expansion coefficient is similar to the specific heat determined by DSC. A typical feature of this mode is that the probe exerts only a very small force (e.g., 0.01 N) on the sample. This mode replaces nowadays the old dilatometry technique.

Compression mode (A): in this mode, the force applied to the sample is high.

Penetration mode (B): in this mode, the determination concerns the softening point of a sample. This is usually done by using the *ball-point* probe.

Stretching or tension mode (C): the fiber or film accessory is used to perform measurements in tension. This allows determining changes in length due to shrinkage or expansion.

Three-point bending mode (D): this mode is ideal for studying the elasticity of stiff samples (e.g., fiber-reinforced polymers)

Swelling (E): many substances swell when they are in contact with liquids. This leads to changes in volume or length that can be measured using the swelling accessory.

Volume expansion (F): an appropriate accessory measures volume changes of liquids. For volume measurements, a liquid with a known coefficient of dilatation contains the sample.

TMA probes and deformation modes for specific applications are shown in Fig. 8.19.

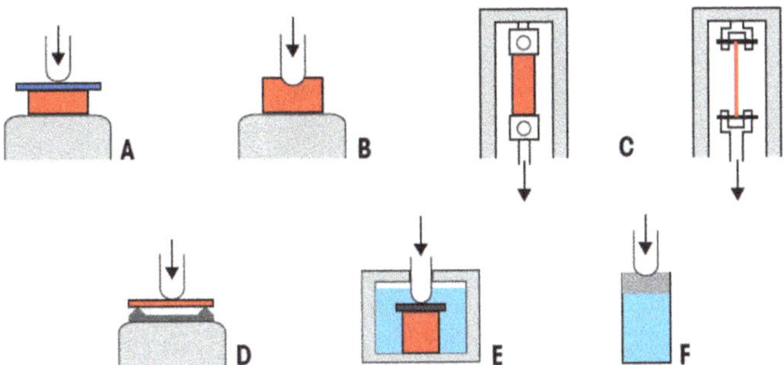

Fig. 8.19: Various types of TMA probes and deformation modes (Property of Mettler Toledo): (a) dilatometric mode; (b) penetration mode; (c) tension mode; (d) 3-point bending mode; (e) swelling; and (f) volume expansion mode.

8.5.1.1 Examples of TMA analyses

T_g determination by TMA on glasses of gold thiophenolate coordination polymers
Sometimes by DSC some glass transitions cannot be highlighted very well. The example below shows the determination of glass transition with more accuracy by thermomechanical analyses then DSC on glasses of gold thiophenolate coordination polymers (CP). These coordination polymers glasses are transparent and luminescent and were obtained, as large pellets, from a simple mechanical pressure of the amorphous powder [10]. The three gold-based CP glasses represented in Fig. 8.20 are composed of thiophenolate [Au(SPh)]n, phenylmethanethiolate [Au(SMePh)]n and phenylethanethiolate [Au (SEtPh)]n. The longer the alkyl chain between the thiolate and the phenyl (Ph) ring, the more transparent the glass formed. The glass transitions, measured by TMA (in compression mode, at 3 °C min^{-1} and a load of 0.1 N with a 3 mm ball-point quartz probe on a *METTLER TOLEDO* TMA/SDTA840^{TM}) (Fig. 8.21), occurred at lower temperatures for CP with longer alkyl chain. The gain in color (yellow) and transparency is directly related to the addition of the -CH_2- motif (i.e., increases of the alkyl chain length) and the flexibility of the ligand. When heated to higher temperatures, these "glass" compounds crystallize and then decompose, as obtained by DSC (*not shown here*). These gold-based CP glasses are a better alternative to transparent and luminescent lanthanide-based materials, banned by the European Union.

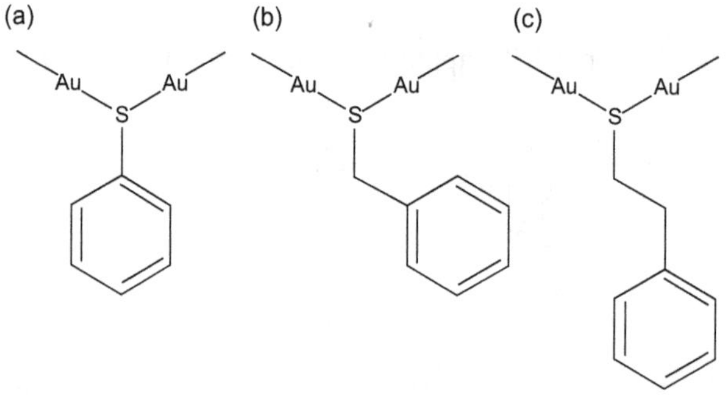

(a) (b) (c)

Fig. 8.20: The three gold-based compounds: (a) $[Au(SPh)]_n$, (b) $[Au(SMePh)]_n$, and (c) $[Au(SEtPh)]_n$.

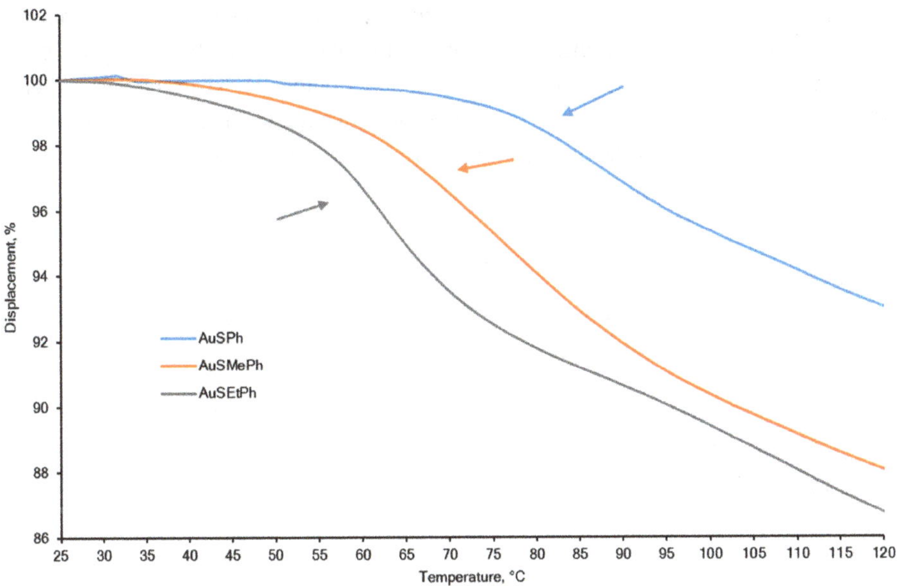

Fig. 8.21: TMA curves for the three gold-based compounds. The arrows show the presence of T_g (reproduced from ref. [10] with permission from the Royal Society of Chemistry).

Determination by TMA of the thermal expansion coefficient of a dental composite

Linear or thermal expansion coefficient can be obtained with TMA by using the following formula:

$$\alpha = \frac{1}{L_0}\frac{\Delta L}{\Delta T}, \quad \text{in } \mu\text{m m}^{-1}\text{ K}^{-1} \text{ or ppm K}^{-1} \text{ or } 10^{-6}\text{ K}^{-1} \tag{8.3}$$

where L_0 is the initial length (or thickness) of the sample at a reference temperature, ΔL is the length variation ($L - L_0$) due to change in temperature, and ΔT is the temperature variation ($T - T_0$).

The surface of the sample should be completely plane and free of roughness for a correct determination of linear expansion coefficient by TMA.

In the next example, the expansion coefficient was determined on a dental composite that was elaborated by insertion of triethyleneglycol dimethacrylate (TEGDMA) inside a mesoporous silica [11]. The use of mesoporous silica as filler in composite influences the thermal expansion coefficient (CTE).

Samples like pellets of 4 mm diameter were analyzed by using a load of 0.01 N, under nitrogen, and a temperature range comprised between ambient and 90 °C at a heating rate of 5 °Cmin^{-1}. The TMA results (Fig. 8.22) for the polymer and the polymer composite show curves of linear variation from which the CTE has been deduced: 110 × 10^{-6} K^{-1} (for TEGDMA) and 71 × 10^{-6} K^{-1} (for composite), knowing that silica has a low CTE of about 0.5 × 10^{-6} K^{-1}. Silica represents only 15% of the complete volume. Polymer fraction outside the mesopores is 50%. Only this fraction is involved in the thermal expansion. Indeed, the polymer encapsulated inside the silica pores (35% porosity) does not contribute to the thermal expansion (reduced chain mobility due to confinement). Therefore, the use of mesoporous silica as fillers in composites reduces the thermal expansion coefficient, as TEGDMA inside the mesopores does not change the thermal expansion.

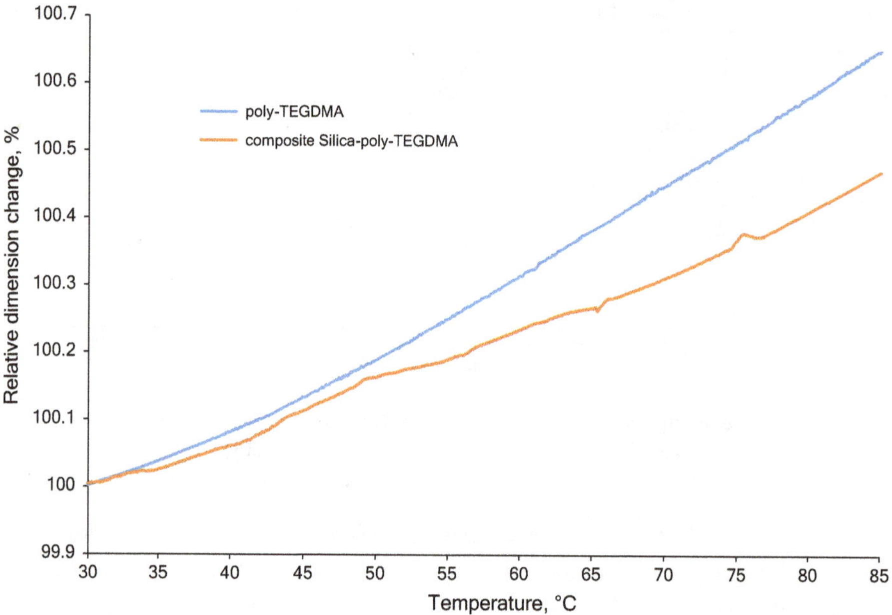

Fig. 8.22: Thermo-mechanical analysis of polyTEGDMA–silica composites (the very small artefacts visible on composite curve are due to the roughness of the sample surface).

TMA to study the thermo-mechanical behavior for poly[B-(methylamino)borazine)]
In the next example, the thermo-mechanical behavior of a series of poly[B-(methylamino)borazine)] has been done in order to set the synthesis temperature (i.e. $T_{thermolysis}$) of these polymers and thus to achieve melt-spinning and produce fine-diameter green BN fibers [12, 13].

TMA curves for poly[B-(methylamino)borazine)] obtained at different $T_{thermolysis}$ ranging from 140 to 200 °C are presented in Fig. 8.23. Three different types of thermomechanical behaviors can be observed. Polymers prepared at $T_{thermolysis} \leq 150$ °C are seen to strongly shrink in thickness as illustrated by the large dimensional change (≈ 85%), the appearance of two stages of softening being most probably caused by different molecular weight species with melting in different temperature ranges. TMA curves for polymers with $T_{thermolysis}$ comprised between 160 and 180 °C show an increasing shape above 120 °C due to the swelling of polymers. Indeed, this phenomenon occurs when low values of load are used (0.1 N, in this study). An increase of $T_{Thermolysis}$ from 140 to 200 °C shifts the softening point to higher temperatures and reduces the capability of the polymer to melt, that is the melting level.

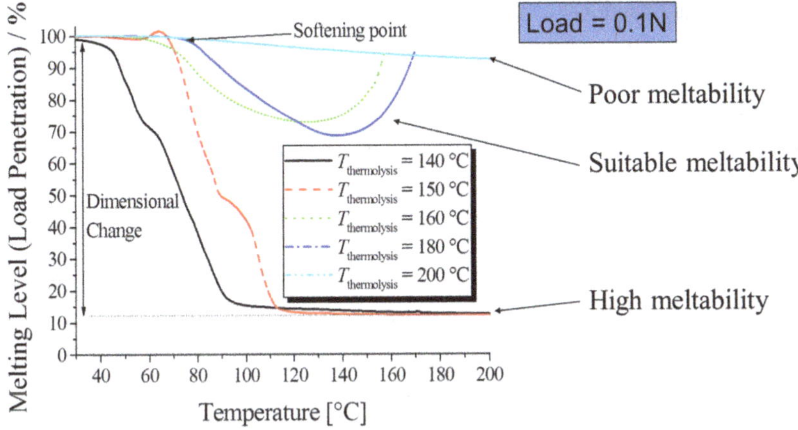

Fig. 8.23: The thermomechanical behavior of a series of poly[B-(methylamino)borazine)] obtained at different synthesis temperatures (reprinted from ref. [12] with permission from Elsevier).

TMA analyses allow to conclude that the polymers with $T_{Thermolysis}$ of 140 and 150 °C were poorly spinnable (due to a too high melting capability), those prepared between 160 and 180 °C were readily spinnable and stretchable (appropriate shrinkage upon heating), and the polymer obtained at 200 °C was not successful according to the absence of dimensional changes upon heating.

8.6 Evolved gas analysis, to know more about the polymer decomposition mechanism

Thermal decomposition of polymers leads to changes in their physical and chemical properties. In some cases, it is necessary to obtain information on the nature of gases emitted in order to understand in detail the various physico-chemical phenomena observed.

When heated or burned, polymers can emit volatile organic compounds (VOCs) that are often toxic, and their impact on the environment must be evaluated.

8.6.1 Principle

Evolved gas analysis (EGA) refers to a family of techniques where the nature and/or amount of gas or vapor evolved is determined (according to the ICTAC; definition approved by IUPAC) [14].

EGA allows identification and/or quantification of gases by coupling TGA, which is a quantitative technique, with an analyzer-like mass spectrometer (MS), Fourier-transform infrared spectrometer (FTIR), GC-MS or micro-GC (μGC). This relies on the quantity of the emitted products (if the heating is done by TGA) to their chemical nature, and thus an overview of what is happening in the material during heating or decomposition can be obtained.

EGA can be done as follows:

- ***On line***: TGA-MS, TGA-FTIR, TGA-GC-MS, TGA-μGC-MS (these couplings are described later in the text);
- ***After or during the heating (or decomposition) of a material,*** as shown just below.

EGA is done either by **bubbling** the emitted gases into a solution to be further analyzed by high-pressure liquid chromatography (HPLC), or by **concentrating** the emitted gases on a sorbent – for example, ***Tenax*** *or* ***activated charcoal*** – followed by either **thermal desorption** or **the extraction of adsorbed gases into a specific solvent** respectively, then GC analysis (or FTIR or MS) to identify the gaseous products.

The choice of the appropriate technique depends on the compounds of interest, their nature and number, or the volatility of compounds, but also on the information sought. To get an overall idea of the nature of the emitted gases, their organic or inorganic character, and the bonds between atoms and the functional groups, the TGA-FTIR is the right coupling.

The TGA-MS coupling, more sensitive than the TGA-FTIR, allows following one or few molecules, but identification of compounds from the same chemical family can not be done without separation. In the case of polyethylene, for example, by using TGA-MS, it is impossible to identify all emitted alkanes, alkenes, or cyclic (and aromatic) compounds. The presence of a gas chromatograph to separate all these compounds is thus essential.

For the analysis of a complex mixture of gases, TGA-GC-MS is more appropriate than the other couplings mentioned above as the GC column can separate more than 100 VOCs. The polarity of the column and its length are very important parameters. For example, with a longer column (e.g. 60 m length compared to 30 m) more compounds will be separated. A non-polar column does not separate well very polar compounds (e.g. compounds with many –OH groups).

For lighter compounds, such as hydrogen, water, carbon dioxide, carbon monoxide, methane, or lighter COV (ethane, ethylene, etc.), and for quantitative analyses, TGA-μGC(-MS) is more adapted than the other cited couplings.

The operating principles of the mentioned couplings are given here with few examples of applications.

For all mentioned couplings, a transfer line is necessary to transport gases from the TGA oven to the gas analysis system. A heated transfer line will prevent condensation of gaseous products through their transport.

8.6.2 Techniques used in EGA

8.6.2.1 TGA-MS coupling

TGA-MS coupling requires a special interface because TGA operates at atmospheric pressure, while MS requires a vacuum of about 10^{-5} mbar. An inert and heated ceramic capillary, bringing into the mass spectrometer only a part (i.e. 1%) of the gases emitted by the TGA in order to maintain the vacuum integrity [15], makes the connection between TGA and MS. This small amount is perfectly adequate as the MS sensitivity is extremely high. The gas coming from the TGA is sucked into the MS due to the drastic decrease of the pressure. Helium is used as carrier gas mainly, but also gases like N_2, air, or O_2 are used. The last two are to be avoided because they reduce the lifetime of the MS filament.

The molecules of gases arriving from TGA are bombarded with electrons, in the case of electron impact (EI) source, in the ionization chamber, and thus fragmented into positive ions, which are then separated (e.g. by quadrupole) according to their mass/charge ratio through a combination of electrostatic and electromagnetic fields. MS provides chemical and structural information like functional groups and side chains.

Measurement modes

Mass spectrometer generally operates in two ways: **SCAN** (monitoring of all ions) and **SIM** (**s**elected **i**on **m**onitoring). In the first one, the ion current is continuously measured as a function of mass-to-charge ratio for a single scan. If all the scans during TGA are joined (with an adapted software), the intensity of each ratio as a function of T/°C and time is obtained. In the **SIM** mode, as its name suggests, one or few selected ions with particular m/z values are monitored. The most abundant mass ions, also called **characteristic ions,** allow searching for a molecule more efficiently (with the *extract ion chromatogram* option). The sensitivity of detection is thus higher, but only few molecules can be followed. For example, if the molecule of interest is benzene as shown below (Fig. 8.24), one or two ions will be selected and monitored. These selected ions (i.e., characteristic ions) show the highest abundances and characterize a molecule. For each molecule, an experimental spectrum with ions of different masses, including the characteristic ions, is obtained and compared with a theoretical spectrum. When the two spectra match well, the identification factor is high (i.e., between 70% and 99%).

TGA-MS for a PVC formulated with and without MoO_3

In this example [16, 17], the addition of 4% molybdenum trioxide (MoO_3) changes the decomposition of *PVC* as shown in the TGA thermograms (upper part) presented in

Fig. 8.24. TGA profiles are slightly different and show a decomposition finishing at lower temperatures in presence of MoO_3.

TGA-MS coupling has been used to follow benzene, whose characteristic ion is 78, as it is emitted during the PVC decomposition. The MS intensity curves (bottom part) for *m*/*z* 78 present a much lower amount of benzene (*and probably other aromatics too*) when MoO_3 is present (Fig. 8.24b): 0.1% versus 2.1% for pristine PVC (Fig. 8.24a). MoO_3 was used as a fume suppressant on the burning polymer. The cyclization and then aromatization processes involved in the usual decomposition mechanism of PVC, after the emission of HCl, are thus highly reduced in the presence of MoO_3, used like cis/trans isomerization catalyst.

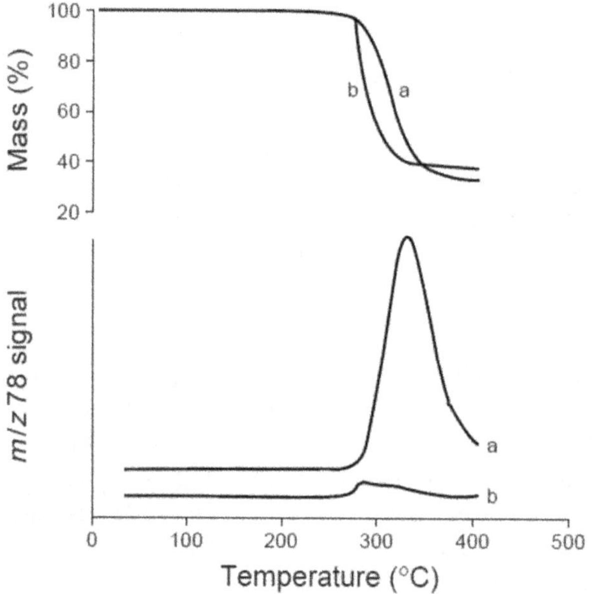

Fig. 8.24: TGA-MS (*m*/*z* 78, corresponding to benzene) curves for a PVC sample formulated with and without MoO_3 (30 mg of sample; 20 °C min^{-1}; under nitrogen; a = PVC, b = PVC + MoO_3) (reprinted from ref. [17] with permission from Wiley).

TGA-MS coupling is not only applied in the field of polymers; various applications exist in the scientific literature, and the technique is nowadays recognized as bringing solutions to many analytical problems.

8.6.2.2 TGA-FTIR coupling

The coupling of TGA with FTIR for gas analysis complements the capabilities of thermal analysis [18, 19]. In particular, it provides a better understanding of the thermal behavior of polymers and information on their composition.

The coupling between TGA and FTIR (Fig. 8.25) allows identifying *on-line* the nature of the volatile products emitted by a substance under a controlled temperature program. Because the emitted gases must be transferred to the FTIR cell without condensing, the coupling requires a heated interface, which consists of

- an oven outlet connection,
- a transfer line heated to 300 °C, and
- an IR cell specific to gas analysis.

The infrared spectra (absorbance as a function of the wavenumber) are recorded continuously. They can then be compared with the gas phase spectra of a library.

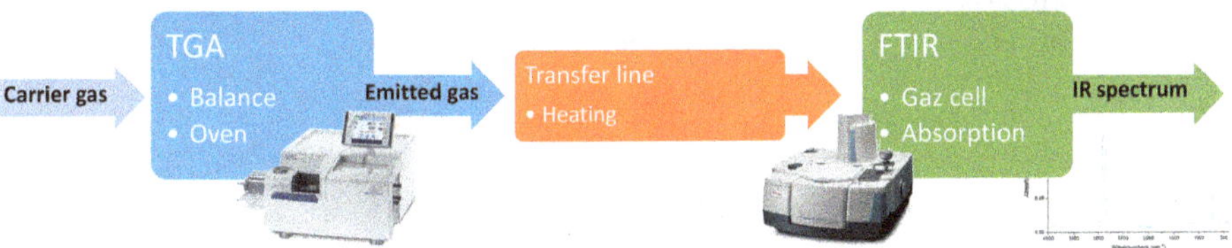

Fig. 8.25: Principle of the coupling between TGA and FTIR.

Infrared principle

Infrared spectroscopy is based on the interaction of infrared light with molecules. Molecules absorb light at specific frequencies that depend on their functional groups. The infrared spectra representing these absorption bands of the atomic groups are characterized by their wavenumber (in cm^{-1}) and their intensity. The part of the electromagnetic spectrum studied in the mid-infrared ranges from 400 cm^{-1} to 4,000 cm^{-1} (Fig. 8.26) [15].

Representation of data

Large amount of data is acquired with the coupling and the representation of the results must be relevant. Several representations are generally used.

Gram-Schmidt

The absorption of each IR spectrum can be integrated over the whole range of wavenumbers. The results are presented as intensity versus time curve. This is called the Gram-Schmidt plot. The Gram-Schmidt is therefore a measure of the sum of the intensities of all the absorption bands and shows the temperatures where gas emission occurs. The trace

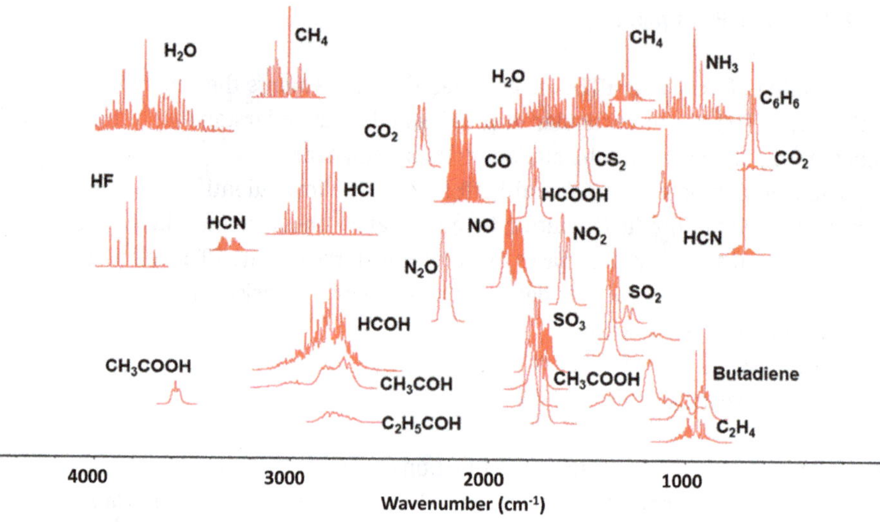

Fig. 8.26: Absorption bands, in the gas phase, characteristic of the main products emitted in TGA-FTIR coupling (reprinted from ref. [15] property of Mettler Toledo).

is similar to the first mathematical derivative of the TGA weight loss (DTG). It does not provide any particular chemical information.

Chemigram

It is possible to display the IR intensity of one or more selected regions (wavenumber in cm^{-1}) as a function of time in order to follow the presence of particular functional groups in the emitted gases. This representation is useful for interpretation purposes. The resulting curves (intensity versus time) are called **chemigrams** or functional group profiles.

Spectrum

An infrared spectrum can be plotted and interpreted at a specific temperature or time. It permits the identification of molecules or mixtures of molecules in the evolved gases.

Examples

Two examples of analyses using TGA-FTIR are given here. For these examples, the analytical conditions are depicted below.

TGA analyses were carried out with a *METTLER TOLEDO* TGA/DSC2^{TM}. Around 20 mg of samples were analyzed by using 70 µL alumina crucibles. The samples were heated from 40 to 750 °C at a heating rate of 10 °C min^{-1}, in dry nitrogen atmosphere at a flow rate of 50 $mLmin^{-1}$.

Infrared spectra were recorded using a Nicolet iS50 FTIR spectrometer from Thermo Fisher Scientific. The spectrometer is equipped with a gas cell, a deuterated triglycine sulfate (DTGS) detector and KBr optics. Background and sample were acquired using 32 scans at a spectral resolution of 4 cm^{-1} from 4,000 to 400 cm^{-1}.

Decomposition of an ethylene-vinyl acetate copolymer (EVA)

Copolymers of ethylene and polar monomers are of great industrial importance and have complementary applications to polyethylene. Even with a small amount of polar comonomer, they have alternative properties to polyethylene. Ethylene vinyl acetate (EVA) copolymer is a copolymer of ethylene and vinyl acetate (VAc). The VAc comonomer reduces melting temperature, crystallinity, and stiffness and improves polarity, adhesive properties, and transparency compared to polyethylene. The versatility of these copolymers results in a wide range of applications for many fields such as electrical insulation (cable jacket, insulation), sporting goods, or solar cell encapsulation for photovoltaic modules.

The three-dimensional (3D) diagram (Fig. 8.27) gives an overall representation of the gases emitted during the TGA. A section perpendicular to the time axis provides an IR spectrum. A section perpendicular to the wavenumber axis provides the absorbance versus time (chemigram).

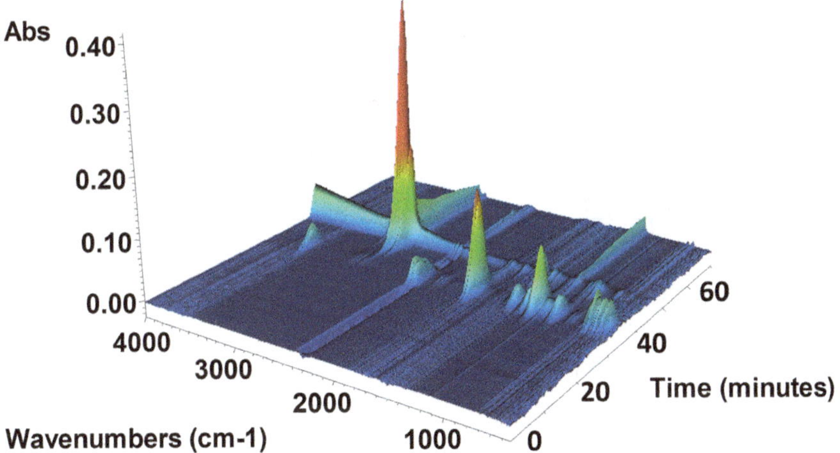

Fig. 8.27: Three-dimensional representation of the decomposition products of EVA. IR spectrum series of the TGA-FTIR measurement. The x-axis corresponds to the wavenumber, the *y*-axis to the time, and the *z*-axis to the absorbance.

The spectra extracted from the 3D representation (Fig. 8.27) at 35 min (345 °C) and 46 min (462 °C) allow identifying the decomposition products as plotted in Fig. 8.28. The analysis of the infrared absorption bands indicates the emission of acetic acid at 345 °C and alkanes at 462 °C [20, 21].

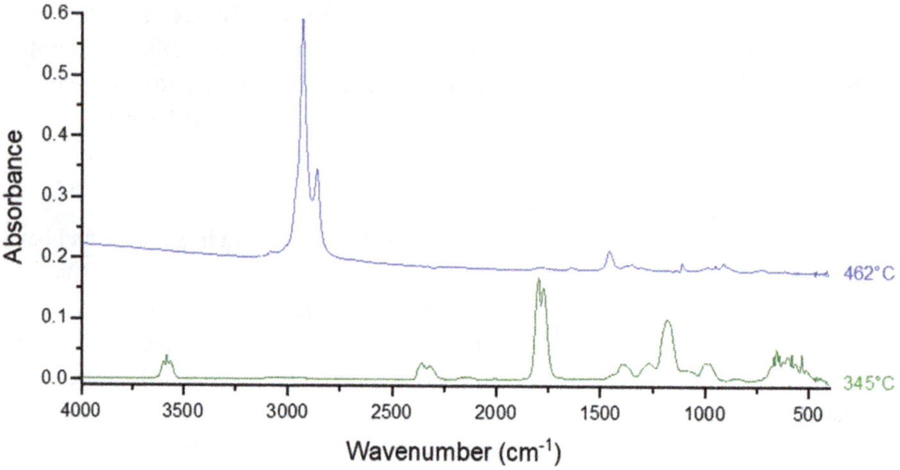

Fig. 8.28: Infrared absorption spectra at 35 min (345 °C, green) and 46 min (462 °C, blue) extracted from the 3D representation.

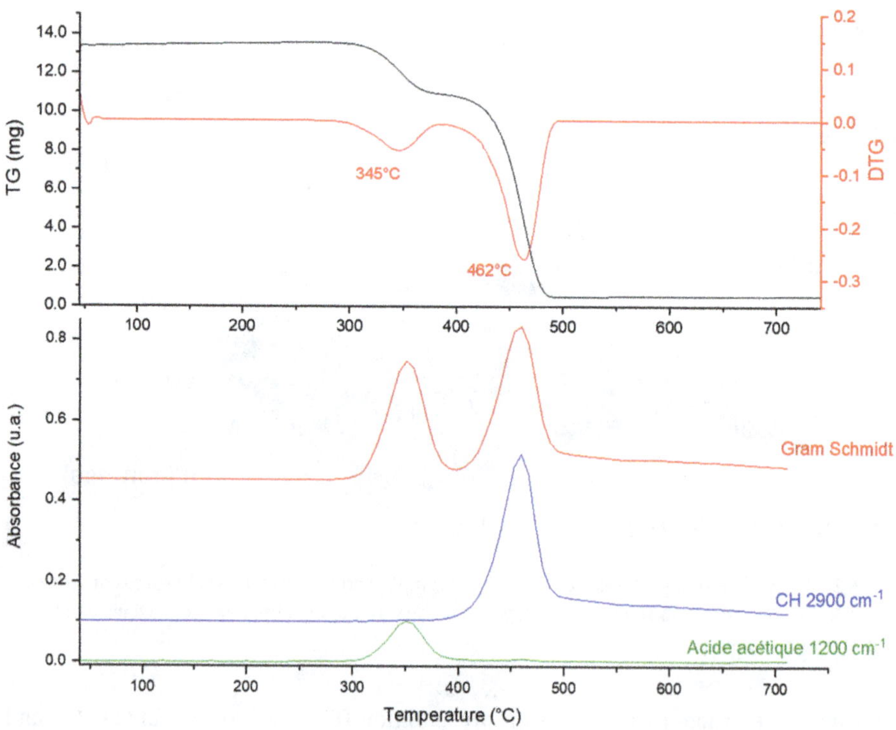

Fig. 8.29: Mass loss curves (TGA, black), DTG and Gram-Schmidt curves (red), and the IR absorption spectra at 1,200 and 2,900 cm^{-1} (chemigrams).

The IR spectrum shows strong bands in the region 3,000–2,850 cm^{-1} due to C–H stretching. C–H scissoring (1,470 cm^{-1}) and C-H rocking (1,383 cm^{-1}) are distinguishable in the spectrum. All these absorption bands indicate the presence of alkane groups.

TGA analysis shows that EVAs are stable up to 300 °C and then decompose into two stages (TGA curve, Fig. 8.29).

The chemigrams at 1,200 cm^{-1} and 2,900 cm^{-1} (Fig. 8.29) show that acetic acid is emitted only in the first step of weight loss, while the polyethylene part of the backbone is degraded in the second step. During the second mass loss step, only the signature of the alkanes appears in the IR spectrum at 2900 cm^{-1}.

Decomposition of PVC

After polyethylene, PVC is the most produced polymer in the world. It has become essential in the fields of building and construction.

The three-dimensional (3D) diagram (Fig. 8.30) gives the overall representation of the gases emitted during the TGA of PVC.

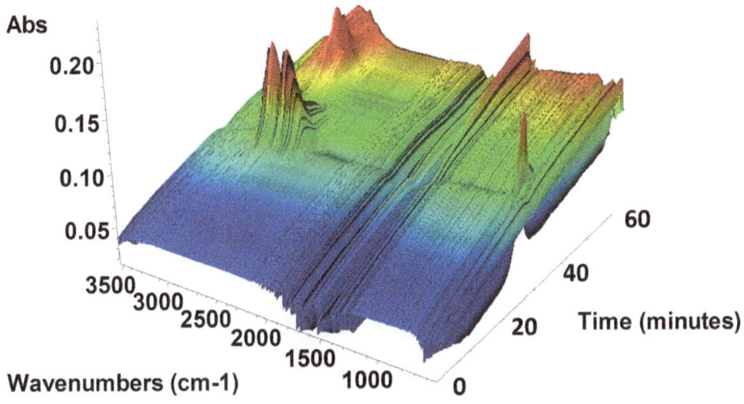

Fig. 8.30: 3D representation of PVC decomposition products.

The IR spectra at 290 °C and 470 °C (Fig. 8.31) show the release of HCl and alkanes respectively. These temperatures correspond to the two steps of the thermal degradation of PVC (Fig. 8.32).

The analysis of the infrared absorption bands (Fig. 8.31) at 290 °C shows the main emission of hydrochloric acid (HCl) corresponding to the wavenumbers between 2500 cm^{-1} and 3200 cm^{-1} [22]. A thin peak at 700 cm^{-1} shows the presence of benzene (C_6H_6) in the gases emitted at 290 °C, which comes from the cyclization of the main chain fragments during the loss of chlorine, followed by aromatization [23, 24].

The IR spectrum at 470 °C, corresponding to the second degradation step, reveals mainly the emission of alkanes at 2900 cm^{-1} (CH stretching) and 1470 cm^{-1} (CH bending).

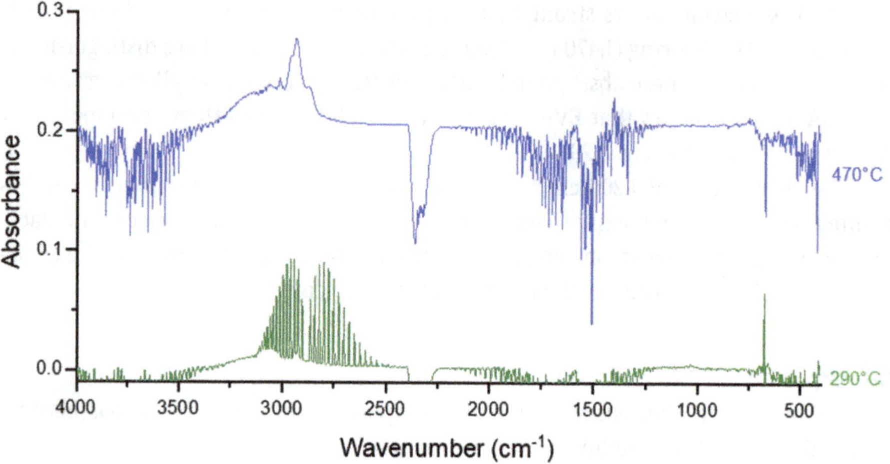

Fig. 8.31: Infrared absorption spectra at 290 °C and 470 °C.

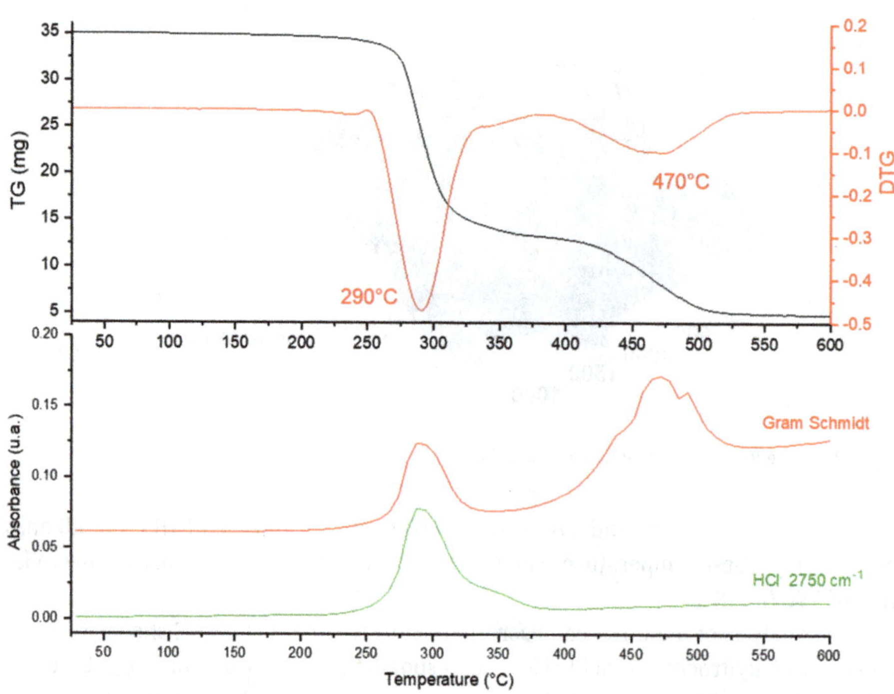

Fig. 8.32: Mass loss curves (TGA, black), DTG and Gram Schmidt curves (red) and chemigram (green) at 2,750 cm^{-1} for PVC.

The chemigram at 2,750 cm^{-1} (Fig. 8.32) shows that the emission of hydrochloric acid continues up to 375 °C (shoulder of the peak towards the highest temperatures).

Limits of TGA-FTIR coupling

The TGA-FTIR coupling can easily give valuable information on the functional groups present in the emitted molecules. However, the presence of molecules in small quantities is sometimes difficult to detect. The sensitivity of TGA-FTIR coupling is lower than for other couplings (e.g. TGA-MS). Moreover, as can be seen from the previous IR spectra, the presence of water has an impact on the readability of the spectra.

During the TGA experiment, only the heating rate allows the separation of the different emitted gases. When there is a large diversity of emitted gases, the IR spectra become complicated to interpret.

8.6.2.3 TGA-GC-MS coupling

GC-MS coupling ensures the separation, identification, and quantification, after a preliminary calibration, of compounds present in a mixture. The online combination of gas chromatography and mass spectrometry (GC-MS) offers multiple analytical possibilities as the two techniques can harmoniously combine their specific advantages. The use of GC-MS coupling covers applications in a wide range of fields, such as perfumery, oenology, the petroleum industry, biology, fine chemistry, and the plastics industry.

Principle of gas chromatography

GC is a technique for separating a mixture of volatile molecules, called “solutes” or “analytes.” The gas mixture, vaporized into the heated injector of GC and then transported through a column containing a liquid or solid substance, constitutes the stationary phase. The transport realized with an inert gas, called “carrier gas”, constitutes the **mobile phase** (e.g., helium). The injector allows both the introduction of the sample into the column of the chromatograph and the volatilization of the analytes. The separation of analytes is based on the difference in affinity of these compounds for the mobile phase and the stationary phase. Capillary columns are simple stainless steel, glass, or fused silica tubes (material inert to the stationary phase and the samples) with an internal diameter between 0.1 and 0.5 mm, and a typical length of several tens of meters, up to 100 m. The inner surface of this tube is covered with a 0.1–5 µm thick film of the **stationary phase**. A capillary column of small diameter, long, with a thick stationary phase and with chemical properties similar to the molecules of the sample typically leads to better separations [25].

The column is enclosed in an oven whose temperature is adjustable (typically between 20 and 350 °C) and programmable. One can work either in "isothermal" or in "gradient", using a certain heating rate, like in a TGA oven. The gradient mode is preferable, as the isothermal method tends to give broad peaks for the most retained species, and therefore a poorer separation. The result is a **chromatogram** with one or more peaks, each peak corresponding to the separated molecule and being characterized by two parameters: **retention time** and **peak area**. By using the same conditions and the same column, the retention time is characteristic of a molecule. Thus, the comparison of retention times can help in the identification of chemical compounds. Moreover, the area of a peak is proportional to the mass of the injected analyte. The proportionality constant, called ***response coefficient***, depends on the chemical species and the detection method used, as well as the injection volume [26].

To identify successfully compounds in the sample, they must be separated efficiently, and the amount of eluted material must be sufficient to be detected by the MS detector.

Principle of mass spectrometry

The **mass spectrometer detector** determines the molecular masses and structure of chemical compounds. Identification is done by analyzing the ions formed from the fragmentation of the substance. Inside the spectrometer, the volatilization of the product, the formation of ions in the gas phase, as well as the separation of the ions according to the ratio of mass m to charge z and their detection take place. The major difficulty in interfacing GC and MS lies in the large pressure difference between the two systems. GC typically operates between 1 and 3 atmospheres (760–2250 Torr; 1 or 2 bars). MS operates at approximately 10^{-5} Torr (mbar) [26]. A good interface should allow both devices to work in their optimal conditions, without altering the sensitivity or the shape of the peaks.

The experimental mass spectrum of a compound is compared with the theoretical spectrum contained in a database, such as NIST. For each identified molecule, an identification quality factor is indicated, in percent. Generally, compounds identified with factors higher than 70% are taken into consideration.

A **TGA-GC-MS** coupling gives plenty of information about a material and its behavior with temperature in terms of mass loss, thermal phenomena, mechanism of degradation/decomposition, nature, and number of the emitted compounds with temperature variation.

Depending on the TGA analysis duration, one or few samplings of gases emitted from the sample are possible with GC-MS. Indeed, the GC-MS method can require 20, 30, or 60 min, in function of the chosen analytical program and the complexity of the sample to be analyzed. Thus, more the analysis time is short and less samplings can be done. If the sample decomposition takes place in a one sharp step of 2 or 3 min (which is the case of a high number of polymers), only one sampling of the emitted gases can be done with GC-MS. To overcome this problem, another similar coupling can be used: **TGA-storage interface-GC-MS.**

The **storage interface** commercialized by SRA Instruments (Lyon/France) is equipped with 16 loops and 2 heated transfer lines at 250 °C, one from TGA oven to the interface entry and the other one from the interface to the GC injector. The loops are also heated at a chosen temperature not exceeding 250 °C. There is therefore a storage step of the emitted gases, which takes place during the TGA analysis, and when the storage is finished, the injection of the content of each storage loop in GC-MS can start. The GC carrier gas (helium) passes through the interface loops and carries their contents into the GC injector (1 loop = 1 GC-MS analysis). Everything is automated so that each storage loop is analyzed with the GC-MS. The loops are ***sulfinert*** treated stainless steel tubes and have a volume of 250 μL each. They are connected to a valve with 16 positions, which is itself connected to two other valves, isolation and injection. All loops can be used if needed. In this way, the evolution of each emitted gas with temperature can be obtained and related to TGA profile. This evolution shows the maxima and minima of gaseous concentrations. With this coupling, three softwares are used, from which two are interconnected (those of GC-MS and storage interface).

Example

Decomposition study of PMMA with TGA-storage interface-GC-MS coupling

The example later shows the analysis of PMMA by using the coupling **TGA-storage interface-GC-MS.** The coupling is presented in Fig. 8.33, with the storage interface between the GC-MS (left side) and the TGA (right side).

PMMA, the most widely used polyacrylic polymer, also known by its trade name, Plexiglas, is a polymer synthesized from methyl methacrylate, its monomer. Figure 8.34 presents its structure. PMMA has very good optical properties (i.e., high light transmittance and transparency) and resistance to ultraviolet and corrosion.

Eleven samplings of the emitted gases have been done during TGA analysis between 200 and 450 °C, which is the thermal decomposition range of PMMA under nitrogen. This polymer decomposes mainly into its monomer (methyl methacrylate), followed by the methacrylic acid. Other VOCs are identified and are emitted over a narrower range between 320 and 410 °C. In Fig. 8.35, TGA decomposition profile of PMMA is superimposed with the major VOC peak areas evolution with temperature. Even if the samplings are spot samplings, we decided to join the points corresponding to the areas of the chromatographic peaks to have a global view of the VOC emission. In fact, the VOCs emitted between two samplings taken with the storage interface are not trapped. However, the samplings are very close to each other in order to obtain a profile of VOC evolution as real as possible. The maxima of concentration of methyl

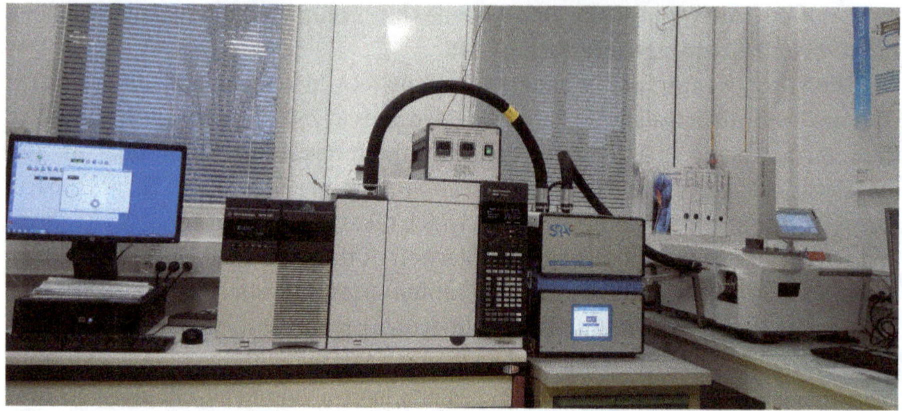

Fig. 8.33: Photo of the TGA-storage interface-GC-MS coupling (with 7890 GC–5977A MS from Agilent Technologies, *METTLER TOLEDO* TGA/DSC3+^{TM} and the storage interface IST16 from SRA Instruments).

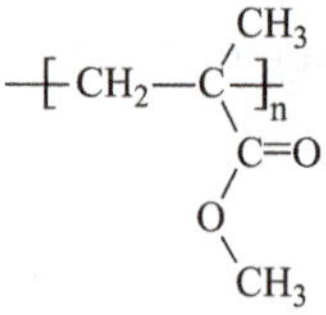

$$-\!\!\left[CH_2-C(CH_3)(C(=O)-O-CH_3) \right]_n\!\!-$$

Fig. 8.34: PMMA chemical structure.

methacrylate (monomer) and methacrylic acid are between 340 and 360 °C. Methacrylic acid starts to be emitted at about 270 °C and its emission finishes at 430 °C, in the same time with that of monomer.

Twelve other compounds were clearly identified; most of them are esters. Among them, acetone and methanol have the highest concentrations. Figure 8.36 shows the evolution profile of the detected VOC throughout the decomposition, except for the monomer and methacrylic acid for reasons of visibility.

End-chain scission with high monomer yield (≥ 90%), process often known as ***unzipping***, is often the predominant decomposition mechanism in polymers with two substituents at the same carbon atom in the repeat unit [27]. In PMMA, the dominant initiation step at low temperatures (around 250 °C) is **end-chain initiation**. At higher temperatures (around 400 °C), the **random-chain initiation** step dominates. The first one involves the breaking-off of a small unit or group at the end of the chain, while the second one involves the breaking of a main chain bond at a seemingly random location, all such main chain bonds being equal in strength. Methyl groups effectively block intramolecular H-transfer, leading to a high monomer yield [27].

Since methyl methacrylate is present in a very high proportion (huge peak area) compared to the other VOCs, it is possible to make an estimate of its proportion

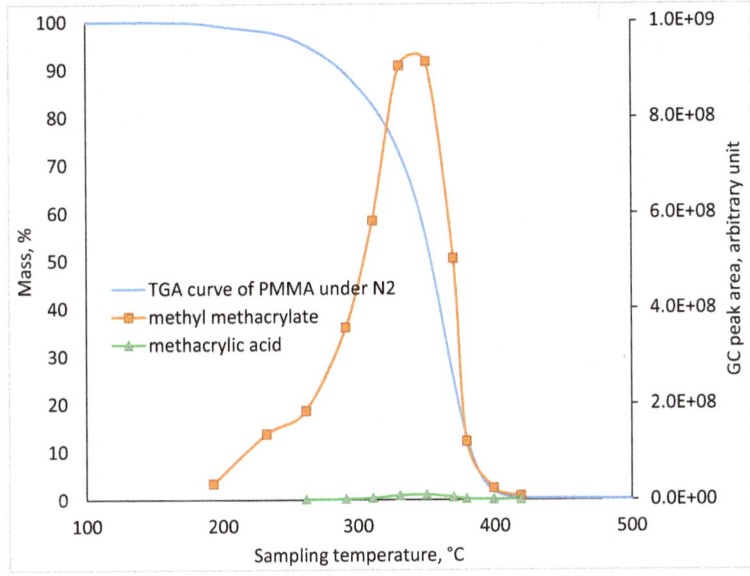

Fig. 8.35: TGA curve of PMMA under N_2 superimposed with the evolution of methyl methacrylate and methacrylic acid (TGA was done with a heating rate of 10°Cmin^{-1}).

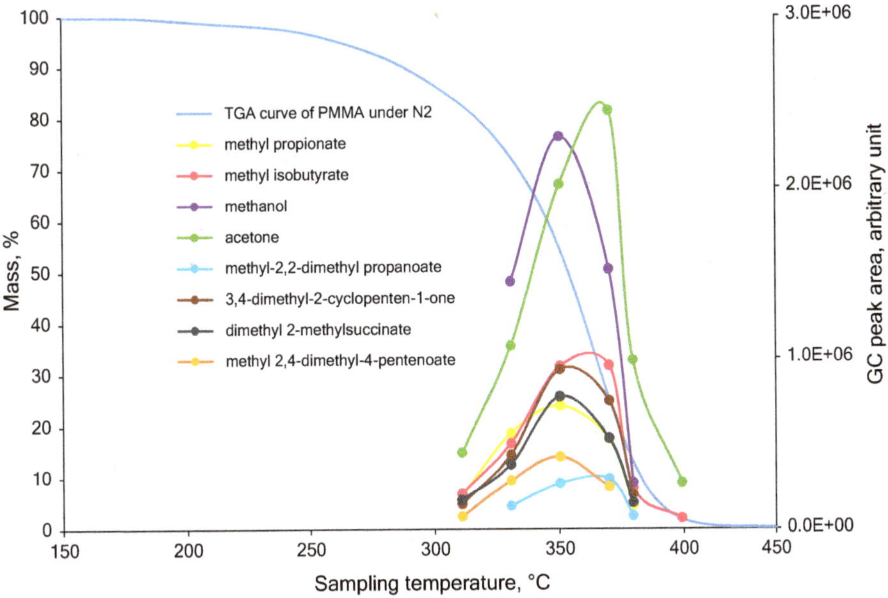

Fig. 8.36: TGA curve of PMMA under N_2 overlaid with the evolution of different identified VOCs, other than the major compounds (methyl methacrylate and methacrylic acid).

among all detected VOCs, by taking into account the sum of all peak areas of the emitted VOCs and by supposing that no other gases (CO_2, for example) have been emitted. Table 8.4 summarizes the results of this estimation. The monomer percentage is estimated to be 98, which is quite close to other values given in the literature [27], comprised between 91% and 98%.

Tab. 8.4: Estimated percentages of VOC identified by TGA-storage interface-GC-MS.

	Sum of all peak areas between 230 and 470 °C (unit area)	Estimated percentages (%)
Methyl methacrylate (monomer)	3,783,288,769	98.4
Methacrylic acid (*second major VOC*)	29,691,434	0.8
Other minor VOC	32,263,417	0.8
All VOC	3,845,243,620	100

8.6.2.4 TGA–micro-GC-MS coupling (TGA-μGC-MS)

Principle

The basic principle of a micro gas chromatograph (μGC) is the same as that of a conventional GC. Differences come from smaller sizes of injector, column, and detector. The miniaturization leads to a decrease in power consumption and a decrease in the analysis time [28].

The analysis time to record one or several chromatograms (if the μGC has several columns) is typically 2–3 min depending on gases to be separated. As an example, if hydrogen is the gas of interest, its separation is less than 1 min.

The μGC is generally equipped with a non-destructive **thermal conductivity detector (TCD)** or **catharometer**, which ensures quantitative results over a wide range of concentrations. TCD operation depends on the different thermal conductivities of various gaseous species. The identification of separated gases is done by retention time or, in the case of complex mixtures, by coupling the μGC with a MS detector.

The μGC can be equipped with several columns, each with its specificity towards the compounds to be separated. Generally, the gases separated with a μGC are **permanent** (H_2, CO, CO_2, CH_4, H_2O, O_2, N_2), **light** (hydrocarbons C1 to C3, light solvent, H_2S, NH_3, COS, etc.), or **medium** (hydrocarbons C4 to C8, BTEX, small acids, furan, etc.) depending on the column nature. The μGC columns have lengths comprised between 4 and 20 meters, and an isothermal temperature value can be chosen between 45 and 160 °C based on the column specificity and the nature of compounds to be separated. As an example, **molecular sieve** column separates permanent gases very well. **PoraPlot-U**, which is a polar porous capillary column, separates CO_2, SO_2, CS_2, H_2S, NH_3,

H_2O, C1–C3, halogenated compounds, ketones, and solvents. Each column is connected to a TCD, and together they form a module, and thus the µGC has one or few modules.

When identification of gases cannot be based only on the retention time, the coupling of the µGC with a mass spectrometer detector (MSD) is necessary. In this case, only one module can be connected to the MSD.

The µGC can be connected to the TGA oven via a transfer line, which can be heated to 80 or 90 °C. The gas flow from the TGA is distributed equally between the modules installed in the µGC [15].

Results presentation

If the µGC is calibrated with the gas of interest (e.g., calibration gas cylinder with 2,000 ppm of H_2 in nitrogen or with 500 ppm methanol in nitrogen), one can plot the concentrations of volatile compounds as a function of temperature or time, together with TGA. In the absence of calibration, the µGC peak area of the compound can be plotted in function of temperature or time. The obtained emission profile can be superimposed with that of TGA. This also gives information on the maxima and minima of concentrations of gases along with temperature. The concentrations measured in µGC are in a broad range, from ppm to 100%.

Evolved gases dilute in the gas flow of TGA. This should be taken into account in the calculation of the concentration of gases for quantitative analyses. In addition, the dilution of gases leads to a weaker detection when traces of gases are emitted from the TGA oven.

Example

EGA of a biomass sample by TGA-µGC

In the example below, two forms of results presentation are shown for a sample of biomass (wood pellets) that has been analyzed by TGA-coupled µGC from ambient up to 750 °C under nitrogen. The µGC used in this study, commercialized by SRA Instruments, was equipped with different columns allowing the separation of the emitted gases as follows: CO, CO_2, CH_4, and H_2 on ***molecular sieve column***, then ethane, ethylene, and H_2O on ***PoraPlot-U column***. Since the µGC was calibrated, the concentration emission/profile of the evolved gases could be plotted against temperature as shown in Fig. 8.37a,b. For reasons of visibility, the evolution curves for CO, CO_2, and H_2O are not shown here.

The first type of profile (Fig. 8.37a) shows the maximum and minimum concentrations for each gas.

The second one (Fig. 8.37b) gives information on the total concentration reached for each gas. Also, we can read the concentrations obtained at a given temperature.

The evolution profiles for water, CO, and CO_2 (not shown here) follow well the TGA mass loss, being emitted from the beginning of the decomposition up to the end.

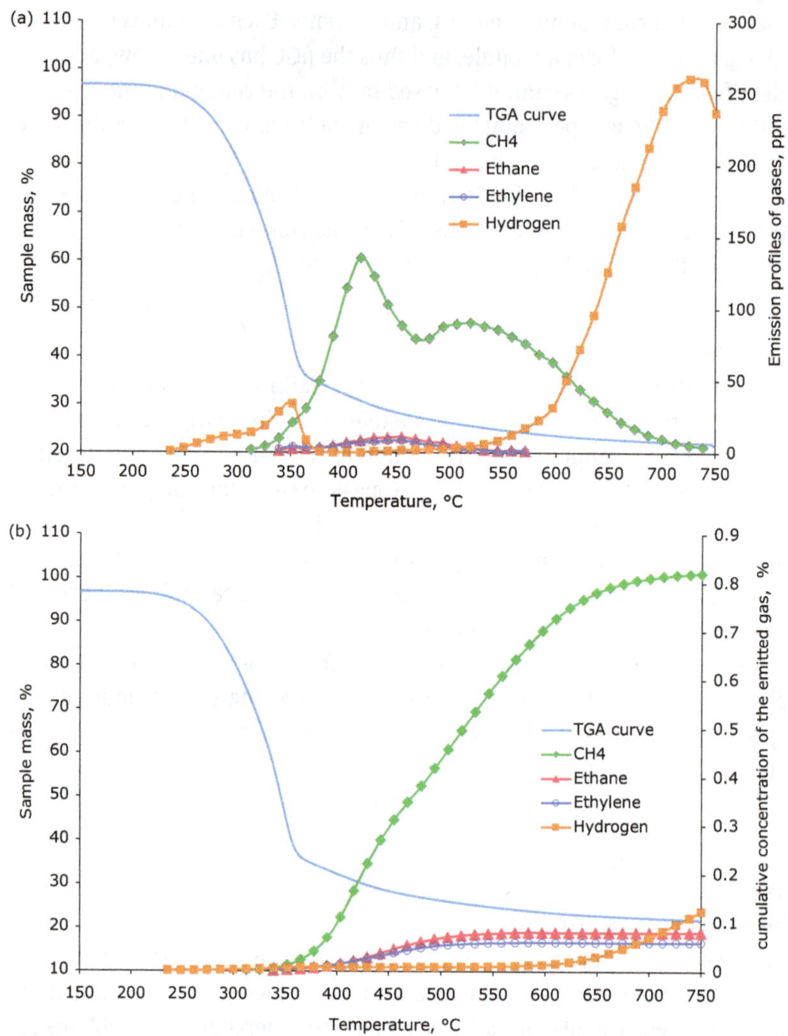

Fig. 8.37: EGA of a biomass sample (wood pellets) by TGA-μGC at 5 °C.min^{-1}: (a) the evolution profile of gas concentrations in ppm with temperature; (b) the profile of the cumulative concentrations of the emitted gases, in %.

This has also been reviewed in the literature [29, 30]. They are emitted in much higher concentrations compared with the other gases shown here (i.e., 25% for water, 10% for CO_2 and 5% for CO). Hydrogen (Fig. 8.37a) starts to be slowly emitted at 230 °C, then its concentration increases more after 600 °C, reaches a maximum at 726 °C, and then decreases. CH_4 is emitted between 300 and 750 °C with an evolution profile in two steps (two maxima of concentration). Ethane and ethylene, emitted between 340 and 570 °C, have the same emission profile with very low concentrations, below 0.1%.

The correlation of the information obtained by TGA (mass variation, DTG and DSC curves) and µGC (nature of compounds with their emission profile) provides a lot of insight into the mechanism of decomposition of a material and an in-depth understanding of the phenomena that take place.

8.6.3 EGA conclusion

To conclude this part, by coupling TGA to MS, FTIR, GC-MS, or µGC-MS instruments clearly enhance the basic information given by TGA alone.

EGA offers the ability to solve a wide range of problems and applications based on the determination of the composition of complex gas mixtures produced from the decomposition of different types of materials.

All these couplings are powerful techniques that yield both quantitative (mass loss) and qualitative (identification) information about the gaseous products released during a TGA measurement. Each coupling has its own specificity. Interpretation of infrared spectra or mass spectra is more complicated when several unknown gases are simultaneously released. TGA-FTIR and TGA-MS are mainly used for simple mixtures of gases when few molecules have to be identified and the other decomposition gases only occur at low concentrations. Another application concerns the detection of residual solvents in active pharmaceutical ingredients. For complex gaseous mixtures, TGA-GC-MS and TGA-µGC-MS are the separation techniques most often used. The last one is more appropriate for small molecules (permanent gases, light solvents, light, and medium VOC) and can be used without MS when the gases of interest are known and already identified by their retention times.

8.7 Conclusion

Thermal analysis and calorimetry provide powerful tools for the characterization of polymers. Thermodynamic transitions, thermal stability, decomposition, and chemical reactions can be accurately identified and quantified over a wide range of temperature.

In this chapter, the most used thermal techniques for the characterization of polymers and different examples of applications have been listed, but this list is not exhaustive. Other thermal analyses techniques, such as dynamic mechanical analysis (DMA), flash DSC, or thermal-microscopy coupled DSC, are also used in the field of polymer characterization.

The development of all these techniques and couplings between different instruments has greatly contributed to the understanding of the physical and chemical phenomena that occur within materials when they are synthesized, shaped, heated, burned, or degraded. Today, these tools are increasingly used in the field of polymer recycling. They undeniably allow a better understanding of the mechanisms involved in the new recycling processes explored in research and development laboratories.

References

[1] Staudinger, H. Über Polymerisation. *Ber. Dtsch. Chem. Ges. A/B*. **1920**, *53(6)*, 1073–1085, doi.org/10.1002/cber.19200530627.

[2] Avérous, L.; Caillol, S.; Cramail, H. Polymères Biosourcés. Principaux Enjeux et Perspectives. *L'actualité Chimique*. **2017**, *422–423*, 68–75.

[3] Rouquerol, J. Les Applications Actuelles de La Calorimétrie. *L'actualité Chimique*. **2019**, *441*, 15–59.

[4] Hale, A. Chapter 9: Thermosets. In *Handbook of Thermal Analysis and Calorimetry, Volume 3: Applications to Polymers and Plastics*. Elsevier Science B. V.: AKRON; 2002, Vol. 3, pp. 295–351.

[5] Gracia-Fernández, C.; Tarrío-Saavedra, J.; López-Beceiro, J.; Gómez-Barreiro, S.; Naya, S.; Artiaga, R. Temperature modulation in PDSC for monitoring the curing under pressure. *J. Therm. Anal Calorim*. **2011**, *106(1)*, 101–107, doi.org/10.1007/s10973-011-1361-8.

[6] Widmann, G.; Riesen, R. *Thermal Analysis: Terms, Methods, Applications*. Hüthig: Heiselberg, 1987.

[7] Grenet, J.; Legendre, B. Analyse Calorimétrique Différentielle à Balayage (DSC). *Techniques de l'Ingénieur*. **2010**, No(*P1205*), 1–27.

[8] Wunderlich, B. *Thermal Analysis*. Academic Press; 1990.

[9] Saddawi, A.; Jones, J. M.; Williams, A.; Wójtowicz, M. A. Kinetics of the Thermal Decomposition of Biomass. *Energy Fuels*. **2010**, *24(2)*, 1274–1282, doi.org/10.1021/ef900933k.

[10] Vaidya, S.; Veselska, O.; Zhadan, A.; Diaz-Lopez, M.; Joly, Y.; Bordet, P.; Guillou, N.; Dujardin, C.; Ledoux, G.; Toche, F.; Chiriac, R.; Fateeva, A.; Horike, S.; Demessence, A. Transparent and luminescent glasses of gold thiolate coordination polymers. *Chem. Sci*. **2020**, *11(26)*, 6815–6823, doi.org/10.1039/D0SC02258F.

[11] Ze Bing, L.; Bois, L.; Grosgogeat, B.; Chassagneux, F.; Toche, F.; Chiriac, R.; Desroches, C.; Colon, P.; Brioude, A. Insertion of Triethyleneglycol Dimethacrylate inside Mesoporous Silica for Composites Elaboration. *Microporous Mesoporous Mater*. **2012**, *160*, 41–46, doi.org/10.1016/j.micromeso.2012.04.055.

[12] Duperrier, S.; Chiriac, R.; Sigala, C.; Gervais, C.; Bernard, S.; Cornu, D.; Miele, P. Thermal behaviour of a series of Poly[B-(Methylamino)Borazine] for the preparation of boron nitride fibers. *J. Eur. Ceram. Soc*. **2009**, *29(5)*, 851–855, doi.org/10.1016/j.jeurceramsoc.2008.07.012.

[13] Chiriac, R.; Sigala, C.; Miele, P. Thermal Analyses Applied to the Preparation of Complex Shaped Ceramics Derived from Preceramic Polymers. In *Design, Processing and Properties of Ceramic Materials from Preceramic Precursors*. Nova Science Publishers, Inc.: New York, 2012, pp. 185–203.

[14] Lever, T.; Haines, P.; Rouquerol, J.; Charsley, E. L.; Van Eckeren, P.; Burlett, D. J. ICTAC nomenclature of thermal analysis (IUPAC Recommendations 2014). *Pure Appl Chem*. **2014**, *86(4)*, 545–553, doi.org/10.1515/pac-2012-0609.

[15] Fedelich, N. *Evolved Gas Analysis*. 2nd EditionCollected Applications Thermal Analysis, *Mettler-Toledo*, 2019.

[16] Charsley, E. L.; Walker, C.; Warrington, S. B. Applications of a new quadrupole mass spectrometer system for simultaneous thermal analysis-evolved gas analysis. *J. Therm. Anal*. **1993**, *40(3)*, 983–991, doi.org/10.1007/BF02546857.

[17] Price, D. M.; Hourston, D. J.; Dumont, F. Thermogravimetry of Polymers. In Meyers, R. A., Ed.; *Encyclopedia of Analytical Chemistry*. John Wiley & Sons, Ltd: Chichester, UK; 2006, p. a2037, doi.org/10.1002/9780470027318.a2037.

[18] Benhammada, A.; Trache, D. Thermal decomposition of energetic materials using TG-FTIR and TG-MS: A state-of-the-art review. *Appl. Spectrosc. Rev.* **2020**, *55(8)*, 724–777, doi.org/10.1080/05704928.2019.1679825.

[19] Materazzi, S.; Vecchio, S. Evolved gas analysis by infrared spectroscopy. *Appl. Spectrosc. Rev.* **2010**, *45 (4)*, 241–273, doi.org/10.1080/05704928.2010.483664.

[20] Marcilla, A.; Gómez, A.; Menargues, S. TG/FTIR Study of the Thermal Pyrolysis of EVA Copolymers. *J. Anal. Appl. Pyrolysis*. **2005**, *74(1–2)*, 224–230, doi.org/10.1016/j.jaap.2004.09.009.

[21] Sultan, B.-Å.; Sörvik, E. Thermal Degradation of EVA and EBA – A Comparison. III. Molecular Weight Changes. *J. Appl. Polym. Sci.* **1991**, *43(9)*, 1761–1771, doi.org/10.1002/app.1991.070430919.

[22] Post, E.; Rahner, S.; Möhler, H.; Rager, A. Study of recyclable polymer automobile undercoatings containing PVC using TG/FTIR. *Thermochim. Acta.* **1995**, *263*, 1–6, doi.org/10.1016/0040-6031(94)02388-5.

[23] Beneš, M.; Plažek, V.; Balek, V. Lifetime simulation and thermal characterization of PVC cable insulation materials. *J. Therm. Anal Calorim.* **2005**, *82*, 761–768, doi.org/10.1007/s10973-005-0961-6.

[24] Matuschek, G.; Milanov, N.; Kettrup, A. Thermoanalytical investigations for the recycling of PVC. *Thermochim. Acta.* **2000**, *361(1–2)*, 77–84, doi.org/10.1016/S0040-6031(00)00549-9.

[25] Grob, R. L.; Barry, E. F. *Modern Practice of Gas Chromatography*. 4th Edition, 2004.

[26] Oehme, M. *Practical Introduction to GC-MS Analysis with Quadrupoles*. Hüthig: Heidelberg, 1998.

[27] Beyler, C. L.; Hirschler, M. M. Thermal decomposition of polymers. *SFPE Handb. Fire Prot. Eng.* **2002**. *2*, pp, 111–131.

[28] Regmi, B. P.; Agah, M. Micro gas chromatography: an overview of critical components and their integration. *Anal. Chem.* **2018**, *90(22)*, 13133–13150, doi.org/10.1021/acs.analchem.8b01461.

[29] Moreno, A. I.; Font, R. Pyrolysis of furniture wood waste: Decomposition and gases evolved. *J. Anal. Appl. Pyrolysis*. **2015**, *113*, 464–473, doi.org/10.1016/j.jaap.2015.03.008.

[30] Slopiecka, K.; Bartocci, P.; Fantozzi, F. Thermogravimetric analysis and kinetic study of poplar wood pyrolysis. *Appl.Energy*. **2012**, *97*, 491–497, doi.org/10.1016/j.apenergy.2011.12.056.

Jyoti Shanker Pandey*, Asheesh Kumar*, and Nicolas von Solms

Chapter 9
Role of calorimetry in clathrate hydrate research

Abstract: This chapter elaborates on the role of calorimetry in gas hydrate studies. Differential scanning calorimetry (DSC) has been considered an innovative tool to measure the thermophysical properties of various gas hydrate systems, which may have applications in energy recovery from marine natural hydrate resources, energy storage, hydrate structure elucidation, and assessment of hydrate inhibitors/promoters. A brief introduction of gas hydrates along with an elaborate discussion on the application of DSC in various hydrate-based phenomena have been presented. Further, challenges and the recent advancements in the high-pressure calorimetry have been outlined.

Keywords: Clathrate hydrates, calorimetry, hydrate inhibitors, dissociation enthalpies, high-pressure micro-DSC

9.1 Introduction

9.1.1 Introduction to gas hydrates and their calorimetric analysis

Gas hydrates or clathrate hydrates are ice-like crystals formed by gas and water molecules under high-pressure and low-temperature conditions. Depending on the guest molecules and thermodynamic conditions, these crystals can materialize in three main structures (i) cubic structure I (sI), (ii) cubic structure II (sII), and (iii) hexagonal structure (sH) [1]. In 1934, Hammerschmidt reported gas hydrates as a nuisance due to their formation, deposition, and plugging in natural gas pipelines [2]. Later in the 1960s, natural gas hydrates were reported as natural gas resources because of vast amounts of hydrate deposits (marine/permafrost) around the world. However, understanding the

*Corresponding authors: **Jyoti Shanker Pandey**, Center for Energy Resource Engineering (CERE), Department of Chemical Engineering, Technical University of Denmark, 2800 Kgs. Lyngby, Denmark, e-mail: jyshp@kt.dtu.dk (JSP)
*Corresponding authors: **Asheesh Kumar**, Upstream and Wax Rheology Division, CSIR-Indian Institute of Petroleum, Dehradun 248005, Uttarakhand, India, e-mail: asheesh.kumar@iip.res.in (AK)
Nicolas von Solms, Center for Energy Resource Engineering (CERE), Department of Chemical Engineering, Technical University of Denmark, 2800 Kgs. Lyngby, Denmark

https://doi.org/10.1515/9783110590449-009

thermodynamic properties of gas hydrates was the major challenge in the 1960s due to the lack of high-pressure thermal analytics [3, 4]. Therefore, to determine the enthalpies of hydrate dissociation, the Clapeyron equation was employed while utilizing the hydrate phase equilibrium pressure-temperature data. However, it was impracticable to determine such properties through calorimetric measurements since gas hydrates are unstable under ambient conditions. Later in 1986–1988, Handa employed an automated Tian-Calvet heat-flow calorimeter to assess the enthalpies of dissociation, compositions, and heat capacities of natural gas hydrates of structure I and structure II hydrate recovered from the Middle America Trench slope sediment off Guatemala and Green Canyon area of the northern Gulf of Mexico, respectively [5]. Later, calorimetric measurements were utilized for various gas hydrate applications (refer to Fig. 9.1), such as (i) to measure the hydrate phase equilibrium conditions of various hydrate formers, (ii) to acquire the thermal properties of tetra-*n*-butyl ammonium bromide (TBAB) hydrate-tetrahydrofuran (THF) hydrate mixture for cold storage application, (iii) to select the hydrate inhibitors having application in flow assurance, (iv) quantification of hydrate-in-oil emulsion stability, and (v) to distinguish the various hydrate structures.

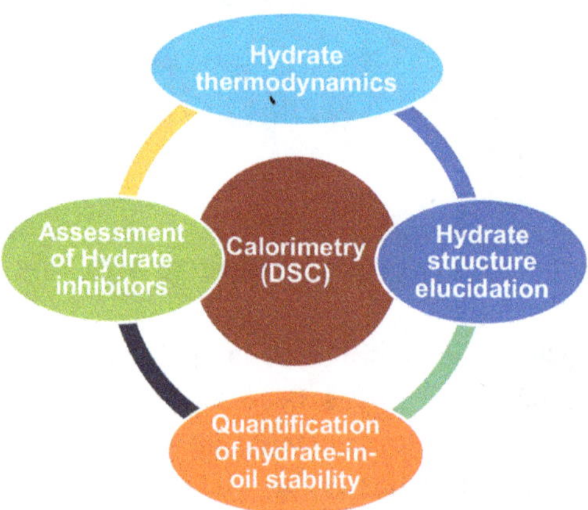

Fig. 9.1: Applications of calorimetry in gas hydrate research.

9.1.2 Introduction to differential scanning calorimetry

Differential scanning calorimetry (DSC) measures the difference in heat flow between a sample and a reference with respect to either temperature or time using a user-controlled temperature profile. DSC analysis makes it possible to understand the influence of many factors. DSC is a relative technique in which measurements are not made in the thermal equilibrium phase because of its dynamic temperature characteristics. In the first study of calorimetric measurement of gas hydrate, the hydrate was prepared ex-situ, and the sample was transferred to the calorimetric cell [5, 6]. This method was challenging due to the unstable nature of the hydrate phase and inaccurate weight measurement. Later, DSC enabled the in situ production of the hydrate by mixing water and gas under high pressure. However, DSC has its pitfalls, such as the lack of stirring, which reduces the contact area between liquid and water, poor gas solubility, high need for supercooling, and uncertainty about the amount of water converted to hydrate, which can, however, be overcome by multiple heating–cooling cycles [7]. These multiple cycles aid in removing the effect of metastable phase and enhance the water to hydrate conversion. Some of the essential DSC application includes

- Phase transitions/phase change such as melting, crystallization, sublimation, and evaporation
- Evaluation of heat capacity and kinetics
- Investigation of adsorption/desorption phenomena

The heat flux between calorimetric block (furnace) and each container is expressed by reference (9.1) and sample (9.2) (refer to Fig. 9.2)

$$\left(\frac{dq}{dt}\right)r = C_r * \frac{dT}{dt} \tag{9.1}$$

$$\left(\frac{dq}{dt}\right)s = C_s * \frac{dT}{dt} + \frac{dH}{dt} \tag{9.2}$$

where

- C_r and C_s are the heat capacity of reference and samples (with the container)
- dH/dt is the thermal power absorbed ($dh/dt > 0$ for endothermic) or released ($dh/dt < 0$ for exothermic) by the sample

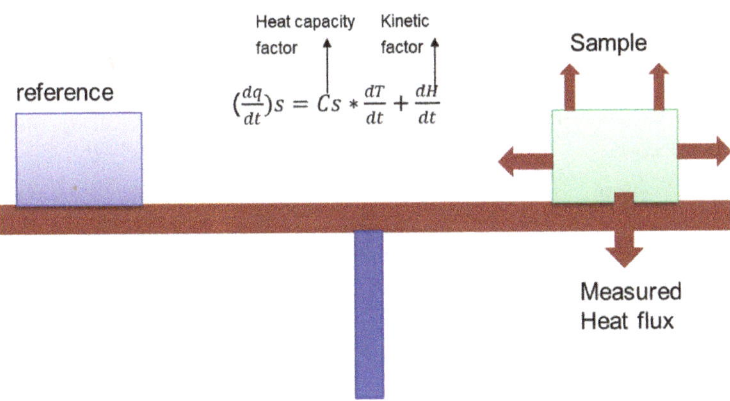

Fig. 9.2: Calorimetry fundamental principle.

9.1.3 Calorimetry-based key measurements

DSC is typically useful to measure the heat associated with phase transition (hydrate dissociation or hydrate formation) and heat capacities. Heat flow in the absence of any phase change corresponds to heat capacity. Enthalpy of formation and dissociation could be measured using DSC to provide valuable information about naturally occurring hydrate dissociation for gas production. The heat capacity quantifies the amount of heat stored in the material or released due to a change in the material phase [8]. Therefore, the heat capacity of clathrate hydrate could provide interesting information about the alignment of host molecules/guest molecules [7]. Many studies have been conducted to measure heat capacity [7–9] and enthalpy of formation and dissociation using heat flow calorimeters and DSCs [10–16].

Another way to calculate the enthalpy of formation and dissociation is to use the pressure–temperature diagram of the system and the Clapeyron equation or Clapeyron approximation. However, the theoretical method has proven to be of limited use in the case of semi-clathrates. A detailed discussion about validity and uncertainty can be found elsewhere in the literature [9]. Error could be due to the low amount of condensed hydrate phase and low gas solubility. Some of the calorimetric measurements and theoretical values (for CO_2 and CH_4 hydrates) are shown in Tab. 9.1.

Tab. 9.1: Dissociation enthalpy of methane hydrates (*n*H = hydration number).

	T (K) and *P* (MPa)	Additives	n_H*	ΔH_{dis} (kJ/mol of gas)	References
For CH_4					
CHFC	273 and 0.1		n.a.	54.19	[11]
	278 and 4.1		5.97	57.65	[12]
	283 and 7.1		5.98	53.24	[12]
	240–273 and 0.036–0.067	0.7A silica	5.94	45.92	[13]
	274 and 0.1		6.38	56.84	[14]
	279–282 and 5		n.a	55.3	[15]
CHF-DSC	273 and 0.1		n.a	51.6	[16]
	281–293 and 5.5–19		n.a	52.21–56.50	[17]
For CO_2					
CHFC	274 and 0.1		7.23	65.22	[14]

H_{diss}, enthalpy of dissociation; CHFC, Calvet heat flow calorimeter; CHF-DSC, Calvet heat flow differential calorimeter.

Some hydrate studies have published heat capacity data too. However, heat capacity is more difficult to measure and have large uncertainty compared to heat associated with phase change measurement [6]. Heat signal associated with heat capacity has a lower amplitude compared to phase change amplitude. Moreover, heat capacity depends on the pressure and temperature conditions due to variation in hydrate compositions. Some challenges associated with measuring heat capacity are [10] (i) the vapour pressure of gas hydrates is temperature dependent, so increasing the temperature will cause the hydrates to dissociate and consequently, the apparent heat capacity will be higher than the actual value, and (ii) the pressure of the free hydrate forming components can significantly affect the measured heat capacity of the hydrates.

9.1.4 Differential scanning calorimetry setup

Two different types of DSC are currently available on the market, Standard DSC and Calvet DSC. The plate-shaped differential thermocouples are available in standard DSC, while thermopile composed of multiple thermocouples are available in Calvet DSC. In standard DSC, contact with the sensor is through the bottom area of the crucibles, while in Calvet DSC, contact with the sensors occurs all around the crucibles. Standard DSC could be used up to 1,600 °C, while Calvet DSC offers very high sensitivity and accuracy. The sample size is much bigger in Calvet than in standard. In Calvet DSC, heat flow measurement becomes quantitative and less dependent on calibrations and on the crucible type. Both types of DSC produce a differential signal between two cells proportional to a heat flux rate, while classical calorimetry measures the heat flux [18].

Moreover, the highly sensitive microdifferential calorimeter (μDSC) uses a semiconductor-type detector (Peltier). It is known for very accurate temperature control of the calorimetric block. However, it is used only for a limited range of temperatures. Therefore, the sample size is intermediate. In μDSC, Peltier elements are used to heat or cool the μDSC calorimetric block, and a separate heat exchanger coupled to the peltier collects the heat from the sides of the μDSC (heat sink). The cells inside the μDSC could be varied as per application, for example, standard cells allow for the transition, isothermal stability, while mixing cells (discussed in the later sections) allow measuring the heat of mixing, wetting (immersion), dissolution, and adsorption.

Some of the main advantages of DSC are (i) dynamic operating mode, (ii) minimal/compensated temperature variation, and (iii) high sensitivity for enthalpy/heat capacity measurements. Despite the advantages, there are some limitations associated to DSC: (i) lack of in situ mixing between two materials and (ii) problem of representativeness and heterogeneity (due to small sample size).

9.1.5 Key influencing factors

A typical DSC curve (also known as DSC thermograms) shows the relationship between the measured heat flux rate dH/dt (on the Y-axis) and the temperature or time (on the X-axis). The DSC technique allows us to measure heat flows, and the temperature measurement corresponds to the phase change (from the heat flow thermograms). The key factors that influence the DSC curve include sample type, experimental condition, and instrumental factor. The sample could be affected by the sample properties, sample size, dilution, sample thermal history, and reference material properties. The critical variation in experimental conditions includes the rate of temperature rise, atmosphere/gas type, and properties. The rate of temperature rise also affects the shape of the DSC curve and phase transition enthalpy. Prolonged heating or cooling rate ensures a quasi-homogeneous temperature profile inside the sample holder and reduces the uncertainty attached to the measurement; however, stepwise heating has suggested a better method to measure the melting temperature than constant rate [19]. In such cases, completion of hydrate melting step could be calculated by the heat signal and the pressure response, which is in good agreement with hydrate modeling software such as CSMGem. Another variation with temperature change includes the two melting runs at different heating rates (0.2 and 0.5 K/min) with the same sample that could be faster than stepwise method and more accurate than single cooling rate experiment [19–21].

9.2 Calorimetric assessment of clathrate hydrate thermodynamics

Understanding the thermodynamic properties of gas hydrate systems is one of the critical factors for various gas hydrate applications/problems. In this direction, high-pressure DSC played an essential role in determining the thermodynamic properties and kinetics of hydrate formation while detecting the phase transitions with respect to time, temperature, and pressure. In this direction, Parlouër et al. [22] reported a good correlation between the methane hydrate dissociation temperature data obtained with DSC and other PVT methods available in the open literature [22]. Figure 9.3 illustrates the typical hydrate dissociation thermograms and the technique used to measure the hydrate onset temperature (dissociation temperature). Refer to the tangent dashed blue line and the cross-over point denoting the hydrate dissociation temperature at 10 MPa (100 bar). The measured dissociation temperatures are plotted against the methane pressure as presented in the figure. It correlates well with the literature hydrate phase equilibrium data.

Moreover, HP-DSC has been widely employed to quantify the hydrate thermal properties (specific heat/heat of dissociation) at natural gas hydrate reservoir pressure and temperature conditions. These properties have offered new insight in predicting the gas production rates during hydrate exploitation from hydrate reservoirs. In this direction, Gupta et al. [17] measured the heat of dissociation for methane hydrate while employing a high-pressure DSC and reported a constant valve 54.44 ± 1.45 kJ/mol gas or 504.07 ± 13.48 J/gm water or 438.54 ± 13.78 J/gm hydrate for a pressure ranging up to 200 bar. Moreover, they observed that the simplified Clausius-Clapeyron equation predictions are not well aligned with the DSC measurements at high pressures; however, the Clapeyron equation predictions corroborated sufficiently with the measured heat of dissociation.

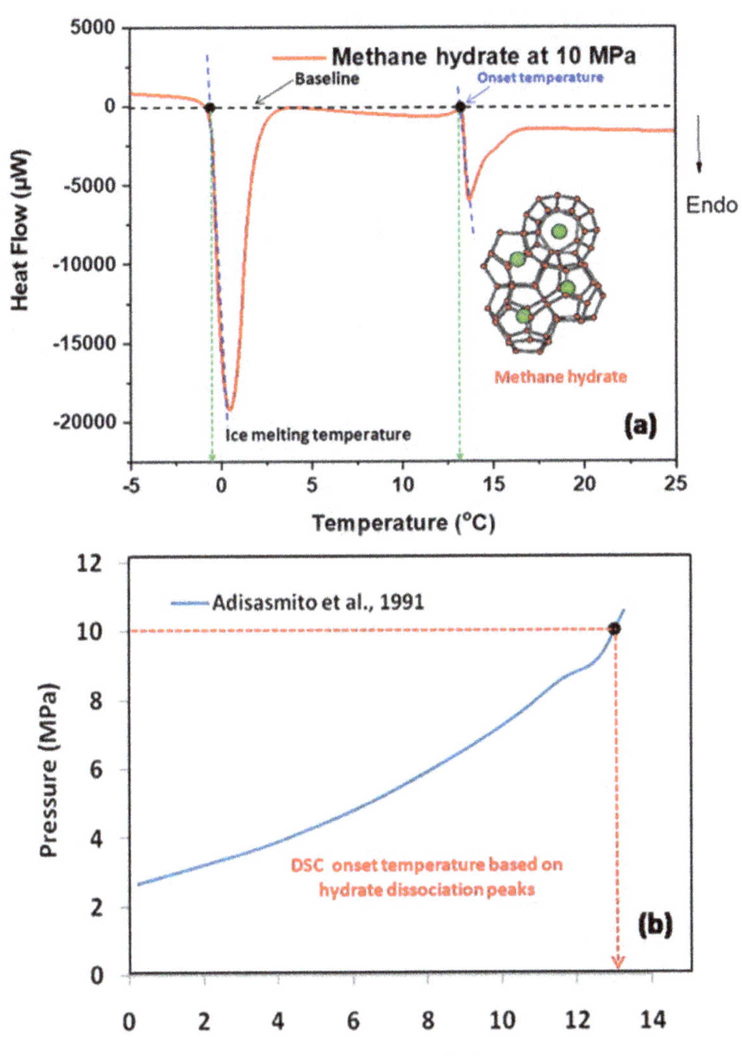

Fig. 9.3: Typical hydrate dissociation thermograms along with the measured phase equilibrium condition of methane hydrate and its comparison with literature PVT data (reprinted (adapted) from [23], copyright (1991), with permission from American Chemical Society).

9.3 Assessment of hydrate inhibitors using calorimetric methods

As discussed in Section 9.1, hydrate formation/deposition inside production and transport pipelines has been a severe problem in the oil and gas industry, which requires a good understanding of pipeline thermodynamics and hydrate inhibitor performance at pipeline conditions. In this direction, high-pressure (HP) DSC has been considered a pioneering tool to probe the statistically reliable time-dependent hydrate formation and decomposition temperature and enthalpy of dissociation, which can be directly applied to investigate the kinetic inhibitor performance. Further, the requirement of a tiny amount of sample (1–10 μL) is another advantage for assessing hydrate inhibitors. Nagu et al. [24] demonstrated the performance of various kinetic hydrate inhibitors and synergists while utilizing an HP-DSC and reported that small amounts of polyethylene oxide (PEO) improved the performance of Luvicap EG (kinetic hydrate inhibitor) in both hydrate nucleation and growth inhibition. Further, they revealed the importance of HP-DSC for selecting potential hydrate inhibitors, which are available in only limited quantities (e.g., antifreeze proteins). Ohno et al. [25] have also utilized HP-DSC to evaluate the performance of chemical/biological kinetic inhibitors in a confined silica gel media. Further, plenty of literature is available in this direction.

9.4 Quantification of hydrate-in-oil stability using DSC

In context to hydrate-based flow assurance problems, entrainment of water/oil droplets and emulsification is one of the critical stages that decide the hydrate formation and deposition in the oil and gas pipelines. Therefore, it is crucial to understand the stability of the emulsified water and its effect on hydrate formation. In this direction, DSC has been utilized as a unique analytical tool to understand the emulsion destabilization from repeated hydrate formation and dissociation cycles (based on endothermic peak area). It is proposed that during hydrate dissociation from the first hydrate formation cycle, the released gas bubbles produce little turbulence, allowing the coalescence of water droplets (refer to Figs. 9.4a and b). Therefore, in the next hydrate formation cycle, less hydrate was formed due to the formation of larger water droplets (DSC peak area reduced). Thus, one can measure the emulsion stability quantitatively based on the integrated area of the hydrate-melting peak of a given cycle. As can be seen in Fig. 9.4c, the integrated area of hydrate dissociation peak in scan 4 (cycle 4) is around half of cycle 2 (scan 2), which indicates a 50% reduction in emulsion stability [22, 26].

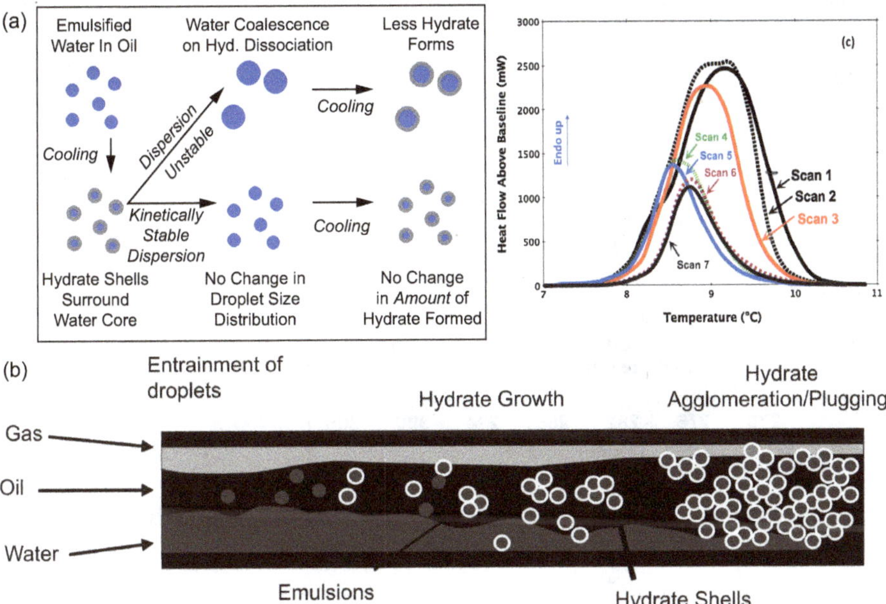

Fig. 9.4: (a) Effect of droplet sizes for kinetically stable/unstable water-in-oil dispersion during repeated hydrate formation–dissociation cycles (reprinted from [27], copyright (2014), with permission from Elsevier). (b) Mechanism of hydrate formation and agglomeration in oil dominated system (reprinted from [26], copyright (2008), with permission from Elsevier). (c) Multiple cycles of DSC hydrate dissociation thermograms for a 30 vol% water-in-oil emulsion.

9.5 Role of calorimetry in hydrate structure elucidation

9.5.1 How to distinguish the different hydrate structures using DSC thermograms

Calorimetry has been considered an innovative instrument to elucidate the hydrate structures or distinguish the ice from hydrate crystals. More precisely, one can easily determine the hydrate structures based on their phase equilibrium conditions (dissociation temperature). For example, Fig. 9.5 exemplifies a hydrate/ice dissociation thermogram of a water-THF-methane system with three endothermic downward peaks wherein the first peak at ~ 273 K can be attributed to ice melting, and the integrated area of the peak could provide an approx amount of ice formed in this system. Further, the second and third endothermic peaks at ~ 283 K and ~ 302 K correspond to structure I (pure methane hydrate) and structure II (mixed methane-THF hydrate). Moreover, an exothermic upward peak after ice melting can be designated to hydrate

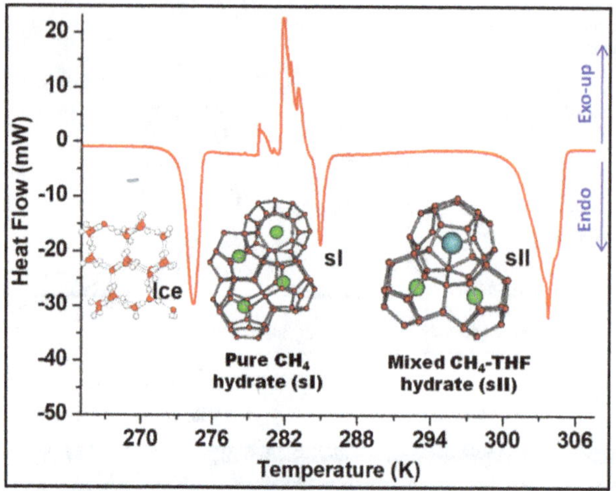

Fig. 9.5: DSC dissociation thermogram to distinguish ice and hydrate structures (sI and sII) (reprinted from [28], copyright (2019), with permission from Elsevier).

re-formation from the melting ice. More detailed information can be found in our publications [28, 29].

9.5.2 CH_4-CO_2 replacement process using a differential scanning calorimeter

To recover the natural gas from marine natural gas hydrate deposits, a unique concept of methane-carbon dioxide (CH_4–CO_2) swapping/replacement was proposed, wherein CO_2 could be sequestered while replacing the CH_4 molecules from hydrate cages. To understand the swapping efficiency, DSC would offer fertile information. In this direction, Lee et al. [30, 31] investigated the CH_4–CO_2 swapping phenomenon while measuring the thermal behavior/ΔH_d values of mixed CH_4–CO_2 hydrates and employing the HP μ-DSC. They measured the heat of dissociation while integrating the endothermic hydrate dissociation peak and combining it with the hydration number. During the replacement process, it was observed that the ΔH_d value of CH_4–CO_2 hydrates increased with increase in CO_2 composition in the hydrate phase (refer to Fig. 9.6) while occupying the large cages of sI hydrates. Based on the above calculations, they reported the CO_2 replacement of 68 ± 2%. Ridzy et al. [16] and Kwon et al. [32] have also measured the ΔHd value of CH_4-CO_2 hydrates using DSC and the Clausius-Clapeyron equation. The ΔH_d values of pure CH_4 and pure CO_2 hydrates were 54.1 ± 0.2 and 57.1 ± 0.1 kJ/mol of gas, respectively, and the average ΔHd value of the CH_4–CO_2 hydrates after the CO_2 replacement was measured to be 55.5 ± 0.2 kJ/mol of gas.

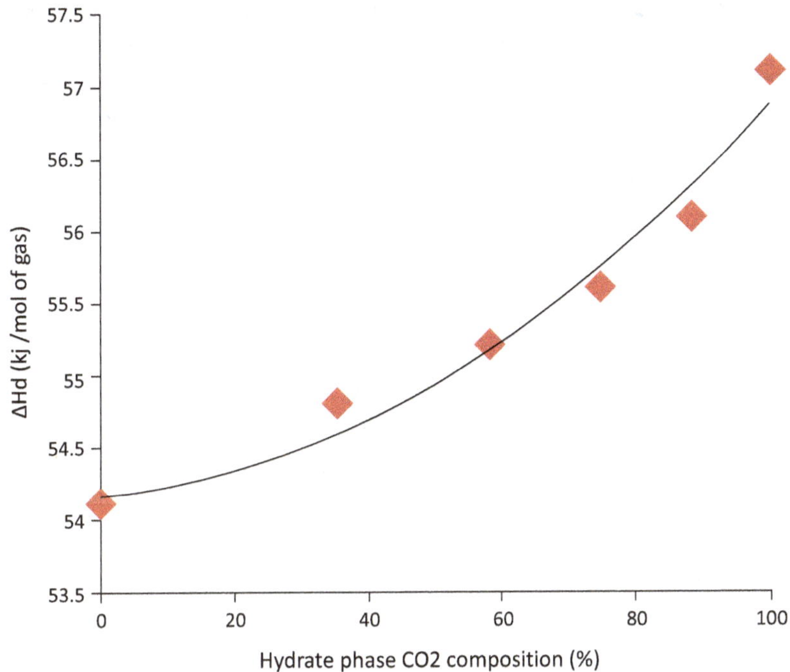

Fig. 9.6: Hydrate dissociation enthalpies (ΔH_d) of mixed CH_4–CO_2 hydrates for various hydrate phase CO_2 compositions (data adapted from [30, 31], copyright (2013), and copyright(2014), with permissions from American Chemical Society and Elsevier, respectively).

9.6 Thermal properties of gas hydrates

For the calorimetric measurement, Handa et al. [11] used Tian-Calbet heat flow calorimetry and calculated that the enthalpy of CH_4 hydrate dissociation into liquid water and gas (h → l + g) was equal to 54.2 kJ/mol. Gupta et al. [17] measured similar values for the enthalpy of dissociation using a DSC and found no significant variations in the measured values for pressure changes up to 20 MPa. In addition to pure CH_4 hydrate, the mixing of other guest molecules such as CO_2, C_2H_6, or C_3H_8 with CH_4 gas molecules was found to alter the dissociation enthalpy wherein molar enthalpy depends on the cage occupancy (CH_4 occupied small cages, while large cages were occupied by larger guest molecules) [16].

Interestingly, it was observed that when CH_4 hydrates formed in the presence of porous silica gel, the dissociation enthalpy decreased to 45.9 kJ/mol [13], which was lower than the value in the presence of the bulk phase (54.2 kJ/mol) [9]. This observation was later disputed by Anderson et al., who showed little change in dissociation enthalpy between bulk medium and porous medium (10–30 nm), also confirmed by the Clausius-Clapeyron equation under different pressure and temperature conditions [33]. The presence of a non-frozen liquid layer and pore wall was suggested as a reason for the apparent decrease in ΔH [34]. CO_2 hydrate dissociation enthalpy calculations show more significant uncertainty than CH_4 hydrates due to the considerable uncertainty associated with dp/dt above the ice point [35]. Experimental measurements and values derived from models could differ as thermodynamic models are sensitive to phase equilibrium, assumptions, and pressure–temperature conditions. Therefore, scientists rely on calorimetry measurements such as DSC as tools/methods preferred to calculate hydrate dissociation enthalpy [36].

Self-preservation has been studied in detail by Stern et al. [37, 38]. Under self-preservation, hydrates dissociate much more slowly at $T < 273$ K due to the ice-shielding effect. There are examples of DSC being used to test the self-preservation of CH_4 hydrate. For instance, Giavarani and Maccioni studied self-preservation by varying the ice content in the hydrate at low pressure (1–3 bar) and low temperature (269–272 K) using DSC and confirmed that ice in high concentration reduces the dissociation rate. The dissociation rate was further reduced at slightly higher pressure [39].

9.7 Recent research trend calorimetry applications

In addition to the calorimetric study of CH_4 or CO_2 hydrates, calorimetric studies are also being carried out in other emerging research areas such as refrigeration, gas separation, hydrogen storage and gas storage using MOFs. Table 9.2 summarizes some of the essential research work.

Tab. 9.2: Emerging hydrate based studies using calorimetry.

Application	Chemicals/conditions	Conclusions	References
Refrigeration	CO_2 + THF (tetrahydrofuran) $P = 0.2$–3.5 MPa THF = 3.8–15 wt%	At 280 K, 78.9% decrease in CO_2 equillbrium pressure observed when 3.8 wt% THF is present Dissociation enthalpy of CO_2 + THF hydrates was observed two times higher than that of CO_2 hydrate	[7, 40]
Gas separation	CO_2 + N_2 + TBAB (tetrabutylammonium bromide) TBAB = 0.17–0.35 wt%	A thermodynamic phase diagram for the gas mixture was generated in TBAB and as a function of pressure	[41]

Tab. 9.2 (continued)

Application	Chemicals/conditions	Conclusions	References
Hydrogen Storage	H_2 + THF (tetrahydrofuran) (P = 14.53 MPa, T = 273.15 K), THF = 3.0 mol%	H_2-THF binary hydrate synthesis wasconfirmed, and 1.875 g/L density proved feasible hydrogen storage	[42]
Hydrogen storage	H_2 + TBAOH (tetrabutylammonium hydroxide), P = 0–40 MPa, salt mol % = 0.0083–0.024	H_2 showed a strong stabilization effect dissociation temperature increased as pressure increased due to hydrogen enclathration into the TBAOH hydrate. A phase diagram was generated using DSC measurement and isochoric reactor experiments	[43]
Gas storage using MOFs	CH_4 + MOFs + water MOFs = HKUST, ZIF-8, ZIF-67	CH_4 storage improved – high water to hydrate conversion and reduction in hydrate nucleation time	[44, 45]

9.8 Challenges and recent advancements in calorimetry

As discussed in earlier sections, the major challenge associated with the HP-DSC is the unstirred cells which lead to delay in gas to liquid dissolution and heterogeneity in the hydrate nucleation, especially in emulsion assessments. Moreover, hydrate nucleation and growth in stagnant conditions were found to be significantly delayed.

To overcome these problems, Torré et al. [46] developed a high-pressure macrocalorimetric cell coupled with a mechanical stirrer, which can efficiently provide the thermophysical properties of gas hydrate systems. However, the sensitivity of stirred macrocalorimetric cells was not as good as unstirred micro-unstirred calorimetric cells. Therefore, Torré et al. [46] developed a novel stirred microcalorimetric cell (in situ mechanical stirrer). A stirring shaft was coupled to the motor by a magnetic coupling (refer to Fig. 9.7), which can reach upto a pressure of 150 bar and rotational speed up to 2,000 rpm.

Fig. 9.7: Advancement in HP-DSC: stirred microcalorimetric cell for high-pressure measurements (reprinted from [46], copyright (2020) with permission from Elsevier).

References

[1] Sloan, E. D. Fundamental principles and applications of natural gas hydrates. *Nature*. **2003**, *426*, 353–359, DOI:10.1038/nature02135.

[2] Hammerschmidt, E. G. Formation of gas hydrates in natural gas transmission lines. *Ind. Eng. Chem.* **1934**, *26*, 851–855, DOI:10.1021/ie50296a010.

[3] Sloan Jr., E. D.; Koh, C. A.; Koh, C. A. *Clathrate Hydrates of Natural Gases*; 3rd ed.; CRC Press: Boca Raton, FL.USA; 2007; ISBN 9780429129148.

[4] Aman, Z. M. Hydrate risk management in gas transmission lines. *Energy & Fuels*. **2021**, *35*, 14265–14282, DOI:10.1021/acs.energyfuels.1c01853.

[5] Handa, Y. P. A calorimetric study of naturally occurring gas hydrates. *Ind. Eng. Chem. Res.* **1988**, *27*, 872–874, DOI:10.1021/ie00077a026.

[6] Handa, Y. P.; Hawkins, R. E.; Murray, J. J. Calibration and testing of a tian-calvet heat-flow calorimeter enthalpies of fusion and heat capacities for ice and tetrahydrofuran hydrate in the range 85 to 270 K. *J. Chem. Thermodyn*. **1984**, *16*, 623–632, DOI:10.1016/0021-9614(84)90042-9.

[7] Delahaye, A.; Fournaison, L.; Marinhas, S.; Chatti, I.; Petitet, J.-P.; Dalmazzone, D.; Fürst, W. Effect of THF on equilibrium pressure and dissociation enthalpy of CO 2 hydrates applied to secondary refrigeration. *Ind. Eng. Chem. Res*. **2006**, *45*, 391–397, DOI:10.1021/ie050356p.

[8] Gabitto, J. F.; Tsouris, C. Physical properties of gas hydrates: A review. *J. Thermodyn.* **2010**, *2010*, 1–12, DOI:10.1155/2010/271291.

[9] Deschamps, J.; Dalmazzone, D. Hydrogen storage in semiclathrate hydrates of tetrabutyl ammonium chloride and tetrabutyl phosphonium bromide. *J. Chem. Eng. Data.* **2010**, *55*, 3395–3399, DOI:10.1021/je100146b.
[10] Rueff, R. M.; Dendy Sloan, E.; Yesavage, V. F. Heat capacity and heat of dissociation of methane hydrates. *AIChE J.* **1988**, *34*, 1468–1476, DOI:10.1002/aic.690340908.
[11] Handa, Y. Compositions, enthalpies of dissociation, and heat capacities in the range 85 to 270 K for clathrate hydrates of methane, ethane, and propane, and enthalpy of dissociation of isobutane hydrate, as determined by a heat-flow calorimeter. *J. Chem. Thermodyn.* **1986**, *18*, 915–921, DOI:10.1016/0021-9614(86)90149-7.
[12] Lievois, J. S.; Perkins, R.; Martin, R. J.; Kobayashi, R. Development of an automated, high pressure heat flux calorimeter and its application to measure the heat of dissociation and hydrate numbers of methane hydrate. *Fluid Phase Equilib.* **1990**, *59*, 73–97, DOI:10.1016/0378-3812(90)85147-3.
[13] Handa, Y. P.; Stupin, D. Y. Thermodynamic properties and dissociation characteristics of methane and propane hydrates in 70-.ANG.-radius silica gel pores. *J. Phys. Chem.* **1992**, *96*, 8599–8603, DOI:10.1021/j100200a071.
[14] Kang, S. P.; Lee, H.; Ryu, B. J. Enthalpies of dissociation of clathrate hydrates of carbon dioxide, nitrogen, (carbon dioxide + nitrogen), and (carbon dioxide + nitrogen + tetrahydrofuran). *J. Chem. Thermodyn.* **2001**, *33*, 513–521, DOI:10.1006/jcht.2000.0765.
[15] Nakagawa, R.; Hachikubo, A.; Shoji, H. Dissociation and specific heats of gas hydrates under submarine and sublacustrine environments. *6th Int. Conf. Gas Hydrates (ICGH 2008).* **2008**, 1–4.
[16] Rydzy, M. B.; Schicks, J. M.; Naumann, R.; Erzinger, J. Dissociation enthalpies of synthesized multicomponent gas hydrates with respect to the guest composition and cage occupancy. *J. Phys. Chem. B.* **2007**, *111*, 9539–9545, DOI:10.1021/jp0712755.
[17] Gupta, A.; Lachance, J.; Jr, E. D. S.; Koh, C. A. Measurements of methane hydrate heat of dissociation using high pressure differential scanning calorimetry. *Chem. Eng. Sci.* **2008**, *63*, 5848–5853, DOI:10.1016/j.ces.2008.09.002.
[18] Rycerz, L. Practical remarks concerning phase diagrams determination on the basis of differential scanning calorimetry measurements. *J. Therm. Anal. Calorim.* **2013**, *113*, 231–238, DOI:10.1007/s10973-013-3097-0.
[19] Lin, W.; Dalmazzone, D.; Fürst, W.; Delahaye, A.; Fournaison, L.; Clain, P. Accurate DSC measurement of the phase transition temperature in the TBPB–water system. *J. Chem. Thermodyn.* **2013**, *61*, 132–137, DOI:10.1016/j.jct.2013.02.005.
[20] Kousksou, T.; Jamil, A.; Zeraouli, Y.; Dumas, J.-P. DSC study and computer modelling of the melting process in ice slurry. *Thermochim. Acta.* **2006**, *448*, 123–129, DOI:10.1016/j.tca.2006.07.004.
[21] Kousksou, T.; Jamil, A.; Zeraouli, Y.; Dumas, J. P. Equilibrium liquidus temperatures of binary mixtures from differential scanning calorimetry. *Chem. Eng. Sci.* **2007**, *62*, 6516–6523, DOI:10.1016/j.ces.2007.07.008.
[22] Le Parlouër, P.; Dalmazzone, C.; Herzhaft, B.; Rousseau, L.; Mathonat, C. Characterisation of gas hydrates formation using a new high pressure micro-DSC. *J. Therm. Anal. Calorim.* **2004**, *78*, 165–172, DOI:10.1023/B:JTAN.0000042164.19602.7e.
[23] Adisasmito, S.; Frank, R. J.; Sloan, E. D. Hydrates of carbon dioxide and methane mixtures. *J. Chem. Eng. Data.* **1991**, *36*, 68–71, DOI:10.1021/je00001a020.
[24] Daraboina, N.; Malmos, C.; Von Solms, N. Investigation of kinetic hydrate inhibition using a high pressure micro differential scanning calorimeter. *Energy & Fuels.* **2013**, *27*, 5779–5786, DOI:10.1021/ef401042h.
[25] Ohno, H.; Susilo, R.; Gordienko, R.; Ripmeester, J.; Walker, V. K. Interaction of antifreeze proteins with hydrocarbon hydrates. *Chem. A Eur. J.* **2010**, 16, 10409–10417, DOI:10.1002/chem.200903201.

[26] Lachance, J. W.; Dendy Sloan, E.; Koh, C. A. Effect of hydrate formation/dissociation on emulsion stability using DSC and visual techniques. *Chem. Eng. Sci.* **2008**, *63*, 3942–3947, DOI:10.1016/j.ces.2008.04.049.

[27] Aman, Z. M.; Pfeiffer, K.; Vogt, S. J.; Johns, M. L.; May, E. F. Corrosion inhibitor interaction at hydrate-oil interfaces from differential scanning calorimetry measurements. *Colloids Surf. A Physicochem. Eng. Asp.* **2014**, *448*, 81–87, DOI:10.1016/j.colsurfa.2014.02.006.

[28] Kumar, A.; Kumar, R.; Linga, P. Sodium dodecyl sulfate preferentially promotes enclathration of methane in mixed methane-tetrahydrofuran hydrates. *iScience*. **2019**, *14*, 136–146, DOI:10.1016/j.isci.2019.03.020.

[29] Kumar, A.; Daraboina, N.; Kumar, R.; Linga, P. Experimental investigation to elucidate why tetrahydrofuran rapidly promotes methane hydrate formation kinetics: Applicable to energy storage. *J. Phys. Chem. C.* **2016**, *120*, 29062–29068, DOI:10.1021/acs.jpcc.6b11995.

[30] Lee, S.; Lee, Y.; Lee, J.; Lee, H.; Seo, Y. Experimental verification of methane-carbon dioxide replacement in natural gas hydrates using a differential scanning calorimeter. *Environ. Sci. Technol.* **2013**, *47*, 13184–13190, DOI:10.1021/es403542z.

[31] Lee, Y.; Lee, S.; Lee, J.; Seo, Y. Structure identification and dissociation enthalpy measurements of the CO2+N2 hydrates for their application to CO2 capture and storage. *Chem. Eng. J.* **2014**, *246*, 20–26, DOI:10.1016/j.cej.2014.02.045.

[32] Kwon, T. H.; Kneafsey, T. J.; Rees, E. V. L. Thermal dissociation behavior and dissociation enthalpies of methane-carbon dioxide mixed hydrates. *J. Phys. Chem. B.* **2011**, *115*, 8169–8175, DOI:10.1021/jp111490w.

[33] Andersen, R.; Llamedo, M.; Tohidi, B.; Burgass, R. W. Characteristics of clathrate hydrate equilibria in mesopores and interpretation of experimental data. *J. Phys. Chem. B.* **2003**, *107*, 3500–3506, DOI:10.1021/jp0263368.

[34] Faivre, C.; Bellet, D.; Dolino, G. Phase transitions of fluids confined in porous silicon: A differential calorimetry investigation. *Eur. Phys. J. B.* **1999**, *7*, 19–36, DOI:10.1007/s100510050586.

[35] Anderson, G. K. Enthalpy of dissociation and hydration number of carbon dioxide hydrate from the clapeyron equation. *J. Chem. Thermodyn.* **2003**, DOI:10.1016/S0021-9614(03)00093-4.

[36] Komatsu, H.; Ota, M.; Smith, R. L.; Inomata, H. Review of CO2-CH4clathrate hydrate replacement reaction laboratory studies – properties and kinetics. *J. Taiwan Inst. Chem. Eng.* **2013**, *44*, 517–537, DOI:10.1016/j.jtice.2013.03.010.

[37] Stern, L. A.; Circone, S.; Kirby, S. H.; Durham, W. B. Anomalous preservation of pure methane hydrate at 1 Atm. *J. Phys. Chem. B.* **2001**, *105*, 1756–1762, DOI:10.1021/jp003061s.

[38] Stern, L. A.; Circone, S.; Kirby, S. H.; Durham, W. B. Temperature, pressure, and compositional effects on anomalous or "self" preservation of gas hydrates. *Can. J. Phys.* **2003**, *81*, 271–283, DOI:10.1139/p03-018.

[39] Giavarini, C.; Maccioni, F. Self-preservation at low pressures of methane hydrates with various gas contents. *Ind. Eng. Chem. Res.* **2004**, *43*, 6616–6621, DOI:10.1021/ie040038a.

[40] Martínez, M. C.; Dalmazzone, D.; Fürst, W.; Delahaye, A.; Fournaison, L. Thermodynamic properties of THF + CO2 hydrates in relation with refrigeration applications. *AIChE J.* **2008**, *54*, 1088–1095, DOI:10.1002/aic.11455.

[41] Deschamps, J.; Dalmazzone, D. Dissociation enthalpies and phase equilibrium for TBAB semi-clathrate hydrates of N2, CO2, N2 + CO2 and CH4 + CO2. *J. Therm. Anal. Calorim.* **2009**, *98*, 113–118, DOI:10.1007/s10973-009-0399-3.

[42] Cai, J.; Tao, Y.-Q.; von Solms, N.; Xu, C.-G.; Chen, Z.-Y.; Li, X.-S. Experimental studies on hydrogen hydrate with tetrahydrofuran by differential scanning calorimeter and in-situ Raman. *Appl. Energy.* **2019**, *243*, 1–9, DOI:10.1016/j.apenergy.2019.03.179.

[43] Karimi, A. A.; Dolotko, O.; Dalmazzone, D. Hydrate phase equilibria data and hydrogen storage capacity measurement of the system H2+tetrabutylammonium hydroxide+H2O. *Fluid Phase Equilib.* **2014**, *361*, 175–180, DOI:10.1016/j.fluid.2013.10.043.

[44] Denning, S.; Majid, A. A. A.; Lucero, J. M.; Crawford, J. M.; Carreon, M. A.; Koh, C. A. Metal–organic framework HKUST-1 promotes methane hydrate formation for improved gas storage capacity. *ACS Appl. Mater. Interfaces*. **2020**, *12*, 53510–53518, DOI:10.1021/acsami.0c15675.

[45] Denning, S.; Majid, A. A. A. A.; Lucero, J. M.; Crawford, J. M.; Carreon, M. A.; Koh, C. A. Methane hydrate growth promoted by microporous zeolitic imidazolate frameworks ZIF-8 and ZIF-67 for enhanced methane storage. *ACS Sustain. Chem. Eng.* **2021**, *9*, 9001–9010, DOI:10.1021/acssuschemeng.1c01488.

[46] Torré, J.-P.; Plantier, F.; Marlin, L.; André, R.; Haillot, D. A novel stirred microcalorimetric cell for DSC measurements applied to the study of ice slurries and clathrate hydrates. *Chem. Eng. Res. Des.* **2020**, *160*, 465–475, DOI:10.1016/j.cherd.2020.06.019.

Ljiljana Damjanović-Vasilić

Chapter 10 Thermal methods as a tool for studying cultural heritage

Abstract: The restoration and conservation of cultural heritage artifacts require knowledge of their chemical and structural characteristics, that is physicochemical characterization of the materials that form the artifacts. This chapter presents different applications of thermal methods in the field of cultural heritage. The examples of successful use of the thermal methods for the studies of mortars, stones, ceramics, leathers and parchments, wood, paper, and painting materials are given. The advantages and limitations of these methods are presented through various examples. This review should be useful for the researchers entering this field.

Keywords: cultural heritage, thermal analysis, mortar, stone, ceramics, wood, leather, paper, paintings

10.1 Introduction

The preservation of cultural heritage is one of the important goals of humankind. Traditionally, archeologists, art historians, historians, and restorers most often classify artifacts according to their physical characteristics, decoration, and style. However contemporary approach requires close collaboration between experts from social and natural sciences. The detailed knowledge of the microchemical and microstructural nature of an artifact is critical in finding solutions to problems of restoration, conservation, dating, and authentication of cultural heritage. Nowadays, it is widely accepted that the properties of a particular product are determined by its composition and structure, which is a result of the raw materials used and the production processes applied. Unquestionably, this is applicable to the objects of cultural heritage. Therefore, characterization of materials that artifacts are made of provide information about production technology and knowledge of ancient societies as well as information about trading routes in certain epochs. Because of specific samples, the main goal of research in the field of cultural heritage is application of nondestructive and fast analytical techniques. The amount of material available for analyses is often limited, and this makes replication of measurements impossible. In addition, investigated materials are commonly heterogeneous, making selection of the representative sample

Ljiljana Damjanović-Vasilić, University of Belgrade-Faculty of Physical Chemistry, Studentski trg 12–16, P.O. Box 47, 11000 Belgrade, Serbia, e-mail: ljiljana@ffh.bg.ac.rs

https://doi.org/10.1515/9783110590449-010

a very challenging task as well as correlation between the results obtained independently by different analytical techniques. Current research methodology in this field most commonly employs a multianalytical approach; in addition, used analytical methods should be nondestructive or at least microdestructive.

Methods of thermal analysis have interdisciplinary character and can be successfully applied for preservation of cultural heritage even though they are often used with complementary techniques. Among the first reports of applications of thermal methods in the field of cultural heritage were differential thermal analyses of ancient ceramics [1] and oil paintings [2]. Not many review articles are available, but they clearly show the application of various thermoanalytical techniques in the studies of cultural heritage: thermogravimetry (TG), derivative thermogravimetry (DTG), differential thermal analysis (DTA), differential scanning calorimetry (DSC), TG coupled with DSC (TG-DSC) or DTA (TG-DTA), thermomechanical analysis (TMA), and dynamic mechanical analysis (DMA) [3–6]. The methods of thermal analysis can be generally considered as destructive, but only few milligrams of sample are usually needed for experiments. Also, the samples do not require any pretreatment for the thermal analyses. These facts make them suitable for the studies of objects with historical and cultural value. Obviously, not only the ancient artworks are subjects of investigations but the modern artworks as well. For example, the thermal methods were used for assessment of the conservation state and the prediction of the long term stability of a sculpture created by Francisco Leiro in 1995 made of synthetic polymeric materials [7].

This chapter presents the various applications of thermal methods in cultural heritage. Emphasis is on characterization of different materials which constitute ancient artworks. The aim is to illustrate possibilities of thermal techniques in this field. Certainly, all available reports are not covered in this chapter.

10.2 Characterization of materials that constitute ancient artworks by thermal methods

10.2.1 Mortars

Ancient mortars are composite materials used as joint elements of bricks or stones or as finishing layers (plasters) in different architectural structures. The first recipes for preparation of mortars as well as technical information are given in "De Architectura" by Vitruvius (first century BC). Some of the ancient recipes have been used through centuries, even during the nineteenth century, because of endurance and quality of obtained materials. Mortars are generally mixtures of two phases, a binder and an aggregate, with water and sometimes organic or inorganic additives. Clay, gypsum, and lime were the most commonly used binders, whereas sand is usually used as an aggregate [8]. The oldest records of mortars with clay (mud) as a binder dates back to

6000 BC in Catal Hüyük Turkey [9]. Gypsum was used for applications in Pharaonic Egypt, and lime-based mortars were identified for the first time in terrazzo floor in Canjenü, Eastern Turkey, dated from 12000 till 5000 BC and in flooring of fisherman's huts in Lepenski Vir, Serbia, dated 5600 BC [9]. Nowadays cement is a dominant binder in production of mortars. However, cement-based mortars are incompatible with old systems: they have a short life, and they can even cause enhanced and accelerated destruction of pre-existing material [10]. Therefore, characterization of historic mortars is important for the preparation of compatible repair mortars. Also, knowledge about used raw materials, their provenience and production process provides socio-economic information about past societies. Mortars are also grouped as hydraulic and nonhydraulic based on their ability to set and harden under water or in air, respectively.

TG, DTG, DTA, and DSC techniques have proven to be very successful in the identification of various constituent materials of mortars and for monitoring the reactions associated with controlled heating of mortars. Very often, results of thermal methods are combined with powder X-ray diffraction (XRPD), Fourier-transform infrared (FTIR) spectroscopy, scanning electron microscopy (SEM), and inductively coupled plasma (ICP) spectroscopy.

Lime-based mortars were the most widely used mortars in ancient times. Preparation of lime-mortars starts with the calcination of limestone ($CaCO_3$) at approximately 900 °C. During this reaction (thermal decomposition of calcium carbonate), calcium oxide (CaO), known as quicklime, is formed and carbon dioxide (CO_2) is released. Then, quicklime reacts with water and forms calcium hydroxide ($Ca(OH)_2$), which is called hydrated lime or slaked lime. Calcium hydroxide is one of the main components of mortars during application [4]. When exposed to atmosphere, hydrated lime slowly hardens because of the reaction with CO_2 from the atmosphere. This carbonation reaction results in the formation of $CaCO_3$ and release of water. It is a slow process which is influenced by the ambient conditions (the temperature, humidity, CO_2 concentration) and by the porous structure of the material [10].

Inorganic additives to lime-based mortars can be either natural materials such as volcanic ash or artificial like ceramic powder. The *cocciopesto* mortars are composed of lime, ceramic fragments and/or ceramic powder, and an aggregate. They have been widely used since the Roman age. This type of mortar is classified as hydraulic because calcium hydroxide and silica (SiO_2) and alumina (Al_2O_3) present at brick surface react with water (pozzolanic reaction) to form calcium silicates and calcium aluminates. This type of mortar was used for tanks, wells, baths, aqueducts, and so on [10].

In the case of organic additives for lime-based mortars, oxblood is one of the most commonly used compound. TG-DSC study, combined with XRD and attenuated total reflectance (ATR) FTIR spectroscopy analyses, of hydraulic and air lime-based mortars with or without oxblood addition showed that at the earlier stage, oxblood addition resulted in slower carbonation rate, and formation of amorphous calcium carbonate. After 12 months, similar amounts of carbonates were detected in mortars with and without oxblood [11].

Gypsum mortars, also very often used type of mortars in ancient times, are produced by the addition of water and an aggregate, usually sand, to the calcium sulfate hemihydrate ($CaSO_4 \cdot 1/2H_2O$) and/or soluble calcium sulfate anhydrite ($CaSO_4$). Hemihydrate is obtained by heating of natural gypsum ($CaSO_4 \cdot 2H_2O$), whereas anhydrite is obtained by calcination of natural gypsum at high temperatures [12]. When this type of mortar is applied (e.g., in the wall), calcium sulfate anhydrite can react with water from the mortar and return to gypsum.

TG provides information about the weight loss of the sample monitored as a function of temperature, and DTA and DSC reveal information about thermal transformations that are exothermic or endothermic in nature, such as dehydration, dehydroxylation, oxidation, and decomposition, as well as crystalline transitions. Typical thermal curves for mortars can be divided into four regions: (a) loss of adsorbed water, <120 °C; (b) loss of chemically bound water of the hydrated salts, such as gypsum if present, 120–200 °C; (c) loss of water chemically bound to hydraulic compounds, for example calcium aluminosilicates, 200–600 °C, and (d) loss of carbon dioxide released during the decomposition of carbonates, >600 °C [13, 14].

Based on TG results, ratio of mass loss due to CO_2 (mass loss due to the carbon dioxide content of the carbonated lime between 600 and 900 °C, in weight loss %) and mass loss of H_2O (mass loss due to chemically bound water of hydraulic products between 200 and 600 °C, in weight loss %) can be used to assess hydraulic nature of mixtures, that is, to distinguish the typical lime-based mortars (air-hardening) and the hydraulic type [14, 15]. Accordingly, higher CO_2/H_2O ratio corresponds to less hydraulic lime-mortar. Some authors classified "true" lime mortars as having CO_2/H_2O ratios higher than 10, hydraulic lime mortars between 4 and 10, and pozzolanic mortars (lime mortar with volcanic ash as additive) <3 [12]. In the study of the mortars from the Baths with Heliocaminus, an architectural building in the complex of the Hadrian's Villa in Tivoli, TG-DSC simultaneous analysis has shown typical trend of pozzolanic mortars, and using CO_2/H_2O ratio allows grouping of mortars with greater hydraulic degree (from marble flooring, vault concretes, arriccio plasters, the coating bedding mortars) and with higher CO_2/H_2O ratio the lime lumps [16].

TG measurements can be used to identify gypsum by weight loss of approximately 26.5 wt% as a result of the transformation to anhydrite [17].

If dolomite ($CaMg(CO_3)_2$) is present in lime-mortars, double endothermic peaks at 780 °C and 860 °C are detected [15]. The presence of hydrated magnesia ($Mg(OH)_2$) contributes to greater plasticity of lime mortars used for building, and its dehydration can be identified at 250–280 °C (hydromagnesite, $Mg_5(CO_3)_4(OH)_2 \cdot 4H_2O$), 350–420 °C (magnesia hydrate, $Mg(OH)_2$), while magnesium carbonates ($MgCO_3$) decompose in the range 450–520 °C [15]. As an example, TG and DSC study of ancient mortars from several churches in Italy, combined with ICP and FTIR analyses, successfully identified different carbonate species used as binders in mortar mixtures: (a) only calcite in calcite lime; (b) calcite and magnesite in magnesian lime; (c) calcite, hydromagnesite, and magnesite in magnesian lime [18].

Quartz can be identified by DSC analysis by endothermic peak at about 580 °C, without mass loss, corresponding to $\alpha \rightarrow \beta$ quartz phase transition [15].

The characterization of mortars used for building the ancient San Matteo hospital in the fifteenth century in Pavia (Italy) by application of the calorimetric and thermogravimetric techniques identified two groups of building materials: the first group corresponding to the plasters on the surface and the second group of the mortars sampled inside the walls. The first group is significantly richer in calcite content than the second group. The high concentration of calcite in the plasters was related to the restoration works done at the end of the eighteenth century [19].

In the case study of Venetian mortars used for building dated the sixteenth century, Biscontin et al. show that the presence of halite (soluble salt – sodium chloride) attributed to environmental conditions and pollution resulted in complex four-step decomposition DSC curve, instead of only one-step DSC curve typical for calcium carbonate, and decrease in decomposition temperature. When sodium chloride was rinsed with distilled water, expected one-step decomposition of $CaCO_3$ was detected [14].

The plaster is a name for mortar used as a specially prepared finishing layer that often served as a ground for the wall paintings of historic monuments. Probably the most common technique of wall painting is fresco technique. It entails application of pigments dispersed in water onto fresh, damp plaster. Painting on such a layer has to be finished one day before the plaster hardens. The thermal studies of painting mortars can reveal information about used production technology and can be employed to distinguish between the different types of mortars used in cultural heritage objects [20] as well as to reveal the composition of different pigments forming the wall painting [21].

Anastasiou et al. studied the plasters from wall paintings of Byzantine and post-Byzantine churches situated in the Balkan region and showed that thermal analysis is a reliable method for the classification of lime-based and gypsum-based plasters as well as that pigments penetrate deeper and form thicker layer on lime-based plasters compared to gypsum-based plasters. In addition, distinction between gypsum used as a constituent element (binder) and gypsum present due to environmental pollution was made. In the case of pollution, gypsum can be a product of sulfation reaction, which involves the dry deposition reaction between limestone ($CaCO_3$) and sulfur dioxide (SO_2) gas, in the presence of high relative humidity, an oxidant (usually oxygen), and a catalyst (Fe_2O_3 or NO_2) or can be caused by acid rain due to the presence of H_2SO_4. Then gypsum is primarily detected close to the surface, especially in cracks and voids, and can be easily distinguished from gypsum used as a binder [22]. Typical thermal curves obtained in the thermal analysis of ancient mortars are shown in Fig. 10.1.

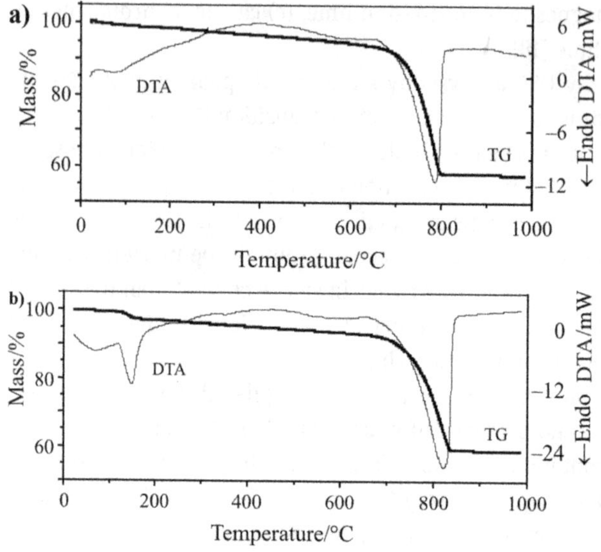

Fig. 10.1: Thermal curves of (a) lime-based plaster and (b) gypsum-based plaster (reprinted from [22] with permission from Springer Nature, copyright 2006).

The endothermic peaks and weight losses at temperatures higher than 750 °C (Fig. 10.1a and b) correspond to loss of CO_2 because of calcite decomposition. Although the pure calcite decomposes at about 840 °C, in the case of mortars (plasters) the lower degradation temperature is characteristic for the CO_2 loss from recarbonated lime [22]. The main difference between thermograms shown in Fig. 10.1 is in the temperature range 120–200 °C: the endothermic peaks with minimum at 132 and 149 °C shown in Fig. 10.1 b correspond to loss of moisture and to the dehydration of gypsum, respectively, allowing identification of gypsum-based mortar [22].

Duran et al. characterize wall painting mortars in ancient cities Herculaneum and Pompeii by TG-DTA analysis, combined with SEM, XRPD, and FTIR techniques. The mortars from Herculaneum were mixtures of lime and silicate compounds, whereas the mortars from Pompeii were obtained by mixing lime with marble grains as aggregate. The presence of highly crystalline marble particles resulted in an increase of carbonate thermal decomposition temperature compared to lime mortars containing silicates. On the other hand, the use of smaller grains and more content of lime produced a shift to lower temperature of the carbonate thermal decomposition [23].

10.2.2 Stones

The stone buildings and monuments are continuously exposed to environmental pollution which results in the formation of grey to black crust on their surface. The fast development of urban environments leads to significant deterioration of outdoor cultural heritage objects. The characterization of the surface layer is very important for the estimation of environmental damage and planning of appropriate restoration work. Thermal methods are successfully applied in the determination of the mechanisms of physicochemical and biological deterioration processes of ancient materials. Degradation patina is usually composed of (a) gypsum, the main product of sulfation reaction between surface materials which contain $CaCO_3$ (sandstones, marble, lime-mortars, etc.) and environmental pollutants (either SO_2 or H_2SO_4) [24]; (b) hydrated oxalates, resulting from reaction of oxalic acid secreted by microorganisms (e.g., algae, lichens, and fungi) with the calcareous stones and mortars [25] or as a result of previous restoration procedures [26]; (c) atmospheric particles, mainly soot particles, embedded within the crust during its formation and responsible for the color of the crust [27]. The atmospheric particles contain heavy metals (such as Fe, Ni, and V) that act as catalyst in reactions that enhance deterioration of stones, such as sulfation reaction. Also, carbon is one of the main constituents of soot particles. The presence of carbon in soot particles is a result of different combustion processes (exhaust emissions from vehicles, forest fires, industrial combustion, domestic heating systems, etc.). It is of interest to determine the amount of carbon from atmospheric particles in black crust in order to assess influence of anthropogenic factor to the damage of stones. Nevertheless, measuring the amount of total carbon in the surface layer is not sufficient because carbon in this layer has different origins. Carbon can be from (i) deposition of atmospheric particles containing elemental carbon and organic carbon compounds; but it can also be from (ii) calcium carbonate from building materials (stones and mortars), (iii) biological weathering because of organisms that produce, among their metabolic secretions, formic, acetic and oxalic acids, (iv) organic compounds (e.g., waxes and oils) used during previous restoration/conservation treatments [28]. Ghedini et al. have shown that DTA and TG methods can be successfully used for distinguishing carbon from different sources in black crust. They have developed analytical methodology based on thermal analysis, preceded by specific chemical treatments, which allows successful differentiation among carbonate, organic and elemental carbon and determination of their quantities [29]. A novel methodology based on the use of TG/DSC and elemental analyses such as CHN (carbon hydrogen nitrogen) has been recently developed to characterize and quantify organic carbon and elemental carbon together with other components present in the black crusts such as gypsum [30].

Dolomite rocks are often used for ancient buildings, and for this type of material the most common degradation layer is composed of hydrated calcium oxalates formed by the activity of living organisms [31]. The thermal behavior of dolomite and hydrated calcium oxalates are well-described. This knowledge can be applied to the study of thermal characteristics of unaltered and altered dolomitic rock samples from cultural heritage monuments. Dolomite has different thermal behavior depending on the atmosphere during heating. Double endothermic peaks typical for dolomite can be better resolved in carbon dioxide atmosphere [32]. Also, grinding of dolomite rocks results in displacement of the first peak to lower temperatures in DTA curves [33]. In the case of hydrated calcium oxalates, three mass-loss steps have been identified in thermogravimetiric analysis, where H_2O, CO, and CO_2 evolved during each step, respectively [25]. Perez-Rodriguez et al. have shown that TG-DTA analysis can be successfully used for the characterization of unaltered and altered dolomitic rock samples, and furthermore for the detection of organic compounds which are products of biological activity at altered surfaces [31].

Franquelo et al. have shown that simultaneous TG, DTG, DTA analyses can identify and quantify constituents of damaged patinas on ancient building, such as the Seville City Hall, but also successfully distinguish different materials used for restorations (mortars, polymeric resins, gypsum) [34].

Clay minerals can be present in stones, either diffused throughout the stone or as coating-filling of void space. Because of their swelling ability they can contribute to the deterioration of monuments. For example, volcanic stone can decompose under weathering processes to form different clay minerals in the stone framework. These clay minerals will lead to stone expanding in contact with water, and even small changes in length can cause serious damage depending on the resistance of the particular stone. In addition, clay expansion can create problems for consolidation of stones during conservation. DTA, TG, and DMA, combined with XRPD, FTIR spectroscopy, and SEM, allowed the identification and quantification of clays in several investigated monuments in the city of Guanajuato, Mexico. Also, the location of clays in the stone structures was determined and related to detect different swelling behavior of investigated materials [35].

TG-DSC analyses, performed by Friolo et al., identified deterioration mechanisms of marble and sandstones used for building two of Sydney's landmarks, the Captain Arthur Phillips Monument at Sydney's Botanic Gardens and Sydney's St. Mary's Cathedral. Acid environment, caused by the presence of SO_2 and CO_2 in air, resulted in the deterioration of marble during sulfation reactions. Also, the sulfation process leads to further destruction of ancient materials because it increases porosity of stone surfaces increasing area for further reactions with environmental pollutants. Contrary, sandstone damage resulted from the destruction of the crystal structure of the clay-based binding material. Changes in crystal structure of the original kaolinite resulted in thermally more stable clay (typical endothermic peak of dehydroxylation of structural water from kaolinite was accompanied with an extra peak at 680 °C) which changed the binding ability of this material [24].

Study by Gatta et al. shows the potential of combining thermal analyses (TG, DTG, and DTA) with chemometric methods for the identification of the marbles' provenance. Thermal decomposition of 16 reference marble samples (of known origin) was investigated, and the values referring to true peak temperatures, mass variations of the main thermogravimetric steps, and TG residues at 1,000 °C, as well as calculated kinetic parameters (activation energy and the log A (A = Arrhenius pre-exponential factor)) were the set of variables used for principal component analysis (PCA). PCA showed grouping of the marbles based on their similarities and successful classification of unknown marble samples [36].

10.2.3 Ceramics

Ceramic objects are well preserved even after several millennia [37]. Therefore, ceramic sherds are usually the most numerous archaeological finds at excavation sites worldwide. The main constituents of ceramics are clays, sand, and feldspars. Clays as well as feldspars can have a wide range of compositions and temperature dependent miscibilities. During firing of raw materials, changes in the mineral assemblage occur as a result of reactions which lead to the formation of new mineral phases and reactions which only produce compositional variations of firing phases [38]. The heterogeneity of samples and presence of amorphous phases additionally complicate the analysis of ceramic samples. Various methods are commonly used for the characterization of ancient ceramics (pottery) such as XRPD, petrography, FTIR and Raman spectroscopy, SEM, X-ray fluorescence (XRF), ICP, optical microscopy, multivariate statistical analysis [39–45].

It is generally accepted that DTA can be used for the identification of clay minerals [46]. Clays are identified based on endothermic signals originating from the dehydration (room temperature to ~ 250 °C) and the dehydroxylation, that is, decomposition of hydroxyl groups and their loss as water occurring in the range 400–650 °C, as well as the decomposition of carbonates, in the case of calcareous clays, in the range 700–800 °C. If present, gypsum shows endothermic effects within the range 120–160 °C. In addition, the formation of new crystalline phases is an exothermic process occurring in the temperature range 900–1,000 °C, whereas endothermic melting is happening at some higher temperature [1]. Dehydroxylation temperature range depends on the type of clay, for example, kaolinite clay dehydroxylates at ~ 450–500 °C, smectite (montmorillonite) clay at ~ 600 °C, illite clay dehydroxylates at ~ 550–900 °C. The processes of dehydration and dexydroxylation result in the formation of meta-clays with pseudo-amorphous structure [47].

Until the second half of the twentieth century, prevailing opinion was that once raw clay is fired, thermal methods cannot be used for the identification of clay minerals because temperatures at which changes occur are usually passed in the original firing [48]. Kingery first showed that DTA analysis can be performed on archaeological ceramic, concluding that "archaeological ceramic samples originally fired in the temperature range bellow 700–800 °C reacquire many of the characteristics of an unfired clay over the millennia" [1]. If the ceramic is fired at temperatures above 900–1,000 °C, constituent clays are altered that DTA indeed cannot provide significant information [1]. During the second firing of ceramics, exothermic reactions and/or reactions with gas release are expected to happen only at temperatures higher than the temperature of the first firing [49].

Immediately after firing, the ceramics begin to rehydrate, adsorbing moisture from the environment. Afterward, slow process of the rehydroxylation of fired clay takes place; meta-clays adsorb water; and recovery of some structural hydroxyl groups occurs. These processes result in the reconstruction of clay minerals [50, 51]. The rehydration and rehydroxylation are processes that result in mass gain.

Figure 10.2 shows representative thermal analysis curves (TG, DTG, and DSC) of the ceramic samples obtained during heating up to 900 °C under nitrogen flow. For production of earthenware, two types of raw clay material were used: calcareous and noncalcareous (containing less than about 5% of CaO) clays [37]. The most obvious difference among thermal curves of noncalcareous and calcareous ceramic is the presence of signal originating from decomposition of carbonates in the temperature range 600–850 °C. Also, the results presented in Fig. 10.2 illustrate that the thermal analyses methods can readily identify and quantify processes of dehydration and dehydoxylation of ancient ceramic.

Based on the fact that the rehydroxylation is a slow process, a method for dating of ancient pottery was suggested [52, 53]. Wilson et al. introduced the use of microbalance for accurate measurement of mass loss of ceramic samples during heating to 500 °C; this mass loss corresponds to ceramic's lifetime water mass gain. Then, the sample was exposed to water vapor to measure its mass gain as a function of time. Rexydroxilation dating is based on the finding that mass gain increases as the fourth root of the time, and this kinetic law is used for the determination of time passed since original firing [53]. Many studies have been made to test or improve rexydroxylation dating method, and they are presented in a detailed review recently published by Drebushchak et al., also containing authors' critical discussion about this dating method [54].

Two main problems that archaeologists want to solve with material scientists are identification of raw materials and their provenance as well as the estimation of the firing temperature. The traditional approach for the estimation of firing temperature and production technology is determination of mineralogical composition of ceramics. Thermal analyses combined with XRPD, SEM, and chemical analysis have been commonly used for the estimation of firing temperature of the ancient ceramics based on mineralogical and chemical composition; usually ancient ceramics are compared with thermally treated potential raw clay material [55–57]. This approach can be critically considered because solid-state reactions, and thus final mineral assemblage of ceramics, are controlled by the temperature and time (the same conversion degree will be achieved for shorter time at higher temperatures and vice versa) as well as by the grain size in the raw materials, their pretreatment, the presence of impurities, and so on [49].

Instead of reconstructing ancient firing procedure, Drebushchak et al. suggested comparison of degree of the thermal transformation among different ancient ceramic samples. They employ thermogravimetric analysis to measure mass loss at dehydration, m_1, in the temperature range from room temperature to 350 °C, and mass loss at dehydroxylation, m_2, in the temperature range from 350 to 600 °C.

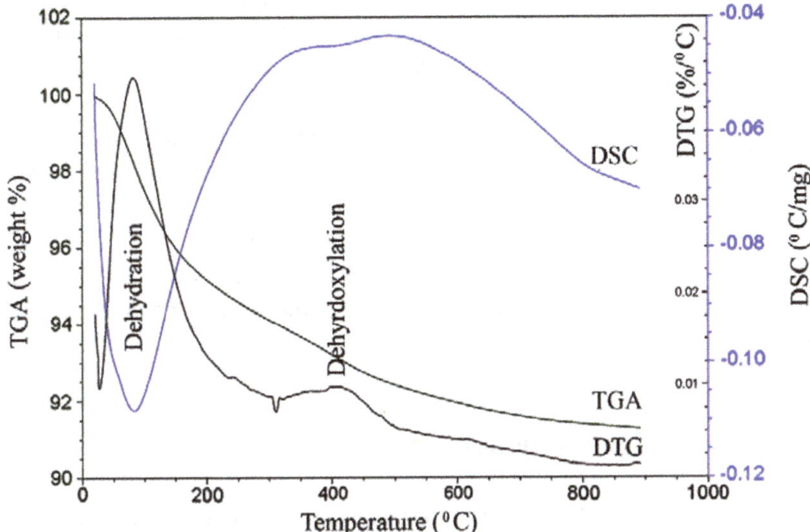

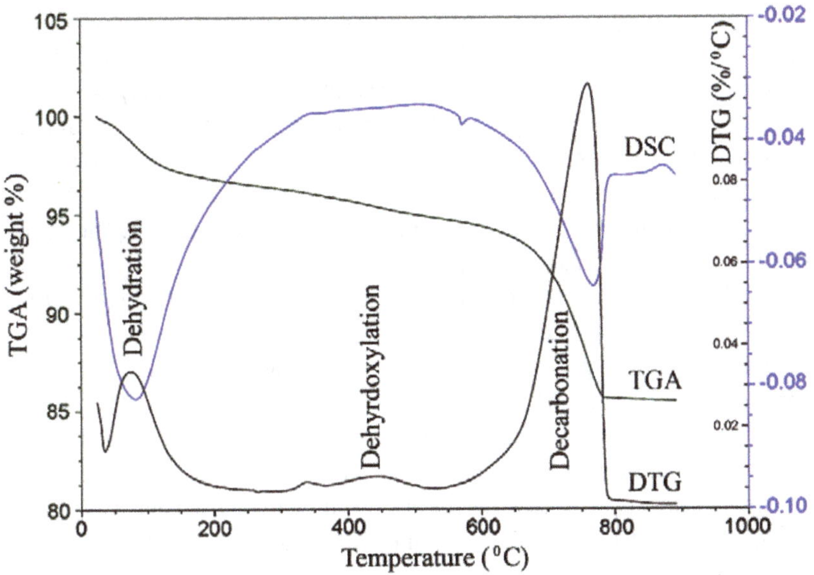

Fig. 10.2: Representative thermal analysis curves (TG, DTG, and DSC) of the pottery samples obtained during heating up to 900 °C under nitrogen flow. (a) Noncalcareous ceramic (chalcolithic cooking vessel, sample GAM-18). (b) Calcareous ceramic (early bronze bowl, sample TA-5) (reprinted from [47] with permission from Elsevier, copyright 2013).

Obtained thermogravimetric results were presented as mass-loss diagram: m_2 versus m_1. The m_2/m_1 ratio is related to degree of transformation of clay in ancient ceramic after original firing and recovery reactions during burial period, whereas the sum of m_1 and m_2 depends on the tempering of ceramic paste. Mild or strong firing condition can be assessed by using the mass-loss diagram [58].

Kloužková et al. studied an archaeological ceramic object, containing kaolinite, by XRF, XRD, TG-DTA, TG-DSC and FTIR spectroscopy. Thermal analysis showed that dehydroxylation of the rehydroxylated kaolinite proceeded at lower temperature compared to the primary kaolinite as the dehydroxylation peak at DSC curve shifted to lower temperature. In combination with results obtained by other used experimental techniques, distinction between primary kaolinite and the rehydroxylated kaolinite was successfully made [59].

10.2.4 Leathers and parchments

Leather and parchment are constituent materials of many important historical and cultural objects (patrimonial objects), such as scrolls, manuscripts, bookbinders, military coats, and so on.

Leather is obtained by the processing of animal skin, most often skin of calves, sheep or goats, and it is one of the oldest biomaterials used by humankind. Tanning of animal skins is a process for production of leather and it was traditionally performed using mimosa plants (by ancient Egyptians), sumac leaves (by people of the ancient Mediterranean) or oak (and pine) bark, nuts, galls and chestnut wood [60]. This process is called vegetable tanning and uses tannins (class of polyphenolic compounds) to increase hydrothermal stability, bacterial resistance and flexibility of leather. The hydrothermal stability mainly depends on the tannin type.

Parchment has been used as the writing material since the second century BC. After Middle Ages, when usage of paper became widespread, parchment was still in use for bookbinders or for special documents. It is obtained by alkaline lime treatment of untanned animal skins, which increases its stiffness [61].

The main component of the animal skin connecting tissue matrix is collagen, a supramolecular fibrillar protein in the form of a triple helix [62]. Consequently, leathers and parchments are collagen-based materials. Tanning results in chemically modified collagen, because of tannin-collagen complex formation, whereas collagen in parchments is chemically unmodified [60].

Deterioration of collagen depends on animal skin origin, for example, of the animal species, its living conditions, age, environment, location on the body, but also on tanning, natural aging, production of parchment (liming and drying), and interaction between ink or binders with collagen. Environmental factors (pollutants such as SO_2 and NO_x, temperature, humidity, UV/VIS light), microorganisms, improper handling and wrong conservation, as well as flooding, fires, earthquakes, wars, and so on are responsible for aging and deterioration of leathers and parchments [61]. Deterioration mechanisms of collagen caused by environmental factors are mainly oxidation and hydrolysis, as well as partial denaturation (gelatization) [63]. For proper restoration and conservation of patrimonial leathers and parchments it is important to determine the degree of deterioration.

Ancient leathers and parchments are commonly studied together with newly made materials and leathers and parchments subjected to accelerated aging for comparison. Accelerated aging involves exposure of leathers and parchments to simulated environmental factors.

Controlled heating of leather or parchment in water is used for assessment of hydrothermal stability of collagen in these materials. It is characterized by shrinkage of leather or parchment because of the denaturation of collagen and it is a good measure of quality of leather and parchment and the degree of their deterioration [64]. During hydrothermal shrinkage the triple helix configuration weakens or dislocates because of cleavage of intermolecular and intramolecular bonds (i.e., hydrogen bonding, hydrophobic bonding and cross-link bridges) [6]. Essentially, denaturation of collagen is transition from the triple helix to a randomly coiled form. Thermal methods, DSC in particular, can be used for measurements of enthalpy changes (ΔH) associated with collagen denaturation as well as the temperature of denaturation (T_d) or shrinkage (T_s) [65]. Value of T_d is influenced by the type of raw material, production method and deterioration processes during lifetime of leathers and parchments; the last one is of importance for studies of ancient collagen-based materials. Figure 10.3 shows DSC curve of new vegetable-tanned leather. As it can be seen, T_d is an extrapolated onset temperature (intersection of the base line and the raising part of the peak). This value is used because it is in the best agreement with conventionally determined shrinkage temperature [65]. When leathers or parchments are damaged the less energy is needed for their thermal denaturation, that is, values of T_d decrease.

Chanine showed that T_d and ΔH must be carefully studied and determined for particular material, because certain cases of parchment deterioration result in unchanged values of T_d but in decrease of ΔH values. In this situation considering only T_d values can lead to false conclusions [65].

DSC curves of the thermal denaturation of leather show higher T_d values compared to parchment. In the case of parchment, only hydrogen bonded water molecules are stabilizing the collagen structure, whereas in leather covalent bonds form intermolecular and intramolecular cross-links due to the inclusion of the tannin bound to collagen. Cross-linking can also be a result of dehydration, UV/VIS light, and gamma rays irradiation. Therefore, even damaged materials can show T_d values comparable to new materials [60].

Cucos et al. studied, by DSC measurements, the effects of 4 years natural aging of new and previously accelerated aged parchments and vegetable-tanned leathers. Decrease in T_d was the largest for new parchments (up to 7 °C), whereas change in T_d values was negligible for strongly aged ones. Values of the enthalpy of denaturation

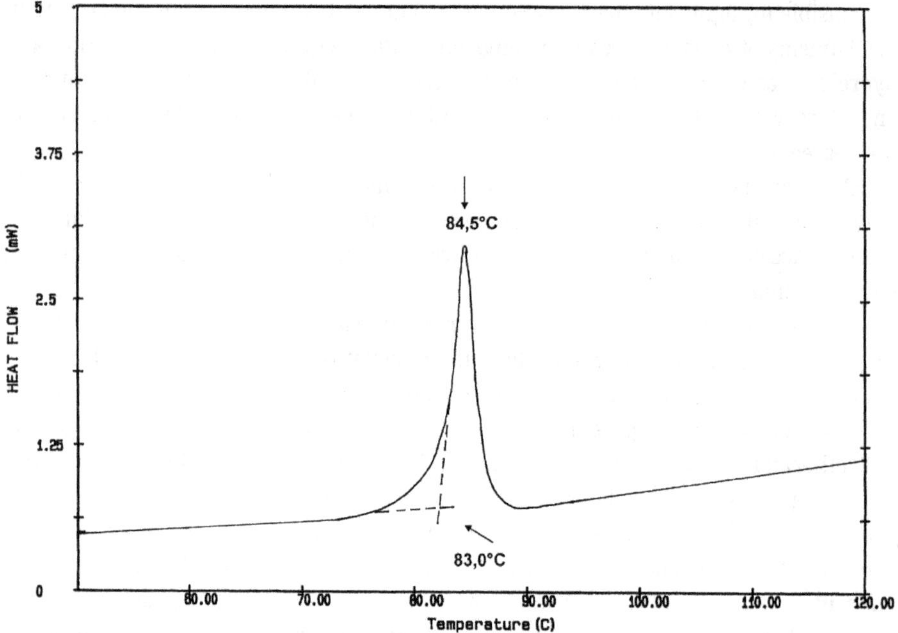

Fig. 10.3: DSC curve of new vegetable-tanned leather (reprinted from [65] with permission from Elsevier, copyright 2000).

decrease substantially with natural aging of parchments. In the case of leathers different behavior was detected. No relation between T_d shift and previous aging was found as well as any significant alteration in the enthalpy of denaturation. Thus, investigated parchments were more susceptible to the effects of natural aging than leather. In addition, this study showed that effect of one day accelerated aging in 2013 on T_d is comparable to 4 years of natural aging. This finding can be used for prediction of effects of natural aging on patrimonial parchments and leathers [66].

Micro DSC analysis can be applied to quantify deterioration of vegetable-tanned leathers using several calorimetric indices of macromolecular change identified for fibrous collagen and deconvolution of the overall DSC denaturation peaks. The temperature at the maximum of endothermic peak is used for classification of different structural domains of collagen: chemically modified, chemically unmodified and gelatinized collagen (i.e. “leather-like”, “parchment-like” and “gelatine-like”) whereas their quantification is based on the enthalpy percent contribution to the overall denaturation enthalpy [67].

Cohen et al. followed shrinkage of leather and parchment, immersed in water, by a standard dynamic mechanical thermal analyzer (DMTA). The differences between historic and modern samples, used as a reference, were detected in the shrinkage rate and the temperature range in which it occurred [64]. The unaged parchment and leather showed a narrow, well defined temperature range (65–70 °C) in which shrinkage occurred. On the other hand, the aged parchment began shrinking when it was immersed in water at 30 °C, and this process continued gradually with increase of temperature [64].

DSC study, in N_2 flow, of collagen-based materials have shown that all investigated materials have one endothermic peak originated from dehydration, whereas new and old parchments and naturally aged leathers (historic leathers) exhibit at least one endothermic peak in the range 126–228 °C. The new vegetable-tanned leathers have shown one peak at a higher temperature (around 243 °C) just before pyrolysis, whereas the majority of recent leathers do not exhibit such a peak. These findings can be used as a criterion for distinguishing between heritage and recently manufactured leather [68].

Combined DSC and TG analysis of ancient, new and artificially aged calf parchments was used to study bulk damage of the investigated materials. The moisture content of collagen-based materials changes depending on the external conditions to which they are exposed, and consequently changes their physical properties. It was found that denaturation temperature, T_d, is directly related to the moisture content which allows a quantitative ranking of the damage experienced by parchment [69].

10.2.5 Wood

The wood has various applications in the field of cultural heritage. This material was used for buildings, furniture, statues, boats, icons, and so on. Archaeological woods are often found underwater as building materials of the boats or as objects carried by the boats or buried in waterlogged land. Waterlogged woods are complex materials for conservation because they can undergo serious damage if dried. Loss of water results in shrinking and deformation of these materials. Therefore, during conservation treatment water from wood cavities is replaced with consolidants. Commonly, consolidants are nontoxic and water-soluble polymer materials, such as poly(ethylene) glycols (PEGs). Polymers with small molecular mass are used when the damage is not serious, whereas for very damaged waterlogged wood application of polymers with large molecular masses is necessary [70].

Deterioration of the wood can be of chemical and biological origin, and results in loss of the main components: cellulose and hemicellulose. Lignin, other constituent of wood, is more resistant, but it also undergoes deterioration.

TG/DTG/DSC study of ancient Egyptian myrrth three was one of the first studies that showed clear distinction between weight losses and degradation peaks of cellulose and lignin indicating that thermal methods can be used for investigation of wood degradation processes during aging [71].

TG and DTG study of a waterlogged wood from relict of a Roman ship dated the first century and fresh wood of the same type (red fir, larch, elm and beech) have shown three distinct processes during controlled heating of the investigated material. The first process was dehydration followed by two successive decomposition steps [72]. For waterlogged wood, decomposition of cellulose was identified at 300 °C, decomposition of lignin at about 400 °C, and ash content as thermogravimetric residue was determined at 650–700 °C. The identical results have been obtained when wood samples were prepared as small cubes or sawdust, revealing that wood samples do not require any pretreatment for thermal analysis. Because dehydration process does not overlap with decomposition of wood constituents, thermograms provide easy and simple way for determination of liquid/solid ratio of archaeological woods [72]. The weight loss up to 150 °C is related to water content and only few milligrams of sample is required for analysis. Therefore, maximum water content in wet wood samples is straightforwardly determined by thermogravimetry [73]. Determination of saturated water content is particularly important because it is an indication of the damage to the inner structure of the wood and of the decomposition processes occurring in these samples [72].

Recent study of the oak wood poles from the prehistoric village of Gran Carro (Lake Bolsena, Italy) tested TG as an alternative method for the chemical characterization of archaeological wood. In the case of reference oak samples, the results obtained for holocellulose are quite reliable, whereas results for the phenolic compounds are more variable. Maximum water content was easily obtained indicating difference in the degree of wood degradation. And for highly degraded wood samples TG was not reliable for determination of either cellulose or lignin content. Comparison of TG results with results obtained by other techniques on waterlogged wood should provide a calibration of thermogravimetric analyses [74].

Vecchio et al. studied five archaeological wood samples: fir, chestnut, poplar, linden and oak; the TG and DSC curves revealed well defined processes which occur during controlled heating of investigated sample: dehydration (endothermic peak up to about 100 °C) followed by two decomposition processes of cellulose and lignin (two exothermic peaks at about 300 and 400 °C). Stability of investigated wood samples was compared based on decomposition temperature and activation energy (determined from kinetic studies) of the first degradation step; oak showed the lowest stability among investigated samples [75].

TG has been successfully applied for quantitative determination of total consolidant content entrapped into the cavities of degraded wood. In addition, DSC was used to investigate mixture of polymers with different molecular masses, used for selective consolidation of waterlogged wood and amount of each polymer was established successfully. Determination of consolidant content in wood after conservation can be used as an estimate of efficiency of performed treatment [70, 76].

Franceschi et al. studied chestnut and fir wood samples which were artificially aged by immersion in deionized and artificial sea water at room temperature and at 40 °C for eight weeks. Also, copper and iron nails have been inserted into wood samples to investigate effect of contact between metals and wood. Untreated wood was used as reference. DSC curves showed that two exothermal peaks of cellulose and lignin combustion are better separated for treated wood samples compared to untreated ones. Also, wood deterioration is more enhanced in the presence of metal; cooper being the more efficient than iron in degradation of lignin and cellulose. Combining results obtained by TG/DTG, DTA and DSC analyses of ancient wood, two different degradation mechanisms are proposed: the first mainly alters crystalline parts of cellulose and the second degrades predominantly amorphous part of the wood [77].

Lime tree is often used for icons and iconostasis found at Romanian churches and monasteries. TG/DTG and DSC study of patrimonial lime tree, during controlled heating in static air atmosphere or oxygen flow, showed that degradation of lime wood occurs through typical three successive processes accompanied by mass losses: dehydration, and exothermal decomposition of cellulose and lignin. Obtained results show that during natural aging mass loss in the first process of thermo-oxidation decreases as well as the ratio between the mass losses in the first and the second processes of thermo-oxidation. The maximum rate of the first process of thermo-oxidation also decreases with aging. It was suggested that these findings can be used as criteria for distinction between patrimonial and forged objects [78].

TG and DTG analysis of lime wood used as a support for 100–200 years old icons and iconostasis show that the characteristic parameters of TG and DTG curves as well as global kinetic parameters of the thermo-oxidations of wood samples decrease with the increase of their deterioration degree. Therefore, these parameters can be used for estimation of painting age [79].

10.2.6 Paper

Paper is a cellulose-based material and its conservation problems are often comparable with the same problems related to wood described in previous paragraph. The thermal methods can be used to determine degree of deterioration and to reveal degradation mechanisms of paper. Typical thermograms of ancient papyrus are shown in Fig. 10.4.

As it can be seen in Figure 10.4, after dehydration process, following two peaks are attributed to decomposition of hemicellulose and cellulose and the decomposition of lignin, respectively [71, 81]. Thermal curves (TG and DSC) show different behavior depending on the different parts of the plants used for production of ancient papyri. These differences can be explained by dissimilar content of hemicellulose, cellulose and lignin in different parts of the plants used for production [80]. For example, ancient Egyptian and Greek-Roman papyri show different thermal curves: peak positions for decomposition of cellulose were slightly different, whereas decomposition of lignin occurred at significantly different temperatures (421.0 °C for the Egyptian and 405.2 °C for Greek-Roman papyri) and had different peak areas. The different shapes of the thermal curves also revealed the presence of starch in Egyptian paper which is

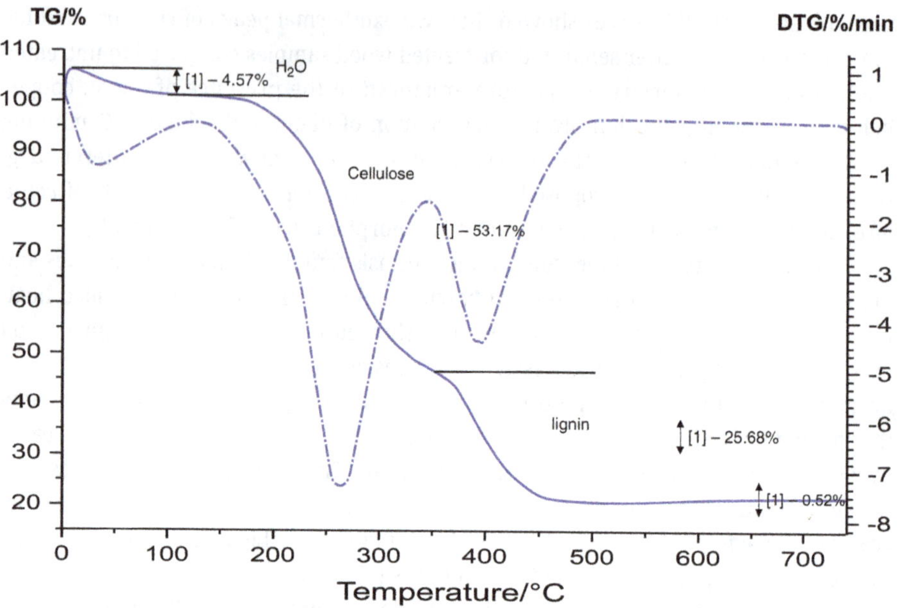

Fig. 10.4: TG and DTG curves of the ancient papyrus; analyses performed in static air, heating rate was 10 °C/min (reprinted from [80] with permission from Springer Nature, copyright 2011).

a naturally occurring mineral in plants used for production, as well as the presence of clay minerals in Greek-Roman paper used during production process [81].

Even though TG is commonly used to quantify lignin in papyrus sheets, recent study compared content of lignin in commercial papyrus sheets determined by TG and two other lignin determination procedures (Klason-lignin and acetyl bromide-soluble lignin). The results revealed large overestimation of the lignin content (~27%) obtained by TG compared to the other methods (~5%), therefore, suggesting that TG should not be used for lignin quantification [82].

Besides cellulose fibers, the one of the most important components of paper are fillers responsible for characteristics of paper and its resistance to aging. The most common fillers used through ages were ground bone ash, white lead, calcium carbonate, gypsum, powdered cuttlebone, wax, zinc oxide, titanium dioxide, clay, kaolin, barium sulfate, calcium sulfate, calcium sulfite, and alum [83]. Usage of fillers that keep pH in the alkaline range, like calcium carbonate, is particularly suitable because it eliminates problem of acid catalyzed decay of cellulose, one of the major degradation mechanisms of paper [84]. Fierascu et al. used thermal methods in combination with other techniques (optical microscopy, FTIR spectroscopy, ICP-atomic emission spectrometry, energy-dispersive X-ray fluorescence, XRPD) for characterization of paper from eight historic documents (testaments, petitions, notices, etc.) dating back to the first half of the nineteenth century. Among eight investigated historic documents, thermal analyses revealed the presence of two different fillers, identified by spectroscopic techniques as $CaCO_3$ and TiO_2 and one sample showed higher amount of hemicellulose and lignin. These findings lead to conclusion that paper used for investigated documents have different origin, in terms of composition and used fillers [83].

Thermal methods are also used for testing the effects of conservation treatment. The cellulose structure weakens during time because of hydrolysis and oxidation processes. Degradation consequently changes the mechanical properties of paper. Current trend in conservation is use of innovative, nontoxic materials compatible with the paper properties. Many new materials have been developed, mainly through sol-gel methods. Recent study investigated ancient paper from the eighteenth century after conservation treatment with hybrid inorganic–organic polyamidoamines (PAAs). Polymerization by polyaddition of organic monomers results in these hybrid materials which serve as protective coatings for paper. The investigated samples were characterized by TG, DSC, SEM-EDX, FTIR, and Raman spectroscopy, as well as static and dynamic-mechanical measurements. Treated materials showed improvement in mechanical and dynamic-mechanical properties, and thermal resistance. Performance of the paper was improved with thickness of the protective material, showing potential of tested material for widespread usage [85].

10.2.7 Painting materials

Preparation of paintings and icons usually means that the ground layer is applied first to prepare the surface of a canvas or a panel, followed by application of a paint layer, which is a mixture of pigments and binding media. Composition of paint layer gives the color quality. Lastly, varnish, which is mainly based on natural resins, is applied as a transparent coating responsible for particular visual effects. Pigments can be organic or inorganic compounds and chronological use of most pigments is known. Consequently, identification of pigments enables indirect dating of painted art objects [86, 87]. The painting materials are often of natural origin, the ancient recipes for preparation of different painting materials are not always known, and in order to achieve particular visual effect artists are mixing various additives even with commercially available paints. Thus, samples taken from paintings and icons are commonly very complex mixtures of different inorganic and organic compounds. In addition, samples underwent aging which caused changes in materials of interest. All this makes analyses of painting materials very challenging, not only because scientists are dealing with mixtures of different classes of compounds, but because it is very difficult to obtain suitable standard materials.

Among the first applications of thermal methods to the characterization of painting materials was the DTA study of samples from oil painting [88]. This study has shown the potential of DTA analysis performed under oxygen atmosphere for dating oil media, not more than 100 years old. Difference in thermal behavior of investigated materials was detected comparing the peak ratio of two exothermic signals observed at 300 and 400 °C.

Various pigments and different binding media were studied by DSC technique, as pure materials and as prepared mixtures, revealing that distinction among binding media based on thermal curves can be made [89]. DSC technique provides direct measurement of heat evolved during exothermic reactions, and the shape of the curves depends on chemical composition of investigated material. The presence of additives can be detected by DSC measurements in oil-containing samples, but it is important to analyze enthalpy values, together with the overall curves shape and the peaks ratios [90].

Natural earths are painting materials widely used from prehistoric times. These materials are inorganic pigments, known as ochres, red earths and boles, siennas (raw and burnt), umbers, green earths, and others. Earthy pigments are clay-rich materials which might be divided according to their coloring agent: it is either some non-clay pigment, for example, iron oxides, or a chromogenous element in the clay structure [91]. Juliá et al. performed TG, DTG, and DTA study of various ochres and siennas and has shown that the detection and quantification of the main components is easily achieved using thermal methods as well as determination of provenience of earthy pigments. Calcium sulfate ($CaSO_4$) was the most common component in the investigated natural earths from Spain, whereas French ochres were basically composed of aluminosilicates. In the case of some ochres the quantification of kaolinite and calcium carbonate was not possible because of the overlapping of peaks, but use of CO_2 atmosphere solved this problem [92]. The ochres used in Australian aboriginal bark painting were characterized by TG and DSC methods, combined with mass spectrometry (MS). Experiments were carried out in both oxidizing atmosphere (air) and nonoxidizing atmosphere (argon). This methodology allowed the identification of kaolinite, quartz, and charcoal as the main components of paint samples. Based on mass losses measured by thermal analysis, the percentage of each component was successfully determined [93].

Thermoanalytical techniques can show differences in thermal behavior of azurite ($Cu_3(CO_3)_2(OH)_2$) and lead tempera paint. Also, comparison of TG curves of the reference samples (mineral-based pigment portlandite in various binders) and samples from different frescoes revealed distinctions in painting techniques used by different authors, Vasari and Zuccari [94].

Binding media are used to provide dispersion of pigments and their cohesion in the paint layer as well as adhesion to the support [95]. Traditional binding media are drying oils, such as cold-pressed linseed oil, stand oil, poppy seed oil, walnut oil, castor oil, sunflower oil. They are mixtures of triglycerides, and have smaller amounts of sterols and vitamins. Drying oils form solid films upon exposure to air as a result of chemical drying, that is oxidation and cross-linking reactions. Natural drying oils are in use nowadays, even though many synthetic products have been available (such as acrylic, alkyd, and vinyl) since the 1930s.

TG/DSC method is a valid tool for studying the thermal and oxidative stability of drying oils. Izzo et al. determined these characteristics for naturally aged oil paints with particular attention to the drying process and the film formation in the early stages, because these processes determine characteristics of the resulting film. TG/DSC analyses have been carried out on unpigmented and pigmented young films (used pigments were cobalt blue ($CoAl_2O_4$), chrome green (Cr_2O_3) and cadmium yellow (CdS)) within 2 years since the oil layers have been painted out. Following formation and transformation of hyperoxide groups, the study showed that (1) after one week, autoxidation of the unsaturated fatty acids present in drying oils takes place; (2) the formation of the film starts after a week and likely in 2 years of curing the film appears to be thermally stable; (3) both processes are accelerated in the case of pigments which contain metals that can act as catalysts (such as cobalt blue) [96].

TG analysis, combined with gas chromatography-mass spectrometry (GC-MS) and direct exposure mass spectrometry (DEMS), was used to study paint layers obtained with linseed oil preprocessed in several ways: water washing, heat treatments, and the addition of driers, with and without heat, as oil was treated in the nineteenth century and different pigments, lead white ($Pb(CO_3)_2 \cdot Pb(OH)_2$), vine black (carbon) and umber (iron oxide with 6–15% of MnO_2 + clay + silica). TG curves obtained under nitrogen flow and mass changes during oxygen uptake registered at a constant temperature (80 °C) showed the different physical behavior of the oil samples. The hydrolyzed, oxidized and cross-linked fractions have been identified and their presence was assigned to the different pretreatments of oil [97]. A similar study showed that composition of aged oil painting is more associated with present pigment and conservation state than oil pretreatment. As a consequence, the ratios between different amounts of fatty and dicarboxlic acid used for distinguishing between drying oils do not appear to be proper analytical approach for identification of oil pretreatment [98].

The presence of metal soaps, complexes formed of metal ions from pigments and saturated fatty acids from the oil binder, in oil paint layers is a growing concern in the conservation of oil paintings. The lead and zinc soaps of palmitic or stearic acid are the most common metal soaps found in oil paintings. DSC study combined with ATR FTIR spectroscopy, using model mixtures of palmitic acid, lead palmitate or zinc palmitate and linseed oil, have shown that the crystallization of metal soaps in oil paint layers is a spontaneous and irreversible process, caused by the very low solubility of metal soaps in oil paints [99].

The proteinaceous materials such as egg, animal glue and casein or milk were also used for centuries as binding materials, particular in the "tempera" technique. The degradation of the proteinaceous binders, ovalbumin (OVA) and casein, and their interactions with pigments azurite ($Cu_3(CO_3)_2(OH)_2$), calcium carbonate ($CaCO_3$), hematite (Fe_2O_3) and red lead (Pb_3O_4) pigments were studied by TG, DSC, FTIR spectroscopy and Size Exclusion Chromatography (SEC). TG/DSC results revealed that the presence of inorganic pigments in tempera paint interact with both proteins inducing a decrease in thermal stability. This trend continues with aging of casein paint replicas.

On the other hand, OVA paint replicas showed increase in thermal stability with aging [95]. Also, the thermal degradation of animal glue (rabbit skin glue) was analyzed by TG, DSC and TG/FTIR techniques, together with mixtures of this glue with the same pigments: azurite, calcium carbonate, hematite, and red lead. The study showed that all the inorganic pigments interact with this collagen-based proteinaceous material (the strongest interaction was with Fe_2O_3) decreasing its thermal stability. Artificial aging with light resulted in additional slight decrease of thermal stability of pigmented paint replicas [100].

DSC as well as dynamic mechanical thermal analysis (DMTA) have proved to be useful to determine state of degradation of painting materials caused by the environmental factors (humidity, light, temperature, gasses such as SO_2 and NO_x) because these techniques are sensitive to chemical changes occurring during aging of painting materials [101]. In order to determine degree of chemical changes of tempera paintings exhibited in art galleries and museums, by DSC and DMTA techniques, test tempera paintings can serve as dosimeters for overall effect of environmental factors. It was shown that the smalt tempera painting, used as dosimeter, can successfully distinguish two sites with controlled and uncontrolled environmental factors [102].

Varnishes are clear, final coatings that are transparent for VIS and UV light. Therefore, they cannot prevent photochemical degradation of paint layers. For centuries natural products, such as gum mastic and since the nineteenth century also dammar resin, have been used as varnishes. These natural products have excellent optical properties, but they yellow over time, and eventually lose their transparency and may crack. Also, upon aging their removal becomes difficult. Modern varnishes are synthetic polymers which are more stable than natural products, but do not have the same appearance. This difference in appearance has been connected to a difference in solution viscosity, which is related to molecular weight. Hence, synthesis and testing of the low-molecular-weight resins became of interest. The resins that are used as varnishes tend to have the glass transition temperatures (T_g) in the range 30–60 °C. Therefore, DSC combined with SEC can successfully measure properties that allow assessment of the suitability of low-molecular-weight resins as varnishes. Maines et al. characterized several commercially available low-molecular-weight resins and determined values of their T_g within the range 30–70 °C, which makes these resins suitable for application as varnishes [103].

The materials used for restoration of canvas painting can be modern materials or prepared by traditional recipes. The requirements for selection of appropriate restoration materials are high stability of color and high thermo-oxidative stability. In practice, the selection of materials is mostly based on the restorer's professional experience and intuition. Simultaneous thermal analysis (DTA and TG) was applied to study the thermo-oxidative stability of baroque oil painting. Among investigated materials, suggested by restores, the fillings prepared by traditional recipes, with the exception of wax, showed good characteristics which make them suitable for restoration of oil painting. The commercial contemporary material Litostucco appeared the least suitable of all of investigated samples; however, its stability can be improved by modification with additives, such as kerotix, ponal, bolus, or linseed oil, in various ratios [104].

The widespread technique for the restoration of canvas paintings, called lining, consists of the application of a new support canvas, glued on the back of the painting. In the second half of the twentieth century, synthetic adhesives, for example, vinyl acetate and acrylate copolymers, were increasingly used by restorers. Degradation of adhesives leads to yellowing, changes in their viscoelasticity, and the formation of volatile organic acids, which can promote the hydrolysis of canvas cellulose and accelerate painting deterioration. Changes in molecular structures can be followed, among other experimental techniques, by thermal methods. Study of synthetic polymeric adhesives (Mowilith® DM5 and DMC2), widely used not only for canvas lining but also for wall painting, wood, paper, or metal, revealed that during degradation changes in molecular structure occur through process of deacetylation followed by a competition between depolymerization and cross-linking. Change detected in molecular weight is very important for restorers because it alters solubility of the adhesives, and in that case traditionally used solvents for cleaning will no longer be effective. In such cases application of microemulsions and micellar solutions is necessary for the removal of the degraded polymers from works of art [105].

10.3 Conclusion

This chapter has presented applications of different thermal methods in the field of cultural heritage, in particular, the characterization of various materials that constitute ancient artworks by these methods. It has been shown that the thermal methods are successfully used for studies of mortars, stones, ceramics, leathers and parchments, wood, paper, and painting materials. Information obtained from these studies is significant for the restoration and conservation of the artwork, reconstruction of "life stories" of certain art objects, verification of their authenticity as well as understanding the level of people's knowledge in certain epochs.

Through selected examples, the advantages and some limitations of application of thermal techniques in the field of cultural heritage have been shown. Even though the number of studies in this area is increasing, particularly in the past decade, there is still potential for wider application of the thermal methods in the characterization of ancient as well as contemporary art objects, whether they are kept in museums and galleries or exhibited outdoors.

References

[1] Kingery, W. D. A note on determination of the differential thermal analysis of archaeological ceramics. *Archaeometry*. **1974**, *16*, 109–112, DOI:10.1111/j.1475-4754.1974.tb01099.x.

[2] Preusser, F. Untersuchung von Werken der Kunst- und Kulturgeschichte mit Hilfe der Differentialthermoanalyse. *J. Therm. Anal. Calorim.* **1979**, *16*, 277–283, DOI:10.1007/BF01910689.

[3] Prati, S.; Chiavari, G.; Cam, D. DSC application in the conservation field. *J. Therm. Anal. Calorim.* **2001**, *66*, 315–327, DOI:10.1023/A101247260.

[4] Pires, J.; Cruz, A. J. Techniques of thermal analysis applied to the study of cultural heritage. *J. Therm. Anal. Calorim.* **2007**, *87*, 411–415, DOI:10.1007/s10973-004-6775-0.

[5] Odlyha, M. Introduction to the preservation of cultural heritage. *J. Therm. Anal. Calorim.* **2011**, *104*, 399–403, DOI:10.1007/s10973-011-1421-0.

[6] Budrugeac, P.; Cucos, A.; Miu, L. The use of thermal analysis methods for authentication and conservation state determination of historical and/or cultural objects manufactured from leather. *J. Therm. Anal. Calorim.* **2011**, *104*, 439–450, DOI:10.1007/s10973-010-1183-0.

[7] Rodriguez-Mella, Y.; López-Morán, T.; López-Quintela, M. A.; Lazzari, M. Durability of an industrial epoxy vinyl ester resin used for the fabrication of a contemporary art sculpture. *Polym. Degrad. Stabil.* **2014**, *107*, 277–284, DOI:10.1016/j.polymdegradstab.2014.02.008.

[8] Carvalho, F.; Sousa, P.; Leal, N.; Simão, J.; Kavoulaki, E.; Lima, M. M.; da Silva, T. P.; Águas, H.; Padeletti, G.; Veiga, J. P. Mortars from the Palace of Knossos in Crete, Greece: A multi-analytical approach. *Minerals*. **2022**, *12*, 30, DOI:10.3390/min12010030.

[9] Elsen, J. Microscopy of historic mortars – A review. *Cement Concrete Res.* **2006**, *36*, 1416–1424, DOI:10.1007/978-94-007-4635-0_10.

[10] Matias, G.; Faria, P.; Torres, I. Lime mortars with heat treated clays and ceramic waste: A review. *Constr. Build. Mater.* **2014**, *73*, 125–136, DOI:10.1016/j.conbuildmat.2014.09.028.

[11] Zhang, K.; Grimoldi, A.; Rampazzi, L.; Sansonetti, A.; Corti, C. Contribution of thermal analysis in the characterization of lime-based mortars with oxblood addition. *Thermochim. Acta*. **2019**, *678*, 178303, DOI:10.1016/j.tca.2019.178303.

[12] Corti, C.; Rampazzi, L.; Bugini, R.; Sansonetti, A.; Biraghi, M.; Castelletti, L.; Nobile, I.; Orsenigo, C. Thermal analysis and archaeological chronology: The ancient mortars of the site of Baradello (Como, Italy). *Thermochim. Acta*. **2013**, *572*, 71–84, DOI:10.1016/j.tca.2013.08.015.

[13] Bakolas, A.; Biscontin, G.; Moropoulou, A.; Zendri, E. Characterization of structural Byzantine mortars by thermogravimetric analysis. *Thermochim. Acta*. **1998**, *321*, 151–160, DOI:10.1016/S0040-6031(98)00454-7.

[14] Biscontin, G.; Pellizon Birelli, M.; Zendri, E. Characterization of binders employed in the manufacture of Venetian historical mortars. *J. Cult. Herit.* **2002**, *3*, 31–37, DOI:10.1016/S1296-2074(02)01156-1.

[15] Moropoulou, A.; Bakolas, A.; Bisbikou, K. Characterization of ancient, byzantine and later historic mortars by thermal and X-ray diffraction techniques. *Thermochim. Acta*. **1995**, *269/270*, 779–795, DOI:10.1016/0040-6031(95)02571-5.

[16] Columbu, S.; Sitzia, F.; Ennas, G. The ancient pozzolanic mortars and concretes of Heliocaminus baths in Hadrian's Villa (Tivoli, Italy). *Archaeol. Anthropol. Sci.* **2017**, *9*, 523–553, DOI:10.1007/s12520-016-0385-1.

[17] Middendorf, B.; Hughes, J. J.; Callebaut, K.; Baronio, G.; Papayianni, I. Investigative methods for the characterisation of historic mortars – Part 1: Mineralogical characterization. *Mater. Struct.* **2005**, *38*, 761–769, DOI:10.1007/BF02479289.

[18] Bruni, S.; Cariati, F.; Fermo, P.; Pozzi, A.; Toniolo, L. Characterization of ancient magnesian mortars coming from northern Italy. *Thermochim. Acta*. **1998**, *321*, 161–165, DOI:10.1016/S0040-6031(98) 00455-9.

[19] Tomasi, C.; Ricci, O.; Perotti, G.; Ferloni, P. Plasters and mortars in the central building of the University of Pavia. *J. Therm. Anal. Calorim.* **2006**, *84*, 33–38, DOI:10.1007/s10973-005-7264-9.

[20] Duran, A.; Robador, M. D.; Jimenez de Haro, M. C.; Ramirez-Valle, V. Study by thermal analysis of mortars belonging to wall paintings corresponding to some historical buildings of Sevillian art. *J. Therm. Anal. Calorim.* **2008**, *92*, 353–359, DOI:10.1007/s10973-007-8733-0.

[21] Perez-Rodriguez, J. L.; Franquelo, M. L.; Duran, A. T. G. DTA and X-ray thermodiffraction study of wall paintings from the fifteenth century. *J. Therm. Anal. Calorim.* **2021**, *143*, 3257–3265, DOI:10.1007/s10973-020-09420-5.

[22] Anastasiou, M.; Hasapis, T.; Zorba, T.; Pavlidou, E.; Chrissafis, K.; Paraskevopoulos, K. M. TG-DTA and FTIR analyses of plasters from Byzantine monuments in Balkan region. *J. Therm. Anal. Calorim.* **2006**, *84*, 27–32, DOI:10.1007/s10973-005-7211-9.

[23] Duran, A.; Perez-Maqueda, L. A.; Poyato, J.; Perez-Rodriguez, J. L. A thermal study approach to roman age wall painting mortars. *J. Therm. Anal. Calorim.* **2010**, *99*, 803–809, DOI:10.1007/s10973-009-0667-2.

[24] Friolo, K. H.; Ray, A. S.; Stuart, B. H.; Thomas, P. S. Thermal analysis of heritage stones. *J. Therm. Anal. Calorim.* **2005**, *80*, 559–563, DOI:10.1007/s10973-005-0694-6.

[25] Perez-Rodriguez, J. L.; Duran, A.; Centeno, M. A.; Martinez-Blanes, J. M.; Robador, M. D. Thermal analysis of monument patina containing hydrated calcium oxalates. *Thermochim. Acta.* **2011**, *512*, 5–12, DOI:10.1016/j.tca.2010.08.015.

[26] Chen, J.; Blume, H.-P.; Beyer, L. Weathering of rocks induced by lichen colonization – A review. *Catena.* **2000**, *39*, 121–146, DOI:10.1016/S0341-8162(99)00085-5.

[27] Sabbioni, C.; Zappia, G. Atmospheric-derived element tracers on damaged Stone. *Sci. Total Environ.* **1992**, *126*, 35–48, DOI:10.1016/0048-9697(92)90482-8.

[28] Riontino, C.; Sabbioni, C.; Ghedini, N.; Zappiaa, G.; Gobbi, G.; Favoni, O. Evaluation of atmospheric deposition on historic buildings by combined thermal analysis and combustion techniques. *Thermochim. Acta.* **1998**, *321*, 215–222, DOI:10.1016/S0040-6031(98)00462-6.

[29] Ghedini, N.; Sabbioni, C.; Pantani, M. Thermal analysis in cultural heritage safeguard: An application. *Thermochim. Acta.* **2003**, *406*, 105–113, DOI:10.1016/S0040-6031(03)00224-7.

[30] Comite, V.; Miani, A.; Ricca, M.; La Russa, M.; Pulimeno, M.; Fermo, P. The impact of atmospheric pollution on outdoor cultural heritage: An analytic methodology for the characterization of the carbonaceous fraction in black crusts present on stone surfaces. *Environ. Res.* **2021**, *201*, 111565, DOI:10.1016/j.envres.2021.111565.

[31] Perez-Rodriguez, J. L.; Duran, A.; Perez-Maqueda, L. A. Thermal study of unaltered and altered dolomitic rock samples from ancient monuments. *J. Therm. Anal. Calorim.* **2011**, *104*, 467–474, DOI:10.1007/s10973-011-1348-5.

[32] Warne, S. S. T. J. Carbonate mineral detection by variable atmosphere differential thermal analysis. *Nature.* **1977**, *269*, 678, DOI:10.1038/269678a0.

[33] Bradley, W. F.; Burst, J. F.; Graf, D. L. Crystal chemistry and differential thermal effects of dolomite. *Am. Mineral.* **1953**, *3-4*, 207–217.

[34] Franquelo, M. L.; Robador, M. D.; Perez-Rodriguez, J. L. Study of coatings by thermal analysis in a monument built with calcarenite. *J. Therm. Anal. Calorim.* **2015**, *121*, 195–201, DOI:10.1007/s10973-015-4432-4.

[35] Reyes-Zamudio, V.; Angeles-Chávez, C.; Cervantes, J. Clay minerals in historic buildings. *J. Therm. Anal. Calorim.* **2011**, *104*, 405–413, DOI:10.1007/s10973-010-1041-0.

[36] Gatta, T.; Gregori, E.; Marini, F.; Tomassetti, M.; Visco, G.; Campanella, L. New approach to the differentiation of marble samples using thermal analysis and chemometrics in order to identify provenance. *Chem. Cent. J.* **2014**, *8*, 35, DOI:10.1186/1752-153X-8-35.

[37] Tite, M. S. Ceramic production, provenance and use – A review. *Archaeometry.* **2008**, *50*, 216–231, DOI:10.1111/j.1475-4754.2008.00391.x.

[38] Riccardi, M. P.; Messiga, B.; Duminuco, P. An approach to the dynamics of clay firing. *Appl. Clay Sci.* **1999**, *15*, 393–409, DOI:10.1016/S0169-1317(99)00032-0.

[39] Damjanović, L.; Holclajtner-Antunović, I.; Mioč, U. B.; Bikić, V.; Milovanović, D.; Radosavljević Evans, I. Archaeometric study of medeival pottery at Stari (Old) Ras, Serbia. *J. Archaeol. Sci.* **2011**, *38*, 818–828, DOI:10.1016/j.jas.2010.11.004.

[40] Damjanović, L.; Bikić, V.; Šarić, K.; Erić, S.; Holclajtner-Antunović, I. Characterization of the early byzantine pottery from Caričin Grad (South Serbia) in terms of composition and firing temperature. *J. Archaeol. Sci.* **2014**, *46*, 156–172, DOI:10.1016/j.jas.2014.02.031.

[41] Damjanović, L.; Mioč, U.; Bajuk-Bogdanović, D.; Cerović, N.; Marić-Stojanović, M.; Andrić, V.; Holclajtner-Antunović, I. Archaeometric investigation of medieval pottery from excavations at Novo Brdo, Serbia. *Archaeometry*. **2016**, *58*, 380–400, DOI:10.1111/arcm.12185.

[42] Perišić, N.; Marić-Stojanović, M.; Andrić, V.; Mioč, U. B.; Damjanović, L. Physicochemical characterization of pottery from Vinča culture, Serbia, regarding firing temperature and decoration technique. *J. Serb. Chem. Soc.* **2016**, *81*, 1415–1426, DOI:10.2298/JSC160823100P.

[43] Gajić-Kvaščev, M.; Bikić, V.; Wright, V. J.; Radosavljević Evans, I.; Damjanović-Vasilić, L. Archaeometric study of 17th/18th century painted pottery from the Belgrade fortress. *J. Cult. Herit.* **2018**, *32*, 9–21, DOI:10.1016/j.culher.2018.01.018.

[44] Stojanović, S.; Bikić, V.; Miličić, L.; Radosavljević Evans, I.; Scarlett, N. V. Y.; Brand, H. E. A.; Damjanović-Vasilić, L. Evidence of continuos pottery production during the late Byzantine period in the Studenica Monastery, a UNESCO World Heritage Site. *Microchem. J.* **2019**, *146*, 557–567, DOI:10.1016/j.microc.2019.01.056.

[45] Damjanović-Vasilić, L.; Bikić, V.; Stojanović, S.; Bajuk-Bogdanović, D.; Džodan, Đ.; Mentus, S. Application of analytical techniques for unveiling the glazing technology of medieval pottery from the Belgrade Fortress. *J. Serb. Chem. Soc.* **2020**, *85*, 1329–1343, DOI:10.2298/JSC200401036D.

[46] MacKenzie, R. C. *Differential Thermal Analysis of Clays*; Mineralogical Society: London; 1975.

[47] Shoval, S.; Paz, Y. A study of the mass-gain of ancient pottery in relation to archeological ages using thermal analysis. *Appl. Clay Sci.* **2013**, *82*, 113–120, DOI:10.1016/j.clay.2013.06.027.

[48] Shepard, A. O. *Ceramics for the Archaeologist. Publication 609*; Carnegie Institute of Washington; 1960.

[49] Drebushchak, V. A.; Mylnikova, L. N.; Drebushchak, T. N.; Boldyrev, V. V. The investigation of ancient pottery. Application of thermal analysis. *J. Therm. Anal. Calorim*. **2005**, *82*, 617–626, DOI:10.1007/s10973-005-6913-3.

[50] Shoval, S.; Beck, P.; Kirsh, Y.; Levy, D.; Gafttand, M.; Yadin, E. Rehydroxylation of clay minerals and hydration in ancient pottery from the 'Land of Geshur'. *J. Therm. Analysis*. **1991**, *37*, 1579–1592, DOI:10.1007/BF01913490.

[51] Shoval, S.; Yadin, E.; Panczer, G. Analysis of thermal phases in calcareous Iron Age pottery using FT-IR and Raman spectroscopy. *J. Therm. Anal. Calorim.* **2011**, *104*, 515–525, DOI:10.1007/s10973-011-1518-5.

[52] Wilson, M. A.; Hoff, W. D.; Hall, C.; McKay, B.; Hiley, A. Kinetics of moisture expansion in fired clay ceramics: A (Time)1/4 law. *Phys. Rev. Lett*. **2003**, *90*, 125503, DOI:10.1103/PhysRevLett.90.125503.

[53] Wilson, M. A.; Carter, M. A.; Hall, C.; Hoff, W. D.; Ince, C.; Savage, S. D.; McKay, B.; Betts, I. M. Dating fired-clay ceramics using long-term power law rehydroxylation kinetics. *Proc. R. Soc. A.* **2009**, *465*, 2407–2415, DOI:10.1098/rspa.2009.0117.

[54] Drebushchak, V. A.; Mylnikova, L. N.; Drebushchak, T. N. Thermoanalytical investigations of ancient ceramics. Review on theory and practice. *J. Therm. Anal. Calorim*. **2018**, *133*, 135–176, DOI:10.1007/s10973-018-7244-5.

[55] Moropoulou, A.; Bakolas, A.; Bisbikou, K. Thermal analysis as a method of characterizing ancient ceramic technologies. *Thermochim. Acta*. **1995**, *269/270*, 743–753, DOI:10.1016/0040-6031(95)02570-7.

[56] Papadopoulou, D. N.; Lalia-Kantouri, M.; Kantirani, N.; Stratis, J. A. Thermal and mineralogical contribution to the ancient ceramics and natural clays characterization. *J. Therm. Anal. Calorim.* **2006**, *84*, 39–45, DOI:10.1007/s10973-005-7173-y.
[57] Ion, R.-M.; Ion, M.-L.; Fierascu, R. C.; Serban, S.; Dumitriu, I.; Radovici, C.; Bauman, I.; Cosulet, S.; Niculescu, V. I. R. Thermal analysis of Romanian ancient ceramics. *J. Therm. Anal. Calorim.* **2010**, *102*, 393–398, DOI:10.1007/s10973-009-0226-x.
[58] Drebushchak, V. A.; Mylnikova, L. N.; Drebushchak, T. N. The mass-loss diagram for the ancient ceramics. *J. Therm. Anal. Calorim.* **2011**, *104*, 459–466, DOI:10.1007/s10973-010-1230-x.
[59] Kloužková, A.; Zemenová, P.; Kohoutková, M.; Mazač, Z. Ageing of fired-clay ceramics: Comparative study of rehydroxylation processes in a kaolinitic raw material and moon-shaped ceramic idol from the Bronze Age. *Appl. Clay Sci.* **2016**, *119*, 358–364, DOI:10.1016/j.clay.2015.11.002.
[60] Carşote, C.; Badea, E.; Miu, L.; Della Gatta, G. Study of the effect of tannins and animal species on the thermal stability of vegetable leather by differential scanning calorimetry. *J. Therm. Anal. Calorim.* **2016**, *124*, 1255–1266, DOI:10.1007/s10973-016-5344-7.
[61] Della Gatta, G.; Badea, E.; Ceccarelli, R.; Usacheva, T.; Mašić, A.; Coluccia, S. Assessment of damage in old parchments by DSC and SEM. *J. Therm. Anal. Calorim.* **2005**, *82*, 637–649, DOI:10.1007/s10973-005-6883-5.
[62] Rosu, L.; Varganici, C. –. D.; Crudu, A. –. M.; Rosu, D. Influence of different tanning agents on bovine leather thermal degradation. *J. Therm. Anal. Calorim.* **2018**, *134*, 583–594, DOI:10.1007/s10973-018-7076-3.
[63] Badea, E.; Della Gatta, G.; Budrugeac, P. Characterisation and evaluation of the environmental impact on historical parchments by differential scanning calorimetry. *J. Therm. Anal. Calorim.* **2011**, *104*, 495–506, DOI:10.1007/s10973-011-1495-8.
[64] Cohen, N. S.; Odlyha, M.; Foster, G. M. Measurement of shrinkage behaviour in leather and parchment by dynamic mechanical thermal analysis. *Thermochim. Acta.* **2000**, *365*, 111–117, DOI:10.1016/S0040-6031(00)00618-3.
[65] Chahine, C. Changes in hydrothermal stability of leather and parchment with deterioration: A DSC study. *Thermochim. Acta.* **2000**, *365*, 101–110, DOI:10.1016/S0040-6031(00)00617-1.
[66] Cucos, A.; Budrugeac, P. The impact of natural ageing on the hydrothermal stability of new and artificially aged parchment and leather samples. *Thermochim. Acta.* **2018**, *669*, 40–44, DOI:10.1016/j.tca.2018.09.006.
[67] Carsote, C.; Badea, E. Micro differential scanning calorimetry and micro hot table method for quantifying deterioration of historical leather. *Herit. Sci.* **2019**, *7*, 48, DOI:10.1186/s40494-019-0292-8.
[68] Budrugeac, P.; Miu, L. The suitability of DSC method for damage assessment and certification of historical leathers and parchments. *J. Cult. Herit.* **2008**, *9*, 146–153, DOI:10.1016/j.culher.2007.10.001.
[69] Fessas, D.; Signorelli, M.; Schiraldi, A.; Kennedy, C. J.; Wess, T. J.; Hassel, B.; Nielsen, K. Thermal analysis on parchments I: DSC and TGA combined approach for heat damage assessment. *Thermochim. Acta.* **2006**, *447*, 30–35, DOI:10.1016/j.tca.2006.04.007.
[70] Cavallaro, G.; Donato, D. I.; Lazzara, G.; Milioto, S. Determining the selective impregnation of waterlogged archaeological woods with poly(ethylene) glycols mixtures by differential scanning calorimetry. *J. Therm. Anal. Calorim.* **2013**, *111*, 1449–1455, DOI:10.1007/s10973-012-2528-7.
[71] Wiedemann, H. G.; Bayer, G. Thermoanalytical study on ancient materials and light it sheds on the origin of letters and words. *Thermochim. Acta.* **1986**, *100*, 283–314, DOI:10.1016/0040-6031(86)87062-9.
[72] Tomassetti, M.; Campanella, L.; Tomellini, R.; Meucci, C. Thermogravimetiric analysis of fresh and archeological waterlogged woods. *Thermochim. Acta.* **1987**, *117*, 297–315, DOI:10.1016/0040-6031(87)88124-8.
[73] Cavallaro, G.; Donato, D. I.; Lazzara, G.; Milioto, S. A comparative thermogravimetric study of waterlogged archaeological and sound woods. *J. Therm. Anal. Calorim.* **2011**, *104*, 451–457, DOI:10.1007/s10973-010-1229-3.

[74] Romagnoli, M.; Galotta, G.; Antonelli, F.; Sidoti, G.; Humar, M.; Kržišnik, D.; Čufar, K.; Davidde Petriaggi, B. Micro-morphological, physical and thermogravimetric analyses of waterlogged archaeological wood from the prehistoric village of Gran Carro (Lake Bolsena-Italy). *J. Cult. Herit.* **2018**, *33*, 30–38, DOI:10.1016/j.culher.2018.03.012.

[75] Vecchio, S.; Luciano, G.; Franceschi, E. Explorative kinetic study on the thermal degradation of five wood species for applications in the archaeological field. *Ann. Chim.* **2006**, *96*, 715–725, DOI:10.1002/adic.200690074.

[76] Donato, D.; Lazzara, G.; Milioto, S. Thermogravimetric analysis: A tool to evaluate the ability of mixtures in consolidating waterlogged archaeological woods. *J. Therm. Anal. Calorim.* **2010**, *101*, 1085–1091, DOI:10.1007/s10973-010-0717-9.

[77] Franceschi, E.; Cascone, I.; Nole, D. Study of artificially degraded woods simulating natural ageing of archaeological findings. *J. Therm. Anal. Calorim.* **2008**, *92*, 319–322, DOI:10.1007/s10973-007-8722-3.

[78] Budrugeac, P.; Emandi, A. The use of thermal analysis methods for conservation state determination of historical and/or cultural objects manufactured from lime tree wood. *J. Therm. Anal. Calorim.* **2010**, *101*, 881–886, DOI:10.1007/s10973-009-0671-6.

[79] Sandu, I. C. A.; Brebu, M.; Luca, C.; Sandu, I.; Vasile, C. Thermogravimetric study on the ageing of lime wood supports of old paintings. *Polym. Degrad. Stab.* **2003**, *80*, 83–91, DOI:10.1016/S0141-3910(02)00386-5.

[80] Franceschi, E. Thermoanalytical methods: A valuable tool for art and archaeology. A study of cellulose-based materials. *J. Therm. Anal. Calorim.* **2011**, *104*, 527–539, DOI:10.1007/s10973-011-1343-x.

[81] Franceschi, E.; Luciano, G.; Carosi, F.; Cornara, L.; Montanari, C. Thermal and microscope analysis as a tool in the characterization of ancient papyri. *Thermochim. Acta.* **2004**, *418*, 39–45, DOI:10.1016/j.tca.2003.11.051.

[82] Bausch, F.; Owusu, D. D.; Jusner, P.; Rosado, M. J.; Rencoret, J.; Rosner, S.; Del Río, J. C.; Rosenau, T.; Potthast, A. Lignin quantification of Papyri by TGA–not a good idea. *Molecules.* **2021**, *26*, 4384, DOI:10.3390/molecules26144384.

[83] Fierascu, R. C.; Avramescu, S. M.; Vasilievici, G.; Fierascu, I.; Paunescu, A. Thermal and spectroscopic investigation of Romanian historical documents from the nineteenth and twentieth century. *J. Therm. Anal. Calorim.* **2016**, *123*, 1309–1318, DOI:10.1007/s10973-015-5089-8.

[84] Fierascu, I.; Avramescu, S. M.; Fierascu, R. C.; Ortani, A.; Vasilievici, G.; Cimpeanu, C.; Ditu, L.-M. Micro-analytical and microbiological investigation of selected book papers from the nineteenth century. *J. Therm. Anal. Calorim.* **2017**, *139*, 1377–1387, DOI:10.1007/s10973-017-6370-9.

[85] Girardi, F.; Bergamonti, L.; Isca, C.; Predieri, G.; Graiff, C.; Lottici, P. P.; Cappelletto, E.; Ataollahi, N.; Di Maggio, R. Chemical–physical characterization of ancient paper with functionalized polyamidoamines (PAAs). *Cellulose.* **2017**, *24*, 1057–1068, DOI:10.1007/s10570-016-1159-8.

[86] Damjanović, L.; Gajić-Kvaščev, M.; Đurđević, J.; Andrić, V.; Marić-Stojanović, M.; Lazić, T.; Nikolić, S. The characterization of canvas painting by the Serbian artist Milo Milunović using X-Ray fluorescence, micro-Raman and FTIR Spectroscopy. *Radiat. Phys. Chem.* **2015**, *115*, 135–142, DOI:10.1016/j.radphyschem.2015.06.017.

[87] Lj., D.; Marjanović, O.; Marić-Stojanović, M.; Andrić, V.; Mioč, U. B. Spectroscopic investigation of icons painted on canvas. *J. Serb. Chem. Soc.* **2015**, *80*, 805–817, DOI:10.2298/JSC140722099D.

[88] Preusser, F. Untersuchung von Werken der Kunst- und Kulturgeschichte mit Hilfe der Differentialthermoanalyse. *J. Therm. Analysis.* **1979**, *16*, 277–283, DOI:10.1007/BF01910689.

[89] Burmester, A. Investigation of paint media by differential scanning calorimetry (DSC). *Stud. Conserv.* **1992**, *37*, 73–81, DOI:10.2307/1506399.

[90] Odlyha, M.; Scott, R. P. W. The 'enthalpic' value of paintings. *Thermochim. Acta.* **1994**, *234*, 165–178, DOI:10.1016/0040-6031(94)85142-5.

[91] Hradil, D.; Grygar, T.; Hradilova, J.; Bezdička, P. Clay and iron oxide pigments in the history of painting. *Appl. Clay Sci.* **2003**, *22*, 223–236, DOI:10.1016/S0169-1317(03)00076-0.

[92] Juliá, C. G.; Bonafé, C. P. The use of natural earths in picture: Study and differentiation by thermal analysis. *Thermochim. Acta.* **2004**, *413*, 185–192, DOI:10.1016/j.tca.2003.10.016.

[93] Thomas, P. S.; Stuart, B. H.; McGowan, N.; Guerbois, J. P.; Berkahn, M.; Daniel, V. A study of ochres from an Australian aboriginal bark painting using thermal methods. *J. Therm. Ana.l Calorim.* **2011**, *104*, 507–513, DOI:https://doi.org/10.1007/s10973-011-1336-9.

[94] Odlyha, M.; Thickett, D.; Sheldon, L. Minerals associated with artists' paintings and archaeological iron objects. *J. Therm. Anal. Calorim.* **2011**, *105*, 875–881, DOI:10.1007/s10973-011-1636-0.

[95] Duce, C.; Bramanti, E.; Ghezzi, L.; Bernazzani, L.; Bonaduce, I.; Colombini, M. P.; Spepi, A.; Biagi, S.; Tiné, M. R. Interactions between inorganic pigments and proteinaceous binders in reference paint reconstructions. *Dalton Trans.* **2013**, *42*, 5975–5984, DOI:10.1039/C2DT32203J.

[96] Izzo, F. C.; Zendri, E.; Biscontin, G.; Balliana, E. TG–DSC analysis applied to contemporary oil paints. *J. Therm. Anal. Calorim.* **2011**, *104*, 541–546, DOI:10.1007/s10973-011-1468-y.

[97] Bonaduce, I.; Carlyle, L.; Colombini, M. P.; Duce, C.; Ferrari, C.; Ribechini, E.; Selleri, P.; Tiné, M. R. A multi-analytical approach to studying binding media in oil paintings; Characterisation of differently pre-treated linseed oil by DE-MS, TG and GC/MS. *J. Therm. Anal. Calorim.* **2012**, *107*, 1055–1066, DOI:10.1007/s10973-011-1586-6.

[98] Bonaduce, I.; Carlyle, L.; Colombini, M. P.; Duce, C.; Ferrari, C.; Ribechini, E.; Selleri, P.; Tiné, M. R. New insights into the ageing of linseed oil paint binder: A qualitative and quantitative analytical study. *PLOS ONE.* **2012**, *7(11)*, e49333, DOI:10.1371/journal.pone.0049333.

[99] Hermans, J. J.; Keune, K.; van Loon, A.; Iedema, P. D. The crystallization of metal soaps and fatty acids in oil paint model systems. *Phys. Chem. Chem. Phys.* **2016**, *18*, 10896, DOI:10.1039/c6cp00487c.

[100] Ghezzi, L.; Duce, C.; Bernazzani, L.; Bramanti, E.; Colombini, M. P.; Tiné, M. R.; Bonaduce, I. Interactions between inorganic pigments and rabbit skin glue in reference paint reconstructions. *J. Therm. Anal. Calorim.* **2015**, *122*, 315–322, DOI:10.1007/s10973-015-4759-x.

[101] Odlyha, M.; Boon, J. J.; van den Brink, O.; Bacci, M. Environmental research for art conservation (ERA). *J. Therm. Analysis.* **1997**, *49*, 1571–1584, DOI:10.1007/BF01983717.

[102] Odlyha, M.; Cohen, N. S.; Foster, G. M. Dosimetry of paintings: Determination of the degree of chemical change in museum exposed test paintings (smalt tempera) by thermal analysis. *Thermochim. Acta.* **2000**, *365*, 35–44, DOI:10.1016/S0040-6031(00)00611-0.

[103] Maines, C. A.; de la Rie, E. R. Size-exclusion chromatography and differential scanning calorimetry of low molecular weight resins used as varnishes for paintings. *Prog. Org. Coat.* **2005**, *52*, 39–45, DOI:10.1016/j.porgcoat.2004.06.006.

[104] Vizárová, K.; Reháková, M.; Kirschnerová, S.; Peller, A.; Šimon, P.; Mikulášik, R. Stability studies of materials applied in the restoration of a baroque oil painting. *J. Cult. Herit.* **2011**, *12*, 190–195, DOI:10.1016/j.culher.2011.01.001.

[105] Chelazzi, D.; Chevalier, A.; Pizzorusso, G.; Giorgi, R.; Menu, M.; Baglioni, P. Characterization and degradation of poly(vinyl acetate)-based adhesives for canvas paintings. *Polym. Degrad. Stab.* **2014**, *107*, 314–320, DOI:10.1016/j.polymdegradstab.2013.12.028.

Vesna Rakić, Steva Lević, and Vladislav Rac

Chapter 11 The application of calorimetry and thermal methods of analysis in the investigation of food

Abstract: This chapter presents the possibilities of using calorimetry and thermal methods of analysis in the field of food investigation. Studying the effect of temperature changes on the physical and chemical properties of raw materials and final products is significant because most manufacturing protocols involve changing the temperature (heating or cooling) over a wide range of values. In addition, temperature changes occur during transport and storage, as well as during final preparations for food consumption. The quality of the final product can be influenced by temperature changes; moreover, temperature changes can affect the functionality, the property of food which is comprehended as very important, nowadays. The chapter gives examples of the application of these methods in the analysis of all main food components (proteins, lipids, and carbohydrates). In addition, the possibilities of their application in the investigation of interactions between food components, and in the research and development related to the newest trends in both food and related technologies, are presented.

Keywords: calorimetry, DSC, TG, ITC, temperature-dependent properties of food

11.1 Introduction

This chapter aims to present the possibility of applying calorimetry and thermal methods of analysis in the fields related to food: food characterization, the control of its quality and safety, as well as in the field of food-processing protocols.

Food, being considered as a source of energy and chemical species having specific roles in the metabolism, can be understood as a set of substances that living organisms eat or drink; liquids taken as drinks are also considered food [1]. It is very well known that the main classes of substances found are carbohydrates, proteins, fats, vitamins and minerals, and dietary fibers. Hence, food samples can contain a large number of chemical species that can be present in all three aggregate states (mixtures of more liquids, liquids with gases or with solids); those substances can be found in a wide range of concentrations, from tents of percents to so-called traces [1, 2]. Hence, it is clear that foods can be very complex systems, what brings us to the fact that food analytics is a real challenge.

Vesna Rakić, Steva Lević, Vladislav Rac, Faculty of Agriculture, University of Belgrade, 11080 Zemun, Serbia

https://doi.org/10.1515/9783110590449-011

Besides, it should be kept in mind that today most of the food consumed is produced by the food industry. Throughout the long history of food production, there have been always two main goals: (i) the transformation of agricultural products (or collected food) into attractive and healthy final product, which is safe for consumption; and (ii) the preservation of food so that it can be later consumed and/or transported. Industrial food production can be global in scale, which includes requirements for extended food shelf life [2, 3]. Therefore, there is a need of constant transformations of production protocols, which consequently requires continuous improvement and adjustment of methods for food analysis.

Most manufacturing protocols (such as sterilization, pasteurization, evaporation, freezing, cooking) involve changing the temperature (heating or cooling) over a wide range of values. In addition, temperature changes occur during transport and storage, as well as during final preparations for consumption. Changes in temperature can cause changes in physical and chemical properties of both the raw materials and the final products (or some of their components); *id est*, the quality of the final product (taste, appearance, texture, etc.) can be influenced [1–3]. Chemical transformations caused by temperature change can be oxidation or reduction, as well as hydrolysis, while the most common physical transformations caused by temperature are evaporation, melting, aggregation, crystallization, or gelation [1, 4, 5]. Moreover, the interactions that exist among food components are affected by temperature changes, as well [1, 6–8]. It is of great importance for food producers to know the impact of temperature on all these processes in order to be able to optimize the conditions of production, transport, and storage, and to control the quality of raw materials and final products.

Due to all mentioned above, it is of great importance for food scientists to have available the analytical techniques that enable to monitor the changes provoked by temperature changes. There are many methods that can be useful in monitoring temperature-dependent food properties. Some of them belong to the group of spectroscopic (UV-visible, IR, NMR spectroscopy) or physical (density, rheology measurement, etc.) techniques. However, as already mentioned, the possibilities of thermal analysis techniques – those that measure changes in some physical properties of materials while a controlled change in temperature occurs – will be presented here. This chapter provides examples of previous applications of thermal methods in the field of food investigation, with the intention to provide insight into the possibilities of these techniques and pave the way for new ideas for their application in the same domain.

11.2 Temperature-dependent properties of foods

In many food samples, changes in temperature cause changes in mass. The mass of the sample can increase or decrease depending on the specific physicochemical process which takes place. Most often, the sample loses mass during heating because volatile components are released (evaporation, drying, and release of gases); however, the absorption of moisture is also possible from the atmosphere, which increases the mass.

The density of pure substances that are not in the process of phase transformation usually decreases with increasing temperature, because the kinetic energy of atoms and molecules increases with increasing temperature, and as a result of their more intensive movement, the space between particles in the sample increases. Knowledge of the changes in density with temperature change is important for engineers who design the production process. For example, density data are important for the choice of material of the containers in which raw materials or finished products are kept, as well as their volume. Similarly, density is a decisive feature if liquid material passes through some pipe. If a phase transformation occurs during the process, the change in density is more pronounced [5].

Rheology studies flow and deformation of some materials. In the case of food, there are many phenomena that involve flow and deformation, like spreading of butter on bread or a stream of liquid from a teapot. The texture of foods, which is very important in the food selection process, is a result of structures formed from food components via complex physical, chemical, and biological changes during processing and storage. Among other properties we sense, there are hardness, elasticity, or stickiness. Moreover, fluid flow properties are very important in the food processing field. In the case of liquids, gels, and solids, their rheological properties can be influenced by temperature and are investigated by food scientists and engineers so as to enable to design food with the desired qualities and to optimize food processing protocols [1, 9, 10].

Phase transformations occur by the action of external factors (temperature) – these are the processes during which the material passes from one physical (aggregate) state to another. In the case of food samples, the most common phase transformations are melting, crystallization, evaporation, condensation, sublimation, and glass transitions [5]. The solid-liquid transformation of the amorphous material is known as glass transition, and it is probably the most important physicochemical characteristics of noncrystalline amorphous solid. More precisely, one amorphous melt can be supercooled to a viscoelastic, "rubbery" state or to a solid, "glassy" state. Foods behave more as syrups rather than rubbers: this kind of transformation is in fact very often in the case of food with low water content (like amorphous sugars, for example). Hence, to know the value of glass temperature, T_g is very important for processing and stability control in the food industry [5, 11]. This transition always occurs below the melting temperature T_M of the material.

A specific transformation that can often be found in foods is gelation – many foods contain components that are capable of forming gel when the food is heated or cooled under appropriate conditions [12]. Usually, gels are three-dimensional networks of aggregated biopolymers or colloidal particles that entrap a large volume of water, resulting in the formation of a "solid-like" structure. The physical properties of gels (appearance – transparent or opaque, water-holding capacity, rheology, and stability) depend ultimately on the type, structure, and interactions of the molecules or particles that they contain – the interactions among molecules being strongly dependent on the temperature. Eggs, starches, jellies, yogurts, and meat products are systems where gelation makes an important contribution to their overall properties. In some foods a gel is formed on heating (heat-setting gels), while in others it is formed on cooling (cold-setting gels). Gels may also be either thermo-reversible (gelatin) or thermo-irreversible (egg-white). Evidently, the gelation temperature (commonly known as the temperature at which a sudden change in viscosity occurs) is an important piece of information, so methods are needed that enable its determination.

The properties and temperature-dependent processes mentioned so far are physical. However, changes in food temperature also cause chemical transformations, that is, changes in chemical properties. The most known change of this type is oxidation, particularly in the case of lipids [13–15]. However, the interactions that occur between the molecules of food components that belong to other groups of substances (proteins, fibers, etc.) are also of great importance [1, 6, 16]. Figure 11.1 shows possible interactions of polyphenols with proteins or cellulose. These compounds of plant origin have attracted a lot of attention due to their antioxidant activities. It is clear that the interactions they accomplish with certain compounds present in food affect this activity, and that is why these interactions are increasingly the subject of research. These interactions, from hydrogen bonds to ionic bridges, are temperature dependent and have the influence on the activity of polyphenolic compounds. Hence, the methods enabling the investigation of such complex events are needed.

11.3 Experimental techniques available to monitor temperature-dependent properties of food: the acquired data

Thermogavimetry (TG) is a technique that enables the monitoring of mass while the temperature is changed in time. Usually, temperature is a linearly increasing function of time, but it can decrease as well, or it can be held at a particular (constant) temperature for a period of time. TG apparatus is a sensitive balance situated within a furnace whose pressure, temperature, and gaseous environment can be carefully controlled – what enables to imitate the various types of processing and storage conditions that a food might be subjected to. Thus, TG enables to determine the changes in mass associated with physical (evaporation, release of gases, condensation, etc.) or chemical (oxidation, hydrolysis) transformations that occur in food samples along heating or cooling.

A **dilatometer** is used to measure the changes of density with temperature. Usually, the subject of interest is the change in fat (or fatty components of food) density with temperature change (during melting or crystallization). This is usually a graduated glass

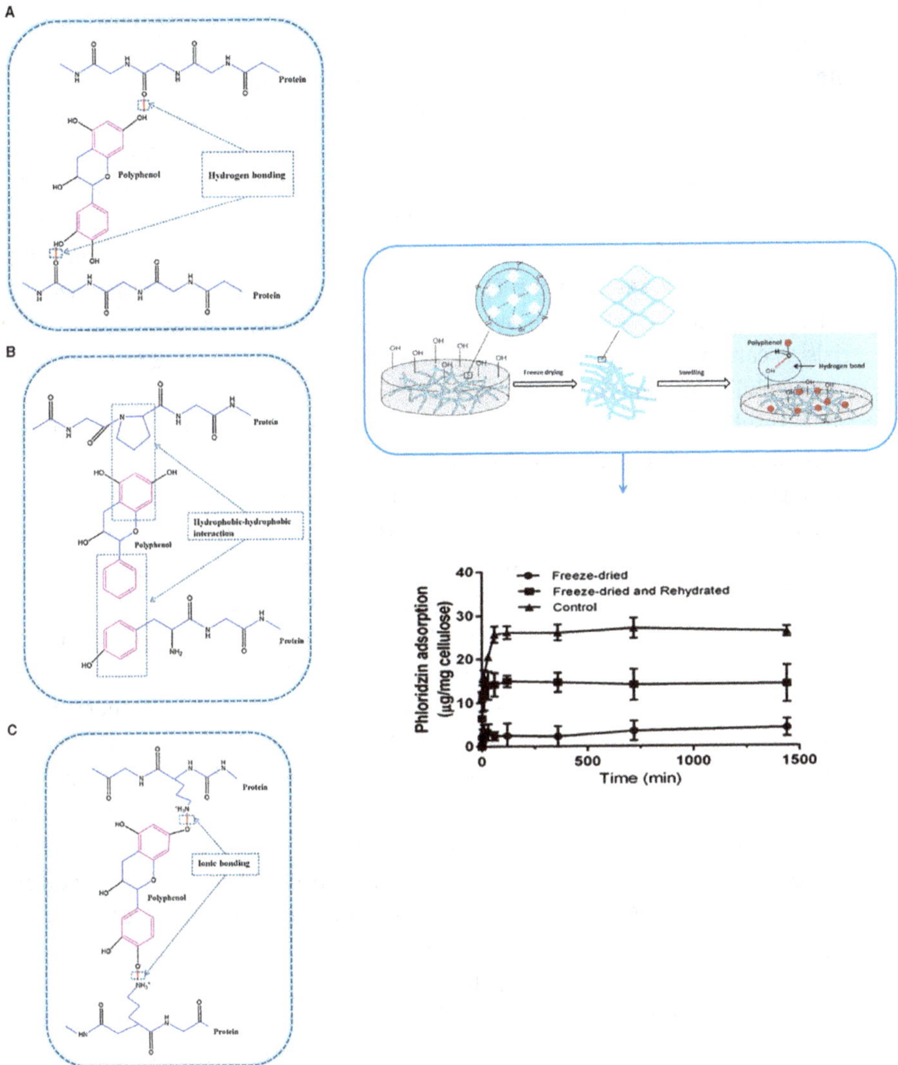

Fig. 11.1: Left: Noncovalent interactions between proteins and polyphenols: (A) hydrogen bonding; (B) hydrophobic-hydrophobic interaction; (C) ion bonding [16]. Right: Interactions of polyphenolic molecules with cellulose (reprinted from [6] with permission from Elsevier, copyright 2017).

U-tube that is kept in a temperature-controlled water bath. The sample, initially put as molten inside this U-tube, is subsequently cooled at a controlled rate, and the change in volume of the sample is measured as a function of temperature. The so-called solid fat index (SFI, which is in fact the solid-to-liquid ratio for some fat, at a temperature of interest) is often the subject of interest of dilatometric measurement. Interestingly, the same data can be obtained much faster using differential scanning calorimetry (DSC) method, which will be elaborated on later.

Rheological thermal analysis techniques (the most known thermomechanical analysis – TMA and dynamic mechanical analysis – DMA) enable to monitor the temperature dependence of certain characteristics such as viscoelastic behavior, expansion coefficient, expansion and shrinkage of fibers ad films, swelling bahavior, softening, and so on. The most common interest in the domain of food science is to determine shear modulus of fatty food, the viscosity of biopolymer solutions, or the shear modulus of biopolymer gels. For example, the TMA method has been used to characterize the rheological properties of Ca-alginate hydrogels, which are important when using these materials as carriers in encapsulation [17]. DMA has been used to investigate protein-based hydrophilic thin films as promising materials for the manufacture of edible food packaging; results related to physicochemical behaviors of hydrophilic films as a function of temperature in isohume conditions have been obtained [18].

Although the mentioned techniques are of great importance in food science, by far the greatest use in this domain is of one of the so-called differential techniques: differential thermal analysis (DTA) and especially that of differential scanning calorimetry (DSC) [1, 2, 5, 19–21]. In both techniques, two measurement cells are situated in a temperature-controlled environment (furnace) that, itself, can be heated or cooled in a controlled manner. In the case of DTA, the apparatus is constructed so as to enable monitoring the temperature difference between a test and a reference material along time, as the identical temperature regime is imposed on both materials. In the case of the DSC method, the calorimeter measures the heat flow into (endothermic) or out of (exothermic) sample undergoing a phase change. The principle of DSC is to keep, for a given temperature program, the sample and reference in the same oven at equal temperature; the signal is recorded as a heat flux. In the case of power-compensated DSC, the electrical power which is required to maintain temperature equality between the sample and the reference material (placed in separate micro-ovens) is equivalent to the calorimetric effect.

Both DTA and DSC can be used to give information related to phase transformations and chemical changes, and for obtaining of data such as the temperature intervals of observed changes and the temperatures where they have maximal values (T_M or T_{peak}, values often related to thermal stability of investigated material). The next important piece of information is the direction of the peak that corresponds to the nature of the transition, being heat absorption (endothermic – like evaporations, melting of solids and denaturation of proteins) or heat releasing (exothermic – like crystallization of carbohydrates and aggregation of proteins) and the determination of inflection points (that are indicative of glass transitions – T_g). Finally, these methods allow the determination of heat effect itself – the value of ΔH can be calculated. A review of the literature shows that the DSC technique has a significantly greater application than DTA for the analysis of food samples. A review of the literature shows that the DSC technique has been more often applied than DTA for the analysis of food samples. The reason is, most probably, that the DTA analysis is designed to allow direct reading of specific temperatures, while in the case of DSC analysis this temperature difference can be converted into a heat-flux difference that allows direct determination of caloric effects (such as melting or crystallization heats). Moreover, DSC instruments are designed to allow working with small sample amounts under suitable conditions (specific atmosphere, pressure, and work at sub-ambient temperatures).

It is of special importance to mention here that DSC enables the determination of heat capacity, a quantity that is a fundamental thermodynamic characteristic of any material. It is a measure of how the material stores additional energy at the molecular level as it is heated. The change in the value of heat capacity for the investigated material that happens during a certain event is a trustworthy indication that a change in the structure of that material has occurred. A simplified way of its determination from DSC measurements involves simple division of heat flow signal by the heating rate and the sample mass. More precisely, it can be done using well-characterized reference material, usually sapphire, as a standard.

The DSC signal is regularly represented as the heat flux plot against time (temperature), but it can also be represented as the value of the heat capacity as a function of time (temperature). If so, the enthalpy change that appears in a certain temperature interval (from T_1 to T_2) can be determined from the integration of the area under the DSC curve:

$$\Delta H(T) = \int_{T_1}^{T_2} C_P dT \tag{11.1}$$

where the integration covers the desired temperature range. The entropy change can be calculated from the area under the plot of C_P/T versus temperature:

$$\Delta S(T) = \int_{T_1}^{T_2} \frac{C_P}{T} dT \tag{11.2}$$

The free energy change ΔG, which indicates whether some process is spontaneous and advantageous for products forming ($\Delta G < 0$) or not ($\Delta G > 0$), can then be calculated using the known thermodynamic equation:

$$\Delta G(T) = \Delta H(T) - T \cdot \Delta S(T) \tag{11.3}$$

Table 11.1 compiles the endothermic and exothermic events that can be monitored during the investigation of food. The table shows the thermal characteristics of individual food components, as well as their significances, *id est*, the feature(s) and the part of the production protocol that are under the influence of these endothermic or exothermic changes. The list of presented features (parts of production protocols), their number, and their variety indicate how important is to know thermodynamic properties of food/food components.

Tab. 11.1: Thermal behavior of main food components.

Food component	Thermal behavior		Significance
	Endothermic	**Exothermic**	
Proteins	Denaturation	Aggregation, crystallization	Thermal stability, functional behavior, storage
Lipids	Melting (polymorphism)	Crystallization, oxidation	Thermal stability, resistance to oxidation, emulsification, storage
Carbohydrates	Melting, glass transition	Crystallization, decomposition	Thermal stability, processing at higher temperatures (e.g., spray drying)
Starch	Denaturation,	Retrogradation, decomposition	Modulation of nutritional properties, i.e., “resistant starch”
Water	Vaporization, sublimation	Crystallization	Process optimization and improved economy. Structure modifications
Yeast	Inactivation	Fermentation	Influence on metabolism and sensorial profile of products
Bacteria	Inactivation	Growth, metabolism	Influence on metabolism and sensorial profile of products; elimination of undesirable bacteria by thermal inactivation (e.g., pasteurization and sterilization)

In addition to the DTA and DSC methods, in which the sample is subjected to a controlled heating regime, isothermal calorimetry also has its application in food science. Isothermal calorimetry is one of the oldest scientific methods for testing matter; however, one very advanced version is used in this domain, so-called isothermal titration calorimetry (ITC), which provides the ability to examine interactions between food components. This is of great importance for understanding the role of individual components in food, especially in terms of their functionality. Functionality of food is manifested by special products that have a positive effect on the mental and physical health of people and that affect the prevention of diet-related diseases [22–24].

Isothermal titration calorimeter can measure enthalpy changes that occur as a result of interactions between different types of molecules. An ITC instrument consists of a reference cell (reference material does not undergo the enthalpy change), a sample cell, and an injector. Small aliquots of the injection solution are sent periodically into the sample solution contained within the sample cell, and the energy required to keep the sample and reference cells at the same temperature is measured as a function of time. The resulting thermogram consists of a plot of heat, Q, versus time (as presented in Fig. 11.2). The nature (exothermic, endothermic), magnitude (area under the curve), and shape of the peaks give the information about interactions between molecules in the injector and in the sample cell [8, 25–29].

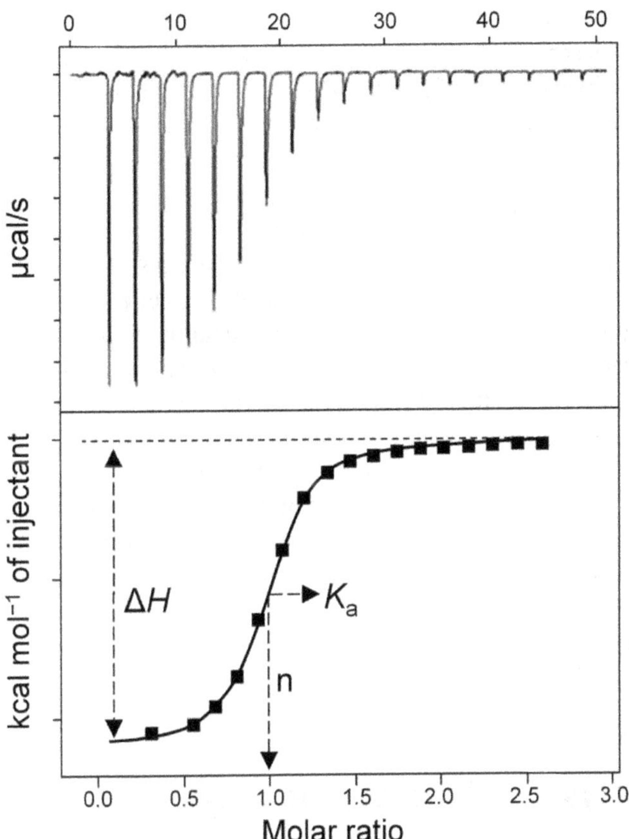

Fig. 11.2: Typical results obtained from isothermal titration calorimetry experiment. Top: the power signal recorded as a function of time, the peaks represent the heats produced upon a ligand aliquot injection into the receptor solution within the calorimetric cell. The peak integration gives the heat evolved. Bottom: Integrated normalized heats from each titration step (corrected by the heats of dilution) give resulting titration curve – the heats of binding as a function of molar ratio of the two components. Affinity constant K_a is calculated as a slope of this curve in its inflection point, ΔH is obtained as a difference between the curve's upper and lower plateaus, while n is the molar ratio of the two components at the curve's inflection point (reproduced from [26] with permission from the Royal Society of Chemistry, copyright 2016).

The example shown in Fig. 11.2 refers to the interaction receptor – ligand [26]. The instrument records the enthalpy change that occurs after each injection as a result of the interaction between the ligand and receptor molecules. The interval between each injection should be long enough to allow the change to go to completion. The area under each peak is proportional to the heat released as a result of the interaction of a single dose of injected substance with the substrate (sample). In addition to measuring the thermal effect itself, it is necessary to check whether the entire amount of injected dose reacted with the sample, because the thermal effect is calculated per mole of injected substance. Finally, we can obtain information about the number of binding sites on the protein and the strength of the binding interactions.

Among all the previously mentioned techniques, TGA and DSC have the greatest application in food science. In addition to the basic techniques, there are also so-called coupled methods – connection of TG or DSC devices with detectors that allow obtaining additional data. First of all, it is very common to connect two thermal methods of analysis – TGA and DSC – a balance is added to the apparatus which consists of sensitive DSC furnace. Then, for systems that require analysis of the gas components that evolve from the sample during heating, there is a possibility to connect devices that allow their analysis (qualitative or both qualitative and quantitative). The gas chromatograph as a detector system is one possibility. The other is to connect the spectrometer (mass spectrometer or FTIR) to be used as a detection system. An important characteristic of DSC measurement is that it enables the characterization of bioactive molecules as they are, without special preparation, which is also valid for the ITC technique. Investigation of significant properties of important food components (macromolecules such as proteins or lipids), such as the melting point but also the strength of the bonds between their segments, is possible using very small amounts of samples. DSC can be used to evaluate the stability of these substances as well as the influence of various factors on it. In the following text, typical examples of the application of these methods in food science will be presented.

11.4 Thermal analysis of food components

11.4.1 Thermal analysis of proteins

The thermal stability of proteins has long been a subject of interest for food chemists. The investigations devoted to proteins' thermal stability and reactivity are of particular importance, especially since the addition of proteins and enzymes to modern food materials has become common in the food industry [29]. Since it became clear that the creation of unique proteins' spatial structures is a process with thermodynamic parameters as decisive, a need for experimental data on heat values associated with intra and intermolecular interactions in which these molecules are involved arose [30]. Measurements that would be performed in very dilute solutions are especially important, since under those conditions it is possible to avoid nonspecific interactions of macromolecules.

Two already mentioned tools are available to gain the information of interest. To acquire the data about proteins' thermal stability, *id est*, to monitor how the native state of protein is changed along the temperature raising, the DSC technique is used. DSC enables to determine the heat associated with temperature change at fixed solvent conditions, and so far, numerous data on protein unfolding processes have been obtained using this method, as shown in many published works and reviews [30–38]. It is a clear indication how important the detailed understanding of the denaturation process is. DSC enables the evaluation of thermodynamic parameters of denaturation (enthalpy and entropy of protein unfolding at the transition temperature can be determined from the area under the heat consumption profile) and thus contributes to the understanding of the relationships between these parameters, and the structure and function of proteins. Another methodology available, ITC, enables measurements to be driven at fixed temperature but associated with the changes in solvent conditions; the results that can be obtained are crucial for understanding the reactivity with various ligands – with other specific macromolecules or with small-molecule effectors [39].

As it is well known that thermal treatment is often applied during processing and industrial production of food, it follows that it is of particular interest to understand in detail the denaturation process, which depends on the structure of the molecule, as well as on numerous external factors. The understanding of mechanisms of protein's folding and unfolding is crucial and gives possibility for controlled use of proteins or some specific food derivatives as ingredients or technological coadjuvants in industrial processes. Unfolding of most proteins occurs as a reversible process at low temperatures (below 70 °C); however, at higher temperatures the process is irreversible. Degradation of the protein structure occurs largely as oxidation of side chains [30] and is usually dependent on protein nature and concentration, pH of the solution, the presence of other chemical species, and the ionic strength [29]. A highly sensitive method such as DSC allows acquiring numerous useful informations through the interpretation of thermograms obtained for different proteins, either in their native states or investigated while the factors that affect their stabilities were varied.

In what follows, some typical DSC results are shown to illustrate how the influence of each of these factors on the thermal stability of proteins can be obtained using this method. When considering the presented results, it should be kept in mind that the sample can be pure protein in a buffered solvent, but also a part of a plant or animal muscle containing proteins. Therefore, the obtained thermograms can be complex profiles with overlapped peaks [19, 31–33, 40–42]. An insight into the thermograms shown in Fig. 11.3 indicates, first of all, that different proteins (whether of animal or plant origin) show DSC profiles in different temperature regions and with different shapes. One can therefore conclude that a different DSC profile may indicate the presence of different proteins in the sample of interest. Figure 11.3 (up) presents the heating profiles of extracted proteins from plants (A – pea, B – chickpea, C – lentil protein); while Fig. 11.3 (bottom) presents three temperature regions where the endothermic peaks assigned to the most abundant meat proteins appear.

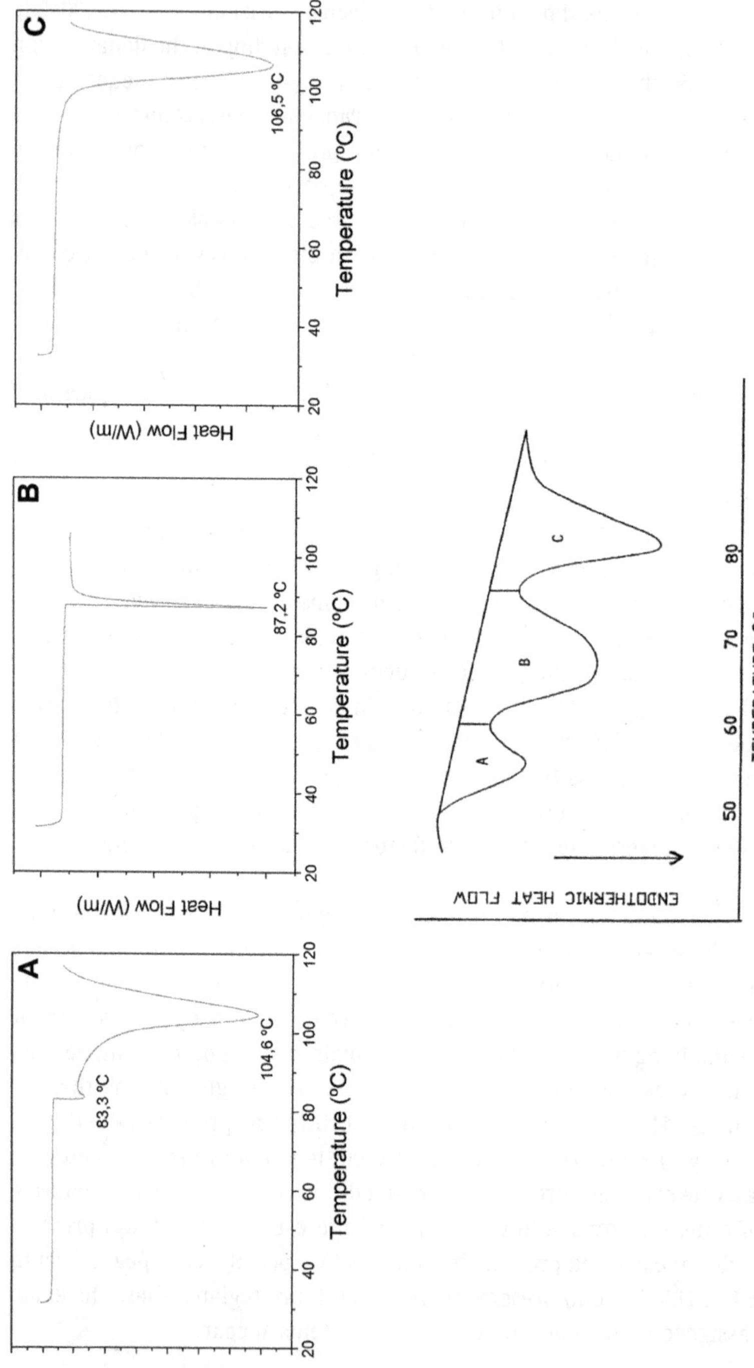

Fig. 11.3: Top: DSC profiles of protein extracts from pea (A), chickpea (B), and lentil (C) (reprinted from [40] with permission from Springer Nature, copyright 2016; bottom – typical DSC result obtained from animal muscle). Three major regions correspond to myosin subunits (A), sarcoplasmic proteins and collagen (B), and actin (C) (reprinted from [41] with permission from Elsevier, copyright 2005).

Although the identification of individual chemical species is not possible using thermal analytical methods, the interpretation of DSC profile of the egg white shown in Fig. 11.4 (left), complementary to the results of other methods, allows us to recognize the presence of specific proteins: conalbumin, ovalbumin, and S-ovalbumin in egg white, as well as the enzyme lysozyme.

Interestingly, S-ovalbumin is a special form of ovalbumin; its concentration can be taken as the indication of freshness of the eggs since it is created by the formation of sulfur cross-links if the pH value changes during storage [1]. Its presence in egg white can lead to the necessity to increase the denaturation temperature by about 9 °C, which explains the higher cooking temperatures in the case of "older" eggs. Lysozyme is an enzyme which shows antibacterial properties; food safety is a reason why extracted lysozyme is often added to some food products [1].

The natural aging of white egg is the explanation of changes that can be noticed in DSC profiles of aged eggs (Fig. 11.4, right). In older samples, changes are not significant in the low-temperature region where denaturation of conalbumin and lysozyme occurs. However, the results obtained in older samples show that the initial ovalbumin is converted mostly to the intermediate or to the S-modification. Apparently, the DSC profile of egg white can serve as an indication of the eggs' freshness.

When considering the DSC profile of a certain protein, the degree of its unfolded part is represented by the area under the thermogram up to some specific temperature; the area under the rest of the peak is proportional to the folded fraction, while the overall area under the DSC profile is proportional to the entire amount of protein present.

It is not only the changing of temperature which may affect the stability of proteins – other factors can provoke the unfolding as well. The addition of acid, for example, leads to a lowering of the pH value, that is to an increase in the concentration of hydronium ions (in an aqueous solution). Consequently, the interactions in the protein are altered – hydrogen bonds and those interactions that involve amino acids become weaker. As a result, as the pH value decreases, the endothermic peaks observed in DSC profiles shift to lower temperature values, what can be seen in the example of thermal denaturation of 11S soybean globulin at different pH values shown in Fig. 11.5.

Alcohols can affect the thermal stability of proteins as well, what is of importance for food technology. Alcohols can decrease the denaturation temperature and, hence, the protein conformational stability. DSC profiles presented in Fig. 11.6 (left) show that DSC peak of 11S broad bean globulin is shifted toward lower temperatures with the increase of ethanol concentration. However, it is important to emphasize that some calorimetric data also reveal that low alcohol concentrations have stabilizing effect for the native protein conformation, what is of interest for food storage [34].

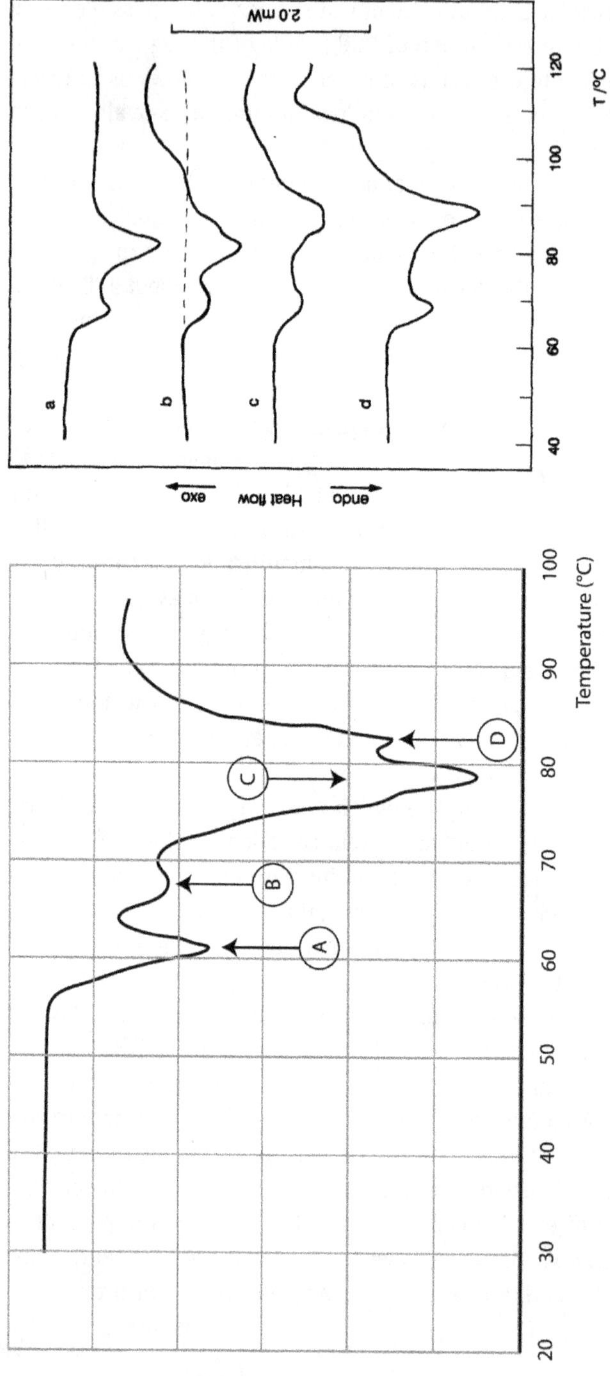

Fig. 11.4: Left: DSC thermogram of fresh egg white. The peaks correspond to A – conalbumin, B – ovalbumin, C – lysozyme, D – S-ovalbumin (reprinted from [1] with permission from IOP Publishing, copyright 2015). Right: DSC traces of egg white: (a) fresh, (b) 4 days, (c) 19 days (d) 34 days; heating rate 2/min (reprinted from [31] with permission from Elsevier, copyright 1992).

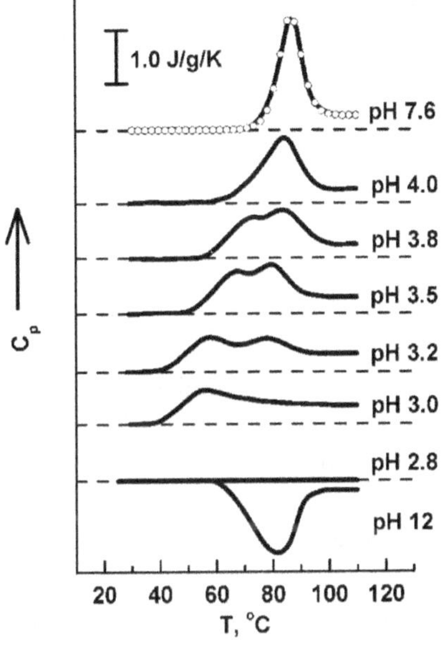

Fig. 11.5: Thermal denaturation of 11S soybean globulin at different pH values (used with permission of John Wiley & Sons – Books from [34], copyright 2009; permission conveyed through Copyright Clearance Center, Inc.).

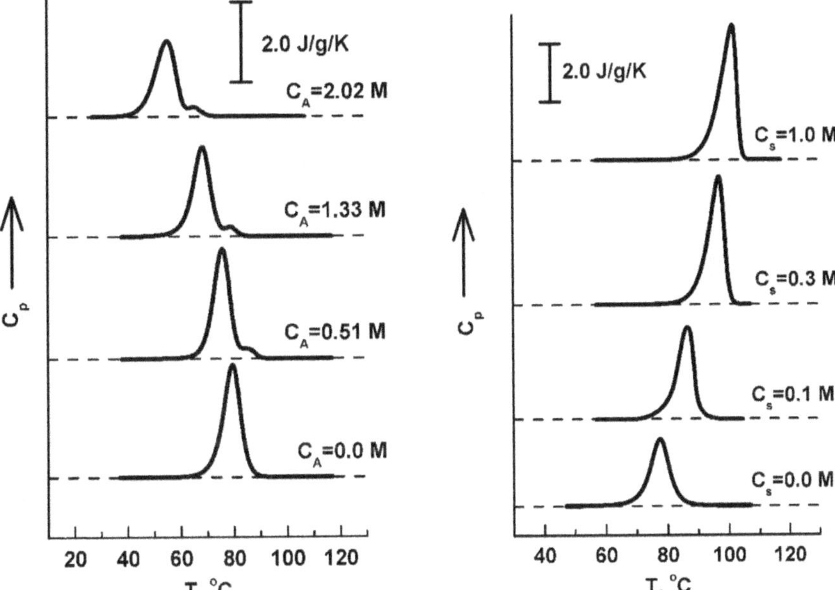

Fig. 11.6: Left: Thermal denaturation of 11S soybean globulin, different ethanol concentrations were added. Right: Thermal denaturation of 11S soybean globulin, different Na^+ concentrations were added (used with permission of John Wiley & Sons – Books from [34], copyright 2009; permission conveyed through Copyright Clearance Center, Inc.).

It is worth noting that the increase in the concentration of some cations can lead to thermal stabilization of the protein, as evidenced by the increase in DSC peak areas and their appearance in higher temperatures regions. As in the case of low ethanol concentrations, the above-mentioned result is of particular importance for food storage. The example of Na^+ concentration effect on 11S broad bean globulin thermal stability is presented in Fig. 11.6 (right) – evidently the increase in Na^+ concentrations shifts peak to higher temperatures. However, it has to be emphasized here that these phenomena are very complex, dependent on many factors, and that the interpretations in some cases are not unambiguous. For example, DSC technique was applied to investigate the influence of different salts ($CaCl_2$, $MgCl_2$, NH_4Cl, NaCl, and KCl) on the thermodynamic stability and reversibility of the lysozyme's folding-unfolding process. Thermodynamic parameters (ΔH, apparent heat capacity change, ΔC_p and the melting transition temperature T_M) were determined; the results of those experiments clearly show that cations destabilize the lysozyme [43].

The products of thermal degradation of protein structure undergo aggregation – a process that often occurs from the very beginning of thermal treatment. These processes have been extensively investigated in the case of, for example, white egg proteins, where, as a result of aggregation, a marked increase in viscosity is observed. Thermal denaturation of proteins is followed by aggregation, which is an exothermic process, and it can be noticed as an exothermic signal that follows denaturation when the sample is heated to temperatures above 100 °C, as shown in Fig. 11.7.

In total, two irreversible processes occur during heating: endothermic unfolding and exothermic aggregation of unfolding products, so the overall DSC signal is created by superposition of these two thermodynamic processes of opposite signs. Both processes – unfolding and aggregation – depend on numerous factors (protein concentration, pH of the solution, ionic strength). As the endothermic and exothermic effects are overlapped, it is not possible to precisely determine the extent of the heat effect that belongs to the formation of aggregates, but it can be concluded that it comprises a smaller part of the total heat involved in the entire unfolding-aggregation cycle. This is a very significant conclusion, since it confirms that thermal treatment of proteins results in a material significantly different from the initial structure.

However, a reliable indication of a change in the structure of any material, including proteins, is a change in its heat capacity which appears as a result of heating. There are many experimental evidences that confirm the change in heat capacity that happen along the heating of proteins. This can be seen directly in the DSC profile (it has been already mentioned that the DSC signal can be displayed as heat capacity versus temperature). Figure 11.8 shows four sets of DSC profiles obtained under the specific values of pH in the case of two proteins and two enzymes. In all these cases the shifts of C_p toward higher values are evident. The partial specific heat capacities of the native and unfolded states are evidently different.

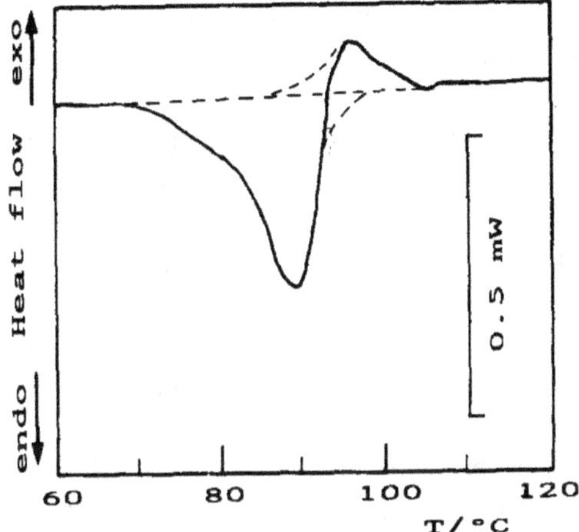

Fig. 11.7: DSC heating profile of commercial ovalbumin in phosphate buffer (pH 6.8) (reprinted from [31] with permission from Elsevier, copyright 1992).

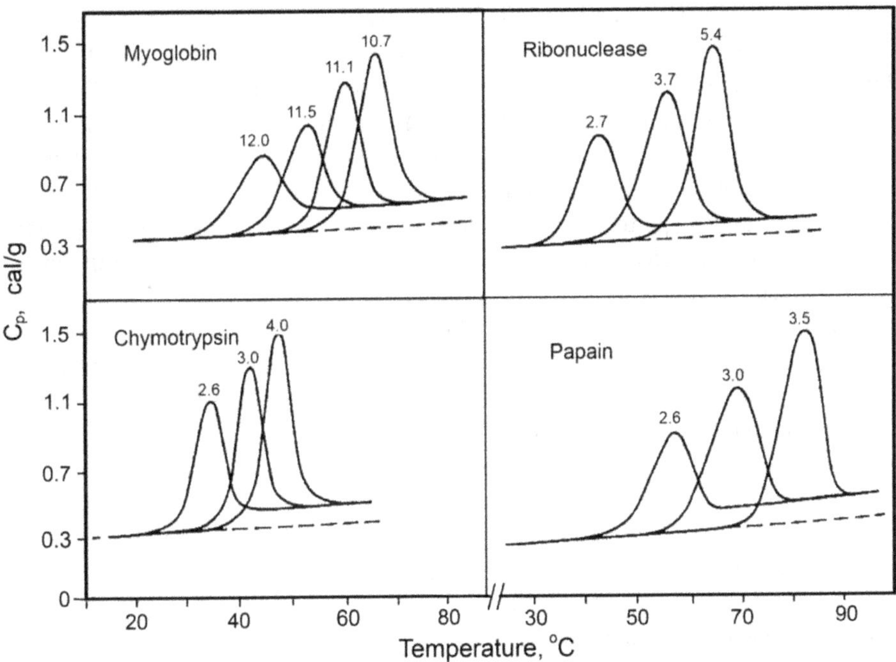

Fig. 11.8: C_p profiles of various globular proteins in solutions having the indicated pH values [44].

These changes in specific heat capacities values are associated with the order-disorder transition (folded to unfolded protein). For the reason of their complex nature, the exact molecular origins of proteins' heat capacity changes have long been a subject of debates. The heat capacity of any protein itself and its change, which appears as a result of denaturation, can be understood as a consequence of hydration effects (protein-solvent and solvent-solvent interactions) and intra-protein interactions – the role of solvent (usually water) being very important. Often, the output from DSC experiment is shown as "excess heat capacity" (sample minus reference) as a function of temperature. Usually, the "buffer baseline" is recorded with both (sample and reference) cells filled with the chosen degassed buffer, using appropriate temperature range and heating rate. Subsequently, a cooled sample cell is filled with protein solution, and heating is repeated using the same experimental conditions. The contribution of protein in the overall heat effect is estimated by subtracting a signal of buffer baseline from the sample signal. Additionally, DSC profile is often expressed as normalized (to a certain amount of protein).

At the beginning of heating, at temperatures below the onset of protein unfolding, a slight increase of baseline is noticed for most proteins. This is explained by the features of water molecule: for the reason of its "hydrogen bond network," water has high specific heat capacity in comparison with any organic substance and is more ordered and tightly packed around the hydrophobic part of protein molecule. With temperature increase, the heat is absorbed, water around protein becomes more like molten (bulk water) – this process increases the heat capacity of aqueous protein solution in this pretransition temperature region. As heating progresses, the protein absorbs energy and starts to unfold, C_p increases abruptly, giving rise to the evident endothermic peak, which can be assigned do unfolding process and with the maximum denoted usually as T_M. Once this process is done, the C_p value decreases at the new baseline, which is usually found at a higher level (positive ΔC_p is noticed).

In the interpretation of this phenomenon of the C_p value increase in the case of proteins during the denaturation process, a large number of proteins were studied using thermal methods; and it was found that this phenomenon is particularly noticeable in globular proteins, that is, they express the highest ΔC_p values. It was found additionally that this heat capacity increment is very specific for a particular protein. Upon further consideration it was observed that the extent of this phenomenon is determined by the overall surface area of the nonpolar groups that are exposed to water during the unfolding process [44]. Finally, the role of water molecules was elucidated – it appears that it is not just a solvent for biological macromolecules, but it is rather a "partner determining their structure." Evidently, calorimetric measurements have a crucial role in all these studies, since they enabled to get direct information on the enthalpy (and entropy) of protein unfolding/refolding from where further conclusions followed.

The possibility of applying calorimetric methods made it possible first to formulate questions and then give answers regarding this very important class of natural macromolecules – proteins. For example, why proteins change from one to another stable conformation, even without adding heat; why denaturation occurs with a relatively small increase in temperature; what is the role of noncovalent interactions; what is the role of the solvent molecule? Obviously, when interpreting the effect of heat capacity increase during denaturation, very important conclusions were reached about the structure and functionality of this crucial group of biomolecules.

Although certainly the results of calorimetric methods were interpreted as complementary to the results obtained by applying other techniques, the effect of increasing the value of ΔC_p during protein denaturation is a good example of the confounding role of calorimetric techniques in the elucidation of such an important phenomenon.

11.4.2 Thermal analysis of fats and oils

Lipids are a very heterogeneous group of compounds of diverse chemical structure and biological origin, including fats, oils, waxes, steroids, and related compounds. The classification can be done according to different criteria such as physical properties at room temperature, polarity, origin, chemical composition and role in the organism, level of complexity, nutritional requirements, and health impact. Based on their physical properties at room temperature, lipids can be divided into oils that are liquid at room temperature and fats that are solid at room temperature (these two groups being important in food consumption and for food industry). They can belong to two big classes if about their origin: they can be of plant or animal origin.

All foods of natural origin contain some amount of lipids. Even in the case of foods of plant origin, lipids can be present at least in seeds, in which their biological role is to be a reserve of energy during germination. There are also foods in which lipids are the dominant component (butter, margarine, and cheese). Basically, natural oils and fats are mixtures of nonpolar molecules – triacylglycerols (triglycerides, TAG) which consist of fatty acids (FA) – chain molecules with an even number of C atoms from 2 to 24. FA molecules with an odd number of carbon atoms appear only in small concentrations in the milk of ruminants [1].

An important property of these materials is that TAG molecules can form spatial arrangements so that ordered structures appear (crystallization occurs); the existence of different crystalline spatial arrangements for the same chemical structure (polymorphism) is also possible – the most known case of polymorphism is certainly that of cocoa butter. In this regard, it is important to point out that fatty acid molecules can contain saturated and unsaturated bonds between C atoms. The presence of cis-double bonds in FA molecules affects the crystallization process – there are fewer defects in crystals formed by FA molecules with saturated bonds than in the case of unsaturated bonds presence. Another important feature of lipids is their melting behavior: the melting temperature of fatty acids can be noticed in the relatively narrow temperature region (~15 to ~80 °C) and increases with the length of the hydrocarbon chain in the FA molecule; while a decrease in melting point is monitored with increasing degree of saturation [45, 46]. The polymorphic forms of one lipid containing the same TAG can exhibit significantly different melting temperatures.

However, it has to be underlined that in natural foods the TAGs contain different FAs that affects the crystal structure, the number of phases, the presence of defects, and finally the melting temperatures. Moreover, the mixture of different TAGs having different FAs can be found in natural food. It can be said that, for most natural foods containing a high percentage of fat, temperature regions rather than the melting point occur and can be observed in DSC thermograms. Furthermore, the thermogram of such samples is created as an overall effect by overlapping the endothermic effects of melting of several types of crystals. The most known example is butter, which is solid at temperatures below 15 °C and liquid at temperatures above 40 °C; but at intermediate temperatures it is semi-liquid, spreadable, and pasty [1]. Hence, a broad distribution of TAGs results not only in a broad temperature region where phase transition occurs but also in special rheological properties assigned to the state in between solid and liquid [5, 47]. For all the aforementioned reasons an important parameter has been introduced to characterize fats: the so-called solid fat content (SFC), which is defined as a percentage of the total fat in the sample which is solidified at a particular temperature [48]. Evidently, this value can vary from 100% at low temperature, where the fat is completely solid, to 0% at high temperatures, where the fat is completely liquid. SFC is an important temperature-dependent property of fats and oils that determines their utility in various applications being a parameter which serves for their characterization and quality control [49]. The common method to determine SFC is NMR spectroscopy – SFC is achieved by measuring both the solid and liquid signals from the NMR free induction decay (FID) of the sample. DSC or DTA can be taken as an alternative method for SFC determination, which has the advantage of obtaining a thermogram in a whole temperature range from one unique measurement. SFC is determined by measuring the heat of fusion successively at specific temperatures; the fraction of fat melted up to those specific temperatures can be determined by reference to the total melting heat (the obtained value should be multiplied by 100) [47].

Figure 11.9 illustrates the difference in melting behavior (presented through the SFC factor) of pure triacylglycerol and an edible oil containing a mixture of many different triacylglycerol molecules [50].

The DSC method enables to collect results related to phase transitions of fatty food and to examine the relations in between their thermal stability and structures. The elucidation of factors that influence melting and crystallization of fats is also a goal of these investigations. Different DSC profiles indicate different chemical composition and/or origin of fats [49, 51, 52]. Animal fats of different origins give different melting profiles. Specific treatments of animal fats will give different melting profiles as well, as can be seen in Fig. 11.10, where the heating profiles of partially hydrogenated soybean oils are presented. Evidently, the origin of fat and the industrial treatment influence significantly the melting profiles; however, we can claim that the temperatures (temperature regions) in which the sample is in solid or liquid state can be read from the DSC thermogram.

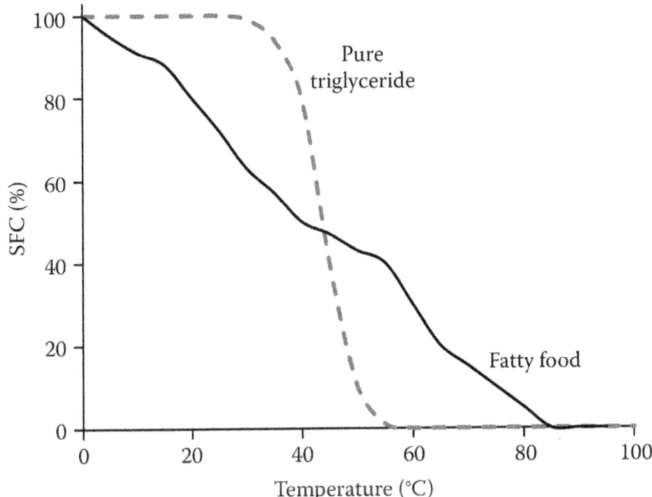

Fig. 11.9: Different melting profiles of pure tryacylglycerol and an edible oil (used with permission of CRC Press from [50], copyright 2016; permission conveyed through Copyright Clearance Center, Inc.).

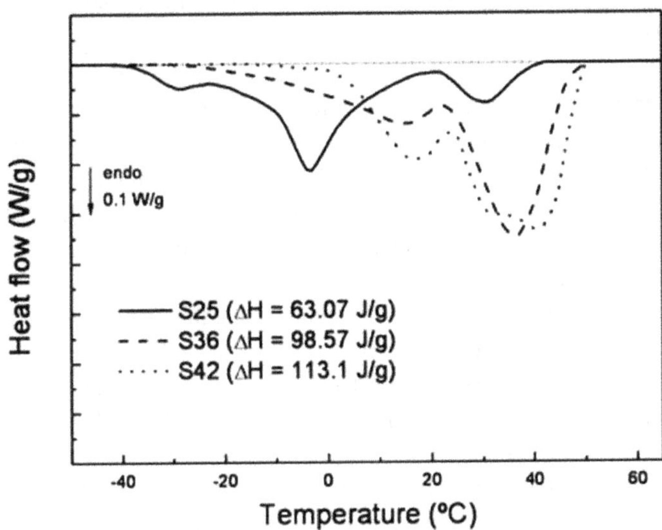

Fig. 11.10: DSC heating profiles of partially hydrogenated soybean oils with different melting points (used with permission of John Wiley & Sons from [49], copyright 2013; permission conveyed through Copyright Clearance Center, Inc.).

In addition to the melting of fat, the crystallization that occurs as a result of cooling is important to be studied and understood in detail. The DSC method, with its associated possibilities of controlled conditions (such as cooling or heating rate, atmosphere, and pressure), is ideal for investigating this process. The understanding of crystallization of fats is important in the case of dairy products (butter or cream, cheese), cacao butter, or coconut oil, for example, since it is related to the overall final products quality. Fat and oil fractionation is the next reason to study crystallization since lipid products with different melting and physical properties are manufactured by crystallization. In addition, triglyceride crystallization is a method to eliminate high-melting compounds from oil, so it remains clear at lower (ambient) temperature [53].

Crystallization curves of milk fat obtained using different cooling rates are presented in Fig. 11.11 (left); two exothermic peaks are noticed. Importantly, different crystallization profiles (the changes in peak shapes, peak areas, and the overlapping degree) are found as a result of cooling rate variation [54]. Figure 11.11 (right) presents cooling profiles of virgin coconut oil blends with two refined oils, soy and safflower oil [55]. Evidently, crystallization of fats can be controlled, which is very important for food industry, the results obtained from DSC experiments being crucial in understanding of this process.

To understand completely the fat crystallization it is necessary to understand its kinetic aspect as well. This is complex phenomena and usually to achieve the overall crystallization kinetics the models such as the Avrami equation is applied, but with the assumptions that the driving force, nucleation and crystal growth rates do not vary, which is often too simplified. It has to be underlined that differential scanning calorimetry has often been used to study the isothermal crystallization kinetics of fats and oils; where the sample is firstly molten and then rapidly cooled to the crystallization temperature; subsequently, the exothermal heat signal induced by the crystallization process is measured as function of time [56–58].

Another important feature of fats and oils can be studied using DSC method is their oxidative stability. Lipid oxidation is a process in which free-radical and nonradical oxygen species react with lipids, causing oxidative destruction of (poly)unsaturated FAs. The mechanism is complex and takes place in three phases. Reactive free (alkyl) radicals (R•), formed by the homolytic decomposition of unsaturated FAs in the first phase, *initiation phase*, further react with oxygen to form peroxide radicals (ROO•); which are responsible for lipid oxidation due to their tendency to accept electrons. During the next phase, *propagation phase*, peroxyl radicals form hydroperoxides with hydrogen from the lipid structure; this leads to further production of radicals and the prolongation of this phase. Hydroperoxides form alkoxy (RO•) or peroxy (ROO•) radicals; subsequently, during the *termination phase*, they react with each other to form stable dimeric products. Volatile hydrocarbons, alcohols and aldehydes (responsible for the smell of oxidized oils) as well as nonvolatile alcohols and ketones are formed as a result of lipid oxidation. Secondary oxidation products, alcohols and unsaturated FAs, also lead to termination products. The result is the formation of viscous materials during

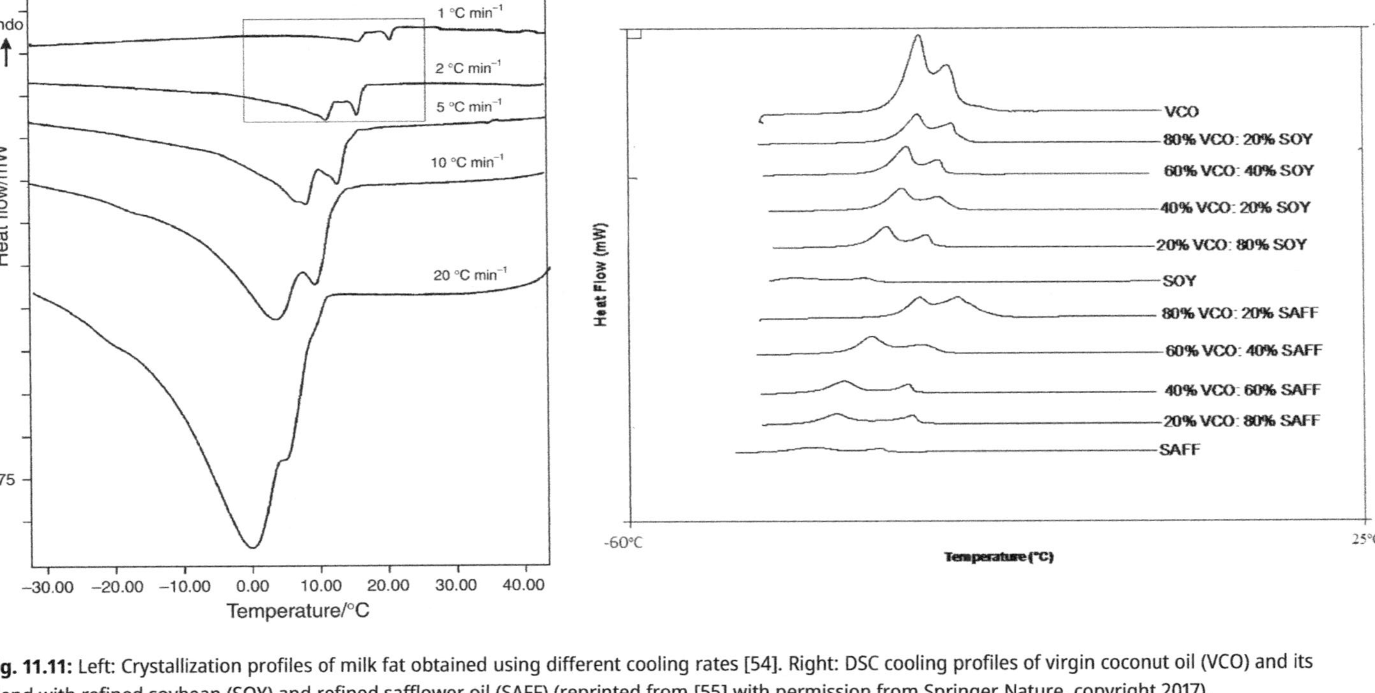

Fig. 11.11: Left: Crystallization profiles of milk fat obtained using different cooling rates [54]. Right: DSC cooling profiles of virgin coconut oil (VCO) and its blend with refined soybean (SOY) and refined safflower oil (SAFF) (reprinted from [55] with permission from Springer Nature, copyright 2017).

polymerization. These polymers are insoluble in fats and represent the final stage of oxidation [59, 60].

Oxidative stability of oil depends mostly on its fatty acid composition, since polyunsaturated FAs oxidize much faster than monounsaturated and saturated FAs, but the stereospecific distribution of FAs in the TAG molecule [61] is also important. It should also be noted that unrefined oils may contain components (oxidation degradation products, free fatty acids, metals, etc.) that contribute to the oxidation. From the other side, they also contain a higher content of natural ingredients with antioxidant properties (tocopherols, carotenoids, phenolic compounds). The shelf life represents the period of time the oil can be preserved from strong oxidation. Therefore, the methods used to determine oil durability are very important. They are based on the accelerated oxidation of oil under the influence of one or more factors that accelerate the process.

Usually, the oxidation is accelerated by heat, in the atmosphere of air. Hence, thermoanalytical methods such as DSC is evidently suitable for the estimation of oxidative stability; and there are indeed literature reports on its application for this purpose. Two approaches exist: the sample must be kept in the oxidative atmosphere (usually synthetic air) either at a constant temperature (isothermal regime) [62–64] or it can be exposed to heating mode (nonisothermal regime) [65, 66]; in both cases, the appearance of an exothermic signal is an indication of an oxidative process. The sample can be kept in closed vessels (under pressure) as well.

Nonisothermal regime was applied in the case of grape-seed oil, where the correlation between the presence of natural antioxidants (total phenolic compounds and α-tocopherol) and oxidative stability has been studied [67]. The coupling of DSC and thermogravimetry was applied. The oxidation onset temperature (OOT) which can be taken as a measure of the end of initiation and a beginning of propagation step is obtained by extrapolation and can be comprehended as a relative measure of oil oxidative stability. In Fig. 11.12, TG signal is evident: coupling of calorimeter with a sensitive thermobalance made it possible to observe a slight mass loss at the beginning of the experiment (loss of small amounts of water and other volatile compounds present in the sample) and then an increase in mass, which was attributed to the consumption of oxygen. It is known from literature that the addition of oxygen to allylic double bonds of FAs causes the mass gain [68]. Both signals (this increase in mass and the exothermic DSC signal) coincide, as can be seen from the figure, and can be attributed to the beginning of the oxidation process.

The OOT values were used to compare the oxidative stability of different oils. In such a way, the oxidative stabilities of grape seed oils obtained from the Cabernet Sauvignon variety by ultrasound assisted extraction were estimated. The oxidative stability of investigated oils was found to be proportional to the duration of ultrasound treatment; since the ultrasounds facilitate the liberation of natural antioxidants (α-tocopherol and phenolic compounds).

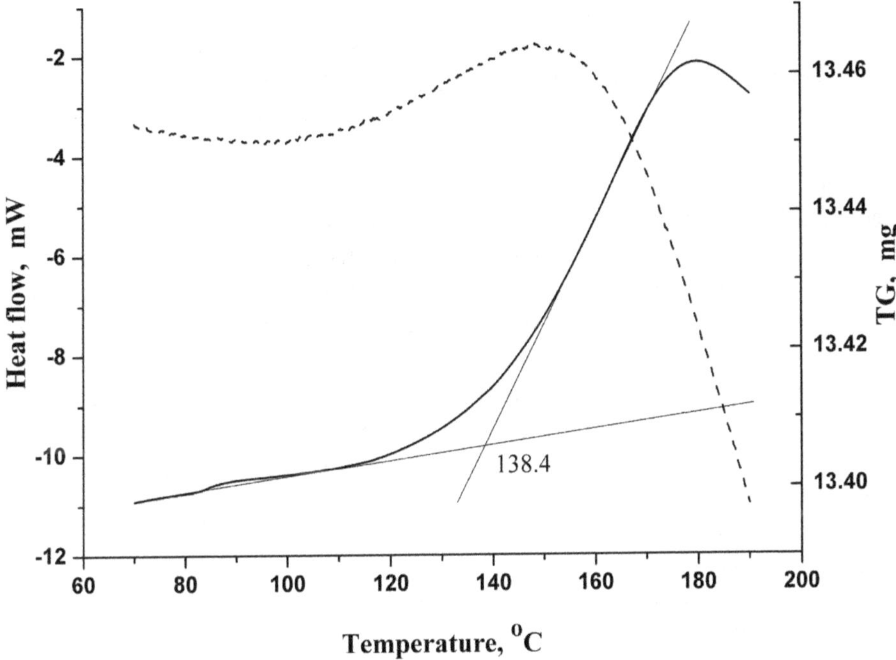

Fig. 11.12: DSC (line) and TG (dots) profiles of nonisothermal oxidation of grape-seed oil. Oxidation onset temperature is taken as the indication of oxidation start (used with permission of John Wiley & Sons from [67], copyright 2014; permission conveyed through Copyright Clearance Center, Inc.).

The DSC method has often been used to achieve another important goal in determining parameters related to the food quality. Furthermore, oils and fats often appear as mixtures made with other lipids, for various reasons: either to increase their oxidative stability or to modify their flavor (for example adding animal tallow to sunflower oil) or simply with the idea of making a cheaper product (adding lard in cocoa butter), which is considered food adulteration [69]. Therefore, there have been many efforts to detect such kind of lipid mixtures, particularly the addition of animal fats (comprehended as contaminants and/or adulterants) in vegetable oils (and in food generally) [70, 71]. These additions are not acceptable for many consumers either for negative nutritional perception related to animal fats consumption or for cultural and religious restrictions. There are also cases of mixing different fats of animal origin (adding tallow to butter, for example) [72, 73], as well as the case of mixing different oils that are not desirable [74]; and there is a need for recognition methods [75].

In the literature, from the pioneering work published in 1992 [76], there are evidences on the application of calorimetric techniques in order of food adulteration recognition: DSC, modulated DSC (DSC techniques where a nonlinear heating or cooling rate is applied to the sample by applying a series of heating or cooling micro-steps followed by an isothermal step), isothermal, or fast calorimetry [75]. Both melting and crystallization profiles have been analyzed. It is particularly important to mention that in these investigations coupled techniques have been applied – usually these are couplings of DSC with some method which enables chemical identification of compounds evolved during heating; the most popular being DSC-Raman spectroscopy, DSC-FTIR (Fourier-transform infrared spectroscopy), and DSC-TG-MS (DSC coupled with thermobalance and mass spectrometer). In the most recent research, DSC coupled with X-ray diffraction technique has been applied. Even though this equipment is primarily applied in materials research and industry, as well as for pharmaceutical purposes, it has been already used to study polymorphic forms and thermal transitions of palm oil products [77] and for the study of polymorphism of extra virgin olive oil [78].

For the purpose of recognizing adulteration, extra virgin olive oil and cocoa butter are the most frequently examined, while in the case of fats of animal origin it is butter, since these foods are the most common subjects of economically motivated adulteration. Olive oil of the highest quality (named extra virgin) is adulterated by cheaper oils (sunflower, soy oil or olive oil of lower quality); while in the case of cocoa and milk butter, the addition of animal lard can be found. In addition, tallow or more water then allowed can be found in milk butter. It can be generally said that the specificities of the calorimetric profiles (the temperature regions of crystallization or melting, T_M values, enthalpy values) of different fats and oils enable the determination of the presence of an unwanted component. However, a statistical approach is often necessary [74]. In the case when the peaks of the two components are far enough on the temperature scale (as in the case of examples presented in Fig. 11.13), the analysis of the results is simpler; however, very often, due to the complexity of thermal profiles, deconvolution of the peaks is required. To conclude, one can state that DSC, either as single or in couplings with other techniques, is very useful in the elucidation of fatty food adulteration.

When considering food containing fats and oils, the next very important application of thermal methods is the investigation of emulsions, that is, their stability. The emulsions are mostly mixtures of two immiscible liquids, one of which is dispersed in the other in the form of small droplets. In the field of food, these systems are extremely important because they are very numerous – they are found in almost all foods, as a majority part (butter) or a minor part (milk). In addition to containing the basic phases that form the emulsion and which themselves have nutritional value, emulsions are very important as carriers of bioactive molecules. For example, fat-soluble vitamins or carotenoids can be protected and carried in a suitable emulsion. There are two main types of emulsions: oil in water (oil droplets dispersed in water – O/W, e.g., milk) and water in oil (W/O – water droplets dispersed in the mass phase, example butter spreads). There are also multiple emulsions of the W/O/W or O/W/O type [81, 82].

Emulsions are thermodynamically unstable systems due to the high surface energy associated with the large contact area between the particles of the water and fat

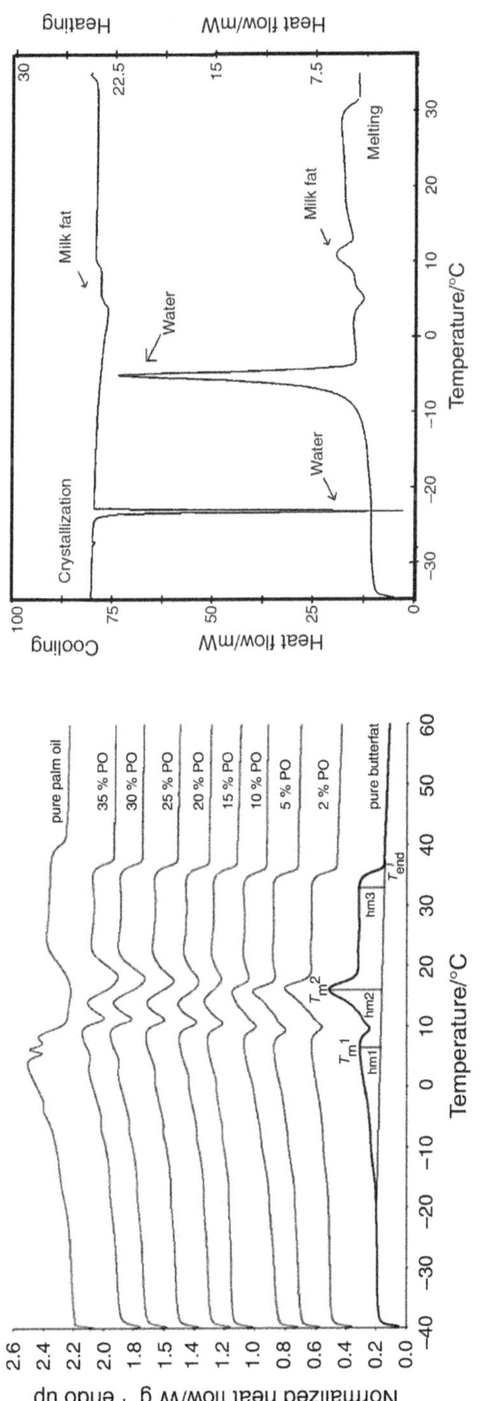

Fig. 11.13: Left: DSC melting profiles of pure milk and its mixtures with palm oil [79]. Right: DSC crystallization and melting of butter containing 15.4% of water (scanning rate 5 °C/min). Both the addition of palm oil (left) and water (right) is evident (reprinted from [80] with permission from Springer Nature, copyright 2012).

phases. The stabilization mechanism is kinetic – strong Brownian movement of particles prevents their separation and sedimentation. In addition, amphiphilic molecules that play the role of emulsifiers exist in emulsions – lecithin or cholesterol being the examples of natural emulsifiers. However, as food is often subjected to heat treatment, it is necessary to elucidate the effects of heating, cooling, freezing, or thawing, which affects the stability and functionality of emulsions. DSC is an ideal method for studying these processes; it is particularly suitable for investigation of crystallization of individual phases present in emulsions [82].

Many phenomena related to the thermal treatment of emulsions were elucidated by the application of the DSC method. First, the crystallization of the fat phase was investigated. Using the DSC method, it was observed that oil dispersed in water crystallizes at a much lower temperature than when it is in the bulk state; the phenomenon is called supercooling (or undercooling). Significantly lower temperature of crystallization was found for hydrogenated palm kernel when it was in the form of an emulsion in comparison with the bulk phase [83]. There is also an effect of added emulsifiers on the temperature and crystallization profile – for example, it has been shown that the addition of diacylglycerol facilitates the crystallization of oil in O/W emulsions [84].

As there is a difference in the crystallization temperature of the oil in the emulsion and in the bulk phase, DSC can be used for the quantification of the separated (free) oil in the emulsion, when the separation occurs as a result of thermal treatment. Figure 11.14 shows the case of confectionary coating fat (CCF) emulsion – from the cooling curves it is evident that, in the first cooling cycle, only one crystallizing peak was detected, indicating that all the oil was in the emulsified state and the emulsion was stable. In the subsequent cycles – the emulsion was heated and cooled again; four cooling-heating cycles were done. Already in the second cooling step, two crystallizing peaks appeared, indicating that one part of the oil was already in the free, separated phase. In the following cooling-heating cycles, the peak assigned to free oil became more and significant, indicating that more oil was separated from emulsion. Importantly, the ratio of the enthalpies of two peaks can be used to estimate the percent of destabilized oil in the emulsion.

In both the O/W and W/O emulsions water crystallization happens. Without going into details here, it can be stated DSC has been widely applied to study water crystallization in the emulsions, and the crystallization of emulsifiers as well. Depending on the emulsion composition and on the emulsifier presence, both water and emulsifier crystallization can influence the stability of emulsions [82, 84]. But, it is very important to underline that DSC technique can be applied to determine the type of emulsion, based on the knowledge that emulsion phases have different thermal behavior in the bulk phase and in emulsified state. This can be done by comparing the thermograms of bulk phase (oil or water) and their corresponding emulsions [84].

Finally, one important role in emulsions is that of proteins that are often present in many food emulsions; it is particularly the case of milk proteins. Proteins mainly have a role of emulsifiers due to their unique amphiphilic character [84, 85–88]. The emulsifying capacity of proteins can be altered under thermal treatment, having in

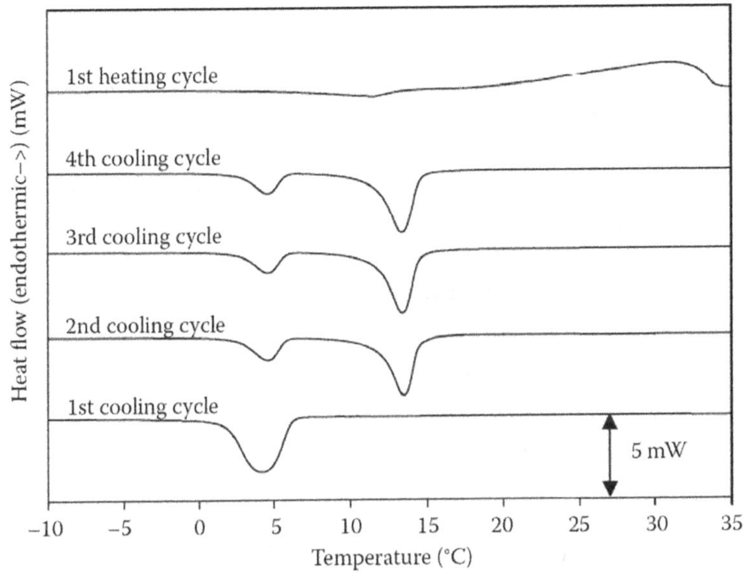

Fig. 11.14: Successive cooling curves of 40% CCF emulsion stabilized with 2 wt% of artificial emulsifier Tween 20 (reprinted from [83] with permission from Elsevier, copyright 2003).

mind that usually above 65 °C the proteins unfold and expose more hydrophobic groups. Unfolded proteins can interact with themselves or with other food ingredients. Whether this will lead to the enhanced emulsifying properties and improved emulsion stability, or to the protein aggregation and the emulsion collapse; is dependent on thermal conditions and other conditions from the surroundings (pH, ionic strength, the presence of other ingredients) [82]. The most important is certainly the protein nature: in studying O/W emulsions of hydrogenated palm kernel oil, DSC was applied to study crystallization/melting behavior of emulsified and no-emulsified samples. For example, it has been shown that total replacement of milk proteins by whey proteins has led to modifications in crystallization behavior of emulsified fat, together with stability against fat droplet aggregation [87].

11.4.3 Thermal analysis of carbohydrates

The third class of nutritionally important compounds comprises carbohydrates. The most abundant is certainly starch, the macromolecules of glucose. Starch makes up a large part of the composition of flour from many grains – wheat, corn, oats, rice – and is also present in potatoes and in other plant and grains. Hence, the most common staple food in many nations, bread necessarily contains starch. In the plants, it has a role of energy storage for germination. In order to provide effective energy storage, starch molecules comprise big number of glucose molecules in a small volume; and such small volume can be achieved if they are packed in crystals. Starch granulae appear in highly crystalline forms, while two types of molecules appear: linear amylose and highly branched amylopectin polymers. Enzymatic cleavage of linear form is easier (by α-amylase), while in the branched form the branching points are cleaved first, by β-amylase. There are other important differences between two forms of starch: they behave significantly differently when dissolving and melting, which determines their physical behavior and all features relevant for cooking and food technology [1]. Fig. 11.15, left, presents typical DSC heating profile of starch together with microscopic pictures of wet potato grains taken at different temperatures.

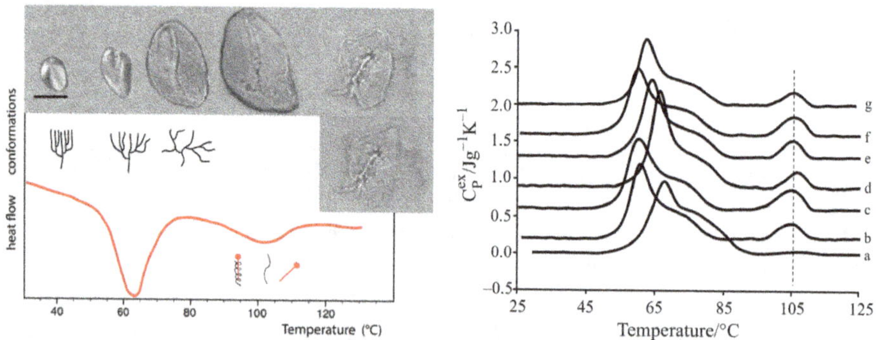

Fig. 11.15: Left: DSC heating profile of starch. Melting point of highly branched amylopectin is ~40 °C, and its structural changes are shown. Microscopic pictures of wet potato grains heated in oil at 30, 65, 75, 80, 120, and 180 °C are presented (reprinted from [1] with permission from IOP Publishing, copyright 2015). Right: DSC profiles of starches obtained from different wheat cultivars, at 50% moisture content. The peak at ~65 °C is related to the starch gelatinization, the second one ~105 °C is related to the dissociation of amylose-lipid complexes (reprinted from [89] with permission from Springer Nature, copyright 2007).

Starch grains melt and dissolve only in the presence of water and heat. The ratio of amylose and amylopectine varies in different plants and grains, even in different cultivars of the same plant, which is reflected in their different heating DSC profiles (Fig. 11.15, right). Starches with high amylose content are interesting because of potential health benefit: it is thought that they are source of resistant starch and that their consumption lowers the incidence of colorectal cancer. Moreover, high-amylose starch can be used for the synthesis of biodegradable packaging materials. Therefore, the determination of the precise amylose content is of interest. Evidently, DSC profile can testify about different starch origins. However, there are also experimental evidences about the possibility to use this method for the determination of amylose content. Rapid and quantitative procedure is enabled by the reversible endothermic transition which is observed for lipid-containing cereal starches and assigned to the melting of helical amylase-lipid complexes. Moreover, there are evidences that the same value can be calculated from the value of exothermic peak obtained upon cooling of gelatinized starch in the presence of phospholipid L-α-lysophosphatidycholine (LPC), which is found to be directly proportional to amylose content. The results are compared with common spectrophotometric iodine-binding procedure [90–92].

Gelatinization is a physical transformation that occurs in starch-containing systems (food) – it is basically swelling of starch granules; hence it requires the presence of water, but also the application of heat, and is essentially an order/disorder transformation. In fact, starch loses its crystalline form during gelatinization. An important parameter for this process is the presence of fats, which exist naturally in many plants; the complexes of fats and starch are very common. All these circumstances exist during cooking, so the gelatinization process occurs normally during thermal processing of food [93].

The process of gelatinization depends on the nature of the raw material (actually on the properties of the starch granule such as composition, morphology, molecular structure, and molecular weight) and the amount of water, the starch/water ratio being a decisive factor. Apart from water and fats, other potentially present components are also of influence – hydrocolloid gluten is of particular importance in the case of dough. Wheat flour, the most often used for bread production, is mainly composed of starch, proteins, and lipids. Upon addition of water, the natural hydrocolloid gluten creates wheat dough, forming a three-dimensional network filled with starch granules. The starch gel and coagulated protein matrix make a surrounding of air bubbles formed during fermentation to form a firm bread structure [94]. Figure 11.16 illustrates the complexity of interactions that can be found in a system like dough.

Evidently, it is very important to study all aspects of gelatinization process. The appropriate methods must be applied to study this complex phenomenon. Since viscosity is altered importantly with the gel forming, rheological methods are involved. Viscosity changes are the result of the swelling of starch granules, and of the increase in the solubility of macromolecules. The obtained results help to understand the important steps in the gelatinization process. Microscopies are nondestructive and allow the observation of visible changes on starch granules. Another method necessary to verify the crystal structure is X-ray diffraction.

Calorimetric methods enable online measurements during the rise in temperature, *id est*, during the gelatinization process itself. In general, during heating, starch molecules undergo the glass transition, then they pass through the phase of gelatinization and formation of amylose-lipid complexes before amorphization occurs, while recrystallization happens as a result of cooling [1]. When investigating all these phenomena, apart from the DSC technique, which most often gives the value of ΔH for gelatinization process, thermogravimetry is also used, as there are significant changes in mass, particularly during gelatinization. Thermogravimetry is an appropriate method to recognize the presence of different steps of mass loosing that appear during heating. TG profiles obtained for four natural starches present four steps in their degradation: first two being assigned to dehydratation and dehydrogenation processes [94]. The authors noticed that starches originated from different cereals exhibited different TG profiles.

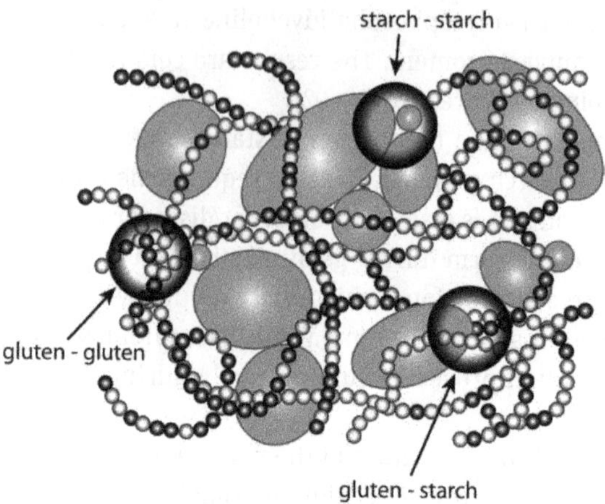

Fig. 11.16: The illustration of possible interactions that can be found in dough (reprinted from [1] with permission from IOP Publishing, copyright 2015).

It is worth noting that application of both techniques (DSC and TG) enables kinetic calculations. There are literature evidences on the application of well-known kinetic models to calculate the kinetic parameters (activation energy) for the gelatinization (Kissinger method) and the degradation of starch (Kissinger, Flynn-Wall-Ozawa, Šatava-Šesták, and Coats-Redfern methods) [95–97].

Finally, it has to be mentioned here that nowadays there is an interest for enzyme-resistant starch (amylose crystals, RS) due to its ability to undergo bacterial fermentation in the large intestine; this is of nutritional significance – RS is comprehendend as a functional food ingredient which can regulate intestinal micro-flora (it has pre-biotic role). Its content in foods is dependent on the type of starch and on the particular processing or cooking conditions. There are literature reports confirming that the degree of starch gelatinization resulting from different processing conditions can affect the RS content of the food products. Since gelatinization process has been extensively studied using thermal methods, and particularly DSC technique, it follows that this technique and the obtained results are crucial for the development of procedures for RS production [89].

11.5 Interactions between food components

The possibility of investigating molecular interactions during which molecular associations are formed can be of crucial importance for the understanding of many physical, chemical, and biological phenomena. Many methods are used for the purpose of understanding molecular interactions; however, only calorimetric techniques provide insight into thermodynamic and kinetic parameters. Calorimetry enables the use of minimal amounts of sample, while the results can be often extracted from only one experiment. As the subject of this text is the application of thermal methods in the examination of the food system, it should be emphasized that the most common interactions of natural compounds that occur (not only) in food are precisely those with proteins. All proteins have the ability to "recognize" and make connections with specific molecules such as other proteins, polyphenolic compounds, amines, amides, peptides, alkaloids, terpenes; but also with molecules – active components of medicines [26]. Obviously, the understanding of food functionality is currently very interesting. However, it should be emphasized that it is actually the motive of designing drugs (based on understanding the structure of active molecules, as well as the thermodynamic parameters of protein binding with those molecules) a decisive factor for performing a large number of calorimetric experiments. The binding of ligand and protein (or other macromolecule) will occur only if the resulting complex is more stable than the initial system. Binding can occur between the ligand and the natural (folded) protein, which will further stabilize the natural state of the protein. If binding of ligand to an already unfolded protein happens, the natural state of the protein will be destabilized.

Two methods belonging to the group of thermo-analytical techniques can be used to investigate the above-mentioned interactions: differential scanning calorimetry and isothermal titration calorimetry. DSC method is particularly suitable for determining whether the interaction between the ligand and the macromolecule has led to stabilization or not. If the ligand preferentially binds to the natural state of the protein (or other macromolecule), the denaturation temperature of the resulting ligand-protein complex will be higher if compared with the temperature at which the protein itself is unfolded. A typical study of this type is an example of selected carbohydrates' influence on the stability of bovine proteins [98]. Glucose, sucrose, and oligosahharide inulin were used as protective agents during freeze-drying. Indeed, the results on thermal protein properties obtained using DSC (denaturation temperature and enthalpy) demonstrated that endothermic transition shifted to higher temperatures, thus confirming the stabilizing effect in the order: inulin > glucose > sucrose.

However, probably the most important contribution of DSC technique in this field of food science is from the study of Maillard reactions. A set of consecutive reactions that start with a reaction between reducing sugars and amino acids or proteins under the influence of heat and going through a large number of steps during which different products can be formed are called Maillard reactions. The components that start the chain of reactions (thus giving further a network of possible reactions) are necessarily found in food, so these reactions that take place during thermal processing (but also to a certain extent at room temperature during food storage) are inevitable. As a result, various compounds are formed, some of which are undesirable and can affect the color, taste, nutritional, and health value of food. For example, if asparagine and reducing sugars are present together (like in the case of potato), one of the resulting compounds in the set of Maillard reactions will be acrylamide, a potential carcinogen for humans which can contribute to many diseases, including those with genetic component [99]. Many efforts have been made to prevent or reduce the occurrence of such undesirable components. One of the possible strategies is to change the initial food composition (adding components such as salts or other proteins) in order to change the mechanism of Maillard's reactions and direct them toward the formation of different products. The application of DSC enabled to study the thermal behavior of a chosen model system undergoing Maillard reactions and to differentiate between reaction conditions and components. In one early work [100], a simplified system containing glucose and glycine was chosen for investigation. In order to clear up the nature of endo- or exothermal peaks, DSC runs were performed either in nitrogen flow or in hermetically sealed crucibles. It is known that a Maillard reaction between glucose and glycine is a consecutive three-step reaction of glucose transformation to 3-deoxyhexosulose, which reacts with glycine to form the unspecified intermediate which gives, by polymerization, un-desired brown melanoidins. Of course, the mechanism is more complex in real food systems where a number of various parallel and/or consecutive reactions can appear. The obtained DSC results revealed that the mechanism of reaction (and consequently, the obtained thermal profiles) is strongly affected by the environmental conditions during analysis, that is the presence of oxygen or the increase of pressure. It was found that the presence of nitrogen and high pressure of air inhibited reactions such as caramelization and formation of melanoidins.

In one of the newest available researches, the reaction of sugars with asparagines was studied – the addition of chickpea protein was applied as one possible strategy to modify the reaction pathway and to reduce the content of acrylamide during baking at 180 °C [101]. The obtained results have revealed both the role of sugar type and the role of protein: the addition of chickpea protein to the carbohydrate-asparagine model system resulted in the greatest reduction in acrylamide content for fructose and lowest reduction for sucrose. The role of DSC was in elucidation of the crystallinity degree of sugars – the lowest one was found for fructose, which means that its uniformity and availability for asparagine were lower than in other sugars. Another contribution of DSC was in finding that sugars' melting points increased with the addition of chickpea protein.

In the case of ITC, stepwise interaction between ligand and protein is carried out, and each injected ligand amount reacts with the protein up to the equilibrium is reached. A generalized description would include the titration of interacting components (ligand and receptor). In such a way the direct assessment to the thermodynamic parameters is enabled: the heat measured for each particular injection is, at constant pressure, related to the enthalpy change. Thus, ITC enables the information about the number of binding sites on the protein (or other substrate) and the strength of the binding interactions. Other thermodynamic parameters characterizing the binding process can be also reached. The limitations of ITC become significant in the cases, where (i) the binding constant is extremely high or binding is extremely slow so the equilibrium cannot be reached in a "real" time of experiment; (ii) the ligand is only soluble in some specific organic solvents. The problem with organic solvent can arise if the forming of ligand/protein complex has very low heat of binding which then can be "covered" by the heat of dilution. Another potential problem is if high amount of solvent is necessary to dissolve sufficient amount of ligand [25, 102, 103]. However, the most important limitation is related to affinity range which is experimentally accessible, and it is directly related to the amount of receptor molecule. It has been established that the following relationship must be satisfied [25]:

$$0.1 < K_a[M]_t < 1,000$$

where $[M]_t$ is the total concentration of receptor in the cell. There are three possibilities: the affinity in between two chemical species is very low ($K_a < 10^4\ M^{-1}$), moderate ($10^4\ M^{-1} < K_a < 10^8\ M^{-1}$), and very high ($K_a > 10^8\ M^{-1}$). The shape of titration curve will depend on affinity. Figure 11.17 illustrates the influence of binding affinities on the shape of ITC curves. The moderate affinity is preferable, and in the case of very low/high affinity, it can be achieved by changing experimental condition (for example – temperature).

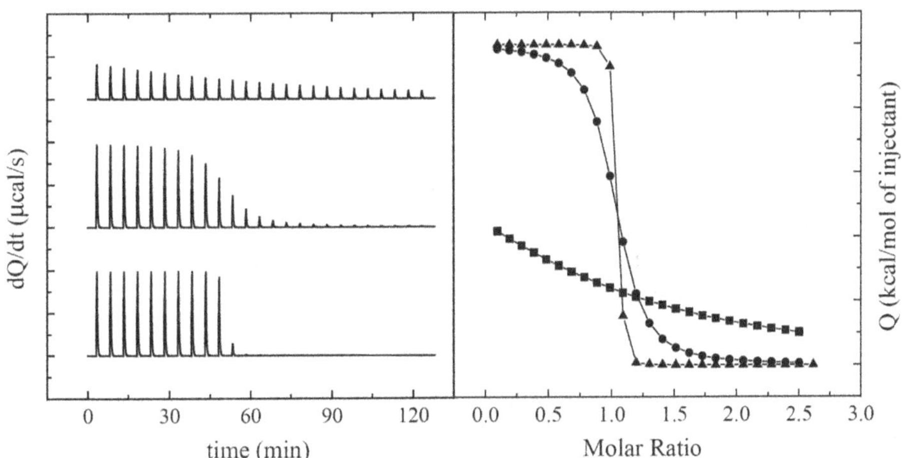

Fig. 11.17: Illustration of the binding affinity influence on the shape of a titration curve. Three titrations simulated using the same parameters (concentrations of reactants and binding enthalpy), but different binding affinities are shown: low (squares), moderate (circles), and high affinities (triangles) (reprinted from [25] with permission from Elsevier, copyright 2005).

The potential of applying the ITC technique in the field of food research is big. Numerous systems have been studied so far. As an example of the investigation related to ligand-protein system, the interactions of coacervate (β-lactoglobulin with albumin from bovine serum) and sodium alginate can be mentioned. It was found that the interactions are spontaneously exothermic at pH = 4.2 [104]. Another research was devoted to study the binding interactions of green tea (GT) flavanoids (catechin) and milk proteins. The results of ITC analysis have shown that the interaction between (+)-catechin and β-casein is endothermic. Hydrophobic bonds between polyphenols and milk proteins have been revealed, the hydrophobic sites on the surface of protein molecules being the place of interaction. These results are significant since green tea flavonoids, which exhibit antioxidant and anticarcinogenic activities could be potential food additives. Hence, the elucidation of their interactions with milk proteins is necessary [105]. Finally, calcium-milk protein interactions were thermodynamically characterized by titration of cow's milk protein with calcium chloride using ITC. Based on the consideration of thus obtained results, as well as other applied methods, it was concluded that the interaction could be rather described as a counterion H^+ to Ca^{2+} exchange than electrostatic [106]. All mentioned results are important not only because of the possibility of practical application in the food industry they enable. Furthermore, such a deep understanding of food components reactivity enables to clarify and improve food functionality.

11.6 Complementary application of thermal methods of analysis with other techniques

11.6.1 Case study 1: chocolate

Chocolate is a product obtained from sugar and materials derived from cocoa beans. It can be described as a suspension of cocoa particles and sugar in a continuous fat phase of cocoa butter. The production of chocolate requires the application of complex technological operations and strict control of their conditions in order to achieve the desired properties of importance: melting behavior, rheological, and textural properties. Chocolate belongs to those materials that have rather the temperature region where melting happens than a precise melting point; when heated, they first soften, and only then melt. The content, composition, and polymorphism of cocoa butter have the greatest influence on the melting behavior and the overall properties of chocolate; therefore, it has been extensively studied using appropriate techniques. In this essential ingredient of chocolate, more than 95% of the lipids present are triacylglycerides (TAGs); among them, the largest amount are 1,3-dipalmitoyl-2-oleoylglycerol (POP), 1-palmitoyl-2-oleoyl-3-stearoylglycerol (POS), and 1,3-distearoyl2-oleoylglycerol (SOS). In addition, small amounts of mono- and di-acylglycerides, polar lipids, free fatty acids, and fat-soluble compounds can be found. During crystallization, TAG molecules are arranged in space into solid regular systems – crystals. In these molecules, different angles between chemical bonds inside the same molecule forming crystal can be found, thus forming different crystal forms, which means that solid TAGs as well as some fatty acids show polymorphism. Polymorphic forms diffract X-rays differently; they also differ in crystal size, crystal form stability, and melting point. It follows that the consistency, texture, plasticity, and other physical properties of fats, and among them cacao butter will depend on the type of TAG crystal [107–

109]. The differences in physical features can be studied using relevant analytical methods such as DSC, NMR, FTIR, and XRD. Wide-angle X-ray diffraction is a standard methodology to study crystal polymorphism of cacao butter, while DSC is the most commonly used method to investigate the processes of its melting and crystallization [110–112]. Most TAGs exist at least in three polymorphic forms usually denoted as α, β′, and β; their molecular packing structures being assigned to hexagonal, orthorhombic, and triclinic crystal forms respectively [5, 113]. Their thermodynamic stabilities, the packing density, the melting point, and the melting enthalpy increase in consecutive order.

Six unique crystal packaging TAG forms have been found in cacao butter and denoted either after the Roman numerals (I, II, III, IV, V, and VI) or the Greek letters: sub-α, α, β', and β, in order of increasing thermodynamic stability [108, 114, 115]. It should be noted that the labeling of these polymorphs is not standardized either in the literature or in the industry. It should be noted as well that instead of six [denoted differently in the literature – γ(I), α(II), β′(β″, β_2', or III), β′(β_1' or IV), β(β_2 or V) and β (β_1 or VI)] sometimes five polymorphs [γ(I), α(II), β′ (IV), β(V) and β(VI)] are mentioned in the literature. In this latter case, polymorph III is considered as a mixture of two forms: (II and IV) [114].

It is interesting to underline that the most thermally stable form of cocoa butter (VI) is not the most desirable in the final products. The desired form is that one which is molten around 31 °C (form V), thus enabling good sensory and texture properties – easy melting and acceptable sensory mouthfeel. Form VI appears if chocolate is heated above the melting point of form V and then cooled without tempering; under these conditions the liquid TAG phase allows the migration of fat to the surface where it recrystallizes and causes fat “bloom” – the appearance of gray (light) fields on the chocolate surface. Therefore, since the presence of V-form enables to obtain the desirable sensory attributes and a long shelf life, in the production of chocolate and related products, control of crystallization process is crucial. To achieve these goals, it is necessary to apply a tempering process which includes a controlled changing of temperature under simultaneous shearing of the chocolate mass. The temperature programs are adjusted and optimized according to the type of chocolate mass being processed (i.e. dark, milk, or white chocolate) [108, 112].

DSC is the method that gives the most information about the effects that heating/cooling has on raw materials and chocolate as a final product. The role of the DSC technique in studying the thermal treatment of chocolate (during production or storage) as well as in investigating the influence of composition (presence of functional additives, aromas, etc.) on thermodynamic and sensory properties is great. There are numerous published results that confirm this statement: DSC was used to study the effects of different factors on melting/crystallization like thermal history of chocolate [109, 110, 116], lecithin [107], free fatty acids [117], and aromas [118]. Further, DSC results enabled to clarify the effects of particle sizes distribution [107] or the addition of crystal promoters like triglycerides [119] on melting/crystallization. Kinetics of these processes has been studied as well. The appearance of fat bloom was another subject of interest [109, 113]. The resulting thermograms are often composed of several overlapped endothermic (in the case of melting) peaks; hence, deconvolution is sometimes needed in order to resolve them in individual contributions [111].

However, it should be emphasized that very often the complementary approach is necessary and that, apart from DSC technique, the methods that enable fingerprint confirmation of specific crystal structures are needed to confirm the presence of certain polymorph. If about cocoa butter, the XRD technique has been used for a very long time for this purpose [108, 109]; however, there are published results of the application of NMR and FTIR spectroscopy as well, and quite recently also the results of Raman spectroscopy [120].

As an illustration of the above-mentioned, Fig. 11.18 (top) shows the overall DSC heating/melting profile of the cocoa butter polymorphs that can be found in chocolate [121], while Fig. 11.18 (bottom) confirms that the XR diffractograms are specific to each of the individual polymorphs [122]. It is very important to underline that the results obtained by these methods make a significant contribution both to quality control and to the design of new products in highly sophisticated confectionery industry.

11.6.2 Case study 2: encapsulation

For the reason of growing demands for the production of functional food, there are increasing needs for new processes and procedures in modern food technology. Food is expected to have a preventive role against diseases, and generally a beneficial effect on physical and mental health. Therapeutic effects of food are also desirable. Therefore, it is necessary that the food contains bioactive components that should be delivered upon consumption in a stable state. However, bioactive components (vitamins, probiotics, polyphenols, omega-3-fatty acids, etc.) are often sensitive to unavoidable factors such as light, oxygen, heat, or moisture. Therefore, it is necessary to protect them in order to be safely incorporated into a food system.

Encapsulation is a methodology designed and developed in a way to enable the introduction of bioactive components but also essential oils and aromas into food in a protected form: it involves entrapping a functionally active, core material into a matrix of an inert material. Tiny particles or droplets are surrounded by a coating wall, or embedded in matrix (homogeneous or heterogeneous) so to form small capsules. The encapsulated material is known as core or active material, fill or internal phase, while the matrix material is described as coating material, wall, capsule, membrane, shell, or carrier material [123, 124].

There are conditions that the matrix, core material, and obtained encapsulates must meet: (1) they must be edible, natural ingredients possibly obtained using solvent-free methods; (2) bioactive materials should be stable as incorporated in matrix, minimal impact on the organoleptic properties should be tolerable; (3) the interaction with other food ingredients should be avoided, as well as degradation of active

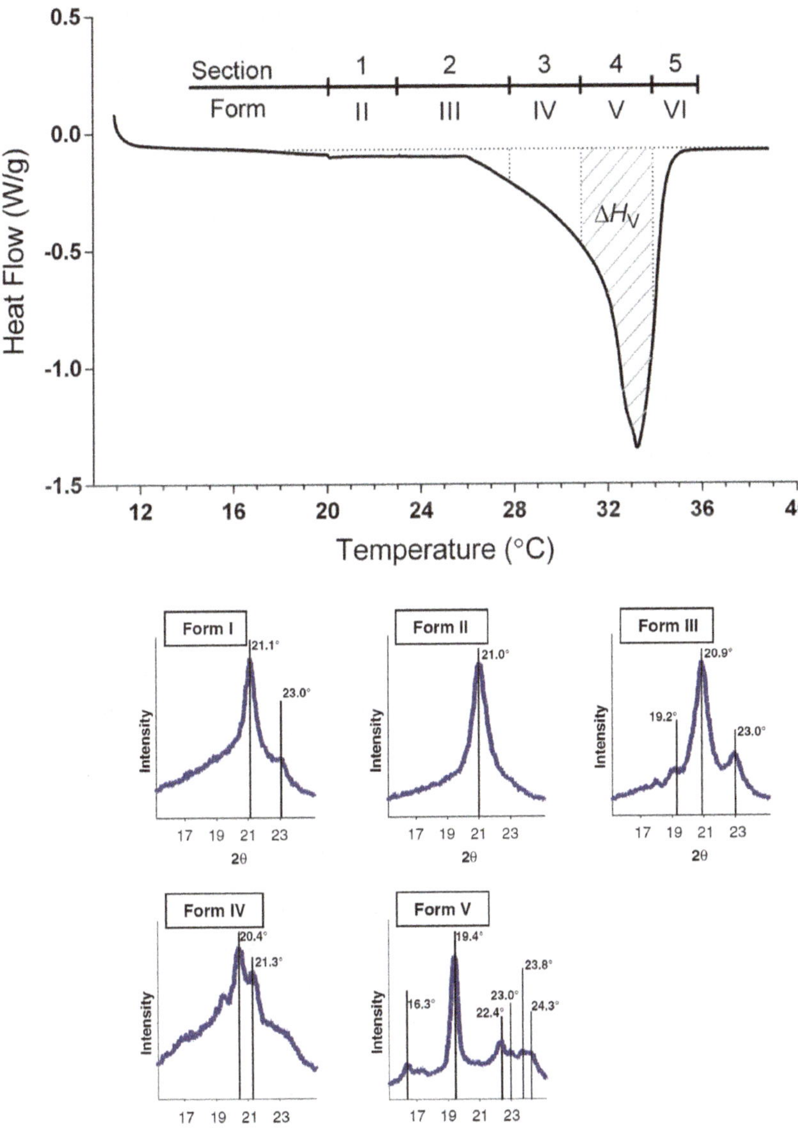

Fig. 11.18: Top: DSC heating profile showing how each temperature section is associated with a particular polymorph (reprinted from [121] with permission from Springer Nature, copyright 2016). Bottom: XRD patterns of cacao butter polymorphs. It should be noted that most probably form IV is contaminated with a small amount of form V (used with permission of John Wiley & Sons from [122], copyright 2009; permission conveyed through Copyright Clearance Center, Inc.).

compounds due to temperature, light, or pH; (4) the encapsulated component should be delivered upon consumption [123]. It is also preferable if procedure can be easily scalable to industrial production. Numerous encapsulation techniques exist, some of them, like spray drying, already applied at the industrial scale. It should be emphasized that the chemical interaction between the active component and the carrier are not desirable, while physical interactions that exist in encapsulate must be strong enough to keep it stable until consumption and weak enough to release the active ingredient from the product when needed. It follows that the choice of matrix is a very important point.

From the aforemetioned, it is evident that it is crucial to provide adequate methods for the characterization of the core material and the matrix as well as the final encapsulate. Microscopy (optical, SEM) and the methods that analyze the structure of the material (XRD, FTIR) are most often used for this purpose. However, as food is very often the subject of thermal treatment, the application of thermal analysis methods is of great importance. Examples of characterization of encapsulated aromas reflect the use of thermal methods in combination with the other mentioned techniques. The role of thermal methods such as DSC or TG or coupled techniques is to check the thermal stability of the obtained encapsulates.

For example, the aqueous extract from endemic species (*Pterospartum tridentatum* – carqueja) was encapsulated within alginate and alginate-inulin microbeads; significant content of polyphenol compounds was found in this extract, and it was protected by encapsulation. The results obtained by TG analysis have shown that microbeads containing extract within alginate as matrix are more stable up to 150 °C in comparison to the empty counterpart. Alginate microbeads containing 20% of inulin turned out to be the best matrix; they provided a slightly delayed extract release compared to plain alginate forms [125].

Another study has shown that Ca-alginate beads are suitable carriers for embedding of aroma D-limonene in order to keep its thermal stability. Thermogravimetry coupled with mass spectrometry was used to study the stability of immobilized D-limonene and to reveal the steps in thermal release of the immobilized flavor. The results pointed that most of the immobilized D-limonene remained intact inside Ca-alginate matrix during the applied temperature regime (30–200 °C), thus revealing a desirable effect of immobilization – thermal stability of core materials achieved within the applied temperature range [126]. D-Limonene was also immobilized in calcium alginate/polyvinyl alcohol matrix using "freezing-thawing" method in order to assure formation of polyvinyl alcohol cryogel structure. TG results again indicated better thermal stability of encapsulated D-limonene in comparison with free aroma [127].

TG analysis was also applied, together with other appropriate techniques (SEM and FTIR) in order to choose the optimal encapsulation system for peppermint essential oil so to enable its placement into the ice cream as a model food product. Ca-alginate was found to be the most suitable matrix – the analysis of FTIR spectra has shown that essential oil and carrier most probably made just a simple mixture in encapsulates. The insight into the results related to thermal stability of free and encapsulated peppermint essential oil that are shown in Fig. 11.19 reveals that encapsulated essential oil was released at elevated temperatures, in the broader temperature regions and in a more controllable manner compared to free peppermint essential oil. It was concluded that encapsulated peppermint essential oil is most probably more suitable for thermal food processes than free essential oil [128].

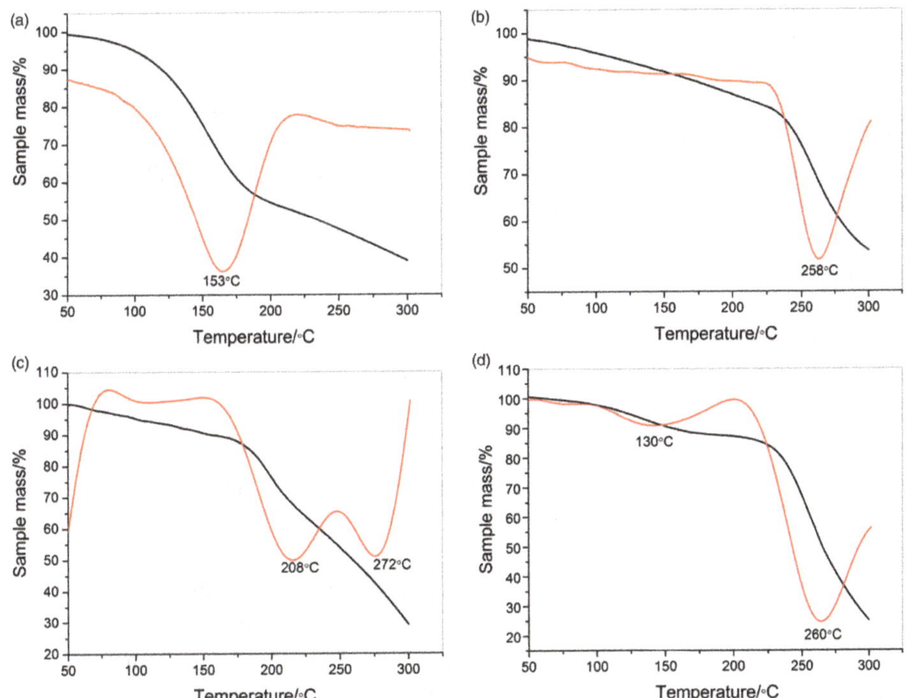

Fig. 11.19: TG analyses of essential oil (a), essential oil encapsulated in Ca-alginate beads (b), gelatine and alginate coacervates (c), and carnauba wax beads (d). Black line, TG curve; red line, differential TG curve (reprinted from [128] with permission from Taylor & Francis (www.tandfonline.com), copyright 2019).

The examples shown confirm that the complementary application of thermal analytical methods with other appropriate techniques is very important in the characterization of encapsulated systems. Evidently, the results obtained using TA methods have a decisive role in determining the thermal stability of these systems, a feature that is often crucial in food processing.

11.7 The role of thermal methods in the newest food technology trends

In recent years, more pronounced and faster changes in food technology trends have been observed. It is expected that in the future efforts will be directed toward topics that are already in the focus of many researchers: sustainable food production, prolonged food shelf life, personal nutrition, and environmental impacts. Indeed, subjects such as alternative protein sources, local foods, nutraceuticals are already in the focus of both academia and industry. A growing concern over environmental impacts of food industry and food waste is noticeable [129].

A "circular economy" approach will consider the re-utilization of food waste products. It will include the consideration of environmental impacts provoked by the accumulation of packaging materials. Hence, there is already, and it will be in the future, a constant need for innovation in the domain of all mentioned topics. As with many other areas of research and development, multidisciplinary approach is inevitable in the food sector as well.

For this reason, growing needs for the application of physicochemical methods for the characterization of matter can be expected. Certainly, methods that can provide answers regarding the thermal stability of either food or materials with which food comes into contact – thermal analysis methods – will be necessary and useful when studying new topics in this area, as they have been until now. In this place, a brief presentation of very recent results, obtained in the last few years, related to the application of thermal analysis methods in the characterization of new materials for food packaging is given.

Plastic has became one of the world's fundamental problems since it is a hazardous material and very challenging to decompose, even though its use in everyday life, particularly as a packaging material, is enormous. Therefore, there is nowadays a growing interest in using bio-based and biodegradable polymers as packaging materials. Such environmentally friendly biopolymers can be either synthesized from bioderived monomers or extracted from biomass or even industrial waste. In the field of food production, transport and storage in the so-called active food packaging (AFP) are especially the focus of interest, since they are designed to increase the shelf life and protect the food quality. Furthermore, degradable films obtained from biopolymers are combined with functional additives (antioxidants or antimicrobial agents) so their transfer from packaging material to food product becomes possible [3, 130, 131]. Besides, there is a growing interest for edible packaging.

However, it has to be mentioned here that biopolymer-derived materials often lack the necessary characteristics such as strength, flexibility, or barrier properties. For that reason there are more and more studies on biopolymer (mainly proteins, lipids, and polysaccharides) usage to obtain degradable materials for food packaging that would satisfy the necessary criteria. Thermal methods of analysis (TMA and DMA) are applied in testing the mechanical properties of these biopolymers. For example, the application of DMA-RH method allowed the analysis of the dynamic mechanical properties of hydrophilic calcium-caseinate films as a function of temperature in isohume conditions [132]. DSC and coupled techniques (like TG/DSC) are used to determine (improve) the thermal properties of bio-based packaging materials. Most often, composite materials are studied and the mentioned TA methods enable to monitor melting profiles (the positions of glass transition and melting temperature, or temperature regions) as well as to find the values of enthalpies associated with those transitions. For example, there are literature reports on alginate-based materials investigations: the blend of sodium alginate, sodium carboxymethyl cellulose, and gelatin was investigated [133]. Further, edible sodium alginate-based films were prepared with fruit apple puree and three kinds of vegetable oils (rapeseed, coconut, and hazelnut oil). The results show that visual appearance was improved in comparison with pure sodium alginate, while DSC results proved stability of these composite edible materials [134]. Another study has shown that composite edible films done with different proportions of pectin, alginate, and whey protein concentrate exhibit good thermal stability, with degradation onset temperatures higher than 170 °C [135].

Multidisciplinary approach is commonly used in studying such kind of systems. The case of gelatin-chitosan edible films proved this statement: the color, transparency, light transmission, mechanical strength, thermal stability, crystalline structures, molecular interactions, and microstructure were assessed using appropriate techniques [136]. The obtained DSC profiles were interpreted in parallel with XRD results so to understand the thermal behavior of particular structures. Finally, it has to be noted that many complex edible materials of this type are starch-based [137, 138]. The same approach is applied: the optical, mechanical, thermal, and physicochemical properties are studied.

In all mentioned cases, the results obtained by applying the selected thermal methods were interpreted as complementary to other applied analytical methods and contributed to an improved understanding of the investigated systems, and finally contributed to the development of these modern materials important for the food industry.

11.8 Conclusions

Thermal methods of analysis comprise a group of techniques that, through decades of development, have become a powerful tool for studying a large number of very different phenomena. They make possible to obtain a large number of parameters useful in the characterization of very different systems. Although some of them are old methods, thanks to the new improved versions that have been developed over decades, as well as the new techniques that have been introduced, it became possible to apply thermal methods in new, mutually divergent areas of research. Among other things, TA methods are now widely used in food-related fields. Importantly, they are useful both in the field of research and innovations, and in sectors such as transport, storage or food safety. Advantages of these techniques are numerous. Usually, small amounts of sample can be used without the need of any chemical modification or labeling. Direct measurements of parameters accessible by specific thermal technique are possible. A wide variety of different materials can be investigated. The samples can be complex; even the purity and interactions among components can be studied. Importantly, we can choose experimental setup very similar to some real (for example industrial) process. The obtained results can be treated as complementary with the results of other techniques. However, we have to keep in mind that, because of system complexity the overlapping of obtained signals is expected, which makes it complex to interpret the results. Nevertheless, the possibilities offered by these methods certainly guarantee the further expansion of their application in the food sector.

References

[1] Vilgis, T. A. Soft matter food physics – The physics of food and cooking. *Rep. Prog. Phys.* **2015**, *78*, 124602, DOI:10.1088/0034-4885/78/12/124602.

[2] Kaletunc, G. Calorimetric Methods as Applied to Food: And Overview. In *Calorimetry in Food Processing: Analysis and Design of Food Systems*; Kaletunc, G., Ed.; USA Whiley-Blackwell: Ames, Iowa; 2009, pp. 5–14, ISBN-13: 978-0-8138-1483-4.

[3] Robertson, G. L. Packaging and Food and Beverage Shelf Life. In *The Stability and Shelf Life of Food*; 2nd ed.; Subramaniam, P., Ed.; Elsevier; 2016, pp. 77–106, ISBN: 978-0-08-100435-7.

[4] Leyva-Porras, C.; Cruz-Alcantar, P.; Espinosa-Solis, V.; Martinez-Guerra, E.; Pinon-Balderrama, C. I.; Martinez, I. C.; Saavedra-Leos, M. Z. Application of differential scanning calorimetry (DSC) and modulated differential scanning calorimetry (MDSC) in food and drug industries. *Polymers.* **2019**, *12*, 5, DOI:10.3390/polym12010005.

[5] Roos, Y. H.; Drusch, S. *Phase Transitions in Foods*; 2nd ed.; Elsevier; 2016, ISBN: 978-0-12-408086-7.

[6] Liu, D.; Martinez-Sanz, M.; Lopez-Sanchez, P.; Gilbert, E. P.; Gidley, M. J. Adsorption behaviour of polyphenols on cellulose is affected by processing history. *Food Hydrocoll.* **2017**, *63*, 496–507, DOI:10.1016/j.foodhyd.2016.09.012.

[7] Wu, X.; Fu, B.; Ma, Y.; Dong, L.; Du, M.; Dong, X.; Xu, X. A debittered complex of glucose-phenylalanine Amadori rearrangement products with cyclodextrin: Structure, molecular docking and thermal degradation kinetic study. *Foods.* **2022**, *11*, 1309, DOI:https://doi.org/10.3390/foods11091309.

[8] Perozzo, R.; Folkers, G.; Scapozza, L. Thermodynamics of protein-ligand interactions: History, presence and future aspects. *J. Recept. Signal Transduct. Res.* **2004**, *24*, 1–52, DOI:10.1081/rrs-120037896.

[9] Ahmed, J.; Ptaszek, P.; Basu, S. Food Rheology: Scientific Development and Importance to Food Industry. In *Advances in Food Rheology and Its Applications*: Elsevier; 2017, pp. 1–6, ISBN: 978-0-08-100431-9.

[10] Nazir, A.; Asghar, A.; Aslam Maan, A. Food Gels: Gelling Process and New Applications. In *Advances in Food Rheology and Its Applications*: Elsevier; 2017, pp. 335–353, ISBN: 978-0-08-100431-9.

[11] Roos, Y. H. Glass transition temperature and its relevance in food processing. *Annu. Rev. Food Sci. Technol.* **2010**, *1*, 469–496, DOI:10.1146/annurev.food.102308.124139.
[12] Kakino, Y.; Hishikawa, Y.; Onodera, R.; Tahara, K.; Takeuchi, H. Gelation factors of pectin for development of a powder form of gel, dry jelly, as a novel dosage. *Form. Chem. Pharm. Bull.* **2017**, *65*, 1035–1044, DOI:10.1248/cpb.c17-00447.
[13] Yanishlieva-Maslarova, N. V. Inhibiting Oxidation. In *Antioxidants in Food, Practical Applications*; Pokorny, J.; Yanishlieva, N.; Gordon, M., Eds.; Cambridge, England, Woodhead Publishing Limited and CRC Press; 2001, pp. 22–70, Woodhead Publishing, ISBN 1 85573 463 X CRC Press ISBN 0-8493-1222-1.
[14] Shaidi, F.; Zhong, Y. Lipid Oxidation: Measurement Methods. In *Bailey's Industrial Oil and Fat Products*; 6th ed.; Shahidi, F., Ed.; John Wiley & Sons, Inc; 2005, pp. 357–385, ISBN 0-471-38460-7.
[15] Micic, D. M.; Ostojic, S. B.; Simonovic, M. B.; Krstic, G.; Pezo, L. L.; Simonovic, B. R. Kinetics of blackberry and raspberry seed oils oxidation by DSC. *Thermochim. Acta*. **2015**, *601*, 39–44, DOI:10.1016/jtca.2014.12.018.
[16] Li, M.; Ritzoulis, C.; Du, Q.; Liu, Y.; Ding, Y.; Liu, W.; Liu, J. Recent progress on protein-polyphenol complexes: Effect on stability and nutrients delivery of oil-in-water emulsion system. *Front. Nutr.* **2021**, *8*, 765589, DOI:10.3389/fnut.2021.765589.
[17] IIjima, M.; Hatakeyama, T.; Nakamura, K.; Hatakeyama, H. Thermomechanical analysis of calcium alginate hydrogels in water. *J. Therm. Anal. Calorim.* **2002**, *70*, 807–814, DOI:10.1023/A:1022252102869.
[18] Bonnaillie, L. M.; Tomasula, P. M. Application of humidity-controlled dynamic mechanical analysis (DMA-RH) to moisture-sensitive edible casein films for use in food packaging. *Polymers*. **2015**, *7*, 91–114, DOI:10.3390/polym7010091.
[19] Biliaderis, C. G. Differential scanning calorimetry in food research. A review. *Food Chem.* **1983**, *10*, 239–265, DOI:10.1015/0308-8146(83)90081-X.
[20] Farkas, J.; Mohácsi-Farkas, C. Application of differential scanning calorimetry in food research and food quality assurance. *J. Therm. Anal. Calorim*. **1996**, *47*, 1787–1803, DOI:10.1007/BF01980925.
[21] Schiraldi, A.; Fessas, D. Calorimetry and thermal analysis in food science. An updated review. *J. Therm. Anal. Calorim.* **2019**, *138*, 2721–2732, DOI:10.1007/s10973-019-08166-z.
[22] Roberfroid, B. M. What is beneficial for health? The concept of functional food. *Food Chem. Toxicol.* **1999**, *37*, 1039–1041, DOI:10.1016/s0278-6915(99)00080-0.
[23] Roberfroid, M. Global view on functional foods: European perspectives. *Br. J. Nutr.* **2002**, *88(S2)*, S133–S138, DOI:10.1079/BJN2002677.
[24] Menrad, K. Market and marketing of functional food in Europe. *J. Food Eng.* **2003**, *56*, 181–188, DOI:10.1016/S0260-8774(02)00247-9.
[25] Campoy, A. V.; Freire, E. ITC in the post-genomic era . . .? Priceless. *Biophys. Chem.* **2005**, *115*, 115–124, DOI:10.1016/j.bpc.2004.12.015.
[26] Calies, O.; Daranas, A. H. Application of isothermal titration calorimetry as a tool to study natural product interactions. *Nat. Prod. Rep.* **2016**, *33*, 881–904, DOI:10.1039/c5np00094g.
[27] Johnson, C. M. Isothermal Titration Calorimetry. In *Protein-ligand Interactions. Methods and Applications.*; Daviter, T.; Johnson, C. M.; McLaughlin, S. H.; Williams, M. A., Eds.; 3rd ed.; Springer Science+Business Media, LLC, part of Springer Nature; 2021, pp. 135–160, DOI:10.1007/978-1-0716-1197-5,, ISBN 978-1-0716-1196-8.
[28] Ghai, R.; Falconer, G.; Collins, B. M. Applications of isothermal titration calorimetry in pure and applied research – Survey of the literature from 2010. *J. Mol. Recognit.* **2012**, *25*, 32–52, DOI:10.1002/jmr.1167.
[29] Eric Plum, G. Calorimetry of Proteins in Dilute Solution. In *Calorimetry in Food Processing: Analysis and Design of Food Systems*; Kaletunc, G., Ed.; Ames, Iowa, USA Whiley-Blackwell; 2009, pp. 67–86, ISBN-13: 978-0-8138-1483-4.

[30] Privalov, P. L. Mucrocalorimetry of Proteins and Their Complexes. In *Protein Structure, Stability, and Interactions*; Shriver, J. W., Ed.; Humana Press, a part of Springer science – Business Media; 2009, pp. 1–39, DOI:10.1007/978-1-59745-367-1.

[31] Rossi, M.; Schiraldi, A. Thermal denaturation and aggregation of egg proteins. *Thermochim. Acta.* **1992**, *199*, 115–123, DOI:10.1016/0040-6031(92)80255-U.

[32] Ferreira, M.; Hofer, C.; Raemy, A. A calorimetric study of egg white proteins. *J. Therm. Anal. Calorim.* **1997**, *48*, 683–690, DOI:10.1007/BF01979514.

[33] Fitzsimons, S. M.; Mulvihill, D. M.; Morris, E. R. Denaturation and aggregation processes in thermal gelation of whey proteins resolved by differential scanning calorimetry. *Food Hydrocoll.* **2007**, *21*, 638–644, DOI:10.1016/j.foodhyd.2006.07.007.

[34] Grinberg, V. Y.; Burova, T. V.; Tolstoguzov, V. B. Thermal Analysis of Denaturation and Aggregation of Proteins and Protein Interactions in a Real Food System. In *Calorimetry in Food Processing: Analysis and Design of Food Systems*; Kaletunc, G., Ed.; Ames, Iowa, USA Whiley-Blackwell; 2009, pp. 87–115, ISBN-13: 978-0-8138-1483-4.

[35] Farber, P.; Darmawan, H.; Sprules, T.; Mittermaier, A. Analyzing protein folding cooperativity by differential scanning calorimetry and NMR spectroscopy. *J. Am. Chem. Soc.* **2010**, *132*, 6214–6222, DOI:10.1021/ja100815a.

[36] Johnson, C. M. Differential scanning calorimetry as a tool for protein folding and stability. *Arch. Biochem. Biophys.* **2013**, *531*, 100–109, DOI:10.1016/j.abb.2012.09.008.

[37] Seeling, J.; Schonfeld, H. J. Thermal protein unfolding by differential scanning calorimetry and circular dichroism spectroscopy Two-state model versus sequential unfolding. *Rev. Biophys.* **2016**, *49*, e9, DOI:10.1017/S0033583516000044.

[38] Mazurenko, S.; Kunka, A.; Beerens, K.; Johnson, C. M.; Damborsky, J.; Prokop, Z. Exploration of protein unfolding by modelling calorimetry data from reheating. *Sci. Rep.* **2017**, *7*, 16321, DOI:10.1038/s41598-017-16360-y.

[39] Privalov, P. L. Thermodynamic problems in structural molecular biology. *Pure Appl. Chem.* **2007**, *79(8)*, 1445–1462, DOI:10.1351/pac200779081445.

[40] Ladjal-Ettoumi, Y.; Boudries, H.; Chibane, M.; Romero, A. Pea, chickpea and lentil protein isolates: Physicochemical characterization and emulsifying properties. *Food Biophys.* **2016**, *11*, 43–51, DOI:10.1007/s11483-015-9411-6.

[41] Tornberg, E. Effects of heat on meat proteins – Implications on structure and quality of meat products. *Meat. Sci.* **2005**, *70*, 493–508, DOI:10.1016/j.meatsci.2004.11.021.

[42] Colombo, A.; Ribotta, P. D.; Leon, A. E. Differential Scanning Calorimetry (DSC) studies on the thermal properties of peanut proteins. *J. Agric. Food Chem.* **2010**, *58*, 4434–4439, DOI:10.1021/jf903426f.

[43] Stavropoulos, P.; Thanassoulas, A.; Nounesis, G. The effect of cations on reversibility and thermodynamic stability during thermal denaturation of lysozyme. *J. Chem. Thermodyn.* **2018**, *118*, 331–337, DOI:10.1016/j.jct.2017.10.006.

[44] Privalov, P. L.; Robinson, C. C. Role of water in the formation of macromolecular structures. *Eur. Biophys. J.* **2016**, *46*, 203–224, DOI:10.1007/s00249-016-1161-y.

[45] Aquilano, D.; Sgualdino, G. Fundamental Aspects of Equilibrium and Crystallization Kinetics. In *Crystallization Processes in Fats and Lipid Systems*; Garti, N.; Sato, K., Eds.; New York, Marcel Dekker Inc; 2001, pp. 1–52, ISBN: 0-8247-0551-3.

[46] Kaneko, F. Polymorphism and Phase Transitions of Fatty Acids and Acylglycerols. In *Crystallization Processes in Fats and Lipid Systems*; Garti, N.; Sato, K., Eds.; New York, Marcel Dekker Inc; 2001, pp. 53–98, ISBN: 0-8247-0551-3.

[47] Belitz, H. D.; Grosch, W.; Schieberle, P. *Food Chemistry 4th Revised and Extended Edition*; Berlin Heilderbeg, Springer Verlag; 2009, DOI:10.1007/978-3-540-69934-7.

[48] Walstra, P. *Physical Chemistry of Foods*; New York Marel Dekker; 2003.

[49] Marquez, A. L.; Perez, M. P.; Wagner, J. R. Solid fat content estimation by differential scanning calorimetry: Prior treatment and proposed correction. *J. Am. Oil Chem. Soc.* **2013**, *90*, 467–473, DOI:10.1007/s11746-012-2190-z.

[50] McClements, D. J. *Food Emulsions. Principles and Techniques*; 3rd ed.; Boca Raton USA CRC Press; 2016, ISBN-13:978-1-4987-2669-6.

[51] Marikar, N.; Alinovi, M.; Chiavaro, E. Analytical approaches for discriminating native lard from other animal fats. *Ital. J. Food Sci.* **2021**, *33*, 106–115, DOI:10.15586/ijfs.v33i1.1962.

[52] Dahimi, O.; Rahim, A. A.; Abdulkarim, S. M.; Hassan, M. S.; Shazamawati, B. T.; Hashari, Z.; Siti Mashitoh, A.; Saadi, S. Multivariate statistical analysis treatment of DSC thermal properties for animal fat adulteration. *Food Chem.* **2014**, *158*, 132–138, DOI:10/1016/j.foodchem.2014.02.087.

[53] Adelman, R.; Hartel, R. W. Lipid Crystallization and Its Effect on the Physical Structure of Ice Cream. In *Crystallization Processes in Fats and Lipid Systems*; Garti, N.; Sato, K., Eds.; New York, Marcel Dekker Inc; 2001, pp. 329–355, ISBN: 0-8247-0551-3.

[54] Tomaszewska-Gras, J. Melting and crystallization DSC profiles of milk fat depending on selected factors. *J. Therm. Anal. Calorim.* **2013**, *113*, 199–208, DOI:10.1007/s10973-013-3087-2.

[55] Srivastava, Y.; Semwal, A. D.; Sajeevkumar, V. A.; Sharma, G. K. Melting, crystallization and storage stability of virgin coconut oil and its blends by differential scanning calorimetry (DSC) and Fourier transform infrared spectroscopy (FTIR). *J. Food Sci. Technol.* **2017**, *54*, 45–54, DOI:10/1007/s13197-016-2427-1.

[56] Foubert, I.; Dewettinick, K.; Vonrolleghem, P. A. Modelling of the crystallization kinetics of fats. *Trends Food Sci. Technol.* **2003**, *14*, 79–92, PII: S0924-2244(02)00256-X.

[57] Foubert, I.; Vereecken, J.; Sichien, M.; Dewettinck; Fredrick, E. Stop-and-return DSC method to study fat crystallization. *Thermochim. Acta.* **2008**, *471*, 7–13, DOI:10.1016/j.tca.2008.02.005.

[58] Himawam, C.; Starov, V. M.; Stapley, A. G. F. Thermodynamic and kinetic aspects of fat crystallization. *Adv. Colloid Interface Sci.* **2006**, *122*, 3–33, DOI:10.1016/j.cis.2006.06.016c.

[59] Yanishlieva-Maslarova, N. V. Inhibiting Oxidation. In *Antioxidants in Food, Practical Applications*; Pokorny, J.; Yanishlieva, N.; Gordon, M., Eds.; Cambridge England, Woodhead Publishing Limited; 2001, pp. 22–70.

[60] Schaich, K. M. Lipid Oxidation: TheoreticalAspects. In *Bailey's Industrial Oil and Fat Products*; 6th ed.; Shahidi, F., Ed.; Wiley-Interscience; 2005, pp. 269–356, ISBN 0-471-38460-7.

[61] Shahidi, F.; Zhong, Y. Lipid Oxidation: Measerement Methods. In *Bailey's Industrial Oil and Fat Products*; 6th ed.; Shahidi, F., Ed.; John Wiley & Sons, Inc.; 2005, pp. 357–385.

[62] Kowalski, R.; Wawrzykowski, J. Effect of ultrasound assisted maceration on the quality of oil from the leaves of thyme. *Thymus Vulgaris L. Flavour Frag. J.* **2009**, *24*, 69–74, doi.org/, DOI:10.1002/ffj.1918.

[63] Tan, C. P.; Che Man, Y. B.; Selamat, J.; Yusoff, M. S. A. Comparative studies of oxidative stability of edible oils by differential scanning calorimetry and oxidative stability index methods. *Food Chem.* **2002**, *76*, 385–389, doi.org/, DOI:10.1016/S0308-8146(01)00272-2.

[64] Ixtaina, V. Y.; Nolasco, S. M.; Tomas, M. C. Oxidative stability of chia (*Salvia hispanica L.*) seed oil: Effect of antioxidants and storage conditions. *J. Am. Oil Chem. Soc.* **2012**, *89*, 1077–1090, doi.org/, DOI:10.1007/s11746-011-1990-x.

[65] Adhvaryu, A.; Erhan, S. Z.; Liu, Z. S.; Perez, J. M. Oxidation kinetic studies of oils derived from unmodified and genetically modified vegetables using pressurized differential scanning calorimetry and nuclear magnetic resonance spectroscopy. *Thermochim. Acta.* **2000**, *364*, 87–97, doi.org/, DOI:10.1016/S0040-6031(00)00626-2.

[66] Simon, P.; Kolman, L. DSC study of oxidation induction periods. *J. Therm. Anal. Calorim.* **2001**, *64*, 813–820, doi.org/, DOI:10.1023/A:1011569117198.

[67] Malicanin, M.; Rac, V.; Antic, V.; Antic, M.; Palade, L. M.; Kefalas, P.; Rakic, V. Content of Antioxidants, Antioxidant Capacity and Oxidative Stability of Grape Seed Oil Obtained by Ultra Sound Assisted Extraction. *J. Am. Oil Chem. Soc.* **2014**, *91*, 989–999, DOI:10.1007/s11746-014-2441-2.

[68] Knothe, G. Dependence of biodiesel fuel properties on the structure of fatty acid alkyl esters. *Fuel Process Technol.* **2005**, *86*, 1059–1070, doi.org/, DOI:10.1016/j.fuproc.2004.11.002.

[69] Gunstone, F. Introduction: Modifying Lipids – Why and How?. In *Modifying Lipids for Use in Food*; Gunstone, F. D., Ed.; Cambridge England, Woodhead Publishing Limited; 2006, pp. 1–8, ISBN-13: 978-1-85573-971-0.

[70] Marikkar, M. N. DSC as a Valuable Tool for the Evaluation of Adulteration of Oils and Fats. In *Differential Scanning Calorimetry Applications in Fat and Oil Technology*; Chiavaro, E., Ed.; Boca Raton USA CRC Press; 2016, pp. 27–48, ISBN-13: 978-1-4665-9153-0.

[71] Marikkar, J. M. N.; Dzulkifly, M. H.; Nor Nadiha, M. Z.; Che Man, Y. B. Detection of animal fat contaminations in sunflower oil by differential scanning calorimetry. *Int. J. Food Prop.* **2012**, *15*, 683–690, DOI:10.1080/10942912.2010.498544.

[72] Marrikar, N.; Alinovi, M.; Chiavaro, E. Analytical approaches for discriminating native lard from other animal fats. *Ital. J. Food Sci.* **2021**, *3*, 106–115, DOI:10.15586/ijfs.v33i1.1962.

[73] Nilchian, Z.; Ehsani, M. R.; Piravi-Vanak, Z.; Bakhoda, H. Comparative analysis of butter thermal behavior in combination with bovine tallow. *Food Sci. Technol. (Campinas)*. **2020**, *40*, 597–604, doi.org/, DOI:10.1590/fst.32019.

[74] Jafari, M.; Kadivar, M.; Keramat, J. Detection of adulteration in Iranian olive oils using instrumental (GC, NMR, DSC) methods. *J. Am. Oil Chem. Soc.* **2009**, *86*, 103–110, DOI:10.1007/s11746-008-1333-8.

[75] Islam, M.; Belkowska, L.; Konieczny, P.; Fornal, E.; Tomaszewska-Gras, J. Differential scanning calorimetry for authentification of edible fats and oils – What can we learn from the past to face the current challenges?. *J. Food Drug Anal.* **2022**, *30*, 185–201, DOI:doi:/10.38212/2224-6614.3402.

[76] Dyszel, S. M.; Baish, S. K. Characterization of tropical oils by DSC. *Thermochim. Acta.* **1992**, *212*, 39–49, DOI:10.1016/0040-6031(92)80218-L.

[77] Zaliha, O.; Elina, H.; Sivaruby, K.; Norizzah, A. R.; Marangoni, A. G. Dynamics of polymorphic transformations in palm oil, palm stearin and palm kernel oil characterized by coupled powder XRD-DSC. *J. Oleo Sci.* **2018**, *67*, 737–744, doi:/, DOI:10.5650/jos.ess17168.

[78] Barba, L.; Arrighetti, G.; Calligaris, S. Crystallization and melting properties of extra virgin olive oil studied by synchrotron XRD and DSC. *Eur. Food Res. Tech.* **2013**, *115*, 322–329, doi:/, DOI:10.1002/ejlt.201200259.

[79] Tomaszewska-Gras, J. DSC coupled with PCA as a tool for butter authenticity assessment. *J. Therm. Anal. Calorim.* **2016**, *126*, 61–68, DOI:10.1007/s10973-016-5346-5.

[80] Tomaszewska-Gras, J. Detection of butter adulteration with water using differential scanning calorimetry. *J. Therm. Anal. Calorim.* **2012**, *108*, 433–438, DOI:10.1007/s10973-011-1913-y.

[81] McClements, D. J. *Food Emulsions Principles, Practices, and Techniques*; 2nd ed.; Boca Raton, USA, CRC Press; 2004, ISBN: 0-8493-2023-2.

[82] Miao, S.; Mao, L. DSC Application to Characterizing Food Emulsions. In *Differential Scanning Calorimetry Applications in Fat and Oil Technology*; Chiavaro, E., Ed.; Boca Raton USA CRC Press; 2016, pp. 243–272, ISBN-13: 978-1-4665-9153-0.

[83] Palanuwech, J.; Coupland, J. N. Effect of surfactant type on the satility of oil-in-water emuklsions to dispersed phase crystallization. *Colloids Surf. A: Physicochem. Eng. Asp.* **2003**, *223*, 251–262, DOI:10.1016/S0927-7757(03)00169-9.

[84] Wardhono, E. Y.; Pinem, M. P.; Wahyudi, H.; Agustina, S. Calorimetry technique for observing the evolution of dispersed droplets of concentrated water-in-oil (W/O) emulsion during preparation, storage and estabilization. *Appl. Sci.* **2019**, *9*, 5271, DOI:10.3390/app9245271.

[85] Cornacchia, L.; Roos, Y. H. Lipid and water crystallization in protein-stabilised oil-in-water emulsions. *Food Hydrocoll.* **2011**, *25*, 1726–1736, DOI:10.1016/j.foodhyd.2011.03.014.

[86] Hoffmann, H.; Reger, M. Emulsions with unique properties from proteins as emulsifiers. *Adv. Colloid Interface Sci.* **2014**, *205*, 94–104, DOI:10.1016/j.cis.2013.08.007.

[87] Relkin, P.; Ait-Taleb, A.; Sourdet, S.; Fosseux, P. Y. Thermal behavior of fat droplets as related to adsorbed milk proteins in complex food emulsions. *J. Am. Oil Chem. Soc.* **2003**, *80*, 741–746, DOI:https://doi.org/10.1007/s11746-003-0766-1.
[88] Tanglao, E. J.; Nanda Kumar, A. B.; Noriega, R. R.; Panzalan, M. E.; Marcelo, P. Development and physico-chemical characterization of virgin coconut oil-in-water emulsion using polymerized whey protein as emulsifier for vitamin A delivery. MATEC Web of Conferences **2019**, *268*, 01002. EDP Sciences, DOI:10.1051/matecconf/201926801002.
[89] Wasserman, L. A.; Signorelli, M.; Schiraldi, A.; Yuryev, V.; Boggini, G.; Bertini, S.; Fessas, D. Preparation of wheat resistant starch. *J. Therm. Anal. Calorim.* **2007**, *87*, 153–157, DOI:10.1007/s10973-006-8209-7.
[90] Mestres, C.; Matencio, F.; Pons, B.; Yajid, M.; Fliedel, G. A rapid method for the determination of amylose content by using differential-scanning calorimetry. *Starch.* **1996**, *48*, 2–6, DOI:10/1002/star.19960480103.
[91] Toro-Vazqueza, J. F.; Gomez-Aldapab, C. A.; Aragon-Pina, A.; de la Fuentea, E. B.; Dibildox-Alvaradoa, E.; Charo-Alonso, M. Interaction of granular maize starch with lysophosphatidylcholine evaluated by calorimetry, mechanical and microscopy analysis. *J. Cereal Sci.* **2003**, *38*, 269–279, DOI:10.1016/S0733-5210(03)00026-2.
[92] Polaske, N. W.; Wood, A.; Campbell, M. R.; Nagan, M. C.; Pollak, L. M. Amilose determination of native high-amylose corn starches by differential scanning calorimetry. *Starch.* **2005**, *57*, 118–123, DOI:10.1002/star.200400368.
[93] Schirmer, M.; Jekle, M.; Becker, T. Starch gelatinization and its complexity for analysis. *Starch.* **2015**, *67*, 30–41, DOI:10.1002/star.201400071.
[94] Naito, S.; Fukami, S.; Mizokami, Y.; Hirose, R.; Kawashima, K.; Takano, H.; Ishida, N.; Koizumi, M.; Kano, H. The effect of gelatinized starch on baking bread. *Food Sci. Technol. Res.* **2005**, *11*, 194–201, DOI:10.3136/fstr.11.194.
[95] Pigłowska, M.; Kurc, B.; Rymaniak, L.; Lijewski; Fuc, P. Kinetics and thermodynamics of thermal degradation of different starches and estimation the OH group and H_2O content on the surface by TG/DTG-DTA. *Polymers.* **2020**, *12*, 357, DOI:10.3390/polym12020357.
[96] Taghizadeh, M. T.; Abdollahi, R. A kinetics study on the thermal degradation of starch/poly (vinyl alcohol) blend. *Chem. Mater. Eng.* **2015**, *3*, 73–78, DOI:10.13189/cme.2015.030402.
[97] Liu, Y.; Yang, L.; Zhang, Y. Thermal behavior and kinetic decomposition of sweet potato starch by non-isothermal procedures. *Arch. Thermodyn.* **2019**, *40*, 67–82, DOI:10.24425/ather.2019.130008.
[98] Campderros, M. E.; Rodriguez Furlán, L. T.; Lecot, J.; Padilla, A. P.; Zaritzky, N. Stabilizing effect of saccharides on bovine plasma protein: A calorimetric study. *Meat. Sci.* **2012**, *91*, 478–485, DOI:10.1016/j.meatsci.2012.02.035.
[99] Parisi, S.; Luo, W. *Chemistry of Maillard Reactions in Processed Foods*; Cham, Switzerland, Springer; 2018, DOI:10.1007/978-3-319-95463-9.
[100] Manzocco, L.; Nicoli, M. C.; Maltini, E. DSC analysis of Maillard browning and procedural effects. *J. Food Process Preserv.* **1999**, *23*, 317–328, DOI:10.1111/j.1745-4549.1999.tb00388.x.
[101] Miskiewicz, K.; Rosicka-Kaczmarek, J.; Nebesny, E. Effects of chickpea protein on carbohydrate reactivity in acrylamide formation in low humidity model systems. *Foods.* **2020**, *9*, 167, DOI:10.3390/foods9020167.
[102] Wiseman, T.; Williston, S.; Brandts, J. F.; Lin, L. N. Rapid measurement of binding constants and heats of binding using a new titration calorimeter. *Anal. Biochem.* **1989**, *179*, 131–137, DOI:10.1016/0003-2697(89)90213-3.
[103] Du, X.; Li, Y.; Xia, Y. L.; Ai, S. M.; Liang, J.; Sang, P.; Ji, X. L.; Liu, S. Q. Insights into protein–ligand interactions: Mechanisms, models, and methods. *Int. J. Mol. Sci.* **2016**, *17*, 144, DOI:10.3390/ijms17020144.

[104] Gorji, E. G.; Waheed, A.; Ludwig, R.; Toca-Herrera, J. L.; Schleining, G.; Gorji, S. G. Complex coacervation of milk proteins with sodium alginate. *J. Agric. Food. Chem.* **2018**, *66*, 3210–3220, DOI:10.1021/jafc.7b03915.

[105] Yuksel, Z.; Avci; Erdem, Y. K. Characterization of binding interactions between green tea flavanoids and milk proteins. *Food. Chem.* **2010**, *121*, 450–456, DOI:10.1016/j.foodchem.2009.12.064.

[106] Canabady-Rochelle, L. S.; Mellema, M.; Sanchez, C.; Banon, S. Thermodynamic characterization of calcium-milk protein interaction by isothermal titration calorimetry. *Dairy Sci. Technol.* **2009**, *89*, 257–267, DOI:10.1051/dst/2009006.

[107] Afoakwa, E. O.; Paterson, A.; Fowler, M.; Vieira, J. Characterization of melting properties in dark chocolates from varying particle size distribution and composition using differential scanning calorimetry. *Food Res. Int.* **2008**, *41*, 751–757, DOI:10.1016/j.foodres.2008.05.009.

[108] Afoakwa, E. O. *Chocolate Science and Technology*; 2nd ed.; Chichester, West Sussex, UK, Wiley Blackwell; 2016, ISBN: 978-1-1189-1378-9.

[109] Beckett, S. T. *The Science of Chocolate*; 2nd ed.; Cambridge UK, RSC Publishing; 2008, ISBN: 978-0-85404-970-7.

[110] Fessas, D.; Signorelli, M.; Schiraldi, A. Polymorphous transitions in cocoa butter. *J. Therm. Anal. Calorim.* **2005**, *82*, 691–672, DOI:10.1007/s10973-005-6934-y.

[111] Declerck, A.; Nelis, V.; Danthine, S.; Dewettinck, K.; de Meeren, P. Characterisation of fat crystal polymorphism in cocoa butter by time-domain NMR and DSC deconvolution. *Foods.* **2021**, *10*, 520, DOI:10.3390/foods10030520.

[112] Ioannidi, E.; Risbo, J.; Aaroe, E.; van der Berg, F. W. J. Thermal analysis of dark chocolate with diferential scanning calorimetry – limitations in the quantitative evaluation of the crystalline state. *Food Anal. Methods*. **2021**, *14*, 2556–2568, DOI:10.1007/s12161-021-02073-6.

[113] Lonchampt, P.; Hartel, R. W. Fat bloom in chocolate and compound coatings. *Eur. J. Lipid Sci. Technol.* **2004**, *106*, 241–274, DOI:10.1002/ejlt.200400938.

[114] Le Reverend, B. J. D.; Fryer, P. J.; Bakalis, S. Modelling crystallization and melting kinetics of cocoa butter in chocolate and application to confectionery manufacturing. *Soft Matter.* **2009**, *5*, 891–902, DOI:10.1039/b809446b.

[115] Le Reverend, B. J. D.; Fryer, P. J.; Coles, S.; Bakalis, S. A method to qualify and quantify the crystalline state of cocoa butter in industrial chocolate. *J. Am. Oil. Chem. Soc.* **2010**, *87*, 239–246, DOI:10.1007/s11746-009-1498-9.

[116] Stapley, A. G. F.; Tewkesbury, H.; Fryer, P. J. The effects of shear and temperature history on the crystallization of chocolate. *J. Am. Oil Chem. Soc.* **1999**, *76*, 677–685, DOI:10.1007/s11746-999-0159-3.

[117] Muller, M.; Careglio, E. Influence of free fatty acids as additibes on the crystallization kinetics of cocoa butter. *J. Food Res.* **2018**, *7*, 86–97, DOI:10.5539/jfr.v7n5p86.

[118] Ray, J.; MacNaughtan, W.; Chong, P. S.; Vieira, J.; Wolf, B. The effect of limonene on the crystallization of cocoa butter. *J. Am. Oil Chem. Soc.* **2012**, *89*, 437–445, DOI:10.1007/s11746-011-1934-5.

[119] Rosales, C. K.; Klinkesorn, U.; Suwonsichon, S. Effect of crystal promoters on viscosity and melting characteristics of compound chocolate. *Int. J. Food Prop.* **2017**, *20*, 119–132, DOI:10.1080/10942912.2016.1147458.

[120] Bresson, S.; Rigaud, S.; Lecuelle, A.; Bougrioua, F.; El Hadri, M.; Baeten, V.; Courty, M.; Pilard, S.; Faivre, V. Comparative structural and vibrational investigations between cocoa butter (CB) and cocoa butter equivalent (CBE) by ESI/MALDI-HRMS, XRD, DSC, IR and Raman spectroscopy. *Food Chem.* **2021**, *363*, 130319, DOI:10.1016/j.foodchem.2021.130319.

[121] Winkelmeyer, C. B.; Peyronel, F.; Weiss, J.; Marangoni, A. G. Monitoring tempered dark chocolate using ultrasonic spectrometry. *Food Bioproc. Tech.* **2016**, *9*, 1692–1705, DOI:10.1007/s11947-016-1755-5.

[122] Pore, M.; Seah, H. H.; Glover, W. H.; Holmes, D. J.; Johns, M. L.; Wilson, D. I.; Moggridge, G. D. In-situ X-Ray studies of cocoa butter droplets undergoing simulated spray freezing. *J. Am. Oil Chem. Soc.* **2009**, *86*, 215–225, DOI:10.1007/s11746-009-1349-8.

[123] Đorđević, V.; Balanč, B.; Belščak-Cvitanović, A.; Lević, S.; Trifković, K.; Kalušević, A.; Kostić, I.; Komes, D.; Bugarski, B.; Nedović, V. Trends in encapsulation technologies for delivery of food bioactive compounds. *Food Eng. Rev.* **2015**, *7*, 452–490, DOI:10.1007/s12393-014-9106-7.

[124] Timilsena, Y. P.; Haque, M. A.; Adhikari, B. Encapsulation in the food industry: A brief historical overview to recent developments. *Food Nutr. Sci.* **2020**, *11*, 481–508, DOI:10.4236/fns.2020.116035.

[125] Balanc, B.; Kalusevic, A.; Drvenica, I.; Coelho, M. T.; Djordjevic, V.; Alves, V. D.; Sousa, I.; Moldao-Martins, M.; Rakic, V.; Nedovic, V.; Bugarski, B. Calcium-alginate-inulin microbieads as carriers for aqueous Caraqueja extract. *J. Food Sci.* **2016**, *81*, E65–E75, DOI:10.1111/1750-3841.13167.

[126] Levic, S.; Solevic-Knudsen, T.; Pajic-Lijakovic, I.; Đorđevic, V.; Rac, V.; Rakic, V.; Pavlovic, V.; Bugarski, B.; Nedovic, V. Characterization of sodium alginate/D-limonene emulsions and respective calcium alginate/D-limonene beads produced by electrostatic extrusion. *Food Hydrocoll.* **2015**, *45*, 111–123, DOI:10.1016/j.foodhyd.2014.10.001.

[127] Levic, S.; Rac, V.; Manojlovic, V.; Rakic, V.; Bugarski, B.; Flock, T.; Krzyczmonik, K. E.; Nedovic, V. Limonene encapsulation in alginate/poly (vinyl alcohol). *Procedia Food Sci.* **2011**, *1*, 1816–1820, DOI:10.1016/j.profoo.2011.09.266.

[128] Yilmaztekin, M.; Levic, S.; Kalusevic, A.; Cam, M.; Bugarski, B.; Rakic, V.; Pavlovic, V.; Nedovic, V. Characterisation of peppermint (*Mentha piperita L.*) essential oil encapsulates. *J. Microencapsul.* **2019**, *36*, 109–119, DOI:10.1080/02652048.2019.1607596.

[129] Nevarez-Moorillon, G. V.; Zakaria, Z. A.; Prado-Barragan, L. A.; Aguilar, C. N. Editorial: New trends in food processing: Reducing food less, waste, and the environment impact. *Front. Sustain. Food Syst.* **2022**, *6*, 856361, DOI:10.3389/fsufs.2022.856361.

[130] Mesgari, M.; Aalami, A. H.; Sathyapalan, T.; Sahabkar, A. A comprehensive review on the development of carbohydrate macromolecules and copper oxide nanocomposite films in food nanopackaging. *Bioinorg. Chem. Appl.* **2022**, 7557825, DOI:10.1155/2022/7557825.

[131] Krasniewska, K.; Galus, S.; Gniewosz, M. Biopolymers based materials containing silver nanoparticles as active packaging for food applications–A review. *Int. J. Mol. Sci.* **2020**, *21*, 698, DOI:10.3390/ijms21030698.

[132] Bonnaillie, L. M.; Tomasula, P. M. Application of humidity- controlled dynamic mechanical analysis (DMA-RH) to moisture-sensitive edible casein films for use in food packaging. *Polymers.* **2015**, *7*, 91–114, DOI:10.3390/polym7010091.

[133] Wang, L.; Campanella, O.; Patel, B.; Lu, L. Preparation and sealing processing of sodium alginate based blending film. *Math. Probl. Eng.* **2015**, 895637, DOI:10.1155/2015/895637.

[134] Kadzinska, J.; Janowicz, M.; Brys, J.; Ostrowska-Ligeza, E.; Esteve, M. Influence of vegetable oils addition on the selected physical properties of apple-sodium alginate edible film. *Polym. Bull.* **2020**, *77*, 883–900, DOI:10/1007/s00289-019-02777-0.

[135] Dalla Rosa, M.; Chakravartula, S. S. N.; Soccio, M.; Lotti, N.; Balestra, F.; Siracusa, V. Characterization of composite edible films based on pectin/alginate/whey protein concentrate. *Materials.* **2019**, *12*, 2454, DOI:10.3390/ma12152454.

[136] Cai, L.; Shi, H.; Cao, A.; Jia, J. Characterization of gelatin/chitosan ploymer ilms integrated with docosahexaenoic acids fabricated by diferent methods. *Sci. Rep.* **2019**, *9*, 8375, DOI:10.1038/s41598-019-44807-x.

[137] Sultan, M. T.; Aghazadeh, M.; Karim, R.; Abdul Rahman, R.; Johnson, S. K.; Paykary, M. Effect of glycerol on the physicochemical properties of cereal starch films. *Czech J. Food Sci.* **2018**, *36*, 403–409, DOI:10.17221/41/2017-CJFS.

[138] Singh, A.; Pratap-Singh, A. Development and characterization of the edible packaging films incorporated with blueberry pomace. *Foods.* **2020**, *9*, 1599, DOI:10.3390/foods9111599.

Index

"effective" acidity and basicity 126
"intrinsic" acidity and basicity 125
μGC 286
1-butanol 136, 141
1-propanol 136, 141
5-(hydroxymethyl)furfural 128

absorption 275, 277
absorption heat pump 225–226
accelerated ageing 324
accessible surface area 53
accumulated intermediates 23
acetate adsorption onto alumina from aqueous solutions 80–81
acetaldehyde 148
acid strength 135
acid-base properties 152
acrolein 141
acrolein yield 151
acrylamide 373
acrylonitrile adsorption 171
activation energy 10, 260
activation energy (E_a) 103–104, 111
active food packaging 382
activity decay 12
adhesives 333
adiabatic calorimetry 234–235
adiabatic shields 235, 238
adsorbate layer 21
adsorption 187, 196, 198
adsorption enthalpies 173
adsorption from solutions 47–48, 52–54, 76–81
adsorption kinetics 54
adsorption microcalorimetry 123
adsorption and micellization of surfactants 69–70
adsorption properties 5
aggregation 348, 356
alumina 148
aluminate spinels 151
ammonia 125, 131, 135, 143, 149
amorphous 247
amount of solute adsorbed 53
amphoteric character 133
amphoteric solids 148
amylopectin 370
amylose 370
ancient ceramic 319–320
anion-modified metal oxide. *high-temperature carbonate* 9
antioxidant 344, 364
apparent molal enthalpy 64–65
application 289
applications 293
Arrhenius plots 20
asparagine 373
atmospheric particles 317
aurichalcite 7
autocatalysis 16
automatic injection mode 55

B3LYP 169
backbone 279
baseline 256
basis set 165
benchmarking 164, 167–168, 170, 177
benzene 274, 279
benzene adsorption 171
benzoic acid 126
benzonitrile adsorption 171
binary surface complexes 81
binding media 329–330
bioethanol 136
biomass 287
biomass valorization 123
bioplastics 247
biosourced 247
blank 256
bond 284
branching 254
Brønsted acid sites 145
Brønsted and Lewis acidity 130
brucite-type precursor 26
butter 366

cadmium adsorption onto zeolite from aqueous solutions 77–79
calcination kinetics 10, 12
calcination of Cu/Zn hydroxocarbonates 8
calcination temperatures 9
calcium oxide 150
calibration 55, 256, 287
calorimeter 231–232, 235–237, 240–241, 250

https://doi.org/10.1515/9783110590449-012

calorimetric cell 50, 61, 231–233, 235–236, 239–240, 294
calorimetry 183, 193, 196, 231–236,239–240, 293, 341
calorimetry technique 125
Calvet 257
carbohydrates 369
carbon 313–314, 317, 331
carbon deposition 28, 34
carbon dioxide 145
carbon filaments 34
carbon-based adsorbates 21
catalytic cycle 3
catharometer 286
cation exchange pathway 79
cellobiose hydrolysis 134
cellulose 325–329, 333
cement-based mortar 313
CeO_2/ZrO_2 142
ceramics 311–312, 319–320, 333
ceria/niobia catalysts 132
chain 284
characteristic ions 273
characterization 289
charged interface 61–63
charging process 186
chelation of metal ions 73
chemical energy storage 6
chemical memory 6
chemical reaction 188
chemical speciation 78
chemigram 279
chemigrams 276
chocolate 376
cholesterol 368
Clapeyron equation 296
classification 248
Clausius–Clapeyron equation 302
clay minerals 318–320
closed systems 186
CO adsorption 168–171, 174
CO/CO_2 hydrogenation 21
cobalt adsorption onto alumina from aqueous solutions 80–81
cocciopesto mortar 313
cocoa butter 366, 376
coke deposition 31
collagen 322
column 272, 281, 287
competitive adsorption 77–79
composite 188, 195, 202, 206, 213
concentration 287
condensation 343
condensed 296
confidence interval 14
consolidant 325
constant pH and ionic strength conditions 61–63
conventional catalysts 2
conversion 130, 134, 139, 252
cooperative adsorption 80–81
copolymer 257
coprecipitated tungsta-zirconia 134
coprecipitation 6
coupling 289
CO_x yield 152
crosslinking 251
crotonaldehyde 150
crystallinity 254, 256
crystallization 10, 258, 342, 376
crystallization kinetics 362
Cu surface area 19
Cu/ZnO catalysts for methanol synthesis 6
Cu-ammonia complexes 176
cumulative enthalpy 55, 66, 70, 76–77
cumulative enthalpy of dilution 66, 68, 70–72
cumulative enthalpy of displacement 66, 72
curing 251, 253
CuZn alloy 18
cyanate ester 263
cycle life span 113

databases 164
– ADS41 164
– database for adsorption bond energies 164
– GMTKN55 164
– Main-Group Chemistry DataBase 164
– SBH10 164
deactivation by coking 34
decomposition 261, 289
degradation patina 317
degree of deterioration 323, 327
dehydration of cations 79
dehydrogenation 97, 99, 103, 108, 111, 114
denaturation 348, 351, 356
density of acid sites 133
dental composite 269
desorption process 193
DFT theory 160

differential enthalpy of displacement 75–76
differential heats of adsorption 126, 131, 133, 135, 143, 149
differential scanning calorimetry 239
differential scanning calorimetry (DSC) 91, 102, 107, 113, 293
differential scanning calorimetry – DSC 346
differential thermal analysis – DTA 346
diffusion coefficients 19
dilatometry 267
dilute aqueous solution 47, 53
dilution experiment 67–69, 72–73
dispersion correction methods
– D2 174
– D3 163, 174–175
dispersion correction" t "*See* van der Waals interactions 163
displacement process 66, 72
dissociation constant 75
dolomite 317
doping 99
dough 371
DRM" t "see dry reforming of methane 34
dry reforming of methane 24
DSC 193, 197, 201, 211, 351, 356, 362, 373–374, 383
DSC 250
DTA 360
DTA-MS 33
DTG curve 26
dynamic mechanical analysis – DMA 346

edible packaging 382
egg-white 353
electric heat pumps 224–225
electron enrichment of the surface 144
emulsifiers 368
emulsions 366
encapsulated aromas 380
encapsulation 378
endothermic and entropy-driven 79, 81
endothermic peaks 301
energy 250
energy recovery 293
energy storage density 184
energy storage materials 212
enthalpies of dissociation 294
enthalpy 103–104, 109, 193–194, 226, 228, 230–233
enthalpy change 51, 349
enthalpy of dilution 68, 70–72
enthalpy of displacement 66, 72, 77–81
Enthalpy of formation 296
enthalpy of micellization 71
entropy 109
environmental conditions 315
epoxy 251
equilibrium bulk solution 52
ethanol 139
ethanol dehydration 139
ethylene 139
ethylene-vinyl acetate 277
evolution profile 284
evolved gas analysis 271
excess properties 228
excess thermodynamic functions 64
exchange-correlation energy E_{xc} 161
exchange-correlation functionals 160
exothermic 184–185, 196
exothermic peak 264
experimental conditions 98
experimental procedure 231–232, 235, 238
experimental setup 238

fats 359
fatty acids 359
fibers 270
fillers 328
firing temperature 320
flexibility 268
flour 371
food 341
food adulteration 365
food components 347, 350
food industry 382–383
food storage 353
food waste 382
formaldehyde 148
fragmentation 282
free carbonate ions 11
freestanding Cu particles 16
fresco technique 315
fructose dehydration 128
FTIR spectroscopy 145
fugacity 228–229
fume suppressant 274
functional additives 382
functional groups 275
functionals with van der Waals correction
– BEEF-vdW 163, 170

– optB86b-vdW 163, 170
– optPBE-vdW 170–171
– PBE-dDsC 171

gas- and liquid-phase reactions 123
gas chromatography 281
gas hydrate studies 293
gas hydrates 293
gases concentration 287
gas-phase calorimetry 125
gelatinization 371
GGA 162, 163, 169, 171–172, 174
Gibbs free energy change 67
glass 268
glass transition 253
glass transition temperature 332
glass transitions 343
glycerol dehydration 141
gradient 282
Gram-Schmidt 275
graphite-like carbon 35
guarded hot plate 201, 211
Guerbet reaction 136
gypsum 314
gypsum mortar 314

halocarbons 227
Harkins and Jura procedure 53
Hartree-Fock approach 160
heat 250
heat capacity 183, 202, 206, 230, 234–240, 294, 295, 356
heat flow 294
heat flux 347
heat of reaction 183
heat released 195
heat storage 212
heat storage capacity 193
heat transfer 186, 201, 204, 206
heating rate 264
heteropolyacids 148
high-pressure calorimetry 293
high-resolution (HR) (S)TEM ((scanning) transmission electron microscopy) 28
high-temperature calcination 29
high-temperature carbonate 9
H-MFI zeolite 129
hot disk 196, 206, 208, 212
hot guarded plates 212
hot wire 205, 212
hot-wire 193, 204
HSE03 170
Hubbard term 175
hybrid GGA/meta-GGA 162
hydrate inhibitors 294
hydrate phase equilibrium 294
hydrate structures 294
hydrated calcium oxalates 317
hydration 196, 211
hydraulic 313–314
Hydrogen carbonate species 145
hydrogen energy vector 92
hydrogen storage capacity 97, 99, 108
hydrogen storage properties 92, 95, 112, 114
hydrogenation 107–108, 111
hydromagnesite 25
hydrotalcite 7
hydrotalcites 137
hygroscopic 196–197, 201, 206, 211

in situ DSC 37
in situ IR 23
in situ microscopy or spectroscopy 81–82
in situ thermal analysis 3
in situ thermogravimetry 21
in situ XRD 9, 18
incremental titration 54, 61–62
individual adsorption isotherm 56–58
infrared spectrum 276
injector 283
interfacial enthalpy 64, 66
intermediate 16
ionic double layer 48, 66
ionic radius 151
IR analysis 11
IR spectrum 277, 279
iron molybdate catalyst 150
isochoric heat capacity 237
isolation of acetaldehyde 150
isothermal titration calorimetry – ITC 348
isotherms 34, 47, 50, 54
isotope effects 37
ITC 374
ITC thermal profiles 59–61

Jacob's ladder 160
Jander-type diffusion 10

kinetic 259, 264
kinetic activity 135
kinetically stabilized intermediate 16
kinetics 104, 108, 111
Kissinger method 103–104

La_2O_3/ZrO_2 142
lack of stirring 295
Langmuir equation and adsorption from solutions 47–48
large uncertainty 296
laser flash 193, 210–212
latent heat storage 184
lattice parameter 18
LDA 169–170
leather 322
lecithin 368
length 269
Lewis acidic sites 144
Lewis and Brønsted acidity 174
lignin 325–329
lignocellulosic 264
lime-based mortar 313–314
lipid oxidation 362
lipids 359
liquid absorption 188
liquid-phase calorimetry 126–127
London-dispersion interactions" t "*See* van der Waals interactions 163
lower amplitude 296
LSDA 161
lysozyme 353

M06-L 169
magnesia 148
magnesium (Mg) 95, 108
Maillard reactions 373
marble 314, 316–318
mass 342
mass spectrometer 282
mass transfer 186, 205, 212
material inertia 196
material phase 296
material properties 209
mechanical 266
melting 261, 343
melting point 376
mesopores 270
meta-clays 319–320
meta-GGA 162, 169–171
metal complexation 73–75
metal dispersion 31
metal soaps 331
metastable phase 295
methane decomposition 34
methanol 139
Mg substitution 27
MgH_2 96–97, 99, 103–104, 108, 111, 113
MgO rocksalt structure 26
microbalance 320
microkinetic model 167
microscopy 380
mineral oxide surfaces 62
miniaturization 286
mixed oxide catalysts 142
mixed oxides 148
mixture 272
mixture of methanol and ethanol 148
mobile phase 281
modern artwork 312
molality units 51
monodentate carbonate 11, 13
monomer 284
morphology 260
mortar 312
multiple heating–cooling cycles 295

Nafion 129
natural aging 323, 327
NEXAFS spectroscopy 16
NH_3 adsorption calorimetry 173
Ni atoms 29
Ni content 26
Ni particles 29
$NiAl_2O_4$ 27
Ni-based catalysts 34
Ni-containing catalyst 25
NiO agglomerates 29
niobic oxide 129
niobium phosphate 129
NiO-MgO solid solutions 28
Ni-rich catalyst 34
nondestructive 311
nonhydraulic 313
nonlinear regression 10
NO_x abatement technology 175
number of basic sites 147

O_2 uptake 33
oil painting 312, 331
oils 359
Olive oil 366
one-site or multi-site binding models 75
open system 185–186
oxidation 364
oxidation of magnesium 98
oxidative coupling of alcohols 146
oxidative propane dehydrogenation (ODP) 37
oxidative stability 362, 365
oxidic support 21
oxygen diffusion 19

painting age 327
painting materials 329
paper 327
papyrus 327–328
parchment 322
partial molal enthalpy 64, 70
particle mobility 31
particle size 33
particle size distribution 33
PBE 169, 172
PBE0 170
peak 256
peak deconvolution 26
Peltier elements 297
permanent 286
petrochemistry 246
pH 356
pH adjustment 62–63
phase change measurement 296
phase transitions 37
pH-dependent electrical charge 62
phenylethylamine (PEA) 126
plane-wave 165
plaster 315–316
plasticizer 253
plastics 246
Pluronic 261
PMMA 283
pollution 315–316
polymorphism 359, 376
polyphenolic compounds 344, 373
potassium pentavanadate 37
pozzolanic reaction 313
pressure-composition isotherm (PCI) 108, 111
pressure-DSC 253
pressure–temperature diagram 296
probe molecules for acidity
- CH_3CN 174
- $CH_3)_3N$ 174
- CO 173
- NH_3 164–165, 173, 175
- probe molecule 173
- pyridine 173
production technology 311, 315–316
profile 263
protein 350, 358
protein unfolding 350, 358
proteins 373
proteins has 350
Prout-Tompkins equation 16
pseudothermal events 21
pulse thermal analysis 31, 37
pump delivery parameters 51, 52, 55, 59–61
PVA 264
PVC 274, 279
PVP 263
PW91 169–170
pyridine 144

quartz 315

radial heat flow 200–201
random phase approximation 163
ratio of strong acidic to strong basic sites 147
ratio of strong basic to strong acidic sites 132, 140, 151
reaction rates 136
reactive adsorbent 197
reactive gas 264
reactive materials 198, 211
reactive nitrous oxide frontal chromatography 19
reactivity 5
recent advancements 293
reduction peak 14
reduction temperatures 27
refrigerant 225, 227–230, 235–236, 238, 240
refrigerants 227–228
relative apparent molal enthalpy 65
relaxation 251
resin 251
resistant starch 348, 372
rexydroxilation dating 320
Rh-based catalyst 34

rheology 343
RPBE 169, 174

salt 188, 201
SCAC 164–169, 171–172
Schrödinger equation 159
screening 99
selectivity 132, 140–141, 147
self-preservation 304
semicrystalline 247
sensible heat storage 184
separation 281
shrink 270
Silica 150, 269
silica-alumina 129
silica-based catalysts 148
silver catalyst 37
simultaneous 261
single-solute and two-solute solutions 65, 67, 70, 72
single-solute solutions 65, 67
sites per surface area (μmol m^{-2}) 133
softening point 271
solid adsorption 188
solid diffusion 10
solid fat content 360
solid solution 26
solid-liquid interface 47–49, 52, 55, 64, 66–67, 70, 76
solid-state storage system 92
solution depletion method 52, 57, 62–63
solvothermal 258
sorbent 272
sorption 185, 188
sorption thermal storage 186
sorption/desorption 193
specific heat capacity 193, 195, 197, 210–212, 235, 240
spectroscopy 378
spinnable 271
stability 12
starch 369
stationary method 199, 201, 206, 211
stationary phase 281
statistical dispersion 14
steady-state methods 199, 202, 205
strontium adsorption onto zeolite from aqueous solutions 77–79
stirring 54
stock solution 55
stone 316, 318
storage interface 283
storage materials 186, 200
strength distribution of acidic sites 135
strong metal-support interaction 24
strongly bound OH-groups 23
structural and redox stability 30
subsurface oxidation 33
successive injections 55, 61, 63
sulfation reaction 315, 317
sulfur dioxide 125, 131, 143, 149
supported Ni catalysts 24
surface complexes 48, 81
surface intermediate 16
surface sites 152
surface titration 29
synthesis of catalyst 6
synthetic 247
syringe pump 50, 51, 56

tanning 322
technique 289
temperature 341
temperature of denaturation 323
temperature programmed desorption (TPD) 91, 97
temperature-programmed reduction 14, 27
ternary solid-metal-ligand complexes 81
textural properties 376
texture 343
TGA-FTIR 281
TGA-MS 273
TGA-μGC 287
the mass-loss diagram 322
thermal analysis 183, 196
thermal behavior 302
thermal conductivity 183, 193, 198–201, 204, 208, 210–212
thermal decomposition 313
thermal diffusivity 193, 199, 202, 206, 210–211
thermal effusivity 193, 198, 208, 211
thermal energy storage 195, 208
thermal expansion coefficient 269
thermal inertia 195
thermal methods of analysis 341
thermal methods of analysis. *See* thermoanalytical methods 1

thermal properties 206, 211–212, 294
thermal resistance 200, 205
thermal stability 348
thermal stability of proteins 353
thermoanalytical methods 3
thermoanalytical technique 312
thermochemical cycle 4
thermochemical heat storage 183, 205, 212
thermochemical properties 2–3, 5
thermodynamic 259
thermodynamic conditions 293
thermodynamic cycles 225, 230
thermodynamic properties 96, 104, 108
thermodynamics 237
thermogram 55, 60, 301, 348, 354
thermogravimetric analysis (TGA) 91, 97
thermogravimetry 261, 344, 380
thermokinetic analysis 16
thermomechanical analysis – TMA 346
thermophysical characteristics 193, 208, 213
thermophysical propertie 293
thermophysical properties 203, 211–212
thermopile 257
thermoplastic 246, 254
thermoset 251
thiophenolate 268
three-dimensional 277
Tian-Calbet heat flow 303
Tian-Calvet calorimetry 164, 166–167
Tian-Calvet DSC 37
Tian-Calvet heat-flow calorimeter 294
time-dependent hydrate formation 300
titania-based catalysts 142
TPO experiments 35
TPR peak 29–30
TPR profile 30
TPR/TPO cycles 30
transient methods 202–203, 206
transition metals 151
triacylglycerols 359
tungsten oxide supported on zirconia 130
two-solute solutions 49, 58, 72

van der Waals interactions 163, 169–170, 172–175
Van't Hoff method 48, 109, 111
vegetable oils 383
volatile 275
volatility 272
volumetric techniques (Sieverts) 95, 107, 111

water content 196–197
water uptake 196, 198, 207
water vapor pressure 10
water-gas shift 34
waterlogged woods 325
wavenumber 275, 279
weathering 317–318
weighted mean apparent molal enthalpy 65
wood 325
working fluid 223–225, 228–229
WO_x clusters 135

XC functionals 167, 171
XRD technique 378

zeolites
– CHA 168, 174
– FAU 168, 174–175
– MFI 167–168, 175
zeolites acidity 173
zeolites basicity 173
zincian malachite 7
zirconia and titania-based catalysts 144
zirconia-based catalysts 142
Zn diffusion 18
Zn incorporation 18

www.ingramcontent.com/pod-product-compliance
Lightning Source LLC
Chambersburg PA
CBHW080654220326
41598CB00033B/5208